THE PURARI – TROPICAL ENVIRONMENT
OF A HIGH RAINFALL RIVER BASIN

MONOGRAPHIAE BIOLOGICAE

VOLUME 51

Series editor
H.J. Dumont

1983 **DR W. JUNK PUBLISHERS**
a member of the KLUWER ACADEMIC PUBLISHERS GROUP
THE HAGUE / BOSTON / LANCASTER

The Purari —
tropical environment
of a high rainfall river basin

edited by
T. PETR

1983 **DR W. JUNK PUBLISHERS**
a member of the KLUWER ACADEMIC PUBLISHERS GROUP
THE HAGUE / BOSTON / LANCASTER

Distributors

for the United States and Canada: Kluwer Boston, Inc., 190 Old Derby Street, Hingham, MA 02043, USA
for all other countries: Kluwer Academic Publishers Group, Distribution Center, P.O.Box 322, 3300 AH Dordrecht, The Netherlands

Library of Congress Cataloging in Publication Data

Main entry under title:

The Purari : tropical environment of a high rainfall
 river basin.

 (Monographiae biologicae ; v. 51)
 Includes index.
 1. Ecology--Papua New Guinea--Purari River Watershed
(Gulf Province) 2. Diseases--Papua New Guinea--Purari
River Watershed (Gulf Province) 3. Nutrition--Papua
New Guinea--Purari River Watershed (Gulf Province)
4. Ethnology--Papua New Guinea--Purari River Watershed
(Gulf Province) 5. Hydroelectric power plants--Environ-
mental aspects--Papua New Guinea--Purari River Watershed
(Gulf Province) 6. Purari River (Wabo) Hydroelectric
Scheme. I. Petr, T. II. Series.
QP1.P37 vol. 51 574s [508.95'3]   82-25862
[QH186.5]

ISBN-13:978-94-009-7265-0   e-ISBN-13: 978-94-009-7263-6
DOI: 10.1007/ 978-94-009-7263-6

Cover design: Max Velthuijs

Contents

Contributors

Alpers, M.P., P.N.G. Institute of Medical Research, Box 60, Goroka, Papua New Guinea.

Burgin, S., Biology Department, U.P.N.G. Box 320, University P.O., Papua New Guinea.

Chewings, V.H., Division of Land Resources Management, CSIRO, Alice Springs, N.T. 5750, Australia.

Conn, B.J., Department of Crown Lands and Survey, Royal Botanic Gardens and National Herbarium, Birdwood Ave., South Yarra, Vic. 3141, Australia.

Cragg, S., Office of Forests, Forest Products Research Centre, P.O. Box 1358, Boroko, Papua New Guinea.

Dyer, K.W., 6 Lancewood St., Victoria Point, Queensland 4163, Australia.

Egloff, B.J., National Parks and Wildlife Service, P.O. Box 210, Sandy Bay, Tasmania 7005, Australia.

Evesson, D.T., Department of Minerals and Energy, P.O. Box 2352, Konedobu, Papua New Guinea.

Frusher, S.D., Kanudi Fisheries Research, Department of Primary Industry, P.O. Box 417, Konedobu, Papua New Guinea.

Gwyther, D., Marine Science Laboratories, P.O. Box 114, Queenscliff, Vic. 3225, Australia.

Haines, A.K., Resource Management Section, Fisheries Division, Department of Primary Industry, Canberra, A.C.T. 2600, Australia.

Hall, A.J., Wellcome Research Fellow, MRC Environmental Epidemiology Unit, Southampton General Hospital, Southampton S09 4XY, U.K.

Hazlett, D.T.G., Department of Medical Microbiology, Faculty of Medicine, University of Nairobi, P.O. Box 30588, Nairobi, Kenya.

Irion, G., Senckenberg-Institut, Schleusenstr. 39A, 2940 Wilhelmshaven, West Germany.

Jozan, M., Center of Health Sciences, University of California, Los Angeles, California 90024, U.S.A.

Kaiku, O., Papua New Guinea Museum and Art Gallery, P.O. Box 635, Port Moresby, Papua New Guinea.

Kisokau, K.K., Office of Environment and Conservation, Central Government Offices, Waigani, Papua New Guinea.

Lambert, J.N., 141, Marlborough Cresc., Sevenoaks, Kent, U.K.

Liem, D.S., 7104 Blanchard Dr., Derwood, Maryland 20855, U.S.A.

Marks, E.N., Queensland Institute of Medical Research, Bramston Terrace, Herston, Brisbane, Queensland 4006, Australia.

Paijmans, K., Division of Water and Land Resources, CSIRO, P.O. Box 1666, Canberra, A.C.T. 2601, Australia.

Pain, C.F., School of Geography, University of N.S.W., P.O. Box 1, Kensington, N.S.W. 2033, Australia.

Pernetta, J.C., Biology Department, U.P.N.G., Box 320, University P.O., Papua New Guinea.

Petr, T., FAO, Fisheries Department (FIR), 00100 Rome, Italy.

Pickup, G., Division of Land Resources Management, CSIRO, Alice Springs, N.T. 5750, Australia.

Poraituk, S.P., Papua New Guinea Museum and Art Gallery, P.O. Box 635, Port Moresby, Papua New Guinea.

Stevens, R.N., Fisheries Department, P.O. Box 22, Port Vila, Vanuatu.

Thom, B.G., Department of Geography, University of N.S.W., Royal Military College, Duntroon, A.C.T. 2601, Australia.

Toft, S., Law Reform Commission, P.O. Wards Strip, Papua New Guinea.

Ulijaszek, S.J., 55, Radford Boulevard, Nottingham NG7 3BQ, U.K.

Warrillow, C., P.O. Box 6181, Boroko, N.C.D., Papua New Guinea.

Wood, A.W., P.O. Box 426, Ingham, Queensland 4850, Australia.

Work, T.H., Center of Health Sciences, University of California, Los Angeles, California 90024, U.S.A.

Wright, L.D., Coastal Studies Unit, Geography Department, University of Sydney, N.S.W. 2006, Australia.

Foreword

One of the major river systems of our country, the Purari River, finds its outlet to the sea in the Gulf of Papua on the southern coast of Papua New Guinea. All highlands provinces contribute to this mighty river: the Erave of the Southern Highlands Province joins with the Kaugel and Wahgi Rivers (Western Highlands), the Tua River (Simbu), and Asaro and Aure Rivers of the Eastern Highlands Province to make the Purari the third largest river in P.N.G. Unlike its rivals, the Fly and the Sepik, the distance between its escape from the mountains and its entrance to the sea is short. After winding its way mostly through deep gorges flanked by high mountains, the river leaves the foothills of the southern slopes of the central cordillera barely eighty kilometers from the sea.

The energy potential of such a river is enormous. Could the waters be utilised in any way to the advantage of the nation? Twelve years ago the Electricity Commission of Papua New Guinea proposed an answer to this question: the building of a dam across the river in the Wabo area of the Gulf Province. The generation of vast quantities of hydro-electric power could be fed into a national distribution grid and heavy industries could be established in the Gulf Province and other suitable localities to benefit from this power.

Should such a project go ahead with the blessing of the nation? In attempting to answer this question, however, many more questions were raised. It was obvious that any such scheme should fit in with the traditional environmental awareness of the Papua New Guinea people. Not only would it affect the people in the immediate vicinity of the dam; such a mighty project could indeed have a considerable impact on the development of this country. But should it be incorporated into then-current government policies? Were there laws requiring an Environmental Impact Statement prior to deciding on such a large project? How could a new nation, a developing country with no large scientific institution, cope with the funding and manpower needed to assess the proposal?

The answers to these questions are history now, but they provide us with a timely lesson on the value of the old saying 'the more haste, the less speed'. It would have been easy to say yes to the project and let the problems sort themselves out along the way. Certainly the need for two-way communication was apparent early, and the establishment of the necessary dialogue soon made both the village people and the high-technology proponents aware of each other, both as people and as owners of resources and skills. This led on one hand to grass roots studies such as the nutritional value of food gathered by the dwellers in the villages of the delta, and on the other hand to esoteric studies such as the mechanism of silt deposition leading to a theoretical computer model.

At the same time the establishment of the Office of Environment and Conservation formalised the government's recognition that no such project should go ahead without an examination not only of the economic balance sheet associated with it, but also of its social and environmental implications. Writing now, twelve years later, it seems hard to believe that these questions were not seriously considered in the very first stages of planning. However, the traditional environmental awareness of the villager has shown a remarkable leap to the modern concept that it is necessary to prepare an Environmental Impact Statement for such large projects and, further, that the user should pay.

The value of the Purari project has been to demonstrate that despite a shortage of technical expertise and unlimited funds it has been possible to gather considerable information at minimal cost. Investigations, often expeditionary in nature, were matched by highly technical analyses, and the result has been a high standard of achievement with a relatively low monetary input. The establishment of the position of co-ordinator of Purari studies within the Office of Environment and Conservation led to a large number of necessary studies of all aspects of the proposal. Apart from using the local expertise, specialists from a number of countries were called upon and assisted; these came from the Institute of Medical Research in Queensland, the University of Sydney in New South Wales, from the CSIRO Division of Water and Land Resources in Canberra, all in Australia; from the Centre for Health Sciences of the University of California in USA; from the CSIR at Taupo in New Zealand; from the Senckenberg Institute at Wilhelmshaven in the Federal Republic of Germany. Financial support from a variety of international and national agencies was forthcoming and is here gratefully acknowledged.

Although government policy now places more emphasis upon the development of smaller projects throughout the country in an effort to maintain a balance between rural and urban areas, the Purari River project must be seen as an innovator in Papua New Guinea. It has firmly established the Office of Environment and Conservation as part of the planning process; it has drawn attention to the need for and Environmental Impact Statement as part of the planning process; indeed it has provided a model for the planning process itself, one which I hope will be reflected in the approach to major developmental projects of the future.

Karol Kisokau
Director,
Office of Environment and Conservation,
Central Government,
Papua New Guinea

Introduction

T. Petr

Purari River is the third largest river of Papua New Guinea, discharging 2607 m³/sec into the sea. It is situated in a high rainfall, humid tropical environment, with a catchment of 33 670 km² extending from the highest mountain of Papua New Guinea, Mount Wilhelm (4510 m), to the Gulf of Papua. The river basin embraces all climatic zones, with the highest mountain having occasional snow cover, through high montane forests, densely populated highland valleys, and poorly populated low montane and lowland forests and swamps, to the coastal mangroves. The rainfall is high, the highest being in the foothills where it reaches over 8000 mm per annum. It is this high precipitation and run off which attracted the attention of planners with the promise of large quantities of cheap hydroelectric power, if a dam were to be built on the Purari. A dam planned for Wabo in the foothills was projected to have an installed capacity of 2160 megawatts (for basic data on Wabo Dam and reservoir see Appendix A). This is about six to seven times more than the total currently installed capacity in the country, of which 155 MW serves the gigantic Bougainville copper mine on Bougainville Island. The Wabo project was to supply large quantities of cheap energy to power-intensive industries, such as aluminium smelting.

In the early 1970s it was decided to carry out an engineering feasibility study for a dam on this river. This study, which was executed jointly by the Australian Snowy Mountain Engineering Corporation, Japanese Nippon Koei, and the Government of Papua New Guinea was completed in 1977. However, the final decision on whether to launch the scheme was still subject to environmental studies. The high environmental awareness of the country has been expressed through the National Executive Council which established that 'an ecological study of the effect of the dam and the overall hydroelectric project must be carried out at an early stage, and certainly before any commitments are entered into to build any such works'. Such a decision was of historic importance, as it has paved the way for similar studies to become an integral part of preparatory planning and environmental studies for other large development projects, such as the Ok Tedi gold and copper mine in the Western Province. Thus, the major connecting theme of this monograph is the Wabo hydroelectric scheme.

To study the whole catchment would constitute an enormous task. The major environmental impacts were to be expected in the area of the projected impoundment and downstream of it. But this was also an area with the most difficult access, and worst weather conditions. In comparison with the relatively easily accessible and therefore better known highlands, the lowlands of the Purari were one of the poorest known environments in Papua New Guinea. The study area is completely covered in tropical forest, has very high rainfall, and is infested with malaria and arbovirus-carrying mosquitoes. There is a population density of only perhaps one person per 5 km^2. Until 1976 access was possible only by floatplane, motorized vessel, helicopter or paddled dugout canoe, the last means of transport still being the most common one. Only in 1976 was an airstrip completed at Wabo, which now receives one small plane per week when weather allows.

The difficult field conditions, and financial and manpower constraints required a special approach. For financial reasons it was decided not to get involved with a package deal approach to environmental studies, which was estimated in 1976 to cost some 2.5 million U.S.$. There was no specialized institute in the country which would be able to execute such studies, and government departments, together with the available institutes and universities had other urgent tasks. As the final decision on the project was still distant, there was some justified hesitation on behalf of the government to spend a large sum of money, which would have to come from the relatively small annual government budget. It was therefore decided to employ a coordinator who would direct the project, have both organizational and scientific responsibilities, and be backed by man-power available in the country. Suitable specialists would however stay on the payroll of their own institutions and departments and undertake individual, usually short-term studies. Only when no specialist was available within the country outside assistance was sought. In this way the cost of the studies was kept to a minimum.

The programme of environmental studies prepared in 1974 by the Office of Environment and Conservation, and later on modified and expanded by a United Nations Development Programme assisted consultant group, was again revised, and subsequently the environmental studies were initiated. Between 1977 and 1982, a succession of two environmental managers, funded initially from the Commonwealth Technical Aid Programme, and later on by the New Zealand assistance scheme, coordinated some 45 research projects. The results of these studies appeared in 21 technical reports (see Appendix B) edited by Dr T. Petr or by Dr A.B. Viner. These studies, together with some additional reviews, have been edited into the present monograph.

The overall objective of the studies was to assess the possible environmental impact of the Wabo Dam on the river and the riparian population. The areas of the future industrial complex and port were left out, mainly because these were not clearly identified. Although most environmental studies concentrated on the area of the future impoundment, and the river below the dam, eventually the studies encompassed a great diversity of aspects of the whole Purari River basin environment.

Tropical rivers situated deep in humid tropical forests are still poorly understood. The Purari River was no exception. Dr G. Pickup, who investigated sediment transport and hydrology, says that even when gauging stations are present, it is quite normal for 20 – 30% of records to be missing and only a few stations have information for a period of more than 10 years. But even less information was available for other environmental aspects. Logistic problems are great, and investigating such areas is costly.

Most baseline data for this book were collected during short expeditions. The project could not afford the establishment of a permanently manned field base, for which it would also be difficult to find a well-qualified person. This did not prevent numerous research teams and individuals from braving the difficult conditions of the project area to collect data for this book. In fact, most researchers found such working conditions challenging and stimulating.

The studies of the people in and around the area of the projected impoundment were given first priority. The poorly known group of the Pawaia was studied for their health and nutritional status, history and contacts, and social aspects of their life. Vectors of malaria and other diseases were also studied, as these contribute greatly to the poor health situation both at Wabo and in the Purari Delta. Nutrition and health studies were extended into the delta, and sago production, fishery and wildlife given special attention as they contribute the most to the diet of the people. Crocodile populations were assessed as they are among the major sources of cash.

Amongst the studies of the physico-biological environment, sediment transport was given high priority. A reservoir life-span depends on the sediment input, and Wabo was to receive 40 million m^3 sediments per year. The retention of sediments would be high as only the smallest particles were to be discharged downstream. Considerable attention was also paid to water and sediment quality in the catchment rivers, with the aims of establishing whether the present human population affects the river and obtaining baseline data on

XV

water and sediment physico-chemistry and mineralogy as a reference for future studies. The impact of the sediment retention would be felt downstream and studies were therefore undertaken to better understand the deltaic and coastal morphology, mangroves, and subsistence and commercially important aquatic biota.

Although much of the catchment is still forested, the high human population density in the highlands, through its deforesting and gardening activities, as well as waste disposal, already evolves some pressure on land and water. Many soils have a protective layer of volcanic ash, preventing their rapid erosion, but plans for utilization of marginal soils are being prepared, which if realized, could speed up the erosion processes. The climate, geology of the catchment, its plant cover and soils character, together with the human impact, determine the water quality and sediment transport. The nutrient transport would then determine the aquatic productivity of future reservoirs. These topics are reviewed and discussed in the relevant chapters of this monograph.

Although an in-depth judgement of the impact of the taming of the Purari River on the multitude of facets of the environment will require still more studies, it is clear that the social life of the Pawaia would be most dramatically affected, not so much by the technical aspects of the scheme as by their sudden full exposure to the modern age. Physical impacts of the scheme would be numerous, although probably much less than those of similar schemes constructed on densely inhabited, highly productive river systems. The Purari now contributes about ten percent of the total sediment input to the Gulf of Papua, and these sediments would be lost to this system. Such loss would probably gradually result in a change in aquatic productivity. This effect could be felt also west of the Purari Delta, in the complex interconnected deltaic system of numerous rivers, where sediment from the Purari is transported and deposited.

When the environmental studies were nearing their completion, it was becoming evident that the Wabo scheme would not be going ahead in the near future. But it was felt that the information collected during these studies should be preserved in a book to make it more easily accessible to specialists in various disciplines, including environmentalists, social scientists, planners, and decision makers, who in the future may face similar situations. The book should also contribute to a better understanding of humid, high rainfall tropical environments, little affected by human activities.

The editor, on behalf of all contributors to this book, wishes to express his high appreciation of the assistance to this project by the Government of Papua New Guinea through the Office of Environment and Conservation and the Department of Minerals and Energy.

I. Physical-chemical environment

1. Introduction to the Purari River catchment

C.F. Pain

1. Introduction

The Purari catchment, of 33 670 km², occupies a considerable portion of the
central and southern part of mainland Papua New Guinea (Fig. 1). It lies between the Kikori and Vailala rivers, and its headwaters comprise the basins of
the Erave, Kaugel, Wahgi, Asaro and Aure rivers. The main divide between
the north and south flowing rivers of the New Guinea island runs along the
Hagen Range, the Sepik-Wahgi Divide, and the Bismarck Range. These three
mountain ranges form part of the watershed of the Purari catchment (Fig. 1).
Mount Wilhelm (4510 m), on the Bismarck Range, is the highest mountain in
Papua New Guinea, while Mount Giluwe (4367 m), the second highest mountain in Papua New Guinea, lies within the Purari catchment.

Most aspects of the physical environment of the Purari River catchment are
dealt with in other chapters in this volume. In this introductory chapter,
therefore, these things are discussed only briefly. Instead, a brief picture of
the development of the catchment is presented, concentrating on its resources
and the way in which their relative importance has changed over time.

2. From hunting and gathering to tillage

In prehistoric times the resources of the catchment were seen in terms of their
usefulness to a stone age society.

The first people to arrive in the Purari catchment were hunters and
gatherers who probably made their way along the upper boundary of the
forest before 30 000 years ago (Hope and Hope 1976). At this time the montane grassland was much more extensive because of the Pleistocene lowering
of vegetation belts on the New Guinea island. The tree line stood at about

Petr, T. (ed.) The Purari – Tropical environment of a high rainfall river basin
© *1983, Dr W. Junk Publishers, The Hague / Boston / Lancaster*
ISBN-13: 978-94-009-7265-0

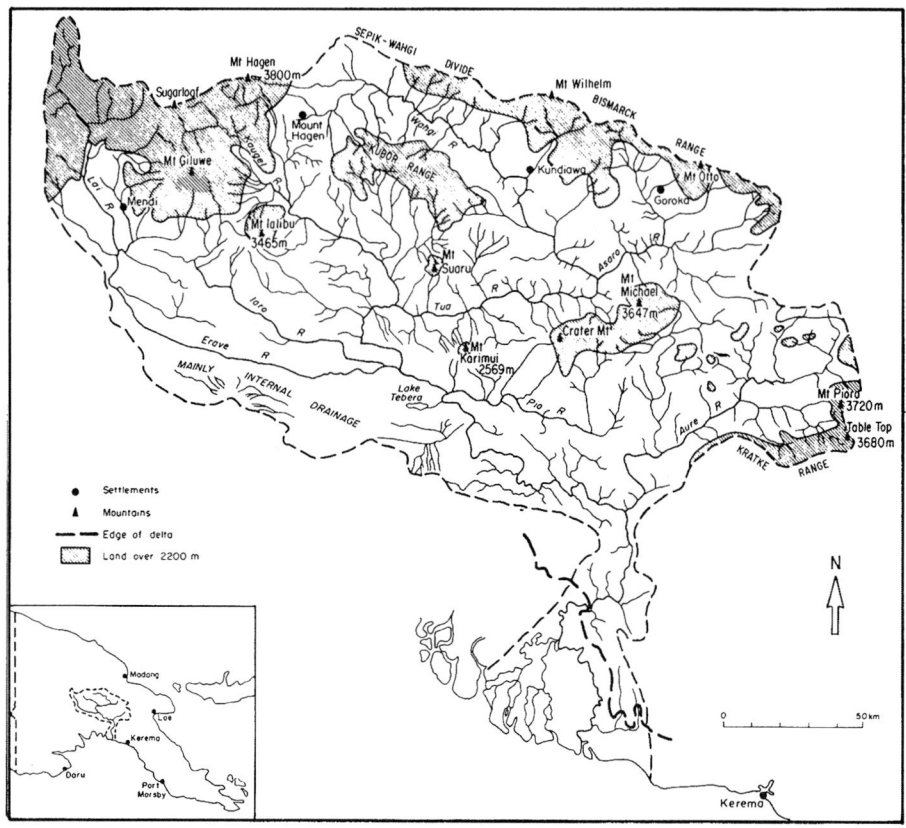

Fig. 1. The Purari River catchment, showing main rivers, mountains, and land above 2200 metres above sea level. The latter coincides approximately with the area covered with alpine grassland in the late Pleistocene.

2100 – 2400 m above sea level. Moreover, it has been suggested that the hunters and gatherers who were the first settlers in the area probably found the forest – grassland boundary a much more productive source of food than either the forest or the grassland alone. In addition, Hope and Hope (1976) point out that even today the area above the tree line contains many trade routes, illustrating the ease of movement in these areas compared with the montane forests (see also Hughes 1977). In the Purari catchment trade routes still pass over Mount Giluwe, and the Kubor Range, as well as the Bismarck Range.

Thus, Hope and Hope (1976) suggest, pre-agricultural man could have established himself more easily in the treeline – grassland ecotone than in the

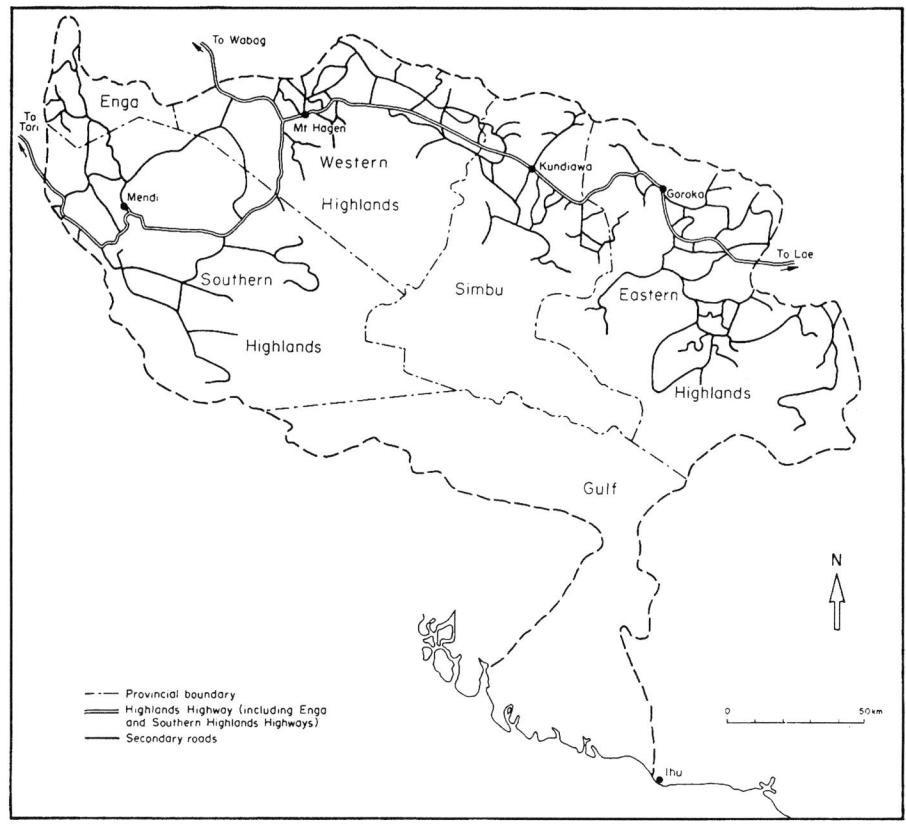

Fig. 2. The Purari River catchment, showing provincial boundaries and the road network.

lower forests. They would also have found these areas more convenient for travel. The first arrivals, then, were using the high altitude grasslands as a travel route, and were looking for the basic resources of food and shelter, both easily obtained from the treeline-grassland ecotone. As stated by Hope and Hope (1976:40),

> 'Hunting has probably been the most important reason for visits to or occupation of the treeline ecotone, both now and in the past. Ecotones, the transition zones between diverse communities, such as forest and grassland, often have a greater absolute number of species, and greater population densities of these species, than do the flanking communities.'

During this early period of occupation along the tree line, people must also have been moving into the areas surrounding the coast. However, in the Purari catchment this must have been a small area, even with the greater land

3

area exposed due to a lower sea level (Fig. 2). These coastal settlers would have been using coastal resources, and particularly fish.

Beginning at least as early as 9000 years ago, people began practicing agriculture in the Purari catchment. The evidence for this comes from Kuk, in the upper Wahgi Valley, where succession layers in the Kuk swamp demonstrate a number of stages of agricultural use of the swamp accompanied by elaborate water control systems (Golson 1977). There is thus a long history of agriculture in the Purari catchment, with both shifting cultivation and sedentary tillage being practised right to the present day. Soil and land thus became very important resources in the catchment from a very early date. Because of environmental differences, sedentary agriculture tends to be confined to the highland valleys, while shifting cultivation occurs in the hill country at lower altitudes (see, for example, Wood 1979).

Other resources were also important in prehistoric times. Hughes (1977) records occurrences of salt, pigments and stone in the Purari catchment that were extensively traded throughout the highlands.

3. European contacts and control

The arrival of Europeans in significant numbers brought a change in the view of resources in the Purari catchment. The European view can be largely summarised by four words, gold, government, missionaries and plantations. Hughes (1977) provides a summary of contacts in the area, while West (1966) makes a number of relevant points about early colonial development.

There were a number of contacts with outsiders as early as the 17th century, but the first influx of outsiders in any numbers along the coast came with the arrival of prospectors in Port Moresby in the 1870s, followed by missionaries and the government (at that time British) in the 1880s and 1890s. When the Australian Government assumed control in the 1900s, the Papuan area, including the lower Purari catchment, was seen as a potentially rich and profitable possession whose resources were available, and should be used (West 1966). To this end land was made readily available on a leasehold basis and plantations became established. In the Purari catchment contact and government and missionary influence had spread in the Kerabi and Erave areas by the mid 1920s. However, it was not until the 1930s that the highland valleys were opened up. The Wahgi Valley was first visited by Europeans in 1933, with the patrols by Taylor and the Leahy brothers, the former representing the government, and the latter prospecting for gold. Earlier, in 1930, Leahy and Dwyer had travelled down the Asaro, Tua and Purari Rivers in the first patrol

4

by Europeans to cross the Purari catchment from north to south.

Missionaries soon followed the government into the highland valleys, and in the late 1930s government administrative posts were set up. Following the Second World War, government and mission influence was consolidated, an airstrip and road network established, and the highlands valleys became the centre of a developing coffee- and tea-growing industry. There was also considerable effort spent on searching for mineral resources, an effort which still continues.

The arrival of the Europeans did not change the traditional view of resources in the catchment, but rather added a new viewpoint. The expectation of mineral wealth brought Europeans into the area, but this was not really fulfilled. Instead, the people of the area became a major resource, for both missionaries, plantation owners, and the government. These three groups had rather different intentions, it is true, but they were basically after the same thing, people either to swell their congregations or to work as labourers. The government in addition had the responsibility of introducing European-style law and order, and developing the communications network. The latter consisted largely of airstrips initially, with later development of roads. Some effort was also put into development of the agricultural capacity of the local people, which helped satisfy their needs for a way into a cash economy. Papua New Guineans began producing significant quantities of coffee in the highlands.

4. The present – planned and integrated development

Between 800 000 and 900 000 people live in the Purari catchment. Although there was a census in 1981, precise figures cannot be given because the catchment boundaries do not coincide with provincial boundaries (Fig. 2) or census division boundaries. Most of these people live in the five highland provinces, with only 9 – 10 000 living in that part of Gulf Province which lies within the Purari catchment.

There is a distinct separation between the highlands and the coastal lowland areas of the Gulf Province, with a zone between about 200 and 800 metres above sea level where there are very few people living. The central highland areas have dense populations and intensive agricultural systems. Outside these areas shifting agricultural systems operate under conditions that do not allow periods of cultivation of more than a few years at most. The 200 – 800 m zone is in a sense a barrier between the two parts of the catchment. This barrier is exemplified by the road network (Fig. 2). There is no road link between the highlands and the lowland areas of the Gulf Province.

Within the Purari catchment, there is a considerable effort being devoted to a systematic reappraisal of the resource base of the area. Three provinces with land within the catchment (Enga, Southern Highlands, and Simbu) are now in the implementation phase of integrated rural development projects, the basic aims of which are to assess the subsistence base of the rural economy, and to find ways of improving the way of life of the people. These projects have involved assessment of soil resources (e.g. Wood 1979; Pain 1982), studies of population and land use (e.g. Wood and Allen 1980), and assessment of forest resources.

There has also been renewed activity in the area of mineral exploration, and the major rivers are now being considered as possible resources, with a hydro-electric power scheme being constructed on the Kaugel River, and detailed feasability studies completed on the proposed hydro dam at Wabo, on the main Purari River. This latter proposal is, of course, the *raison d'être* behind the present volume.

5. Conclusions

During man's occupance of the Purari catchment there has been considerable shift in emphasis on the resources of the catchment. Some resources have been important throughout the time of man's occupation, although the emphasis has changed. Land has always been sought after, although its use varies over time and from place to place. These changes have come about as a result of the changing contacts with surrounding areas, both inside and outside Papua New Guinea. For example, in prehistoric times, the status of a resource depended on whether or not it was useful to the inhabitants of the catchment. After contact with Europeans, however, other items in the catchment, not regarded as resources by the original inhabitants, became more important. The extreme example of this is the use of rivers for electricity generation. The prehistoric peoples could not have conceived of getting energy from rivers in this manner, yet now there is a serious proposal not only to harness this energy, but in the case of the Wabo dam, to use this energy largely outside the borders of the catchment. Thus, while some resources in the catchment are still of local importance only, others have come to be regarded as resources available to Papua New Guinea, and even to areas beyond Papua New Guinea. With this change in emphasis on resources in the catchment, the people in the catchment have become dependent on events and people outside the catchment. For example, the early hunters and gatherers who made the first forays into the catchment, and set up the first societies there, were not dependent on anyone outside their own community. This continued down to the agricultural societies, although with the development of trade routes, some dependence on outside factors was introduced. Now, however, the ideas of energy seekers from Australia and Japan, and the destruction of Brazil's coffee crop by frost, as well as a large number of other external factors, all have an influence on the way the resources of the Purari catchment are perceived, and used.

References

Golson, J., 1977. The making of the New Guinea highlands. In: J.H. Winslow (ed.), The Melanesian Environment, pp. 46–56. Australian National University Press, Canberra.

Hope, J.H. and G.S. Hope, 1976. Palaeoenvironments for man in New Guinea. In: R.L. Kirk and A.G. Thorne (eds.), The Origin of the Australians, pp. 29–54. Australian Institute of Aboriginal Studies, Canberra.

Hughes, I., 1977. New Guinea stone age trade. Terra Australis 3. Australian National University Press, Canberra.

Pain, C.F., 1982. Enga soils: reconnaissance soil map and explanatory notes. Enga Yaaka Lasemana, Suppl. Volume. National Planning Office, Waigani, Papua New Guinea.

West, F.J., 1966. The historical background. In: E.K. Fisk (ed.), New Guinea on the Threshold, pp. 3–19. University of Pittsburgh Press.

Wood, A.W., 1979. The effects of shifting cultivation on soil properties: an example from the Karimui and Bomai Plateaux, Simbu Province. P.N.G. Agric. J. 30: 1–9.

Wood, A.W. and B.J. Allen, 1980. Land use and population studies for rural development in the Southern Highlands. Australian Geographer 14: 308–310.

2. The climate of the Purari River basin

D. T. Evesson

1. Introduction

The climate of the Purari River basin, which extends on the island of New Guinea between latitudes 5°S and 8°S and longitudes 143°E and 146°E, is determined by its position within the humid tropics. In Papua New Guinea, the overall character of the climate is that of high constant temperatures, high rainfall and high humidity.

Papua New Guinea (PNG) is affected by two major airstreams, the north-westerlies and the southeasterlies. These streams are controlled by the relative positions and intensities of that part of the subtropical high pressure belt which lies across Australia and the Tasman Sea, and the intertropical convergence zone (ITCZ), a belt of low pressure cells which roughly coincides with the thermal equator (Löffler 1979).

When the intertropical convergence has moved south of PNG, in the wake of the sun's apparent movement southwards towards the Tropic of Capricorn, a northwesterly airflow is experienced over PNG (Fig. 1). This season lasts from about December to March. At that time, the high pressure belt moves to the south of the Australian continent. Following the sun's movement north again the intertropical convergence moves over PNG and a southeasterly airstream develops between this system and the high pressure belt, which gradually moves into central Australia latitudes. This season generally lasts from May to October (Fig. 2). The months of April and November can be considered as changeover periods from one windstream to the other as the ITCZ passes over PNG. These periods are commonly known as the doldrums because winds are generally light and variable.

These seasonal effects are extremely important in the climatology of the Purari River catchment. Weather systems, such as tropical low pressure systems, surface and upper level low pressure troughs from southern latitudes,

Petr, T. (ed.) The Purari – Tropical environment of a high rainfall river basin
© *1983, Dr W. Junk Publishers, The Hague / Boston / Lancaster*
ISBN-13: 978-94-009-7265-0

10

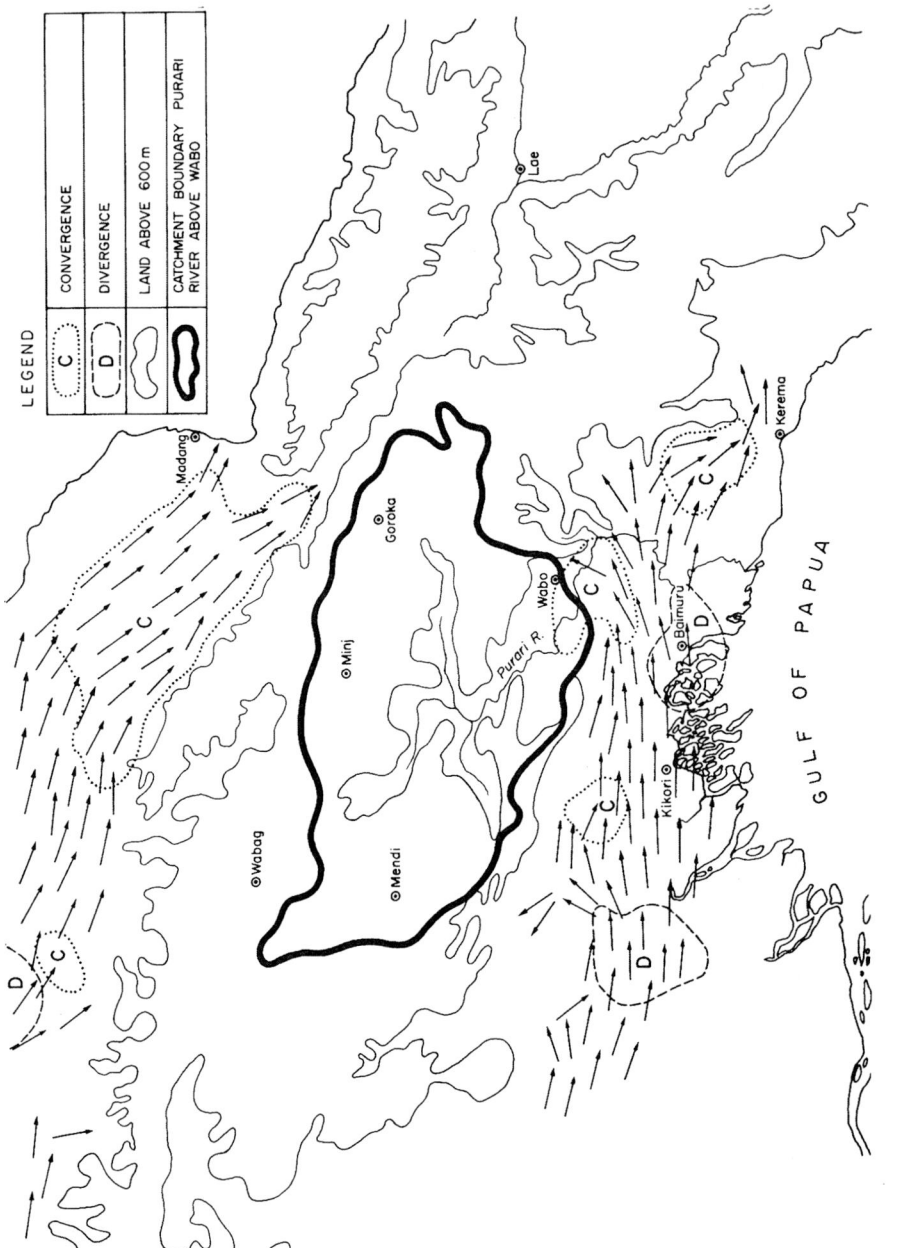

Fig. 1. Topographical convergence with northwesterly airflow (December – March).

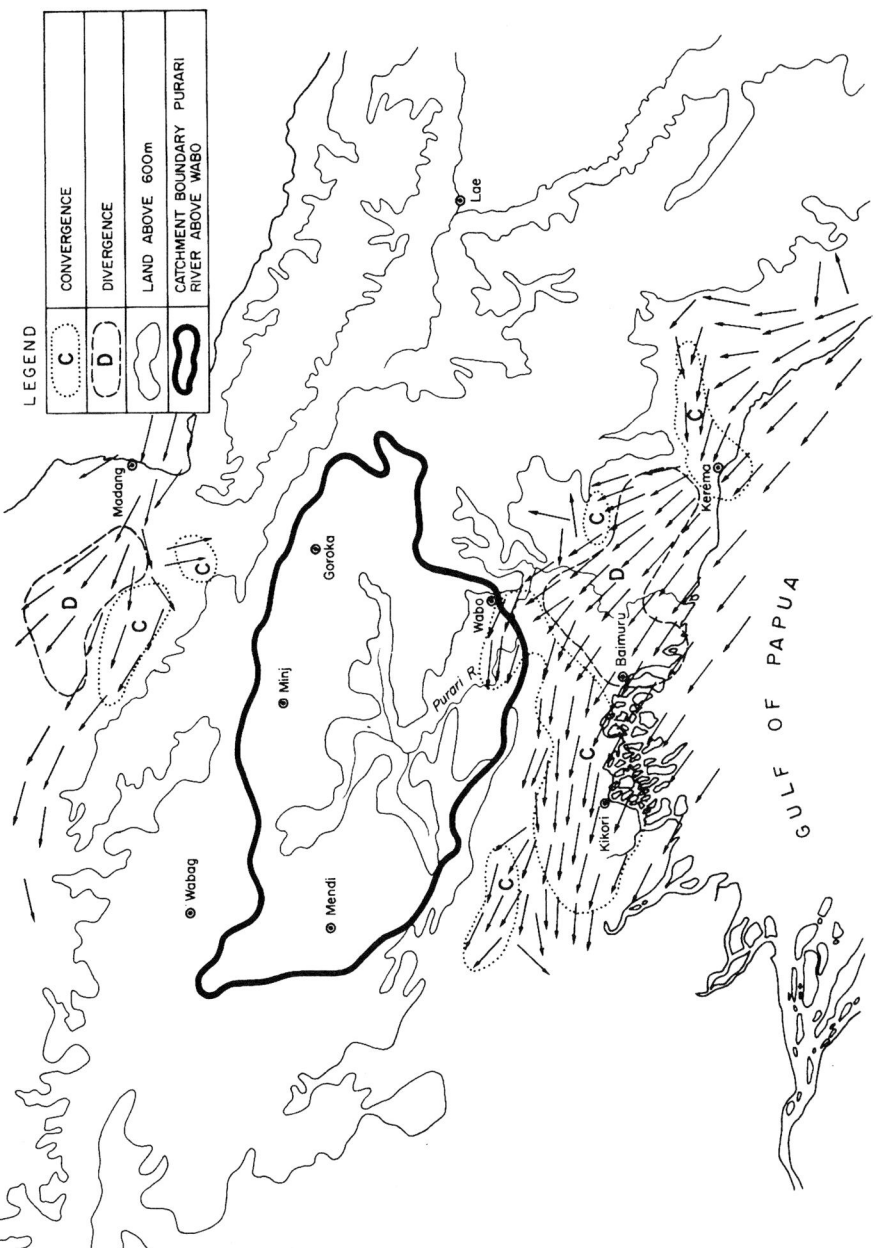

Fig. 2. Topographical convergence with southeasterly airflow (May – October).

11

at times influence the weather over the Purari basin. Low and middle level cloud can be generated, leading to rain showers, and winds can be strong and gusty.

The Purari basin includes the highest PNG mountain, Mt. Wilhelm (4510 m a.s.l.), with occasional snow showers, and it extends to the Gulf of Papua, bordered by extensive mangrove swamps. The climate of the Purari catchment is thus very much dependent on topographical relief and its alignment with respect to the major air streams. Local effects are the main determinant with respect to such elements as cloud cover, rainfall and wind. Temperature, humidity and evaporation are generally influenced more by the above elements than directly by local effects.

2. Clouds

Convergence which is produced by interaction of air pressure systems or by the shape of the land, causes air to rise; it is a common reason for cloud formation in a hilly and mountainous relief. The higher the land exposed to an airstream the more often a cloud is likely to form over it.

This is due to certain topographical features which act as effective barriers to other features further downstream. When cloud forms on a mountain, the air travelling down the leeside will warm by compression at the same rate as it cooled when rising, but its moisture charge will have been depleted by the cloud-forming process. As it ascends the next ridge there will be less moisture available for cloud formation and, generally speaking, the condensation level will be higher. That is, the cloud base will be higher and the cloud size smaller. The stage can be reached when the moisture content of the air has dropped so much that cloud will not form on a particular ridge. This is the major reason for the rainfall distribution in the Purari basin, with the lowest hills receiving highest rainfall, while the highest mountains have a relatively low precipitation.

Some nocturnal showers and thunderstorms over the Purari basin are caused through radiational cooling leading to the growth of cumulus or cumulonimbus clouds. In valleys of the Purari basin clouds form fairly commonly as a result of cool air flowing down slopes of high mountains. When mixing with warmer air layers, condensation leads to stratus cloud formation. Such clouds can have a base height of from 30 meters to about 500 meters with a thickness generally less than 300 meters. It can persist until after sunrise, depending on its thickness and the extent of cloud above it. Mean morning and afternoon cloud cover for two major seasons is shown in Fig. 3.

12

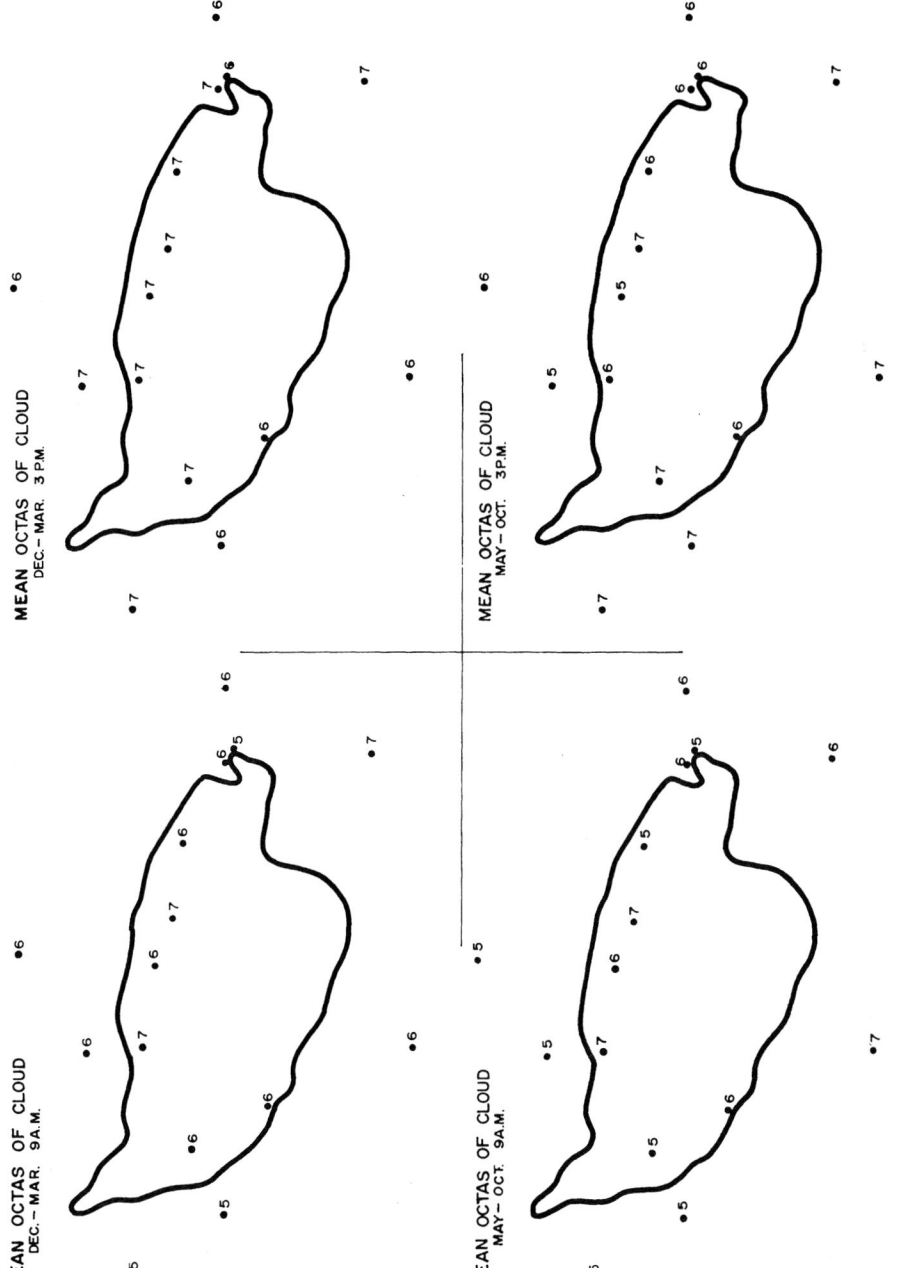

MEAN OCTAS OF CLOUD
DEC.-MAR. 3 P.M.

MEAN OCTAS OF CLOUD
MAY-OCT. 3P.M.

MEAN OCTAS OF CLOUD
DEC.-MAR. 9A.M.

MEAN OCTAS OF CLOUD
MAY-OCT. 9A.M.

Fig. 3. Mean morning and afternoon cloud cover for two major seasons.

13

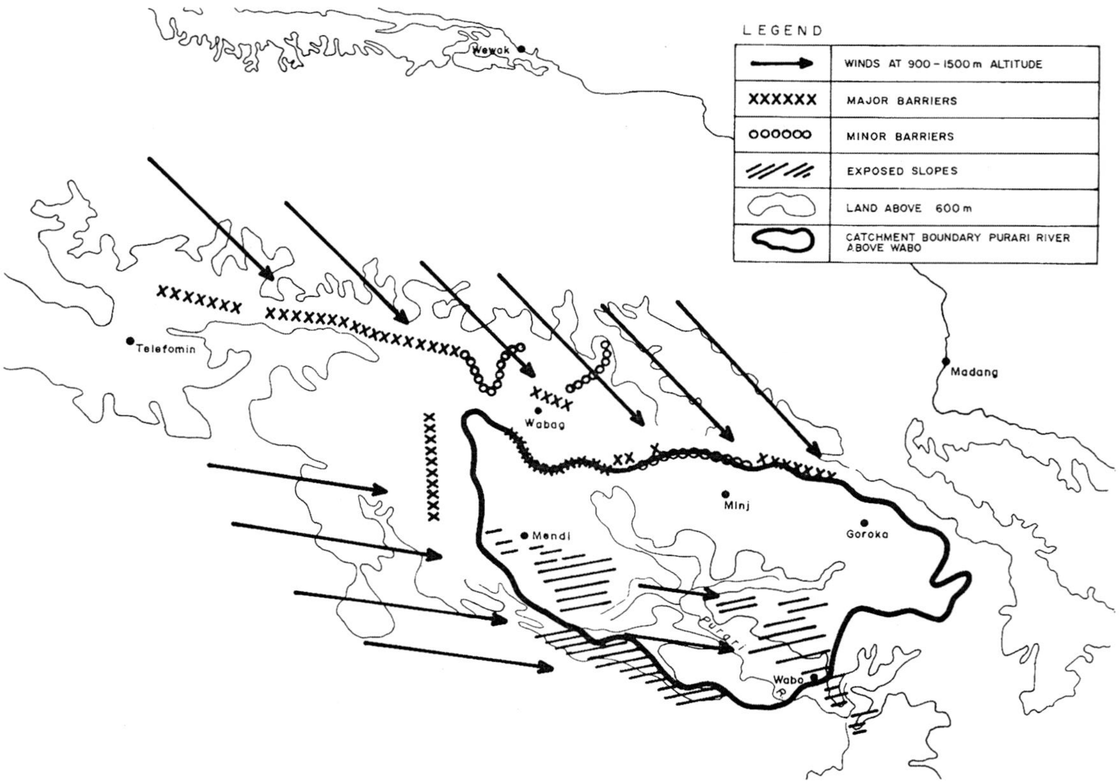

Fig. 4 Mountain barriers to northwesterly airflow.

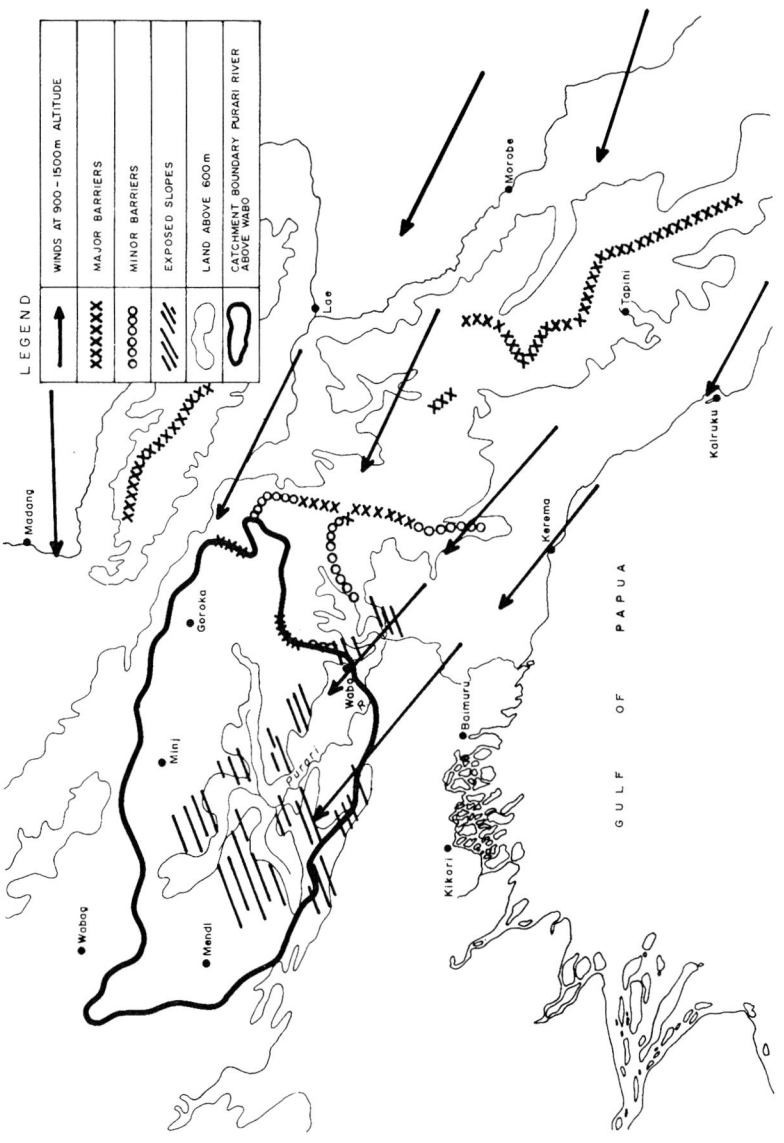

Fig. 5. Mountain barriers to southeasterly airflow.

Fog formation is common. Fogs occur in valleys of the highlands above 1000 m almost every day of the year, generally lasting from about 7 a.m. to 9 a.m. with a few persisting longer. The frequency of fogs generally decreases with decreasing elevation. Marshy and lake areas, however, would have greater incidence than other places of the same elevation. The Wabo dam site area has a fairly high incidence of fog.

Away from the coast, the rapid fall in temperature in late afternoons allows condensation to occur with fog or low stratus cloud formation. Air moving with up-valley or the start of down-valley winds, given the right circumstances of temperature and relative humidity, can also lead to formation of clouds and fogs.

3. Rainfall

Generally speaking, those areas subject to the most cloud receive the most rain. Most rain over the Purari basin in both seasons is produced by a combination of topographical effects and thermal or nocturnal convective processes. The major exception occurs in the Wabo area as will be discussed later on.

Because of the varied effects of the two major airstreams, it is best to consider rainfall from a seasonal viewpoint. Figure 6 shows average isohyets for the northwest season (December to March). The major feature is clearly the rain shadow effect over the northern and central parts of the catchment. The northwest stream is captured by the Central, Schroder and Bismarck ranges to the northwest and north of the basin, and its moisture load is well depleted by the time it moves into the watershed area.

That part of the northwesterlies which flows to the south of the central cordillera of the New Guinea island tends to travel parallel to the ranges, thus limiting uplift effects. However, the lowest part of the catchment, in the Wabo area, does receive high rainfall. Some topographical uplift occurs, but mainly low level convergence is responsible for these rainfall peaks. Comparison of Fig. 1, 4, and 6 illustrates this point.

The isohyetal average areal rainfall for the Purari catchment upstream of Wabo was found to be 1400 mm for the northwest season.

Comparison of Fig. 7, representing the southeast season (May to October), with Fig. 6 clearly shows that the highest and lowest rainfalls occur roughly in the same areas but that the distribution across the catchment is markedly different. Figures 2 and 5 illustrate the convergence and topographical effects experienced by this airstream. Though the moisture carried by this stream is less

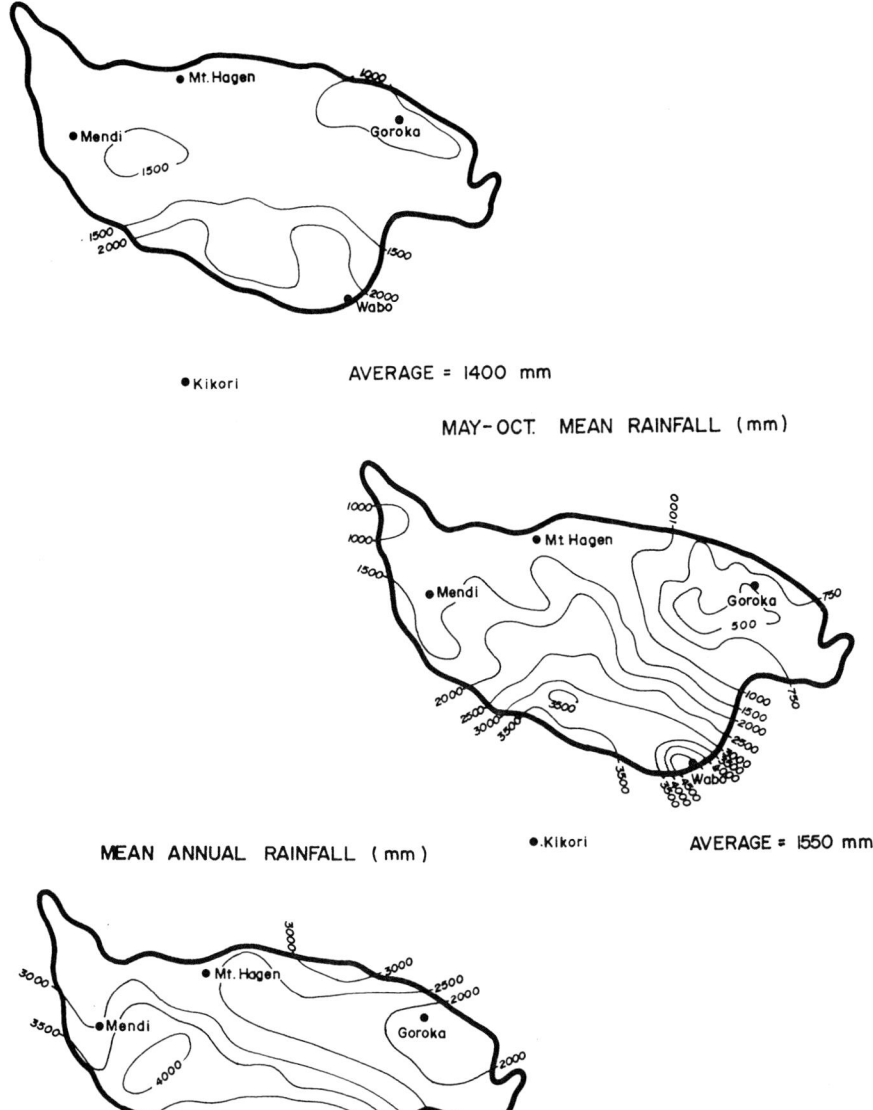

Fig. 6 – 8. Mean rainfall (in mm) in various seasons and annual.

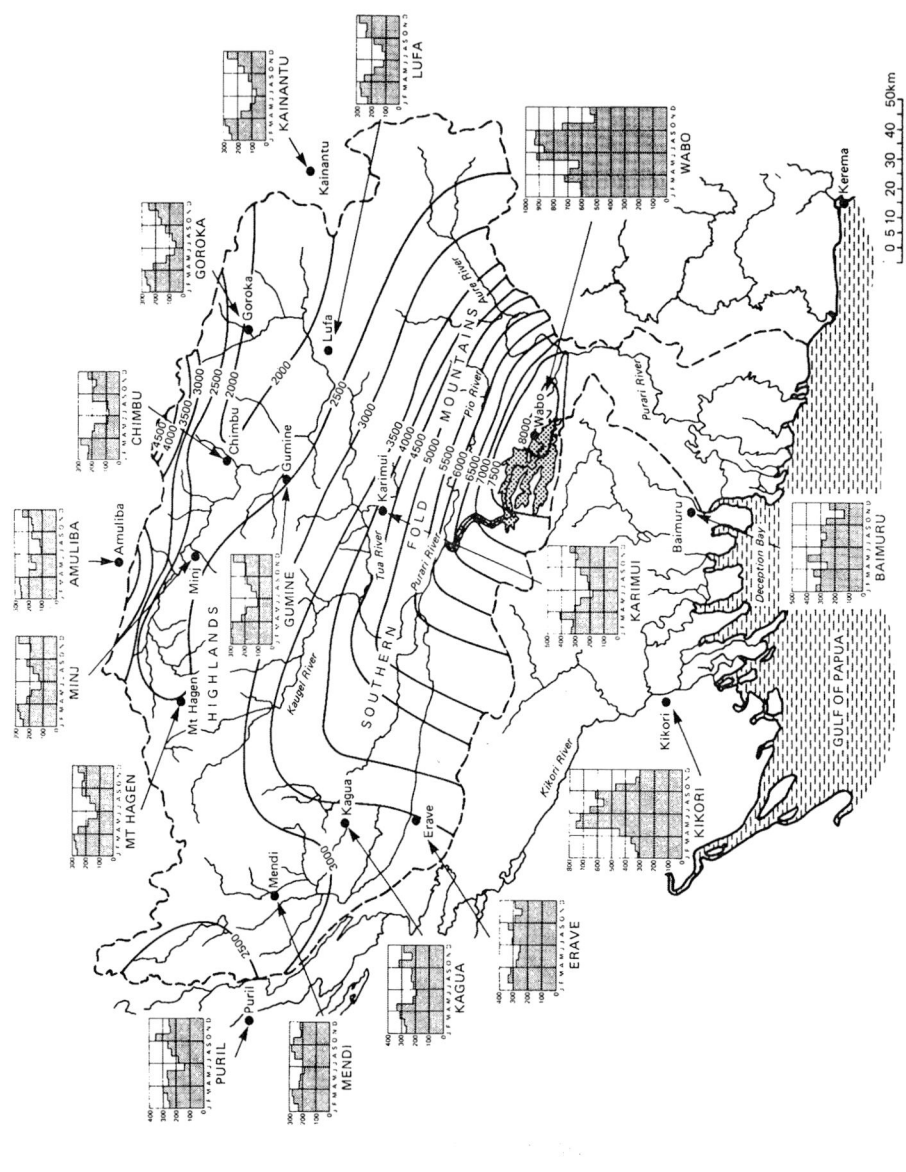

Fig. 9. Monthly rainfall (in mm) data. Isohyets refer to annual rainfalls in mm. Redrawn from SMEC-NK (1977) and data presented by Evesson (1980).

18

than that of the northwesterlies, the very strong physical influences just mentioned are sufficient to cause much more precipitation.

The very strong low level convergence in the Wabo area results in the extremely high rainfall. Thus, the heaviest rainfall occurs in the southeast corner of the catchment and grades towards the northwest corner. The northeast corner, which is protected by the orientation of the topography from the southeasterlies, receives the least rainfall.

The isohyetal average rainfall for May to October for the Purari basin upstream of Wabo was found to be 1550 mm. Annual average isohyets for the Purari basin (Fig. 8) clearly show the influences of the seasonal effects. The isohyetal average for the catchment was 3500 mm.

The onset of either the northwesterlies or southeasterlies varies from year to year. The changeovers generally occur during April and November. Sometimes almost all April is subject to northwesterlies, but other years the southeasterlies may begin already in the first week of April.

Total rainfall during the May to October period over the catchment is higher than, or as high as the total during the March to December period, except over the northeastern corner.

Monthly average rainfalls for selected stations in the catchment are presented in Fig. 9.

4. Winds

Because of the rugged nature of the topography winds have a strong 'local' component. Up-valley winds during the day and down-valley winds overnight are common, and the latter can gust in excess of 60 kilometers in some locations. Where the valleys lay in the direction of the major airstreams, wind speeds could be even higher. Channeling effects are also likely, but no general statement is possible, as each site displays its own individual characteristics.

Mt. Hagen is one of two stations in the Purari basin for which observations are available. The other is Goroka, both stations being situated in highlands. These stations are therefore not representative of the catchment as a whole. At Mt. Hagen, for about 50% of the time at 3 p.m. the wind comes from a southwesterly quarter in both northwest and southeast seasons. The orientation of the valley in which the observation site lies is the reason for this. Between approximately sunrise and 10 a.m., if there is a wind, it is generally less than 10 km per hour. The strongest winds tend to occur about mid-afternoon when thermal effects are at a maximum. Table 1 shows greatest wind gusts, on a monthly basis, at Goroka, and the times of the occurrences.

Table 1. Maximum annual wind gusts in Goroka.

Year	Date/month	Direction	Speed (km/hr)	Time
1965	22/02	E	42.5	15.40
1966	23/03	ENE	57.6	15.30
1967	27/12	NE	53.6	17.10
1968	24/10	SSE	44.6	16.30
1969	21/10	ENE	44.6	14.10
1970	09/01	NE	37.1	16.40

Surface winds in the northwesterly season are, as mentioned earlier, very much dependent on local topography. However, sometimes strong northwesterly winds do reach the higher peaks of the mountains. During such conditions a very significant wave motion is set up in the airstream as it flows over the mountains.

Valleys grading from the northwest to the southeast can expect gusty gully winds at night as the general stream will enhance local nocturnal effects. Similarly, valleys graded in the opposite direction will experience enhanced afternoon up-valley winds.

The equivalent situation to that occurring with the northwesterlies, in relation to valley wind enhancement, is experienced in the southeasterly season. However, mountain wave production then is limited because of the different relationship between the surface air and the air at 1500 to 3500 meters.

The lower parts of the catchment are open to the southeasterlies and will, at times, experience strong winds.

5. Evaporation (evapotranspiration)

No evaporation data is available in the catchment area apart from five years of records from a U.S. Class A pan at Kuk (5°48′S, 144°17′E) with an elevation of 1630 m. The average annual evaporation was 1488 mm. Using temperature, humidity and sunshine and the above data McAlpine et al. (1975) derived estimates of evaporation for a number of places in the Purari basin. Isopleths of these data are shown in Fig. 10.

It is well accepted that evaporation measured from U.S. Class A pans gives an overestimate of actual evaporation from a lake surface. Houman (1973) concluded that lake evaporation would be 0.6 to 0.8 times the measured pan evaporation, depending on the general climate of the area and the time of the

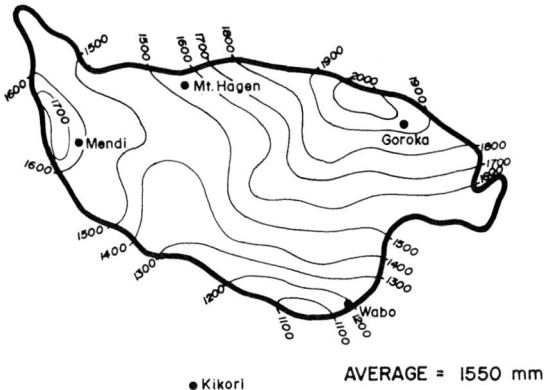

ESTIMATES OF PAN EVAPORATION (mm)

AVERAGE = 1550 mm

Fig. 10. Estimates of pan evaporation (in mm).

year, the lower value being appropriate in hot, dry climates and the higher value for cool, wet ones. For hot, wet conditions, then, a coefficient of 0.7 would seem a logical choice.

The annual average from Fig. 10 for the catchment is 1550 mm. Using 0.7 as the coefficient it can then be estimated that the annual average lake evaporation would be 1100 mm. If only the actual lake area for the proposed dam at Wabo is considered then the value of annual lake evaporation is down to 850 mm. This is because of the large amount of cloud and rainfall in this area.

The average annual amount of water which passes Wabo in the Purari River is 78.3 km^3. This amount represents the total rainfall that has entered the river system either directly, or by running off the land or by percolation through the ground into the banks or beds of the streams. The difference between the rainfall and the average flow represents the catchment losses which are evaporation and possibly percolation out of the catchment or underground storage. A simple water balance equation for the catchment may then be written:

> river flow = rainfall plus inwards percolation minus
> evaporation minus outwards percolation
> minus underground storage

As a significant part of the Purari catchment and the adjacent catchments have very porous rock strata, percolation should be a factor to consider.

Annual average figures for the whole catchment give:

21

river flow (in rainfall equivalent units)	= 2700 mm
rainfall	= 3500 mm
evaporation minus inwards percolation plus outwards percolation plus underground storage	= 800 mm

It seems likely from the physical characteristics of the catchment that percolation has a net inwards value and underground storage over the average can be considered as a net zero. The equation then reads:

evaporation minus inwards percolation = 800 mm, or

evaporation = 800 mm plus inwards percolation

Considering the geology of the strata, the most likely areas of percolation are in the northwestern and western borders of the catchment. Because of the lack of stream gauging information for adjacent catchments in these areas little can be done to assess the overall value of inwards percolation. However, some figures do exist for the upper Strickland catchment. The area of this catchment likely to contribute to percolation is about 6000 km². The average annual rainfall is 3600 mm, the average river flow is 1900 mm, which leaves 1700 mm for evaporation and percolation. From Fig. 10 the average Class A pan evaporation is 1500 mm and, as this will decidedly be an overestimate of the actual evaporation, then the percolation out of the upper Strickland catchment is in excess of 200 mm per 6000 km² area. This would mean possible percolation into the neighbouring Purari catchment of more than 40 mm/unit area. As two to three times this percolation figure for the upper Strickland would not be considered unreasonable, a catchment evaporation for the Purari of the order of 1000 mm would seem likely. This is less than earlier estimates.

6. Radiation

Solar radiation data is not available for anywhere within the Purari catchment. The large amount of cloud over the area would certainly reduce insolation, and Fig. 3 shows the average cloud amounts for 9 a.m. and 3 p.m. on a seasonal basis.

7. Temperature

The yearly ranges of temperature are almost invariably insignificant when compared to the diurnal ranges. Plotting annual average maximum and minimum temperatures against elevation (0 to 2000 m) for the Purari catchment gives a range for minima of approximately 22.8°C to 11.6°C, and for maxima, 30.0°C to 22.5°C (Fig. 11).

Two stations outside the Purari basin (Kikori and Kerema), both near sea level, fit well for minima, but sea breeze effects during daytime give maximum values lower than might have been expected for their elevations.

Since the 'best fit' lines are almost parallel it can be deduced that the average diurnal range of temperature is independent of elevation and tends to be the same for most places, except of course coastal areas affected by sea breezes and, perhaps, also the highest peaks where maximum temperatures are probably reduced due to a greater preponderance of cloud there than over the other parts of the catchment.

8. Humidity

Air humidity is related to air temperature and hence to altitude (Table 2). When comparing stations of differing altitudes, the fluctuations of the relative humidity are quite irregular and are far less pronounced than those of the absolute humidity, which decreases in a consistent manner with increasing altitude. January and July values illustrate the lack of seasonal variation.

Table 2. Average daily humidity.

Station	Altitude (m)	January		July		Annual	
		Relative humidity%	Absolute g/kg	Relative humidity%	Absolute g/kg	Relative humidity%	Absolute g/kg
Kerema[a]	5	84	19.0	92	19.0	88	19.0
Kikori[a]	75	85	19.2	94	17.8	90	17.8
Lake Kutubu[a]	810	79	14.2	82	14.2	79	14.2
Goroka	1565	75	13.8	73	12.0	74	12.0
Kainantu	1565	78	13.8	80	12.0	79	13.0
Aiyura	1570	80	13.8	80	12.0	80	13.0
Tari[a]	1600	79	13.1	77	13.1	79	13.1
Mt. Hagen	1630	81	13.2	84	12.0	81	13.2
Mendi	1675	78	12.3	78	11.6	77	12.3
Wabag[a]	1980	81	12.7	79	11.2	80	11.9

[a]Outside Purari catchment.

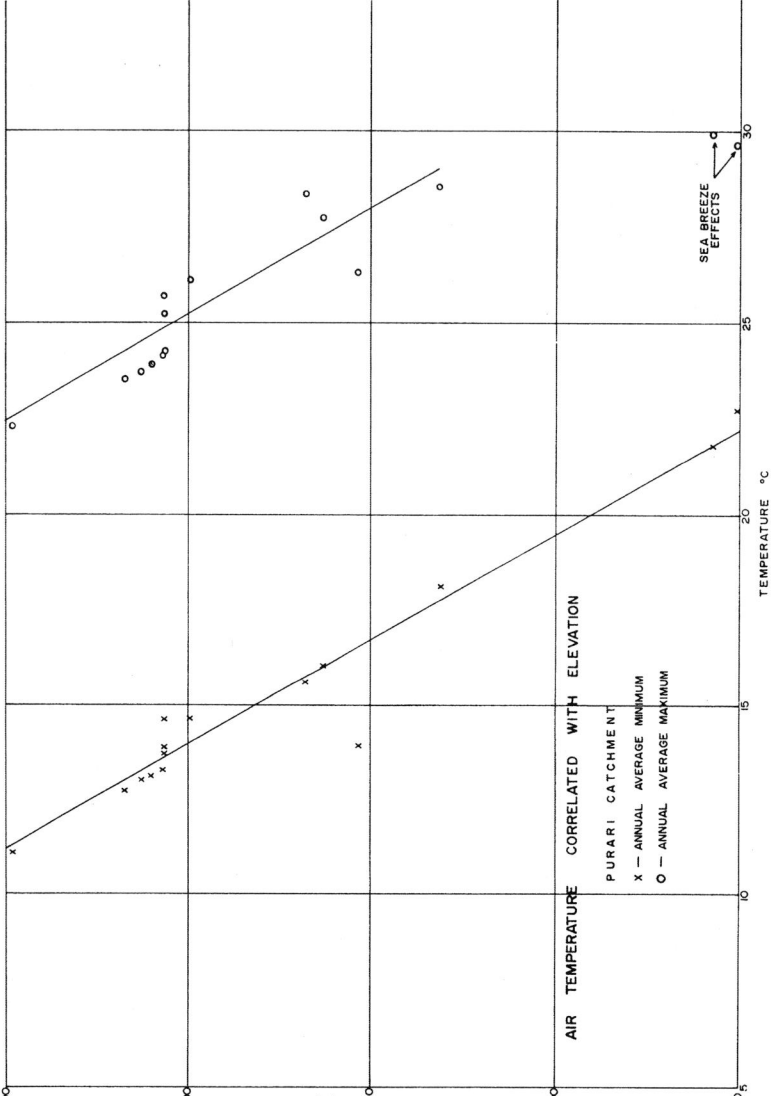

Fig. 11. Air temperature correlated with elevation, as recorded in the Purari catchment.

24

9. Other phenomena

Widespread severe frosts were experienced in the highlands in 1940 or 1941, 1949, 1950, 1962 and 1972, generally above 2000 m elevation. Light to moderate frosts with severe frosts in very localised areas, occur in most years.

Between 1000 m and 2000 m light frosts are common and severe frosts are very rare. Below 1000 m, generally, frosts of any type are unlikely except in some isolated pockets where topographical influences help to produce frost-favourable conditions.

Thunderstorms occur throughout the year and are responsible for a large percentage of the rain received over the catchment. Down-drafts and down-bursts associated with thunderstorms, though rare, can cause major damage to trees, leaving large gaps in the forest cover for a few years. Squalls associated with local winds have been discussed earlier. Hail is possible in the highest areas but is very infrequent.

No statistics are available on lightning strikes, but it is guessed that a reasonable number do occur, especially on higher peaks.

The Purari catchment is unlikely to be directly influenced by tropical cyclones. The closest a tropical cyclone came to the catchment in known history was in January, 1921, when the centre of the storm passed within about 200 nautical miles of Wabo.

10. Summary

The climate of the Purari basin is dictated by the effect of topography on the two main seasonal airstreams, the northwesterlies and the southeasterlies. It varies from a hot and very wet environment over southern parts to a relatively dry and mild environment in the northeastern corner.

The rugged nature of the terrain in the catchment, with many peaks over 3000 m, makes it very difficult to generalise about the climate as one mountain may have a rain shadow on one side and a wet spot on the other.

However, some details are worthy of note. The area is one of the cloudiest places on earth and the rainfall in the Wabo locality (more than 8000 mm per year) rivals the world's highest. Temperatures depend on elevation and range from hot at low altitudes to cold at the highest elevations. Frosts above 1000 m are common and snow occurs on the highest peak, Mt. Wilhelm, though it usually melts during the day.

Below Wabo the Purari River flows onto a coastal plain area forming a substantial delta. Rainfall over this delta region increases northwards in both seasons. Temperatures range from about 22 to 30°C, the maxima being fairly strongly affected by cooling breezes coming from the south during the afternoons at any time of the year.

Acknowledgement

Figure 8 was kindly provided by Dr G. Pickup.

References

Evesson, D.T., 1980. The climate of the Purari River catchment above Wabo. In: A. Viner (ed.), Purari River (Wabo) Hydroelectric Scheme Environmental Studies, Vol. 16, pp. 1–16. Office of Environment and Conservation, Waigani, Papua New Guinea.

Houman, C.E., 1973. Comparison between pan and lake evaporation. Technical Note No. 126. World Meteorological Organisation, Geneva.

Löffler, E., 1979. Papua New Guinea. Hutchinson Group, Richmond, Victoria, Australia.

McAlpine, J.R., G. Keig and K. Short, 1975. Climatic tables for Papua New Guinea. C.S.I.R.O. Div. Land Use Res. Tech. Pap. No. 37. C.S.I.R.O, Canberra.

Snowy Mountains Engineering Corporation – Nippon Koei (SMEC-NK), 1977. Purari River Wabo Power Project feasibility report, 8 volumes.

3. Geology and geomorphology of the Purari River catchment

C.F. Pain

1. Introduction

The Purari catchment exemplifies many of the central features of Papua New Guinea's geology and geomorphology. Foremost of these are the recent geological origin of much of the island, with late Tertiary marine sediments now considerably uplifted and deformed, the intense erosion that has taken place since uplift, and the influence of volcanism. Only small areas of rocks are older than the Tertiary, and most of the catchment did not emerge from the sea until after the Eocene. Since then, however, uplift volcanism and erosion have been rapid, and have led to some spectacular landforms. Large strato-volcanoes, bold limestone scarps and deeply incised gorges are marked features of the landscape. Slopes are predominantly steep and lead down to rivers that almost always flow as swift, muddy streams. The Purari catchment provides a good example of the geology and landforms of tectonically active areas.

This paper aims to provide a synthesis of the geology and geomorphology of the Purari catchment. This synthesis is achieved by description of the geology and landforms, and by historical interpretation of the development of the landscape. It is based on considerable field experience in the highlands, supplemented by map and aerial photograph interpretation and, especially in the lowland areas, the published data of other workers. The delta area is excluded from this paper, since it is covered by Thom and Wright (this volume).

2. Geology

The oldest rocks in the Purari catchment are Upper Paleozoic in age, but Tertiary rocks cover the largest area. The distribution of different rock types is

Petr, T. (ed.) The Purari – Tropical environment of a high rainfall river basin
© *1983, Dr W. Junk Publishers, The Hague / Boston / Lancaster*
ISBN-13: 978-94-009-7265-0

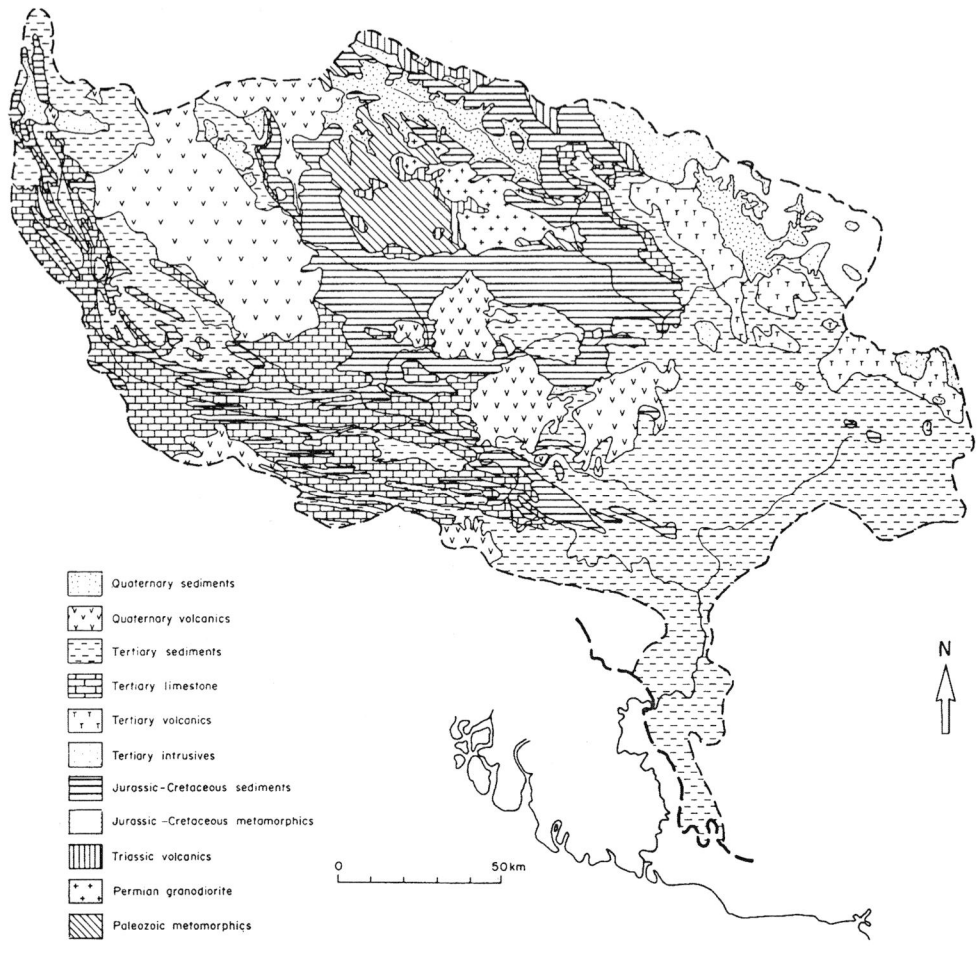

Fig. 1. Simplified geology of the Purari catchment. —— —— —— edge of delta.

shown in Fig. 1, which is derived from Bain and Mackenzie (1974, 1975), Brown and Robinson (1977) and Bain et al. (1972, 1975), together with corrections to 1981 by the present author. On Fig. 1 the rocks of the Purari catchment are divided into 11 map units on the basis of age and rock type. This generally involves combining several named rock units. These map units are discussed before they are placed in their structural setting.

2.1. Rock types (Fig. 1)

2.1.1. Paleozoic metamorphics. The Omung Metamorphics, the oldest rocks outcropping in the Purari catchment, are weakly metamorphosed slates, phyllite and shales. These rocks are comparatively resistant to erosion, and stand high in the Kubor Range.

2.1.2. Permian granodiorite. The Kubor Granodiorite intrudes into the Omung Metamorphics. The granodiorite is coarse grained, and occurs as a number of small unmapped bodies as well as the large area mapped on Fig. 1. The Kubor Granodiorite outcrops only in the area of the Kubor Range. Bain and Mackenzie (1975) consider that an Rb-Sr age of 244 m.y. most probably gives a true indication of the age of the granodiorite.

2.1.3. Triassic volcanics. These fine-grained lavas and tuffs, which include the Kana Volcanics, are found in the western parts of the Kubor Range, and along the Sepik Wahgi Divide. Like the Omung Metamorphics, they are resistant to erosion, and occur in topographically high areas.

2.1.4. Jurassic – Cretaceous metamorphics. This unit includes the Goroka and Bena Bena Formations. The rocks are weakly to moderately metamorphosed, and occur in the area east of Goroka, on the Bismarck range (Fig. 1).

2.1.5. Jurassic – Cretaceous sediments. For most of the area of this map unit, these sediments fall into the Wahgi Group comprising the Upper Jurassic Maril Shale, Lower Cretaceous Kondaku Tuff and the Upper Cretaceous Chim Formation. These rocks are shales, siltstone, sandstone and smaller amounts of volcanics and limestone. In particular the Chim Formation is important because it is a very weak rock unit, suffering severe soil erosion by both running water and massive landsliding (Blong 1981; Blong and Humphreys 1982). The Maril Shale and the Kondaku Tuff are found in areas peripheral to the Kubor Range, while the Chim Formation occurs over a much wider area which includes the southern part of the catchment.

2.1.6. Tertiary intrusives. This mapping unit includes the Bismarck Intrusive Complex and the Michael Diorite. The Bismarck Intrusive Complex, consisting of hornblende diorite, minor gabbro, and a whole range of other intrusive rock types, is found along the Bismarck Range in the northeastern part of the Purari catchment. Michael Diorite is confined to Mount Michael and small unmapped intrusive bodies to the west. The Michael Diorite has an isotopic age of 7.3 ± 0.2 m.y. (Bain and Mackenzie 1974).

2.1.7. Tertiary volcanics. The Yaveufa Formation is composed of agglomerate with interbedded lavas and tuffs, together with derived sediments. It is found southwest of Goroka, and extends eastwards to the edge of the catchment. Rocks from the Yaveufa Formation have isotopic ages ranging from 12.5 – 15 m.y. (Page and McDougall 1970).

2.1.8. Tertiary limestone. Four main formations make up this mapping unit. In the northern part of the Purari catchment there are small areas of Nembi Limestone (shelf environment), Nebilyer Limestone (shelf environment) and Chimbu Limestone (reef environment). However, the major limestone formation is the Darai Limestone (shelf environment) which is found extensively in the south and west of the catchment. The Darai Limestone is Upper Oligocene to Middle Miocene in age, while the other limestone formations are slightly older.

2.1.9. Tertiary sediments. This map unit includes all Tertiary non-limestone sedimentary rocks. The major formations are the Era Beds, the Orubadi Beds, the Movi Beds, the Aure Beds and the Pima Sandstone. The shallow marine Pima Sandstone contains material derived from the Kondaku Tuff and the Chim Formation. The Orubadi Beds are conformable on the Darai Limestone, and the Era Beds in turn are conformable on the Orubadi Beds. The Orubadi Beds tend to be weak and in the Mendi area undergo extensive landsliding. In place they are overthrust by the Darai Limestone.

2.1.10. Quaternary volcanics. There are 8 Quaternary volcanic complexes in the Purari catchment. These are Hagen, Giluwe, Ialibu, Suaru (Au), Karimui, Crater, Duau-Favenc, and Murray. These volcanics are mainly basaltic to andesitic, and contain lavas, agglomerates, tuffs and associated sediments. Löffler, Mackenzie and Webb (1980) summarise available isotopic dates for Quaternary volcanic rocks in the highlands of PNG. The oldest dated rocks come from Giluwe, where lavas near the summit have K-Ar ages of about 1.3 m.y. Bain et al. (1975) report a K-Ar date of 1.1 m.y. from the Kara Plug, southwest of Ialibu. Elsewhere Quaternary volcanics date at between 300 – 200 thousand years, with one K-Ar date of 72 000 being obtained from Crater Mountain (author's unpublished data). However, it seems likely that at least some of the volcanoes in the Purari catchment began activity at the beginning of the Quaternary, while Bain and Mackenzie (1974) suggest that Crater Mountain may have continued activity into the Holocene.

Considerable areas of deposits derived from volcanic mudflows (lahars) surround the major volcanic centres. These are discussed further in a later section of this paper.

30

Volcanic centres in the Purari catchment produced a large amount of volcanic ash (tephra). Because these tephras are widespread, but thin, they are not mapped on Fig. 1. They are up to 20 m thick in places (Pain and Blong 1976) and appear to have been derived from three distinct sources in the highlands (Pain and Blong 1979). Of these sources, two, the eastern and central sources, are in the Purari catchment, while the western source is outside. The central source is Mount Hagen, which produced a large number of tephras from a caldera complex towards the southern end of the Hagen Range. The uppermost tephra from Hagen (Tomba Tephra) is the youngest from a highland source; however, it is greater than 50 000 radiocarbon years in age.

The eastern source is probably centered on Crater Mountain, although there may also be contributions from Mount Yelia, outside the Purari catchment. The western source, tephras from which are found as far east as Mount Hagen town, has not been located, but may include Doma Peaks, Kerewa, Sisa, and perhaps even Bosavi. All these volcanoes are outside the Purari catchment.

Only the extreme southern part of the Purari catchment is likely to have escaped a covering of tephras from Quaternary volcanoes in the highlands.

2.1.11. Quaternary sediments. The areas mapped as Quaternary sediments in Fig. 1 consist of a wide variety of materials including alluvium, fanglomerates, colluvium, talus, lake sediments, and small areas of peat. These sediments occur as both relicts and as the products of present-day sedimentation. They occur mainly in the highland areas of the catchment where they occupy broad valleys formed either by tectonic warping, or more usually as a result of volcanic eruptions damming river valleys (Blong and Pain 1976).

2.2. Structure

A map of faults and folds in the Purari catchment shows that a relatively simple structure, the Kubor Anticline, separates two strongly deformed areas, the New Guinea Mobile Belt and the Papuan Fold Belt (Fig. 2). The New Guinea Mobile Belt, 50 – 100 km wide and 1600 km long, contains most of the major high angle faults and intrusive and metamorphic rocks of Mesozoic or younger age in Papua New Guinea (Bain 1973). In the Purari catchment it is dominated by the Bismarck Fault Zone, with abundant branching faults and shear zones. Bain et al. (1975) note that there is at least 3000 m of vertical dis-

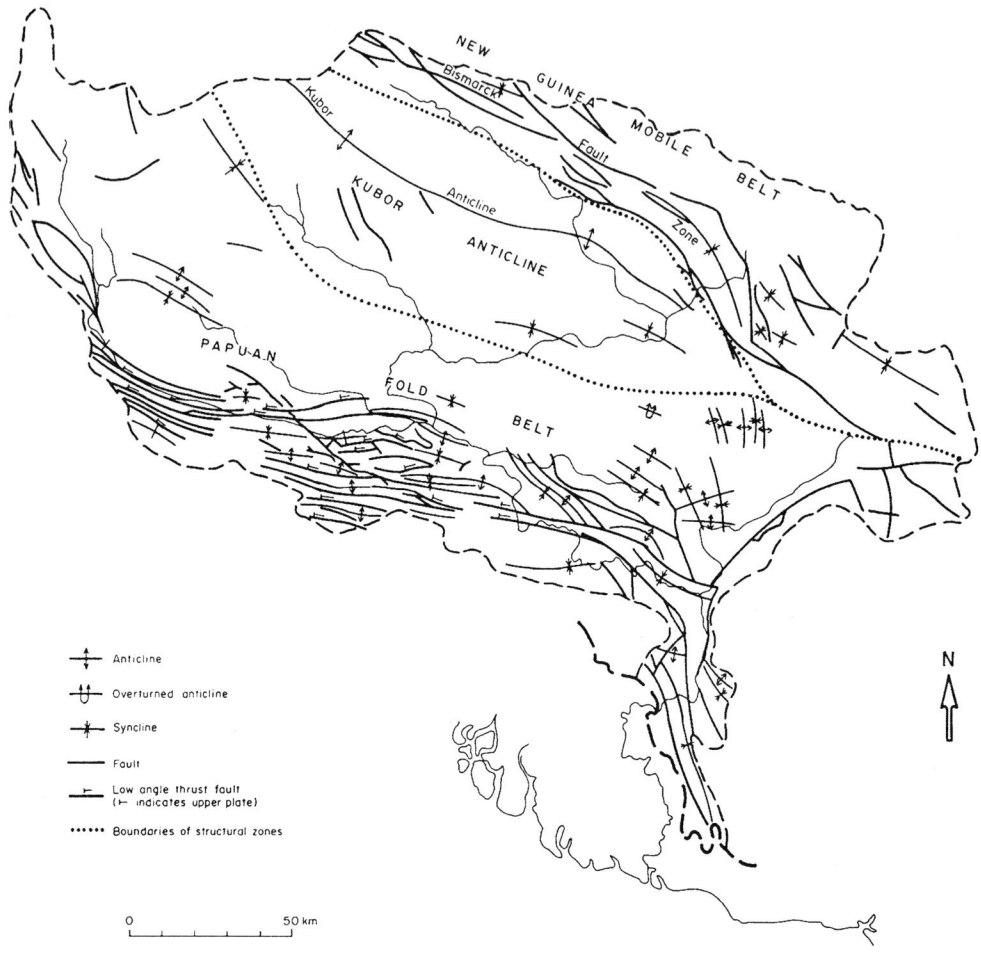

Fig. 2. Geological structure of the Purari catchment.

placement, north side up, spread over the width of the fault zone. Bain et al.
(1975) also note that, although lateral displacement may have occurred, there
is no firm evidence of such movement.

The Kubor Anticline is a broad upwarp 140 km long and up to 65 km wide.
It can be traced from Mount Hagen town in the west to Mount Michael in the
east. The core of the anticline consists of low grade Paleozoic metamorphics
intruded by Permian granodiorite. The anticline is only weakly deformed.

The Papuan Fold Belt is characterised by folding and thrust faulting from
the north. Considerable shortening of the Darai Limestone has taken place,

32

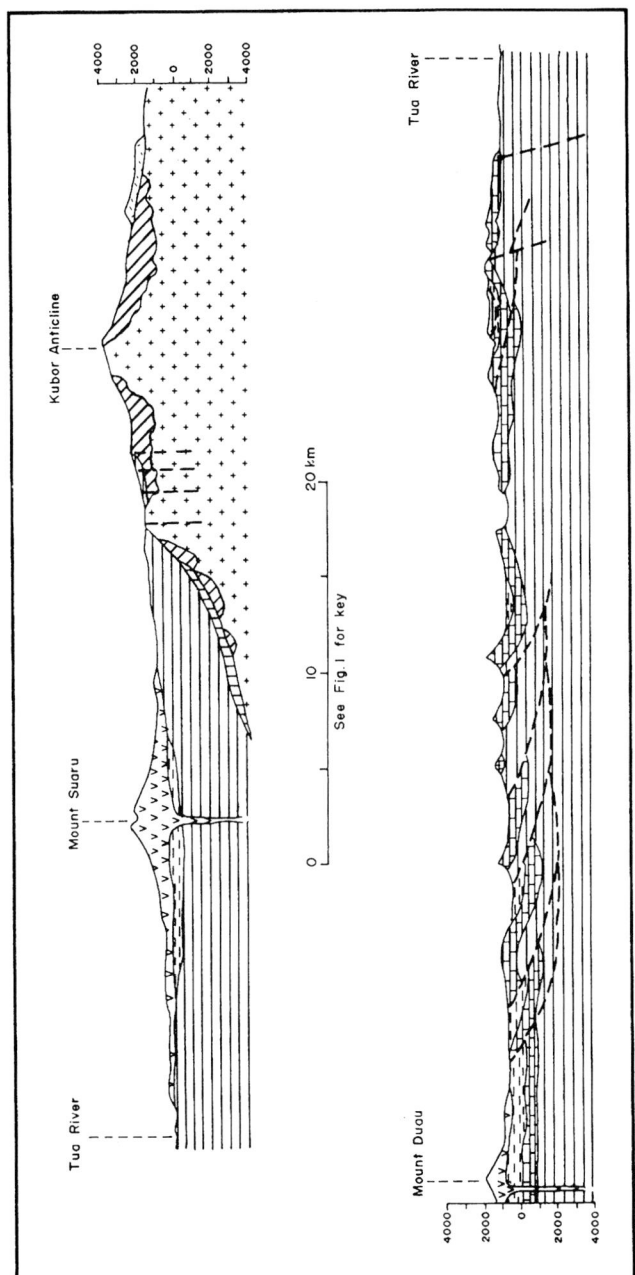

Fig. 3. Geological cross section of the southern fold mountains and the Kubor Anticline.

33

with the limestone being thrust over the Cretaceous Chim Formation (Fig. 3).

2.3. Surficial materials

A section on surficial materials is not often included in a regional description of geology and geomorphology. However, many rivers in the Purari catchment have a turbid appearance, and this discussion is included in a volume concerned primarily with the environmental impacts of the Wabo dam. It therefore seems appropriate to consider the nature of the materials that are most likely to provide the sediments that will eventually fill the lake.

2.3.1. Volcanic ash. Surveys carried out by the CSIRO (Perry et al. 1965; Haantjens et al. 1970) first established the presence of volcanic ash in the PNG highlands, and more recent work (Pain and Blong 1976, 1979; Pain and Wood 1976; Pain unpublished) has established that this ash is spread over much of the highlands. The ash is well bedded, and ranges in thickness from as much as 20 m near Mount Hagen town to as little as 0.5 – 1 m in areas east of Goroka and to the south. It is also strongly weathered, and it provides the parent material for a group of soils that are easily worked and have the ability to withstand long periods of continuous cultivation.

Volcanic ash is present on slopes up to 35° on all except unstable rocks (Blong and Pain 1978). Over much of the Purari catchment, then, the absence of volcanic ash indicates slopes over 35° or unstable underlying rocks. Only in the southern and eastern extremes is it probable that there is little if any cover of volcanic ash, by reason of distance from source.

Because volcanic ash is not easily eroded, areas covered with ash are not important sources of sediment at present.

2.3.2. Colluvium and alluvium. In general there is very little late Quaternary colluvium and alluvium in the Purari catchment. Most of the Quaternary sediments shown in Fig. 1 are covered with volcanic ash. However, the younger colluvium, especially that derived from rocks such as the Chim Formation, and more particularly alluvium, are important sources of sediment in the Purari River system. As Pickup (this volume) points out, river bank collapse contributes important amounts of sediment to channels.

2.3.3. Weathering profiles. Weathering profiles are important sources of sediment in some rock types throughout the catchment. As mentioned above, the Cretaceous Chim Formation and the Upper Tertiary Orubadi Beds are weak,

34

and often sheared, and are easily eroded both by surface wash and by land-slides. These materials produce considerable amounts of sediment (Blong 1981; Blong and Humphreys 1982).

On slopes over 35° over most of the catchment, and in areas of unstable rocks and where volcanic ash is not important, weathering profiles comprise most of the surficial material. Weathering in PNG has been discussed in some detail by Haantjens and Bleeker (1970). For the most part, on steep slopes and in areas where volcanic ash does not cover the surface, the soil and weathering profiles are thin, often less than 1 m. There are no quantitative data on sedi-ment yield from these kinds of profiles in PNG, but qualitative observations suggest that surface wash is more important on these materials than on volcanic ash. This is related to the fact that volcanic ash soils are highly permeable with high structural stability, while soils on other materials are less so.

A pertinent observation here is that of Pickup (this volume) who notes that the average annual sediment load of the Purari River at Wabo is 40 million m^3/yr, while that of the Aure River at its confluence with the Purari is 48 million m^3/yr. This is despite the fact that the catchment of the former is at least 5 times that of the latter. The fact that the Aure River has very little volcanic ash cover, while the Purari above Wabo has a very widespread cover of ash may account for the difference in sediment load, because the Tertiary sediments of which the Aure catchment is composed are particularly suscepti-ble to erosion. Furthermore, the area with least volcanic ash cover in the Purari catchment above Wabo is near Wabo, and Pickup (this volume) sug-gests from the rating curve at Wabo that most of the sediment load in the Purari River is derived from near the dam site. A volcanic ash cover may thus be an important factor in the generally low sediment yields in much of the highland part of the Purari catchment.

3. Geomorphology

Given the wide variety of rock types and the strong structural control, it is not surprising that there is a wide variety of landforms as well. This variety of landforms is briefly described before the geomorphic history of the area is considered.

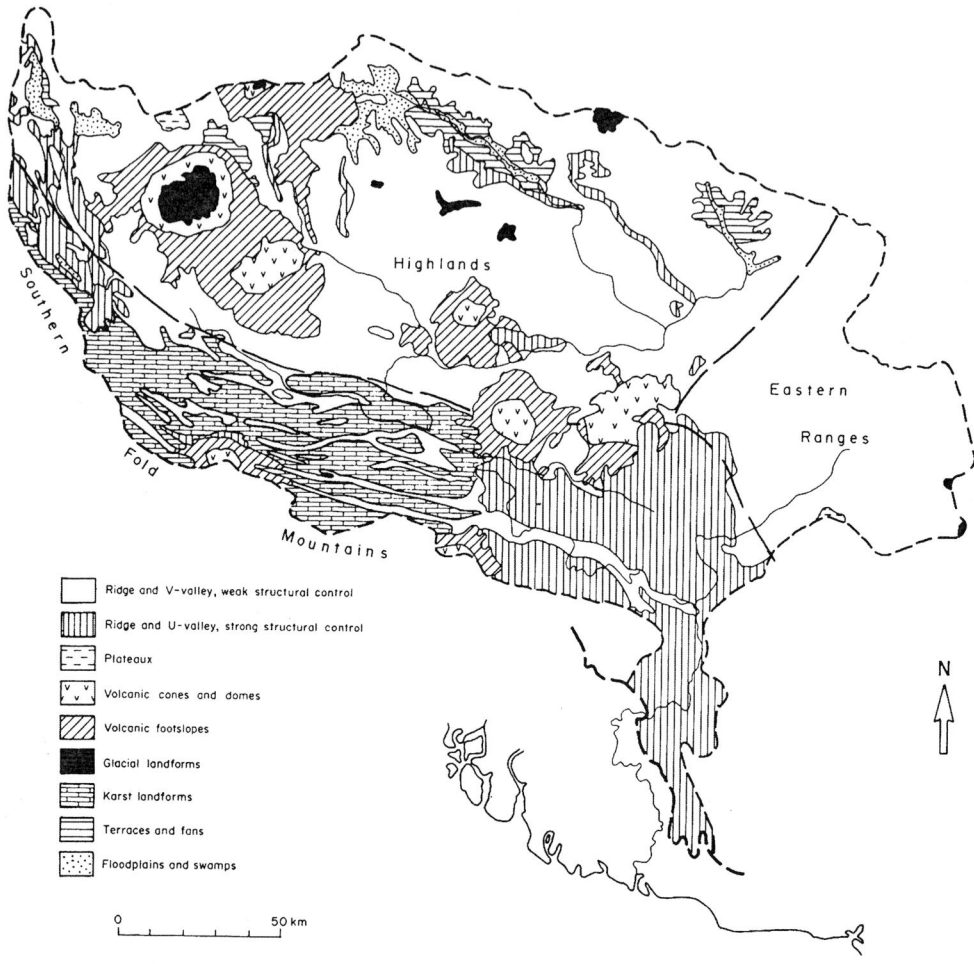

Ridge and V-valley, weak structural control

Ridge and U-valley, strong structural control

Plateaux

Volcanic cones and domes

Volcanic footslopes

Glacial landforms

Karst landforms

Terraces and fans

Floodplains and swamps

0   50 km

N

Fig. 4. Landforms of the Purari catchment.

3.1. Landform types

Nine groups of landforms are defined in the Purari catchment. Their distribution is shown in Fig. 4, which is modified after Löffler (1974).

3.1.1. Ridge and V-valley forms, with weak structural control. There is a great variation in the details of these landforms, but their overriding features are their sharp crests, steep and straight valley sides, and V-shaped valleys. They

36

are the most widespread landforms in the Purari catchment, occurring predominantly in the northern half, but also in smaller areas in the south (Fig. 4). Löffler (1977) points out that these forms are often considered to be the typical landform of the humid tropics. The sharp crests are in places less than 1 m wide, and are rarely rounded. There is thus a general lack of an upper convex slope. The valley sides may be irregular in detail, but the overall pattern is of straight slopes leading from the crest right to the edge of the channel. The valley floors are narrow and V-shaped, with comparatively rare accumulations of depositional material.

Drainage patterns are dendritic, or weakly modified dendritic, and drainage densities and slope angles vary on different rock types. Soft rocks tend to have dense drainage patterns, while more indurated rocks have lower drainage densities. Slope angles on valley sides range from 25° to more than 35°. These landforms show evidence of active downcutting, and their form suggests that slope development is controlled by the rate of weathering of the underlying bedrock. In some places, however, thick covers of volcanic ash on slopes up to 35° show that some other control must be operating. These volcanic ash-covered areas have attributes of relict surfaces, and are discussed further below.

3.1.2. Ridge and V-valley forms, with strong structural control. Strong structural control is evident in the landforms of the southern part of the Purari catchment, in the Papuan Fold Belt, as well as in a few places elsewhere (Fig. 4). Landforms are mainly of the ridge and V-valley form, but ridge and valley directions are controlled by the strike of the underlying rock. In addition slope angles are controlled by the dip of the rock. They thus consist of a series of dip and scarp slopes, forming cuestas and hogbacks with slope angles depending on the dip of the rock. Often these landforms reflect the alternation of soft and hard beds in the underlying rock. However, apart from the strong structural control, the comments on ridge and V-valley forms in the previous section apply also to the landforms described here.

3.1.3. Plateaux. Although plateaux are relatively common in Papua New Guinea, only three small areas are found in the Purari catchment. These are the Sugarloaf Plateau, west of Mount Hagen, a small unnamed plateau between Mount Ialibu and Mount Suaru, and another small plateau in the Kratke Range, south of the Aure River. The Sugarloaf Plateau is formed on volcanic rocks and owes its low relief to flat-lying lavas and volcanic detritus laid down in depressions in the original volcanic surface. Small volcanic cones and domes, of which the Sugarloaf is the highest are a feature of this surface.

The other two plateaux are formed on limestone. All three have low relief and slopes of less than 20°. They are presumably remnants of once larger areas with similar landforms, since all three are surrounded by steep erosional slopes.

3.1.4. Volcanic cones and domes. The eight Quaternary volcanic complexes which lie wholly or partly within the Purari catchment are dominant landforms. Since none of them are now active, they have been eroded so that only parts of their original surfaces are still present. The degree of dissection varies from volcano to volcano, although, with the exception of Mount Giluwe, all are conical strato-volcanoes with slopes up to 30° on the main cones and often steeper on the walls of streams which dissect them. Giluwe is a shield volcano and has convex sides with many small satelitic volcanoes (Blake and Löffler 1971). Crater Mountain is a complex with several main centres forming a basement on which there are a number of small well-preserved cones (Bain et al. 1975). All the others are simple volcanic cones with concave slopes. Some of these volcanoes (e.g. Karimui) still have planaze remnants on their flanks. However, others (e.g. Lalibu) are much more highly dissected. Where remnant surfaces are not preserved on the sides of the volcanoes, ridge and V-valley landforms predominate.

Ollier and Mackenzie (1974), noting that all except Mount Hagen are more deeply incised on the southern side, suggested that this may be because the main rain-bearing winds come from that direction. However, it is possible that depth of dissection also relates to available relief and the retreat of knickpoints, since in general the sides with lower basal altitudes are also the most deeply dissected.

3.1.5. Volcanic footslopes. Footslopes of varying extent surround the eight volcanic cones (Fig. 4). These footslopes are mainly low angle slopes resulting from deposition of volcanic debris during the eruptions (e.g. lavas and agglomerates), or secondary deposition from lahars and alluvial fans. Lahars are particularly important around Hagen and Giluwe. Around Hagen they extend for long distances down tributary valleys of the Kaugel River, as well as to the north, outside the Purari catchment. Within these river valleys they form high, gently sloping terraces. Around Mount Hagen town, there is a well-developed lahar moundfield. Lahar deposits are found underlying terraces in tributaries of the Erave River, and are derived from Mount Giluwe. In fact, lahar deposits from Giluwe have been traced as far downstream as Erave (R.J. Blong, personal communication), although they cannot be mapped at the scale of Fig. 4.

Elsewhere volcanic footslopes and fans surround the volcanic cones. In most places there are large areas of remnant surfaces, indicating that there has not been complete dissection of these footslopes. This landform complex thus consists of flat or gently sloping surfaces of original deposition separated by steep-sided and often deep ravines.

3.1.6. Glacial landforms. Late Pleistocene glacial landforms in the Purari catchment are found on Mount Giluwe, Mount Hagen, Mount Wilhelm, several peaks in the Kubor Range, Mount Piora, and Table Top Mountain. These landforms, best described by Löffler (1972), consist mainly of cirques and small U-shaped valleys descending to altitudes of about 3550 m. Only on Giluwe was glaciation very extensive during the Pleistocene. There an ice cap of 188 km² led to the development of a glaciated plateau and several well-developed U-shaped valleys. On Giluwe there are a series of recessional moraines, but on the other mountains only terminal moraines still exist.

3.1.7. Karst landforms. Most of the southwestern part of the Purari catchment consists of karst landforms (Fig. 4). These landforms, the result of solution of limestone by water, are notable for the lack of any organised surface drainage pattern. Most water sinks into the limestone and is carried away by an underground drainage system.

Although Löffler (1974) notes the occurrence of 4 kinds of karst landforms in PNG, polygonal karst dominates in the Purari catchment. Polygonal karst takes its name from the polygon-shaped depressions that occupy much of this landform type. The polygonal depressions are surrounded by a variety of different shaped hills and towers. In many places the often angular nature of these landforms is somewhat subdued by a cover of volcanic ash, which gives the forms a more rounded appearance (Pain, in press). Polygonal karst is fully described by Williams (1971).

3.1.8. Terraces and fans. In the headwater areas of the Lai, Kaugel, Nebilyer, Nembi, Wahgi and Asaro Rivers there are wide basins filled with fluviatile deposits. Around the margins of these as well as a number of smaller basins there are a series of relict alluvial terraces and fans up to 60 m above the present channels. In some valleys (e.g. the upper Kaugel Valley) flat terraces are the remnants of old lake floors. Fans may, in part, be composed of colluvial materials rather than alluvium, and are often truncated by the valleys of the present channels. Terraces are incised to various depths. The landforms are thus flat to gently sloping surfaces separated by gullies and ravines.

3.1.9. Floodplains and swamps. Floodplains and swamps are well developed in the same areas as the terraces and fans. Floodplains occur along the narrow valleys cut into the terraces and fans, and swamps are found where there is some obstruction to water flow. The Kandep swamps, at the head of the Lai River, are the largest swamps in the Purari catchment, with an area of approximately 450 km². There are also large swamps in the headwaters of the Wahgi River. There is a whole range from comparatively 'dry' floodplains to swamps that become lakes during prolonged rain. In the Kandep swamps, permanent lakes are found in the swamps, and are separated from the main channels by narrow ridges of sediment (Jennings 1963).

3.2. Geomorphic regions

Löffler (1977, Fig. 6) divides PNG into 14 geomorphic regions. Of these, three occur in the Purari catchment (Fig. 4). The southern fold mountains, as the name suggests, are dominated by structurally controlled landforms, mainly in the form of strike ridges, cuestas and hogbacks. This region also contains nearly all the karst landforms of the catchment. Elevations tend to be lower than in the highlands and erosional landforms dominate over depositional forms.

The highlands have larger areas of depositional forms, and are higher, generally lying over 1500 m in altitude. Although there are volcanoes in the southern fold mountains, volcanic forms reach their best expression in the highlands. Dow (1977) makes the observation that the highlands of PNG are a faulted and, inplaces, deeply dissected high plateau. Certainly, with their relatively low relief, broad intermontane valleys, and fairly abrupt transition to surrounding high relief and dominantly ridge and valley landforms, the highlands have a broad plateau-like appearance.

The third geomorphic region in the Purari catchment was called the eastern metamorphic ranges by Löffler (1977). Here, however, this region is simply called the eastern ranges, for in the Purari catchment there are only rare outcrops of metamorphic rocks, the eastern ranges being underlain in the main by Tertiary sediments of the Aure Beds. The eastern ranges consist mainly of ridge and V-valley landforms, with only slight structural control, at altitudes generally below 1500 m, although in places (e.g. Mount Piora) they rise to more than 3000 m and have glacial landforms (Fig. 4).

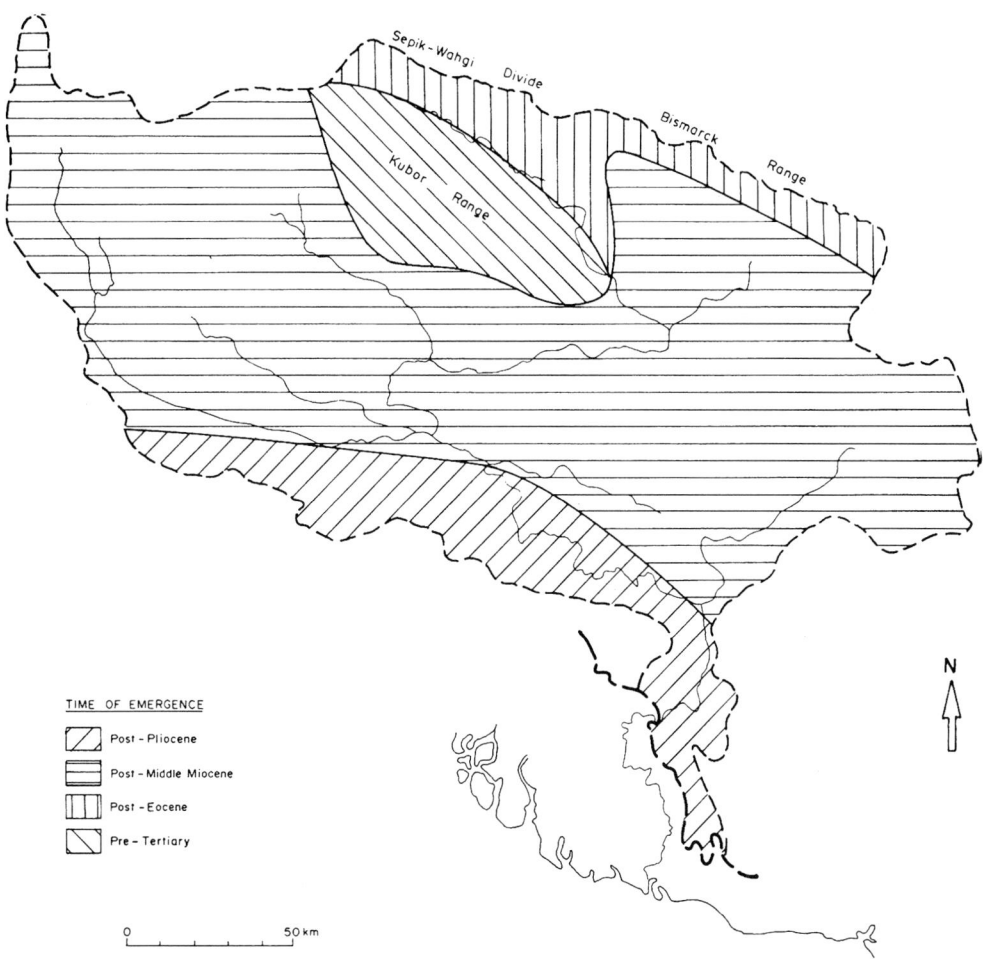

Fig. 5. Stages of emergence of the Purari catchment.

3.3. Geomorphic history

Using source material provided in Dow (1977), a map of the Purari catchment showing age of emergence was prepared (Fig. 5). The Kubor Range, according to Bain et al. (1975), has been land since the middle of the Jurassic. A small part of the Purari catchment, along the Sepik-Wahgi Divide and the Bismarck Range, became land between the Eocene and Middle Miocene, but the major emergence began after the Middle Miocene and intensified during the Pliocene. The southern part of the catchment remained an area of marine sedimentation during the Pliocene. Pliocene sediments such as the Wongop Sandstone and the Era Beds grade from shallow marine sediments to estuarine-deltaic sediments, showing progressive emergence.

Erosional development of the Sepik-Wahgi Divide, the Bismarck Range and the Kubor Range was thus well under way by the Middle Miocene. However, it was not until the emergence of the southern part of the highlands, and the southern fold mountains, that the landforms of the Purari catchment began to take on their present form. The beginning of this uplift and emergence is usually placed in the Upper Tertiary and most workers consider that uplift continued throughout the Quaternary (e.g. Bain et al. 1975; Dow 1977).

It has become usual to discuss the overall development of major landform features in PNG in terms of the interaction of the Australian and Pacific plates during the late Cainozoic (e.g. Löffler 1977; Pain 1978). In the Purari catchment this interaction is supposed to have led to regional uplift, with fault movement in the New Guinea Mobile Belt to the north of the Kubor Anticline, and intense folding and thrust faulting due to compression and gravity sliding to the south. Regional uplift, particularly along the Kubor Anticline and the eastern ranges certainly explains much of the present distribution of landforms in the area, but compression as a mechanism for folding no longer seems adequate to explain so-called fold mountains. It seems more likely that folding deformation of rocks takes place before uplift (Ollier 1981).

In the southern fold mountains uplift was preceded or accompanied by spectacular folding and thrust faulting of Tertiary sediments, mainly the Darai Limestone (Upper Oligocene – Middle Miocene) (Fig. 3) (Jenkins 1974). In the northern part of the fold mountains the Orubadi Beds (Upper Miocene – Pliocene) were involved, while to the south the Era Beds (Pliocene- –Pleistocene) are also involved. This sequence fits with progressive emergence from the north, beginning in the Pliocene and continuing in the Quaternary. Gravity sliding followed by emergence thus moved south with time, and may well be continuing at the present time in the area south of the present coastline, perhaps involving Holocene deltaic sediments as well.

Following emergence, erosion of the uplifted sediments proceeded in the normal way, with structurally controlled ridge and V-valley forms developing on clastic sediments, and karst landforms developing on the limestone. Present drainage lines are probably little different from those that were formed in the initial depressions on the newly emerged land surface.

Although the Kubor Range and the ranges on the divide to the north were land by the Middle Miocene (Fig. 5), the rest of the highlands landform region did not become land until later. It seems likely that the peripheral highland areas, and perhaps the eastern ranges as well, emerged during the late Miocene or early Pliocene during the first stages of regional uplift. Immediately following uplift the highlands were deeply incised and the main watersheds were established before the eruption of the highland volcanoes (Bik 1967).

Warping, differential uplift and volcanism all played a part in the development of the basins that are found along the northern side of the Purari catchment (Fig. 4). In many places downstream from these basins, steep, deeply incised major rivers (e.g. the Wahgi below Kundiawa) suggest that there may have been a period in the development of the highlands when base level was higher than it is today. This is supported by the relict ridge and V-valley landforms reported by Löffler (1977) in small areas of the highlands. However, there is very little evidence left of any hiatus in uplift that may have led to partial planation of the area.

Volcanic activity in the Purari catchment spanning all but the last part of the Quaternary has had a major impact on landforms in the catchment. This impact includes the formation of the major volcanic piles which are a prominent feature of the landscape, and the production and deposition of large amounts of mudflow and tephric material over a considerable portion of the catchment. In addition, volcanic damming of rivers led to the formation of some of the largest highland basins in the area.

Following uplift and incision, and during the last stages of volcanic activity in the highlands, infilling of the highland basins took place. This infilling, by alluvial, colluvial, lacustrine and swamp deposits (Blong and Pain 1976; Pain 1978), took place in various stages and over varying lengths of time. The presence of relict terraces and fans in some basins shows that they were infilled in at least two stages. In the Kaugel Valley infilling was completed shortly after Tomba Tephra was deposited in the area, and the sediments are now being incised, while in the Kandep Basins and in the upper Wahgi Valley, infilling is still continuing.

During the last part of the Pleistocene two periods of glacial activity produced glacial landforms on the highest mountains in the catchment (Löffler 1972). Since then incision and deposition has continued at a slow rate. This is demonstrated by the fact that a large part of the catchment is covered with tephra units more than 50 000 years old, indicating stability over that period.

The geomorphic history of the eastern ranges seems to have been very simple. The area became land sometime during the Pliocene, and has since undergone erosion, possibly episodic, to form the ridge and V-valley landforms that occupy almost all of this region. Localised glaciation is the only interruption to this process that can be demonstrated in the present landscape.

The geomorphic history of the Purari catchment may thus be summarised as being one of uplift, with gravity sliding in the south and faulting in the north, followed by incision, volcanism, and basin formation and infilling.

4. Conclusion

The landforms of the Purari catchment have been developed largely by erosion of rocks that have emerged from the sea since the Middle Miocene. The resulting forms are mainly ridge and V-valley landforms, some of which are strongly controlled by underlying rock structure. Constructional landforms in the catchment include volcanic forms, and fans and terraces of alluvial and colluvial origin. A cover of volcanic ash which lies over most of the catchment has an important influence on sources of sediment that is carried by the Purari River. Only areas which have little or no ash cover contribute large amounts of sediment.

References

Bain, J.H.C., 1973. A summary of the main structural elements of Papua New Guinea. In: P.J. Coleman (ed.), The Western Pacific: Island Arcs, Marginal Seas, Geochemistry, pp. 147 – 161. University of Western Australia Press, Perth.

Bain, J.H.C., H.L. Davies, P.D. Hohnen, R.J. Ryburn, I.E. Smith, R. Grainger, R.J. Tingey and M.R. Moffat, 1972. Geology of Papua New Guinea 1:1 000 000 map. Bureau of Mineral Resources Australia, Canberra.

Bain, J.H.C. and D.E. Mackenzie, 1974. Karimui, Papua New Guinea, Sheet SB/55 – 9. 1: 250 000 Geology Series, Explanatory Notes, Bureau of Mineral Resources Australia, and Geological Survey of Papua New Guinea.

Bain, J.H.C. and D.E. Mackenzie, 1975. Ramu, Papua New Guinea, Sheet SB/55 – 5. 1:250 000 Geology Series, Explanatory Notes, Bureau of Mineral Resources, Australia, and Geological Survey of Papua New Guinea.

Bain, J.H.C., D.E. Mackenzie and R.J. Ryburn, 1975. Geology of the Kubor Anticline, central highlands of Papua New Guinea. Bureau of Mineral Resources Australia Bulletin 155. 106 pp.

Bik, M.J.J., 1967. Structural geomorphology and morphoclimatic zonation in the central highlands, Australian New Guinea. In: J.N. Jennings and J.A. Mabbutt (eds.), Landform Studies from Australia and New Guinea, pp. 26 – 47. Australian National University Press, Canberra.

Blake, D.H. and E. Löffler, 1971. Volcanic and glacial landforms on Mount Giluwe, Territory of Papua and New Guinea, Bull. Geol. Soc. Am. 82: 1605 – 1614.

Blong, R.J., 1981. The Time of Darkness: Local legends and volcanic reality in Papua New Guinea. Australian National University Press, Canberra.

Blong, R.J. and G.S. Humphreys, 1982. Erosion of road batters in Chim Shale, Papua New Guinea. Inst. of Engineers Australia. Civil Eng. Trans. CE24: 62 – 68.

Blong, R.J. and C.F. Pain, 1976. The nature of Highland valleys, central Papua New Guinea. Erdkunde 30:212–217.

Blong, R.J. and C.F. Pain, 1978. Slope stability and tephra mantles in the Papua New Guinea highlands. Geotechnique 23: 206–210.

Brown, C.M. and G.P. Robinson, 1977. Explanatory notes on the Lake Kutubu Geological Sheet. Geological Survey of Papua New Guinea, Report 77/12, and 1:250 000 geological map (preliminary).

Dow, D.B., 1977. A geological synthesis of Papua New Guinea. Bureau of Mineral Resources Australia. Bulletin, 201.

Haantjens, H.A. and P. Bleeker, 1970. Tropical weathering in the Territory of Papua New Guinea. Australian J. Soil Research 8: 157–177.

Haantjens, H.A., J.R. McAlpine, E. Reiner, R.G. Robbins and J.C. Saunders, 1970. Lands of the Goroka–Mount Hagen area, Territory of Papua and New Guinea. C.S.I.R.O. Land Research Series, No. 27. C.S.I.R.O., Melbourne.

Jenkins, D.A.L., 1974. Detachment tectonics in western Papua New Guinea, Bull. Geol. Soc. Am. 85: 533–548.

Jennings, J.N., 1963. Floodplain lakes in the Ka Valley, Australian New Guinea. Geographical Journal 129: 187–190.

Löffler, E., 1972. Pleistocene glaciation in Papua and New Guinea. Z. für Geomorph. suppl. 13: 32–58.

Löffler, E., 1974. Explanatory notes to the geomorphological map of Papua New Guinea. C.S.I.R.O. Land Research Series, No. 33. C.S.I.R.O., Melbourne.

Löffler, E., 1977. Geomorphology of Papua New Guinea. Australian National University Press, Canberra. 195 pp.

Löffler, E., D.E. Mackenzie and A.W. Webb, 1980. Potassium-argon ages from some of the Papua New Guinea highland volcanoes, and their relevance to Pleistocene geomorphic history. J. Geol. Soc. Aust. 26: 387–397.

Ollier, C.D., 1981. Tectonics and landforms. Longmans, London. 324 pp.

Ollier, C.D. and D.E. Mackenzie, 1974. Subaerial erosion of volcanic cones in the tropics. J. Trop. Geography 39: 63–71.

Page, R.W. and I. McDougall, 1970. Potassium-argon dating of the Tertiary $f_{1/2}$ Stage in New Guinea and its bearing on the geological time-scale. American Journal of Science 269: 321–342.

Pain, C.F., 1978. Landform inheritance in the central highlands of Papua New Guinea. In: J.L. Davies and M.A.J. Williams (eds.), Landform Evolution in Australasia, pp. 48–69. Australian National Univ. Press, Canberra.

Pain, C.F., in press. Geology, landforms and landuse distribution on the Nembi Plateau, Southern Highlands Province, Papua New Guinea. University of P.N.G. Occasional Papers in Geography.

Pain, C.F. and R.J. Blong, 1976. Late Quaternary tephras around Mount Hagen and Mount Giluwe, Papua New Guinea. In: R.W. Johnson (ed.), Volcanism in Australasia, pp. 239–251. Elsevier, Amsterdam.

Pain, C.F. and R.J. Blong, 1979. The distribution of tephras in the Papua New Guinea highlands. Search 10: 228–230.

Pain, C.F. and A.W. Wood, 1976. Tephra beds and soils in the Nondugl–Chuave areas, Western Highlands and Chimbu Provinces. Science in New Guinea 4: 153–164.

Perry, R.A., M.J. Bik, E.A. Fitzpatrick, H.A. Haantjes, J.R. McAlpine, R. Pullen, R.G. Robbins, G.K. Rutherford and J.C. Saunders, 1965. General report on lands of the Wabag–Tari area, Territory of Papua and New Guinea, 1960 – 61. C.S.I.R.O. Land Research Series, No. 15. C.S.I.R.O., Melbourne.

Pickup, G., 1983. Sedimentation processes in the Purari River upstream of the delta. This volume, Chapter I, 11.

Thom, B.G. and L.D. Wright, 1983. Geomorphology of the Purari Delta. This volume, Chapter I, 4.

Williams, P.W., 1971. Illustrating morphometric analysis of karst with examples from New Guinea. Z. für Geomorph. 15: 40 – 61.

4. Geomorphology of the Purari Delta

B.G. Thom and L.D. Wright

1. Introduction

The Purari Delta is one of a number of large deltaic complexes which border the Gulf of Papua. Along with the Fly, the Kikori and many other rivers, the Purari drains the western and central highland region of Papua New Guinea (Fig. 1). Upper sections of these rivers are located in highly mountainous terrain (to 4510 m, Mt. Wilhelm) with steeply descending valleys debouching onto a deltaic plain 30 to 50 km wide. Average annual rainfalls ranging from 2000 mm to 8500 mm in the catchment of the Purari result in a mean annual discharge at Wabo of about 2360 m³/sec, carrying 88.6 million m³/year of sediment into the delta (Pickup 1980; this volume). These inputs provide the material for a major deltaic complex of global significance.

Our studies of various deltaic regions of the world, encompassing a variety of climatic, geologic and hydrologic settings, have shown that several factors control delta development (Wright et al. 1974). These include: (i) geologic setting in relation to Quaternary tectonics and sea-level change; (ii) fluvial contribution to the deltaic plain in the form of water and sediment discharge; (iii) climatic and vegetation of the drainage basin and deltaic plain; and (iv) the dynamic regime of the receiving basin (in this case the Gulf of Papua). The receiving basin serves as a sink for sediments and energy discharged by the river. In many areas like the Gulf of Papua, wind, wave and tidal forces in the basin play a major role in molding delta shape. The slope and shape of the subaqueous profile fronting a delta is also important; it affects the rate of horizontal sediment accumulation so that deltas can prograde faster over flat, shallow shelves than over steep ones. Furthermore, wave power which reaches the delta shoreline is as much or more of a function of offshore slope as it is of the deepwater wave characteristics (Wright and Coleman 1973).

The aim of this chapter is to describe the main morphologic and dynamic

Petr, T. (ed.) The Purari – Tropical environment of a high rainfall river basin
© *1983, Dr W. Junk Publishers, The Hague / Boston / Lancaster*
ISBN-13: 978-94-009-7265-0

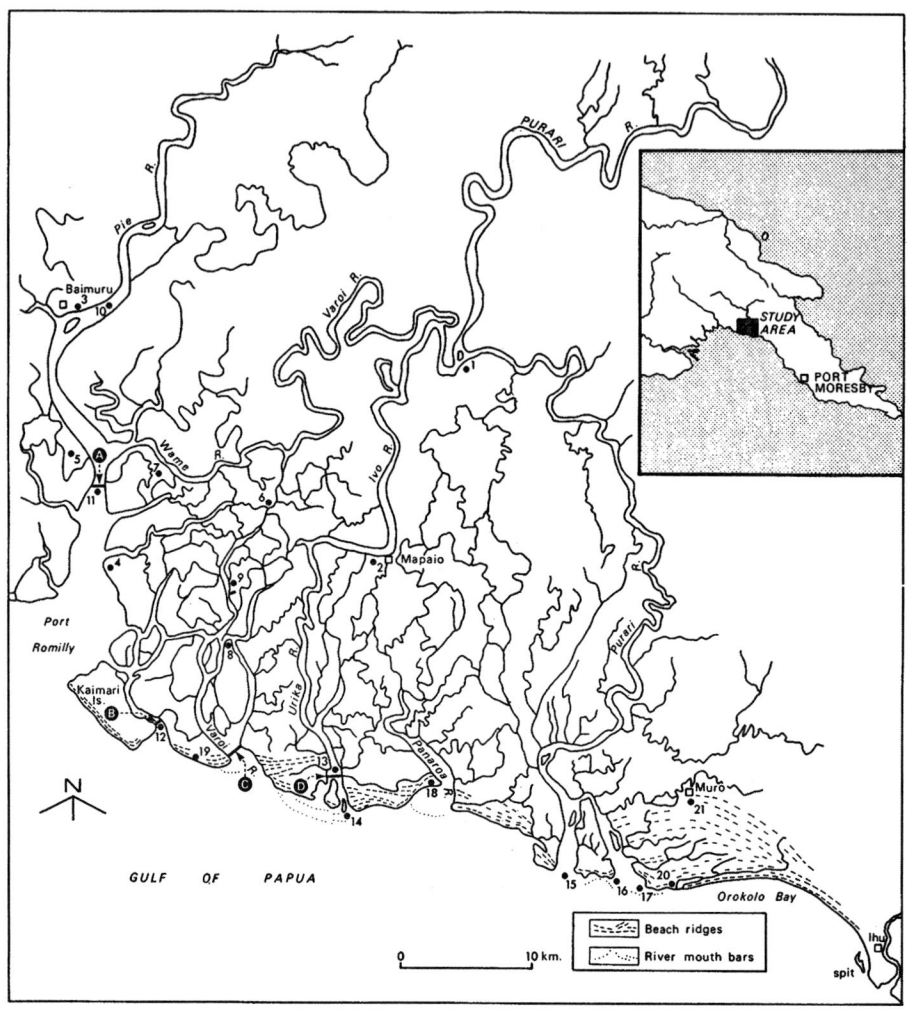

Fig. 1. Purari Delta, Gulf of Papua, showing general morphology and location of samples (Table 1) and channel sections (Fig. 5).

characteristics of the Purari, and to compare briefly these characteristics with other deltas which we have studied. The field investigation on which this is largely based was limited to a two-week period in July, 1979. This study has been supplemented by an analysis of aerial photographs, and by field data collected by others, especially T. Petr.

2. Geologic setting

The stratigraphy of the Purari Delta region has been investigated by deep drilling and seismic reflection for oil exploration. The coastal margin is actively subsiding in association with crustal downwarping of the Papuan basin.

Throughout the Mesozoic and Cainozoic sediments have accumulated in a basin between a relatively stable shallow shelf region to the west and the volcanic belt east of the Purari Delta (Khan 1974). Two north – south trending troughs developed in the Miocene, the Omati and Aure. Over 3000 m of sediments have accumulated in these troughs separated by the 'Wana Swell' (A.P.C. 1969). Near the presently active delta more or less continuous sedimentation persisted in the Eocene, Lower and Middle Oligocene and throughout the Lower Miocene. Limestones, probably of reefal origin, are common in the Lower Miocene, indicating that the western part of the delta at this time was some considerable distance from the present shoreline (Khan 1974). Middle and Upper Miocene sediments are mainly marls and mudstones with some evidence of emergence and erosion.

The Pliocene sediment in the Delta region consists of a thick, brackish, marine and estuarine series of sandstones, mudstones and coal, which conformably overlies the darker fully marine shales of Miocene age . . . The Pliocene sediments were laid down in a very shallow marine embayment extending over the Delta region and they became transitionally terrestrial northwards (ibid, p. 268).

Strong folding occurred in the northern part of the basin in the Middle Miocene to form the embryonic core of the central mountains (Ruxton 1969). Tectonic activity took place on a north-northwest – south-southeast trend. Complementary subsidence continued to the southwest into the Quaternary with volcanic detritus associated with widespread eruptions to the east accumulating within Pleistocene estuarine deposits (Khan 1974).

Figure 2 is a graphic log of Wana No. 1, a drill hole west-northwest of Baimuru on the Era River. Similar results were obtained from Iviri No. 1 to the south (see Khan 1974, p. 273). Of interest to modern deltaic studies, is the relatively thick accumulation of Pliocene and Quaternary sediments. At Wana these units are 1700 m thick with their thickness increasing eastwards indicating marked subsidence in relatively recent times (A.P.C. 1969). Deposition was wholly in shallow water in a marginal marine environment. Fine-grained sediments, such as mudstone, predominate with the proportion of sandstone layers increasing upwards. This implies that modern depositional environments are similar to those which prevailed in this area throughout much of the last 5 million years. The tectonic setting appears to have favoured the preservation of a thick mud record shed from rapidly eroding, uplifted highlands to the north.

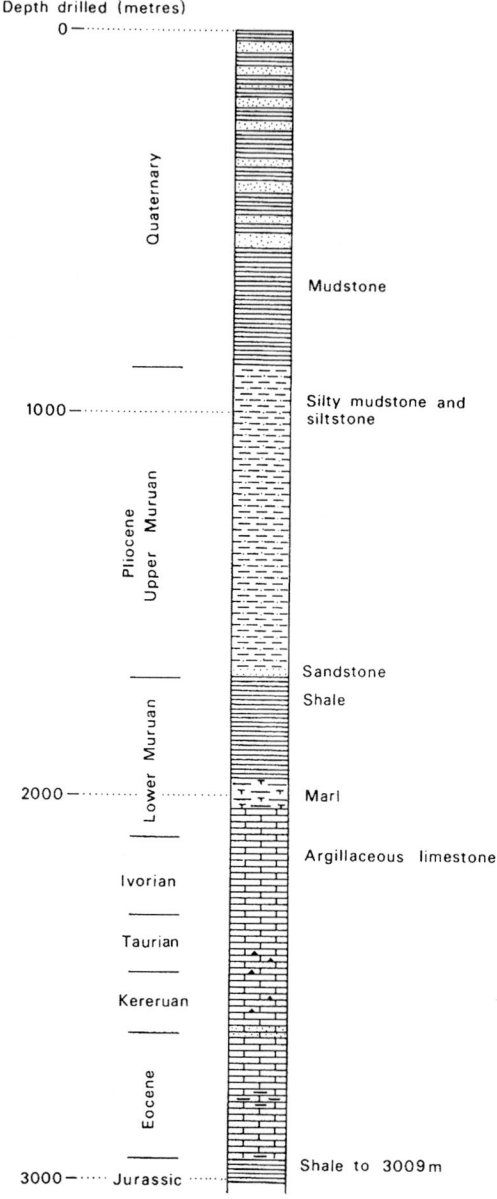

Fig. 2. Log of Wana No. 1 drill hole located west-northwest of Baimuru on the Era River (source: BP Petroleum Dev. Aust., July, 1968).

50

3. Modern dynamic environment

Details of the climatic, ecologic and hydrologic environment of the Purari Delta are discussed elsewhere in this book. It is important to note the following regional characteristics:

The maximum recorded discharge rate of the Purari River between 1961 and 1978 was 10450 m³/sec (Pickup and Chewings, this volume) which is close to half the flood-stage discharge rate of the Mississippi River. The Purari delivers approximately 88.6 million m³/year of sediment to the head of the delta of which 6% is sand, 48% is silt and 46% is clay (Pickup, this volume). Within the delta region, this total water and sediment load is disseminated seaward by way of several distributions, with the largest percentage of the total load being carried by the Wame-Varoi, Ivo-Urika, and Purari channels (Fig. 1). The remainder of the water-sediment load appears to be distributed throughout the delta complex by way of an intricate network of smaller channels or *ehes*.

The delta region experiences semi-diurnal tides with a mean spring range of 3 metres (mesotidal). Tidal transports are thus significant in the lower estuaries of the delta region as well as along the delta front. Strong reversing tidal currents prevail in the estuarine channels which receive relatively small fluvial inputs (e.g. Pie estuary), but tidal influences are also present within active river mouths.

The region is strongly influenced by the southeast trade winds and the waves generated by them. Trade wind-generated waves arriving from the southeast have an average period of 8 – 9 seconds, and average 2 – 4 metres in height during the southern hemisphere winter months (May – September). Wave rider records from Kerema indicate that on an annual basis wave height exceeds 0.25 metres for 90% of the time, 0.7 metres for 50% of the time, 1.5 metres for 10% of the time, and 2.0 metres for 1% of the time. The medium (50%) significant wave height during winter is 1.3 metres as compared to a medium summer wave height of only 0.3 metres.

4. Deltaic morphology

The northern Gulf of Papua between Kerema in the east and Kikori in the west can be divided into three distinct depositional subdivisions. Each will be discussed with the main emphasis being on the central region, the Purari Delta. Marginal delta areas to the east and west are important in that they provide a framework within which the delta proper is situated.

4.1. Eastern region

This region lies immediately east of the Purari River (Fig. 1). Ruxton (1969) has identified four major geomorphic units: littoral plains, alluvial plains, foothills and the hills and mountains of the 'Kukukuku lobe'. In the eastern region, bedrock comes closer to the coast than is the case with the two regions to the west; in fact at The Bluff, half way between Ihu on the Vailala River and Kerema, there is an outcrop of bedrock where waves have produced a well-developed rock platform.

The shoreline of the eastern region is relatively straight being composed of a narrow (1 – 2 km wide), beach-ridge plain. The beach ridges form elongate sand barriers cut by small river channels and tidal creeks. Figure 1 shows a fanning pattern of ridges east of the Purari River. In general, the mouths of the channels and creeks are deflected towards the west. Ruxton (1969) has described tidal flats of various vegetation types depending on salinites from low-lying environments landward of the sand barriers (Alele, Nipa and Murva land systems). Freshwater swamps behind these barriers merge into and are being replaced by accreting alluvial plains (Elala land system, Ruxton 1969, p. 11). River sedimentation is predominantly fine grained with freshwater peats accumulating locally in poorly drained swamps.

The eastern region is strongly affected by waves generated by southeastern trade winds. These waves prevail during winter months. Their oblique approach to the shoreline establishes a net east to west littoral drift pattern, one manifestation of which is the pattern of deflected creek mouths referred to above. The drift also explains the large northwest – southeast trending spit near Ihu (Fig. 1).

There is a broad wave-breaker zone along this coast. Three to four shore-parallel nearshore bars are common. It appears that during winter the beaches attain a highly dissipative state (Wright et al. 1979).

Sands collected on Ihu spit and from Orokolo Bay (Fig. 1) have been sieved. Figure 3 shows the cumulative frequency distribution of two samples revealing a moderately well-sorted, medium to fine sand. In general, grain size decreases westwards. There is little evidence for these sands being delivered to the littoral zone by streams in the eastern region. It is suggested that they were mostly reworked landwards from the inner shelf during the Postglacial Marine Transgression to form the beach-ridge plain. The finer fraction is progressively being reworked alongshore to the west to add to the store of river sands which are accumulating at the mouth of the Purari Delta. Beach profile characteristics and offshore bottom sediment samples are discussed in the Wabo Power Project report (Govt. P.N.G. 1977). Sand-sized sediment forms

52

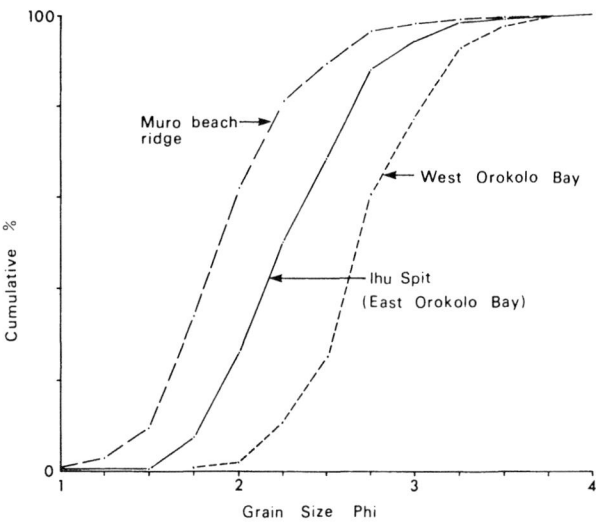

Fig. 3. Cumulative frequency distribution for three sand-sized samples east of Purari Delta. (West Orokolo Bay sample is No. 20 in Table 1; Muro beach-ridge sample is No. 21 in Table 1; see Fig. 1 for location.)

a narrow band less than 1 km wide in Orokolo Bay with clay offshore (ibid, Fig. 22).

4.2. Purari Delta

The central region is 50 km wide and consists of the active and inactive mouths of the Purari Delta. Figure 1 shows the symmetrical pattern formed by the distributary system of the Purari. At the head of the delta, the main stream branches into three major distributaries: the Wame-Varoi, Ivo-Urika, and Purari channels. These distributaries radiate from a point 35 km from the sea. Our limited observations suggest that the Ivo-Urika distributary carries 60 – 70% of the high stage discharge whereas the Purari distributary is dominant during low stage.

The general morphology of the Purari Delta is one of intricate branching distributaries (Fig. 1). The stream pattern has the appearance of a twisted maze with many interconnections, locally known as *ehe* (pronounced 'air-hair'). Except where logs of fallen trees block channels, boat travel is relatively easy from one part of the delta to another. Tidal channels rapidly widen near their mouths to form funnel-shaped entrances. Vegetated middle-ground shoals are common within these river mouths.

The deltaic plain can be divided into two parts, an upper and a lower plain. In the upper plain the main river channels dominate the distributary network. However, smaller interconnected channels occur between the Ivo and the Purari (Fig. 1). Well-developed levees, up to 2 m high occur along the banks of the main channels with villages like that at Mapaio being located on them. This is the freshwater portion of the deltaic plain possessing a dense cover of rain forest except for some open interdistributary swamps. Traces of abandoned, sinuous channels can be observed on aerial photographs in the upper section especially between the Ivo and the Pie estuary. These channels are choked with vegetation under which peat is presumably accumulating.

The lower deltaic plain is characterized by ephemerally brackish water conditions grading to more saline waters near the mouths of the less active distributaries. The brackish water zone is defined by the lowest limit of *Phragmites ? karka* and by the upper limits of *Sonneratia lanceolata* and *Pandanus* sp. (brackish water pandanus). The range of interstitial water salinities near the boundary of the lower deltaic plain is 1 to 7.5‰ (Petr and Lucero, 1979).

River channel salinities have been measured in the lower deltaic plain at two times of the year. During the wettest months salinities no higher than 9‰ were observed. The highest salinities occur in the western part of the delta in the Pie estuary. In November, salinity values more than doubled in many areas. Petr (1980, p. 8) has noted:

The coastal waters between the Vailala River in the east and the Kikori Delta in the west are characterized by less than full sea water salinities which is the direct result of high freshwater input through the rivers, including the Purari, into the Gulf of Papua.

Lowest salinity values occur in the front of the Urika and Varoi river mouths in the centre of the lower deltaic plain. Alele and Aievi Passages of the Purari distributary were not studied in July, 1979, but observations later in the year revealed higher values than those at the Urika mouth (Petr 1980). More saline waters penetrate the funnel-shaped entrances of major channels west of the Purari Delta, including Port Romilly (Pie estuary). Salinites at Baimuru some 30 km from the Gulf have regularly been measured at 7.5‰ (Petr 1980), although in July, 1979, the maximum value recorded here was 3.5‰.

In the lower delta salinity gradients appear to be associated with mangrove zonation patterns (see Floyd 1977, for description of mangroves along the northern Gulf). Tall forests of *Excoecaria* fringed by *Sonneratia lanceolata* and *Nypa* swamps form the inner fringe of brackish-water communities. Adjacent to Port Romilly there exists a mixed *Rhizophora – Bruguiera* community where salinities appear to exceed 10‰ for most of the year. However, near the

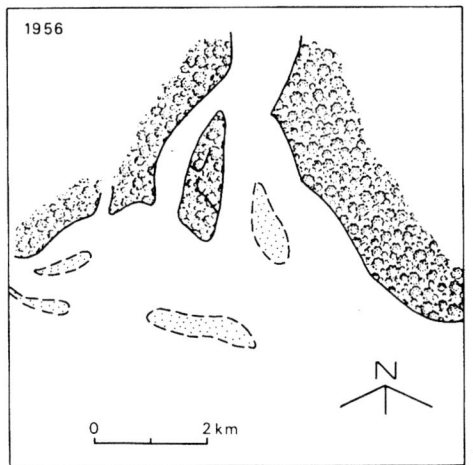
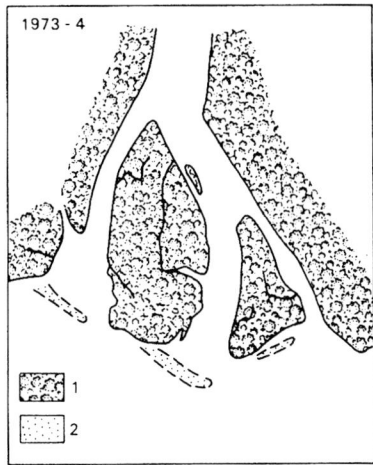

Fig. 4. Sketch maps based on aerial photography of changing river mouth morphology of Urika River. Legend: 1. Vegetated surfaces; 2. Prominent sand bars.

centres of predominant freshwater and sediment discharge the only mangrove which is at all extensive is *Sonneratia lanceolata*.

Active river mouths, like that of the Ivo-Urika, are rapidly changing their morphology. Figure 4 depicts changes in shoal and bar patterns sketched from rather cloudy aerial photographs. In 1973, the mid-channel islands had more than doubled their size compared to 1956. Our observation in 1979 showed further seaward growth of *Sonneratia* behind an accreting river mouth bar. The southeast channel seems to have been a more effective distributary in 1956 than it is today; the western channel is now the main distributary and possesses a sandy bed load in comparison with sandy mud in the southeast channel.

Erosion is locally quite severe along the delta margin away from active river mouths. Tall mangrove forests are being destroyed by wave attack (for instance, west of the mouth of the Varoi River). This impact represents the effect of shifting channels and centres of active sediment discharge through time.

Discontinuous beach ridges occur along the delta front between distributary mouths. Coconuts are locally grown on these low-relief but relatively well-drained sands. Figure 1 indicates the general location of the beach ridges and their truncation by shifting channels and shoreline positions.

4.3. Western region

To the west of the active delta is an extensive complex of tide-dominated funnel-shaped estuaries. Relatively small rivers draining the western highlands feed the heads of these estuaries. The Pie River is the easternmost example although the Wame distributary of the Purari flows into this estuary. The Kikori River is the largest of the rivers in this region. Mangroves are common on tidal flats (Floyd 1977).

From the few available observations it is apparent that these estuaries are partially mixed. Tidal currents dominate the exchange of water and sediments. Linear tidal ridges extend offshore into the Gulf to depths of 40 m.

The western region has many characteristics of deltas dominated by tidal processes with tidal channels widening progressively downstream and becoming moderately sinuous upstream (Wright et al. 1973). Mud-laden waters circulated by tides lead to vertical accretion of mangrove-covered tidal flats.

5. River mouth and estuary morphodynamics

Channels and entrances of the Purari Delta, through which water and sediment are debouched into the sea for subsequent redistribution by waves and currents, are of two dynamic types: (i) active river mouths dominated by seaward-flowing fluvial discharge, and (ii) tidal-dominated estuaries which experience flow reversals over a tidal cycle. The second type includes atrophying river distributaries which may exist near to active river mouths.

5.1. Active river mouths

The Purari River branches at the head of the delta into three distributaries: the Purari to the east, the Ivo-Urika in the centre and the Varoi-Wame to the west. Current profiles and cross-section surveys measured in July, 1979, indicated that the largest fraction of flow was carried by the Ivo-Urika distributary with the Purari distributary second in importance. Flow velocities in the upper Ivo-Urika distributary averaged 1.6 to 2.0 m per second. Velocities in the Upper Purari and Varoi-Wame distributaries respectively averaged 1.4 m per second and 1.2 m per second. The central distributary, the Ivo-Urika, carries a sandy bed load and exhibits downstream migrating megaripples with heights of 0.5 – 1.0 m all the way from the head of the delta to the sea.

The main discharge of the Ivo-Urika distributary enters the sea by way of the Urika mouths (Fig. 4). Salinity – temperature profiles in the mouth and just upstream from the mouth indicate that seawater is completely excluded all the way to the river mouth bar. Zero salinities prevailed throughout the water column in association with a strong downstream flow of fresh river water. Near the mouth, the Urika channel bifurcates at least twice as shown in Fig. 4. Outflow velocities through the main outlet were observed to be in the order of 1 m per second; those through the secondary southeast outlet were in the order of 0.5 m per second. Consistent with its high outflow velocities and sandy bed load, the morphology of the Urika outlet conforms closely to the 'friction dominated' model described by Wright (1977). Characteristic features include extensive 'middle ground' bar deposits, shallow (depth = 2 – 4 metres) (Fig. 5D) and bifurcating channels (Fig. 4).

Wave modifications of river mouth deposits are also pronounced. These take the form of wave-reworked longshore bars backed by swales fronting the middle ground shoals, and elongate river mouth bars across the entrances of the Urika. Wave effects at the Purari outlet are even more conspicuous.

The Purari double outlet is constricted by broad subaqueous levees surmounted by swash bars. It is fronted by broad arcuate bars which are awash at low tide and exhibit ridge and runnel topography. The features at the Purari outlet (Fig. 1) are similar in some respects to those of other wave-dominated river mouths (see Wright 1977; Wright, Thom and Higgins, 1980).

A significantly lower fraction of the total discharge is carried by the Wame distributary which debouches by way of the Wame mouth into the sheltered environment of the Pie estuary. In contrast to the other active river mouths, wave effects are negligible; however tidal flows appear to play a more important role in the lower Wame as evidenced by the funnel shape of the lower 5 kilometres of the channel and the relatively deep cross-section (Fig. 5A) in the adjacent Pie estuary.

The dominant river mouths just described, particularly the Urika and Purari mouths, are the main sources of sand to the delta front. The sinks for these sands appear to be the distributary channels, river mouth bar deposits and the discontinuous interdistributary beach ridges. Collectively the major mouths, together with the mouths of numerous secondary channels, are the sources of an extensive mass of mud-laden fresh water. Aerial reconnaissance indicates that seaward of the outlets over the delta front, the muddy effluents merge to compose a coastal band of fresh to brackish water which remains trapped inshore by waves and winds. This band supports a heavy load of

---

*Swale is a depression between the ridges of a beach-ridge plain.

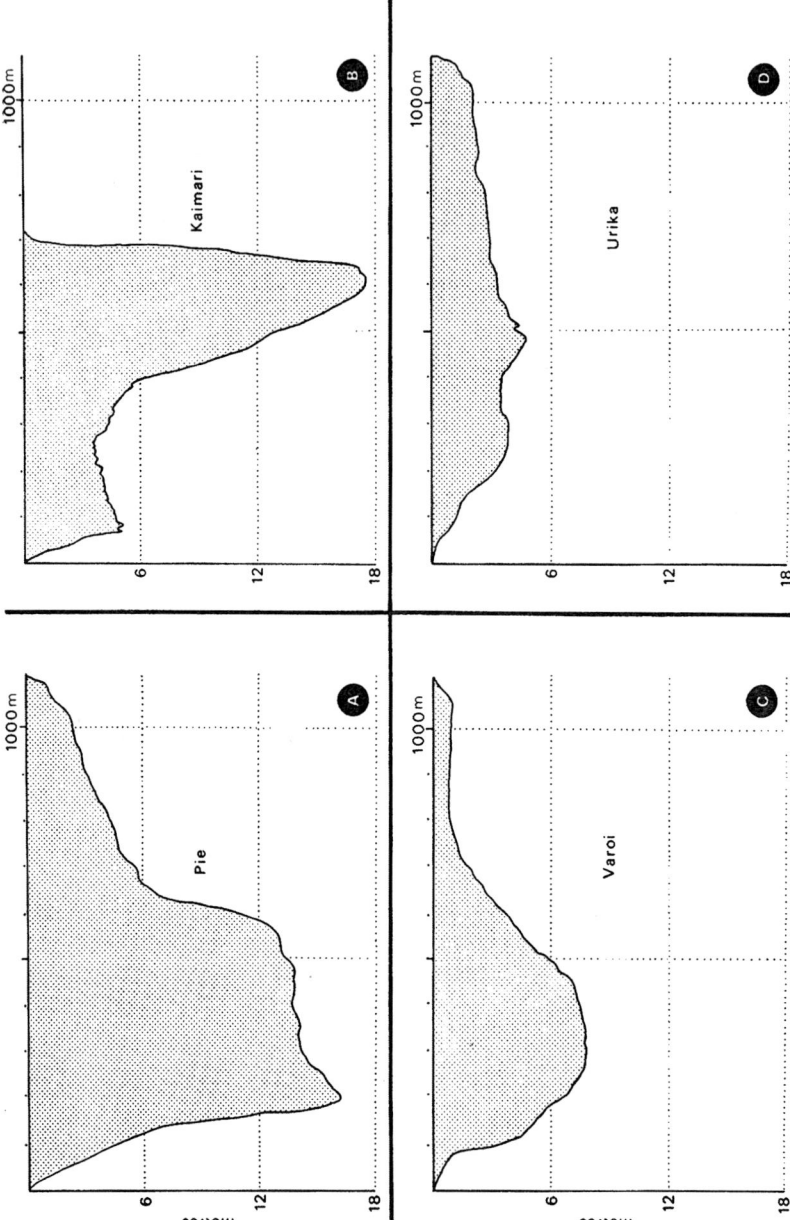

Fig. 5. Channel cross-sections, lower deltaic plain (see Fig. 1 for location).

58

suspended fines, and is particularly noticeable during the southeast trade wind season.

5.2. Tide-dominated estuaries

The remainder of the channels and estuaries of delta region carry a much lower fluvial discharge and appear to be maintained largely by tidal flows. Included in this category are the Panaroa estuary, the Varoi estuary, Kaimari channel (behind Kaimari Island) and the Pie-Port Romilly estuary. The following characteristics distinguish the tide-dominated estuaries from the fluvially-dominated active river mouths in this region:

(i) Bed load transport is small and channel beds are composed largely of mud rather than sand.

(ii) Flow velocities are weaker in the tide-dominated estuaries than in the active distributaries with typical speeds being $0.25 - 0.5$ m per second; flows reverse direction with the tide.

(iii) The tide-dominated estuaries were weakly stratified in July, 1979, with some upstream penetration of brackish water. At the time of our observations salinities even near channel bottoms did not exceed 10‰; however observations by Petr at times of lesser discharge indicated somewhat higher salinities (Petr 1980).

(iv) Bidirectional tidal scour in the absence of continual fluvial input of sediments has led to greater deepening of the tide-dominated estuaries. The downstream reaches of these estuaries are typically 10 metres or more deep as compared to average depths of 3 to 4 metres for the active river mouths. In the case of Kaimari channel, a channel 18 metres deep has been cut into clay (Fig. 5B). The Varoi mouth has a similar cross-sectional shape (Fig. 5C).

(v) At seaward of the entrance to the larger estuaries, such as Port Romilly and the estuaries farther west, linear tidal ridges similar to those described by Wright et al. (1973) are present and are aligned parallel to the apparent direction of tidal flows. Little is known at present of the sediment and tidal flow characteristics of these ridge fields.

6. Deltaic stratigraphy and sediments

No attempt has been made to systematically sample modern environments of deposition in the Purari Delta. Drilling is limited to shallow augering by the authors in the beach-ridge plain near the mission village of Muro, east of the

59

delta, a cored hole in the floodplain of the upper delta at Mapaio, and drill logs obtained for purposes of geophysical exploration on file at P.N.G. Geological Survey (Port Moresby).

Figure 1 shows the location of selected samples collected during the reconnaissance survey of July, 1979. Grain size, organic content and heavy mineral content* are listed in Table 1. Cumulative frequency distributions for beach sands east of the Purari Delta are shown in Fig. 4, and for a sample of beach-ridge sand collected at Muro Mission.

It is evident from Table 1 that the proportion of clay, silt and sand is not only highly variable between environments of deposition, but also within environments. However, some trends can be noted from this table.

Table 1. Sediment properties – Purari Delta.

Sample No.	Environment of deposition	Clay %	Silt %	Sand %	Heavy mineral %	Organic matter %
1	Levee, upper delta	4	18	78	2	6
2	Levee, upper delta	3	14	83	1	7
3	Levee, estuary	44	31	25	16	9
4	Levee, estuary	44	42	14	–	13
5	Backplain, estuary	51	33	16	–	15
6	Backplain, lower delta	38	24	38	–	31
7	Backplain, lower delta	10	54	36	–	11
8	Backplain, lower delta	42	41	17	–	18
9	Levee, lower delta	20	32	48	–	13
10	Channel, upper estuary	60	28	12	–	12
11	Channel, lower estuary	41	26	33	2	11
12	Channel, lower estuary	8	34	58	–	12
13	Channel, active river	1	2	97	9	–
14	River mouth bar – seaward	1	1	98	7	3
15	River mouth bar – seaward	1	–	99	30	3
16	River mouth bar – seaward	1	–	99	22	1
17	River mouth bar – landward	1	–	99	10	2
18	Beach – eroding	–	–	100	21	2
19	Beach – eroding	7	17	76	4	11
20	Beach – accreting	–	–	100	9	2
21	Beach-ridge plain	1	1	98	19	2

*Heavy minerals in the context of this chapter are a mixed group of minerals with a specific gravity above 2.65, e.g., ilmenite, magnetite, rutile. No attempt was made to separate different mineral types.

(i) The proportion of sand in levee sediments is higher in the upper than in the lower delta.

(ii) Backplain environments (i.e. backswamp of tidal flat environments between levees) have high organic contents.

(iii) Heavy mineral percentages are highest in river mouth bar environments and very low in levee, backplain and estuarine channel environments (an exception is No. 3).

(iv) Heavy mineral content of beaches is also high; no analysis has been made as yet of heavy mineral types in any environment although magnetic minerals are common.

(v) Estuarine channels have highly variable grain size proportions and organic content in contrast to the main Urika channel where sand predominates.

(vi) Eroding beaches have variable grain sizes depending on material being eroded, for instance No. 19 is a sample from an eroded backplain environment.

(vii) Sand mineralogy in the delta is quite mixed reflecting an immediate fluvial origin with limited reworking by waves.

The stratigraphy of the delta is little known. In particular, the relationship of wave-reworked sands along the delta front to fluvial deposits requires investigation. A cored hole to 12 m depth at Mapaio in the upper delta (Fig. 1) bottomed in fine sand. Wood from 8 to 8.3 m depth in clay was radiocarbon dated (SUA − 1326); the result of 745 ± 155 14C yrs B.P. indicates accumulation at a time when sea level was lower than present at least in eastern Australia (Thom and Chappell 1975). The estimated elevation of this site is + 4 m MSL (mean sea level). An accumulation rate of approximately 1 m in slightly less than 1000 radiocarbon years is indicated. There is no macro-fossil evidence of marine influence in the sands or clays of this hole.

Drill logs from holes used in geophysical exploration show a consistent pattern. In the lower delta 10 − 15 m of grey clay often rich in organics overlies 'blue sand'. This sand is coarser towards the surface and varies in thickness from 8 to over 40 m. In the upper delta near the junction of the three distributaries, a coarse sand unit exists under 3 m of clay and extends to − 30 m where it overlies 'fine blue sand'. It is not known if there are any discontinuities within these sands.

It is difficult to see how such an extensive and thick sand sheet covered by mud could accumulate under present-day conditions. Assuming that the sand body is laterally continuous then three models are possible.

The first involves the sands building up during a period (or periods) of lower sea level and being deposited by fluvial processes. Under this model, the

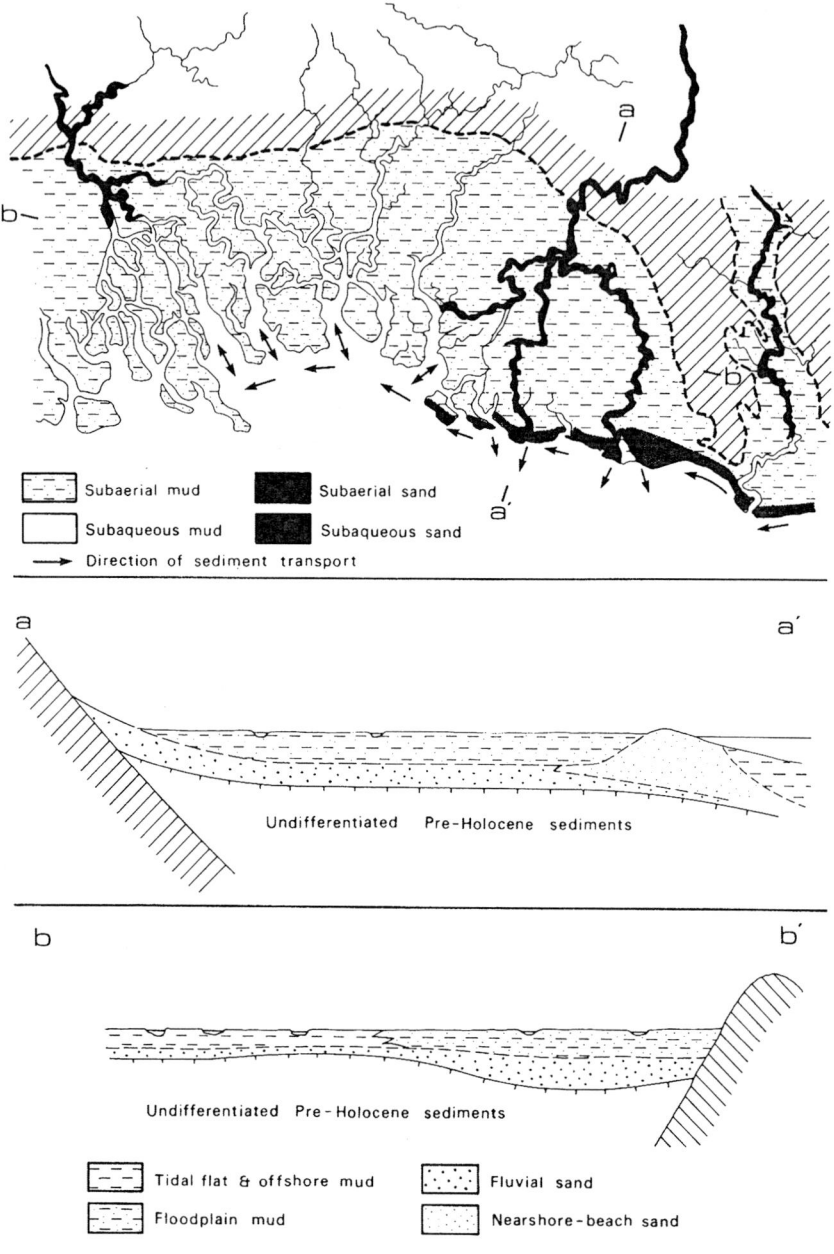

Fig. 6. Generalised depositional model of Purari deltaic complex showing sediment types, stratigraphic units and directions of sediment transport. Information shown here requires further field study for confirmation.

clays formed in backplain/levee environments during the Postglacial Marine Transgression (P.M.T.) behind a receding delta front sand sheet. This sheet later prograded seawards during the subsequent 'stillstand'.

The second model requires the sands to be deposited at the delta front during a marine transgression. Scarcity of shells in modern environments may explain their absence from older analogs. The sands extended landwards during the P.M.T. then prograded seawards during the 'stillstand'. It is difficult to see how this model would explain the thickness of sand extending so far inland.

A third model involves elements of the first two insofar as fluvial sands deposited at a lower sea level underlie wave-reworked river mouth bar and fluvial plain sands in the lower deltaic plain. This model is shown in Fig. 6.

7. Conclusion

The Purari deltaic system is one of the major components of a much larger Gulf of Papua depocentre. The larger-scale depocentre encompasses the sinks for sediments derived from the Fly, Turama, Kikori, Purari Rivers as well as several smaller rivers. It is only the northern part of this depocentre, that is the Purari Delta and neighbouring estuaries to the west, that this chapter is concerned.

The Purari-Kikori deltaic complex exhibits many of the environmental and morphological characteristics of the Ganges-Brahmaputra and Irrawaddy deltas (as described by Wright et al. 1974). In terms of tidal range and wave regime all three complexes are very similar. Although the Ganges-Brahmaputra has a larger discharge than that of the combined Purari-Kikori system, the discharges of the Purari and Irrawaddy are comparable.

With respect to morphology, overall geometry and size, the Purari-Kikori complex is most similar to the Ganges-Brahmaputra complex. In particular, complicated networks of numerous distributaries, funnel-shaped estuaries and interconnecting channels are common to both systems. However if one considers the Purari Delta *per se* as an isolated unit its closest analogy is probably the Irrawaddy Delta. Like the Purari, the Irrawaddy is a tropical delta exhibiting a similar number of tidally-influenced wave-constricted river mouths.

Much work remains to be undertaken on all aspects of deltaic morphology and stratigraphy in the Purari region. The area presents an exciting deltaic complex, not only from the point of view of research on river mouth dynamics under varying seasonal conditions, but also for investigating deltaic stratigraphy and depositional sequences in the Holocene. We have established the existence of dateable materials within backplain sediments, but there is an apparent scarcity of shells from within beach ridges. The Purari region needs to be better understood in terms of location and movement of sand bodies. We have identified two main sites of sand deposition; on the one hand there are the active but unstable river channels, and on the other the modern river mouth bars and adjacent, discontinuous beach ridges (Fig. 6). Muds accumulate in shallow water offshore, in the tidally-dominated channel region to the west, and in the subaerial deltaic plain and mangrove-covered tidal flats.

Critical to the future evolution of the delta front is the construction of the Wabo Dam. Any structure which interferes with the input of sediment to the delta is most likely to result in modification of deltaic morphology. Ly (1980) has recently documented the erosive effects of the Akosombo Dam on the Volta River on the coast of Ghana. A similar sequence of events could occur in the Purari Delta if the input of sands and fine-grained sediments is restricted, thereby leading to extensive delta-front erosion rather than the present-day localised erosion associated with channel switching. It is also likely that such extensive erosion would be accompanied by significant ecological changes. Further studies will be needed to appreciate the implications of dam construction on deltaic geomorphology and ecology.

References

Australasian Petroleum Company Pty. Ltd., 1969. Final report Era – Pie – Purari Seismic Reflection Survey D-1. United Geophysical Corporation Party 130, on file Geological Survey, P.N.G. 22DZ (see also 22CA).

Floyd, A.G., 1977. Ecology of the tidal forests in the Kikori – Romilly Sound area Gulf of Papua. P.N.G. Dept. of Primary Industry, Division of Botany, Lae. Ecology Report, No. 4. 59 pp.

Government of Papua New Guinea, 1977. Purari River Wabo Power Project feasibility report, Vol. 3: Port, industrial and urban development.

Khan, A.M., 1974. Palynology of Neogene sediments from Papua (New Guinea). Stratigraphic boundaries. Pollen et Spores 16: 265 – 284.

Ly, C.K., 1980. The role of the Akosombo Dam on the Volta River in causing coastal erosion in central and eastern Ghana (West Africa). Marine Geol. 37: 323 – 332.

Petr, T., 1980. Purari River environment (Papua New Guinea): a summary report of research and surveys during 1977 – 1979. Office of Environment and Conservation, Waigani, Papua New Guinea.

Petr, T. and J. Lucero, 1979. Sago palm (*Metroxylon sagu*) salinity tolerance in the Purari River Delta. In: T. Petr (ed.), Purari River (Wabo) Hydroelectric Scheme Environmental Studies,

Vol. 10: Ecology of the Purari River catchment, pp. 101 – 106. Office of Environment and Conservation, Waigani, Papua New Guinea.

Pickup, G., 1980. Hydrologic and sediment modelling studies in the environmental impact assessment of a major tropical dam project. Earth Surface Processes 5: 61 – 75.

Pickup, G., 1983. Sedimentation processes in the Purari River upstream of the delta. This volume, Chapter I, 11.

Pickup, G. and V.H. Chewings, 1983. The hydrology of the Purari and its environmental implications. This volume, Chapter I, 8.

Ruxton, B.P., 1969. Regional description of the Kerema – Vailala area. In: Lands of the Kerema – Vailala area, Papua New Guinea. C.S.I.R.O., Melbourne. Land Research Series, No. 23, Part II: 9 – 16 (see also part III, Land sys.).

Thom, B.G. and J. Chappell, 1975. Holocene sea levels relative to Australia. Search 6: 90 – 93.

Wright, L.D., 1977. Sediment transport and deposition at river mouths: a synthesis. Bull. Geol. Soc. Amer. 88: 857 – 868.

Wright, L.D. and J.M. Coleman, 1973. Variations in morphology of major river deltas as functions of ocean wave and river discharge regimes. Amer. Assoc. of Petroleum Geologists Bull. 57: 370 – 398.

Wright, L.D., J.M. Coleman and B.G. Thom, 1973. Processes of channel development in a high-tide-range environment: Cambridge Gulf – Ord River delta. J. Geol. 81: 15 – 41.

Wright, L.D., J.M. Coleman and M. Erickson, 1974. Analysis of major river systems and their deltas: morphologic and process comparisons. Louisiana State University. Coastal Studies Institute Technical Report, 156. 114 pp.

Wright, L.D., J. Chappell, B.G. Thom, M.P. Bradshaw and P. Cowell, 1979. Morphodynamics of reflective and dissipative beach and inshore systems: southeastern Australia. Marine Geol. 32: 105 – 140.

Wright, L.D., B.G. Thom and R. Higgins, 1980. Wave influences on river-mouth depositional process: examples from Australia and Papua New Guinea. Estuarine and Coastal Marine Sci. 11: 263 – 277.

5. Soil types and traditional soil management in the Purari catchment

A.W. Wood

1. Introduction

There has been no comprehensive soil survey of the whole Purari catchment. C.S.I.R.O. Division of Land Use Research have conducted reconnaissance soil and land resource surveys in the northern part of the catchment (Perry et al. 1965; Haantjens et al. 1970), and in the delta area (Ruxton et al. 1969). Information on soil types in the area is also available from generalised soil maps and reports of the whole of Papua New Guinea (Haantjens et al. 1967; Haantjens 1970a; Bleeker 1974; FAO-UNESCO 1976; Bleeker and Healy 1980; Wood 1982a, 1982b) and from investigations by independent research workers, particularly in the highlands (Pain, 1982; Pain and Blong 1976, 1979; Pain and Wood 1976; Wood 1979, in press).

Studies of traditional cultivation and soil management have also been largely restricted to the highlands. Information on soil management is available from Wonerara (Ollier, Drover and Godelier 1971); Okapa and the Fore (Bourke and Allen 1979; Sorenson 1976); the Goroka basin (Howlett 1962); Chimbu (Brookfield and Brown 1963); the Wahgi valley (Powell et al. 1975); the Kaugel valley (Bowers 1968); the Nembi plateau (Allen et al., in press); and the Karimui area (Wood 1979). Wood and Humphreys (1982) have compiled a review of the knowledge of traditional soil management for the whole of Papua New Guinea.

This chapter describes the main soil types in the Purari catchment. Published information is combined with additional data from the author's own field observations on the characteristics and distribution of the main soil types. This is followed by a description of traditional cultivation techniques and their impact on soil properties and erosion.

Petr, T. (ed.) The Purari – Tropical environment of a high rainfall river basin   67
© *1983, Dr W. Junk Publishers, The Hague / Boston / Lancaster*
ISBN-13: 978-94-009-7265-0

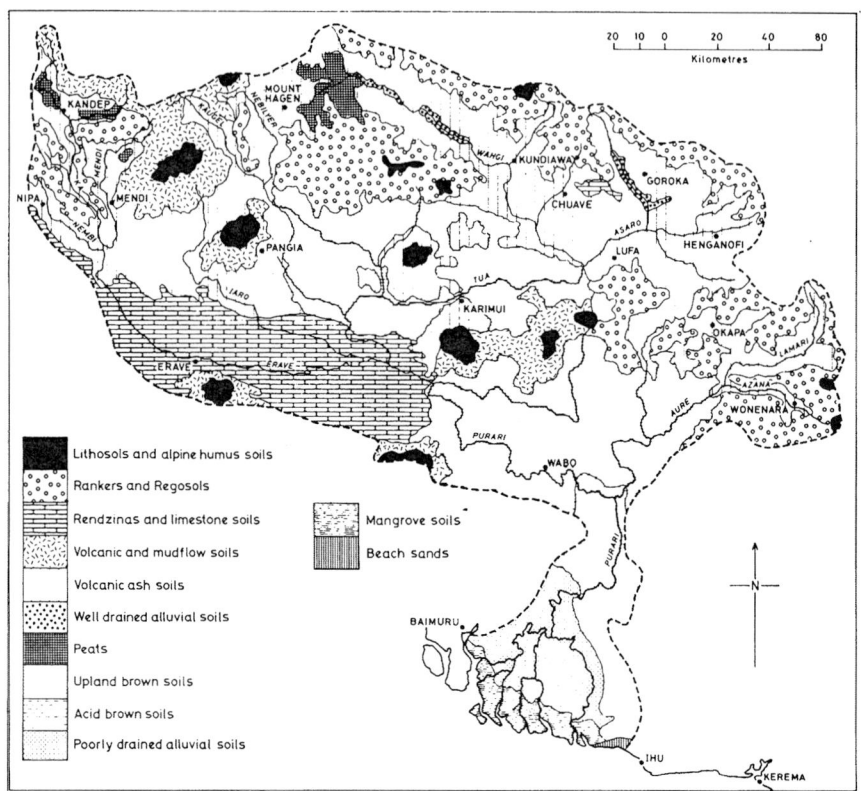

Fig. 1. Soil types of the Purari catchment.

2. Soil types

Since soils are often highly variable over small areas in Papua New Guinea (Bleeker and Speight 1978) it is only possible to present a generalised picture of the distribution of soils on a map of the whole Purari catchment (Fig. 1). The mapping units are associations of soils based largely on the assumption that there is a relationship between landforms, soils and vegetation. Some of the mapping units are quite broad and contain a wide variety of soil types.

Soil nomenclature has been kept as simple as possible given the wide ranging contents and broad readership of this volume. Internationally used soil terms which are not all based on one particular system of soil classification have been used on the map. The names of the mapping units are intended to convey as much information as possible about the nature of the soil and the associated landforms. For equivalent terms, using the American Soil Taxonomy (U.S.D.A. 1975) and the FAO soil nomenclature (FAO-UNESCO 1974), see Table 1.

The descriptions of soil types have been divided for convenience into two groups: those occurring in highland areas above 1000 m; and those of the lowlands.

2.1. Highland soil types

2.1.1. Lithosols and alpine humus soils. These occur on rugged mountain summit areas at 3000 – 4500 m on a variety of rock types including andesite, diorite, granodiorite, and metamorphics. Many of the summit areas have been glaciated, resulting in extensive areas of bare rock or very shallow soils (Löffler 1972). Soil types are mainly shallow Lithosols or Rankers. Deeper Alpine

Table 1. Soil nomenclature using international systems of soil classification.

Map unit		U.S. Soil Taxonomy	FAO system
2.1.1	Lithosols and Alpine Humus soils	Cryandept, Cryaquept, Cryorthent, Cryochrept, Cryumbrept, Cryofolist, Tropofolist	Lithosol dystric Cambisol dystric Histosol
2.1.2	Rankers and Regosols	Humitropepts Eutropepts	Ranker humic Cambisol dystric Cambisol
2.1.3	Rendzinas	Rendoll, Troporthent, Eutropept	Rendzina eutric Regosol
	Deep dark clay soils	Andaquept, Hapludoll	mollic Gleysol
2.1.4	Volcanic and mudflow soils	Hydrandept Humitropept	humic Andosol
2.1.5	Volcanic ash soils	Hydrandept	humic Andosol
2.1.6	Well-drained alluvial soils	Tropofluvent	eutric Fluvisol
2.1.7	Peats	Tropofolist, Tropohemist, Tropofibrist, Troposaprist	eutric Histosol dystric Histosol
2.1.8	Upland brown soils	Humitropept Tropohumult	humic Cambisol dystric Cambisol
2.2.1	Acid brown soils	Dystrochrept	dystric Cambisol
2.2.2	Poorly drained alluvial soils	Hydraquent Tropohemist	gleyic Fluvisol eutric Histosol
2.2.3	Mangrove soils	Hydraquent	eutric Gleysol
2.2.4	Beach sands	Orthopsamment	cambic Arenosol

Humus soils and peats occur in depressions and flatter areas where the cool, wet climatic conditions have encouraged local accumulations of raw organic material.

A sequence of Alpine Humus soils formed on glacial and fluvioglacial deposits on the summit of Mt. Giluwe has been described by Rutherford (1964a). The soils are shallow with a dark brown silty clay to clay surface horizon overlying a well-decomposed black organic clay subsoil. Rutherford attributed the arrangement of horizons to a form of humus illuviation, though the possibility of topsoils being buried by volcanic deposits was not eliminated. These soils are strongly acid, high in organic matter and total phosphorus, with a low base saturation. Feldspar is the dominant mineral in the sand fraction, with a high proportion of amorphous material in the clay fraction.

The different soil types around Mt. Giluwe are arranged in a series of semi-parallel zones (Rutherford 1968). Shallow, poorly developed, highly organic Lithosols occur on the summit, deeper Alpine Humus soils occur below, with poorly drained soils on the fluvioglacial deposits of the upper slopes, and volcanic soils on the lower slopes. Rutherford attributed this concentric arrangement of soil types to a reduction in the intensity of soil-forming processes with altitude, thus following an essentially zonal approach.

Equivalent terms for these soil types in other classification systems are given in Table 1. Humphreys (personal communication) has calculated lapse rates and soil temperatures in the New Guinea highlands and found that soil temperatures of less than 8°C (the limit of 'cry' soils using the American Soil Taxonomy) are restricted to above 3600 m. Cryandepts are the dominant soil type above this altitude.

2.1.2. Rankers and Regosols. They are common on steep sloping, rugged mountains and hills between 500 and 3000 m. Soil profiles are generally less than 70 cm deep due to the instability of most slopes, although deeper Regosols occur where colluvium has accumulated. Soil creep and slumping are common processes with landslides occasionally occurring. Soil textures range from silty clay to clay with varying amounts of weathered rock fragments occurring throughout the profile (Haantjens 1970b). Profiles are well drained having dark brown topsoils, moderately high in organic matter overlying brown to yellowish brown subsoils. Chemically they are acid with a low content of bases, due to constant leaching (Haantjens 1969).

2.1.3. Rendzinas and limestone soils. These have formed only on calcareous rocks, and occur mainly on the limestone in the southwest of the catchment between 1000 and 3000 m. Shallow Rendzinas are the dominant soil type, occurring on the steepest slopes. The Rendzinas are 10 to 50 cm deep, with black to very dark brown topsoils overlying brown to dark grey clay, which rests abruptly on the underlying limestone. Fragments of limestone occur in all horizons, and the influence of the limestone causes pH values to be high particularly in the lower horizons where they may exceed 7. Exchangeable calcium levels are also high (Rutherford and Haantjens 1965; Wood, in press).

Depressions in the limestone and gentle slopes have deeper clay soils, with up to 1 m of highly organic, well-structured topsoil overlying yellowish brown to strong brown clay. The high organic matter content is probably related to poor drainage. Deep dark clay soils have largely developed where the limestone has been mantled by volcanic ash. The soil has formed in a mixture of ash and the insoluble residue from the weathering of the limestone. Inputs of topsoil which have washed down from slopes of dolines or depressions have contributed to the deep topsoils.

2.1.4. Volcanic and mudflow soils. These soils occur on the dissected slopes of major volcanoes between 2000 and 4000 m. They have developed on a variety of volcanic materials including andesitic lava flows, agglomerate, volcanic ash, lahars and mudflows. Profiles are moderately deep with black to dark brown organic topsoils abruptly overlying olive brown to yellowish brown subsoils, often containing fragments of weathered lava. Soil textures range from clay loam to clay. The mineralogy of the clay fraction is dominated by gibbsite and amorphous material, whilst the sand fraction contains mainly feldspars, amphiboles and opaque minerals (Bleeker and Healy 1980).

Much of the volcanic materials on steep slopes have been affected by landslides, mudflows and other forms of mass movement, which were probably initiated by major earthquakes. Pain (1975) has described a widespread mudflow in the Kaugel valley, developed in Late Quaternary volcanic deposits, which occurred between 20 000 and 30 000 years ago on the slopes of Mt. Giluwe. The mudflow soil has a well-developed organic topsoil overlying yellowish brown sandy clay containing large volcanic boulders. The matrix of the mudflow contains mainly feldspars and hornblende in the sand fraction, with allophane and halloysite in the clay fraction.

2.1.5. Volcanic ash soils. From the agricultural viewpoint this is the most important soil type in the Purari catchment. Volcanic ash soils occur in all the major highland valleys, where most of the population of the catchment is located. The soils are formed on deep deposits of volcanic ash (tephra) which overlie a variety of sedimentary rocks including sandstone, shale, limestone

Fig. 2. Volcanic ash soil profile exposed in a road cutting near Kundiawa, showing the dark humic topsoil and the underlying tephra layers.

and mudstone. The volcanic ash has come from three main sources, and at one time must have covered most of the northern catchment (Pain and Blong 1979). On some of the steep unstable slopes, the ash cover has been partially or completely removed by erosion and landslips, particularly where the underlying rock is a fine-grained shale or mudstone. Elsewhere, the ash cover is remarkably stable, even on slopes of up to 35° (Blong and Pain 1978).

Volcanic ash soils are very deep, with thick dark brown to black humic topsoils overlying yellowish brown to strong brown clay subsoils (Fig. 2). The topsoil has a well-developed crumb structure and a very friable consistence even when wet. Chemical properties include a high organic matter content, high cation exchange capacity, low base saturation, and moderate acidity. Available phosphorus is low in all horizons due mainly to fixation by amorphous material (Parfitt and Mavo 1975). They have excellent physical properties which make them highly suited to cultivation. The granular structure, friable consistence, and rapid infiltration of moisture into the topsoil provides ideal conditions for plant growth. Structural aggregates are stable and the amount of soil erosion is low except on steep slopes. The clay fraction contains gibbsite, amorphous material and subsidiary halloysite (Parfitt 1975). Amphiboles and feldspars dominate the sand fraction.

Rutherford (1962) and Haantjens and Rutherford (1964) have suggested that the main reason for the uniformity in soil type throughout the highland valleys is the prevailing cool wet climate. They argue that humic brown clay soils are zonal soils that have formed on a variety of parent materials including volcanic ash, sandstone, greywacke, shale, siltstone and limestone. They distinguish between the soils formed on volcanic ash and those formed on sedimentary rocks, the former having thicker and darker organic topsoils; the latter having firmer, more massive and plastic subsoils. This zonal theory has been disputed by Pain and Wood (1976) who consider that the humic brown clay soils occur only where the sedimentary formations are mantled by thick deposits of volcanic ash. It appears that the extent of volcanic ash deposits has been greatly underestimated in the highlands.

These soils are classified as Hydrandepts using the American Soil Taxonomy and humic Andosols using the FAO classification. Bleeker (1974) termed them humic Acrisols, but most profiles lack the diagnostic characteristics of an argillic horizon. Rutherford (1964b), however, has noted the presence of clay cutans in volcanic ash soils near Nipa.

2.1.6. Well-drained alluvial soils. These occur extensively on floodplains and low terraces in the Asaro and Wahgi valleys, and as narrower deposits in other valleys in the highlands. They are well drained, coarse to medium textured soils with colours ranging from olive brown to dark yellowish brown. Organic matter levels are low but there is a moderate supply of bases and phosphorus.

2.1.7. Peats. Peats are deep organic soils which have developed where organic matter has accumulated under conditions of poor drainage. The topsoil has a dark brown to black peaty clay with a crumb or blocky structure when cultivated. They are very acid with a pH of 4 – 4.5, but when cleared and drained the organic matter breaks down, acidity is reduced and they become very fertile and productive soils (Drover 1973).

2.1.8. Upland brown soils. This association includes a wide variety of soil types occurring on hilly and mountainous country between 1000 and 2500 m. The soils are of variable depth depending on slope steepness and stability. Brown Regosols occur on unstable slopes. Brown Forest soils and Brown Podzolic soils occur on more stable areas. The surface horizons are well developed, dark brown and friable and overlie brown to yellowish brown heavier clay subsoils. Chemical characteristics are variable, but these soils are usually acid, high in organic matter with a low content of bases.

2.2. Lowland soil types

2.2.1. Acid brown soils. These occur on strongly dissected hill country between 0 and 500 m. They are uniform fine-textured soils, yellowish brown to olive brown in colour. They often have a poorly developed A horizon and a clearly defined B horizon, the latter resulting from weathering activity (Bleeker 1969). They are characterised chemically by a strongly acid reaction.

2.2.2. Poorly drained alluvial soils. This association occurs on poorly drained level floodplains and swamps. Fine-textured soils predominate, which are peaty in the permanent swamps and strongly gleyed due to the poor drainage. The water table is always close to the surface and during the wet season, many of the soils are subject to inundation.

2.2.3. Mangrove soils. This association occurs along the tidal mangrove flats at sea level and consists mainly of marine muds and peat. Soils are dark grey and massive silty to heavy clays containing a moderate amount of organic matter. They are usually alkaline and saline due to the marine influence.

2.2.4. Beach sands. These occupy a small area of both ridges in the extreme southeast of the catchment. The soils have little or no profile development with brown structureless loamy sand surface horizons overlying greyish brown sand (Bleeker 1969).

3. Traditional soil management

Methods of traditional soil management have evolved as a response to different ecological conditions. The inhabitants of the highland valleys above 1200 m have the most distinctive and elaborate soil management techniques which they use for sweet potato cultivation. In the sparsely populated highland margins between 500 and 1200 m, and over much of the lowlands, systems of shifting swidden cultivation with long bush fallows are used to regulate soil fertility. In the swamps of the delta region, the subsistence system is based largely on the gathering of sago and other wild foods, with little cultivation of the soil. Examples of soil management practices are described from these three ecological zones in the catchment, and where possible the response of the soils to different management practices is indicated.

3.1. Soil management in the highlands

The Highlanders are semi-permanent cultivators rather than shifting cultivators, and cultivate the same land for many years without lengthy fallows. The highland valleys contain the greatest concentration of rural population in Papua New Guinea, and it is proof of the high natural fertility of highland soils that local population densities in excess of 150 persons/km^2 are supported largely by subsistence cultivation. One of the most striking features of cultivation in the highlands is the apparent permanence of gardens and their neat, orderly arrangement in the landscape. Brookfield (1962, 1964) noted the main features of highland agriculture as being:
- a great dependence on a single root crop, sweet potato
- the absence of food storage so that crops have to be harvested daily from gardens
- the negligible importance of hunting
- the keeping of large numbers of pigs
- the use of systems of enclosure to separate pigs from the food gardens
- the cultivation of gardens in grasslands by means of tillage
- the use of special soil conservation techniques

Highland societies have a highly developed perception of soils and different environments. They can name various rocks and soils (Ollier, Drover and Godelier 1971) and can put forward reasons why some areas are cultivated and others not (Brookfield 1964). The intimate knowledge of the environment and of the suitability of different soil types for cultivation possessed by Highlanders is reflected in many areas by a close adjustment of garden types, crops and cultivation practices to soils. In some parts of the highlands crops are segregated into different types of gardens on the basis of soil type, with sweet potato cultivated in large 'open fields' and other crops grown in 'mixed gardens'. In the Kaugel valley, sweet potato gardens are located on fertile alluvial soils on river bottomlands and terraces, whilst the mixed gardens are located in fallow plots in the valley and along the forest boundary high up the valley sides (Bowers 1968). On the Nembi Plateau, the most fertile soils in depressions and dolines in the limestone and on river flats are used for mixed gardens, with sweet potato gardens confined to the volcanic ash soils on the valley slopes (Allen et al. 1980) (Fig. 3).

Highland groups are also noted for the sophistication of their agricultural techniques. Soil management in the predominantly volcanic ash soils of the

Fig. 3. Cultivation pattern on the Nembi Plateau with mixed gardens in depressions and sweet potato gardens on the slopes.

highland valleys consists of elaborate tillage and soil conservation practices which allow almost continuous cultivation of the land. Soil management has been interpreted as being designed for controlling erosion, soil fertility, surface and subsurface water, and soil temperature (Brookfield 1962; Wood and Humphreys 1982). Most soil management techniques have been developed exclusively for the cultivation of the staple sweet potato crop, so as to alter the soil environment to suit the sweet potato's particular growth requirements.

Tillage methods generally involve the preparation of raised beds consisting of a loose porous bed of soil through which water can percolate, with intervening drainage channels to dispose of excess water. Three types of tillage are used for sweet potato cultivation in the highlands: complete tillage, gridiron ditching*, and mounding (Brookfield and Hart 1971). Complete tillage is used in the steep country of northern Chimbu (Brookfield and Brown 1963), gridiron ditching in central Chimbu (Brookfield 1966) and the Wahgi valley (Powell et al. 1975), small mounds in the Eastern Highlands (Howlett1962), large mounds in the Kaugel valley (Bowers 1968), and elongate mound-like beds on the Nembi Plateau (Allen et al. 1980).

The different tillage techniques are probably local responses to the need for aerating grassland soils and promoting the breakdown of organic matter and release of nutrients, as Clarke and Street (1967) have shown. Brookfield (1962) however suggested a climatic explanation for the distribution of tillage methods, with complete tillage and small mounding for conserving moisture in the drier areas of greatest rainfall seasonality; gridiron ditching where there is less seasonality; and large mounding where there is no dry season but where the frost risk is high.

Tillage increases the depth of topsoil for crop growth and restricts the amount of erosion. During short fallows between sweet potato crops, the soil is particularly vulnerable to erosion, especially when pigs are put into gardens to break up the soil. Pigs are a major agent of erosion due to their rooting activities in gardens and fallows. Landslides are a common form of soil loss on steep slopes. Brookfield (1962) suggested that the use of gridiron ditching in central Chimbu may help control this problem by carrying away excess water and preventing percolation. This is supported by Barrie (1956) who noted that gardens with heavy clay soils on mudstones had a more elaborate network of drains. Gridiron ditching is also thought to reduce rill erosion by dividing the slope into shorter segments (Wood and Humphreys, in press). Cultivation techniques developed explicitly for erosion control have been reported from

*Gridiron ditching consists of a rectangular network of ditches with intervening raised soil beds (Fig. 4).

Fig. 4. Sweet potato gardens in the upper Chimbu with a gridiron arrangement of ditches.

the upper Chimbu valley where small fences, tree logs or low stone walls are constructed across the slope in steep gardens (Brookfield and Brown 1963; Paglau 1982).

Soil erosion is not a spectacular problem in the highlands due to moderate rainfall intensity and highly permeable soils with a high organic matter content. Measurements of erosion from garden plots in northern Chimbu have shown soil losses ranging from 10 tonnes/hectare/year on a 10° slope to 65 tonnes/hectare/year on a 40°C slope (Humphreys 1981, and personal communication). These losses are very low by international standards, but considering the length of time for which garden plots are cultivated continuously in the highlands, which can be up to 100 years on volcanic ash soils, the amount of soil lost during the lifespan of a garden is substantial. Erosion rates are generally low wherever there are volcanic ash soils, even on steep slopes. Where the ash cover is missing however, erosion rates are much greater with some of the highest erosion rates in the world recorded from road batters on mudstones of the Chim Shale formation in Simbu Province* (Blong and Humphreys 1982).

*Simbu Province is the administrative unit encompassing the Chimbu people.

Additional methods for soil fertility improvement are associated with some forms of tillage. Composting is commonly practised where there is large mound cultivation. The compost is composed of the vines from the last sweet potato crop together with grass and weeds from the garden. This is dried and covered with soil to make a mound. The decomposing compost improves soil fertility by increasing the amount of organic matter and releasing nutrients. The structure and moisture-holding capacity of the soil are also improved. In addition, the decomposition of the compost generates heat which raises soil temperatures and reduces the impact of ground frosts (Waddell 1972).

Probably the most elaborate of highland soil management techniques are the methods of field drainage, designed to dispose of excess surface and sub-surface water. This is essential for the swamp cultivation of sweet potato which is better adapted to drier conditions. The active drainage of peaty swamp soils for subsistence cultivation is restricted to only a few areas in the highlands. There are two examples of major swamp management in the Purari catchment: the Kandep basin and the upper Wahgi valley. Archaeological evidence from the Kuk swamp in the upper Wahgi valley indicates that drainage and swamp cultivation techniques were developed as far back as 9000 years ago, and that many of the contemporary soil management practices may have evolved from earlier cultivation methods (Golson 1981).

The main rationale behind soil management techniques in the highlands is firstly to allow the land to be cultivated for as long as possible, to satisfy increasing demand for food from a growing population, and secondly to preserve the soil for future generations. Some parts of the highlands have experienced a permanent reduction in soil fertility. The most vulnerable areas have high population densities and a shortage of arable land, resulting in long periods of cultivation without fallow. Agronomic studies of garden soils on the Nembi Plateau, an area having some of the highest population densities in the highlands, indicate a steady depletion of soil nutrients and an increase in acidity as length of cultivation increases (Wood, in press). The gradual removal of topsoil by erosion and soil creep has been suggested as the major reason for the fertility decline. Intensive research on soil deterioration in the upper Kikori catchment around the Tari basin has shown that some soil types (peat soils, well-drained alluvial soils, and volcanic mudflow soils) do not decline in fertility after up to 200 years of cultivation. Volcanic ash soils however do show a decline in soil fertility and crop yields. The principal cause of soil deterioration is thought to be losses of topsoil by erosion (Wood 1980).

Fig. 5. Cultivated garden near Karimui. Trees have been felled across the slope to restrict erosion.

3.2. Soil management in the highland margins

The highland margins experience some of the highest rainfall in the catchment, with annual totals between 5000 and 10 000 mm. Much of it falls as high intensity storms. Soil loss through erosion and nutrient losses by leaching are thus far greater than in the highlands. The traditional method of land use under these conditions is a form of shifting cultivation or bush fallowing, where the forest is gradually opened up and gardens prepared in the clearings (Fig. 5). Gardens are cultivated for 1 – 2 years and are then left fallow as yields decline.

The response of the soils to cultivation in this environment is very different to that in the highlands. A study of the effects of shifting cultivation on the properties of volcanic ash soils on the Karimui and Bomai plateaux showed that soil fertility declined much more rapidly with cultivation than on similar soils in the highlands (Wood 1979; Wood and Pain 1977). The fertility decline was attributed to a reduction in soil organic matter resulting from vegetation clearance. This in turn caused a reduction in the soil's exchange capacity making the exchangeable bases more susceptible to removal by leaching. The nutrient status of the soil thus declines.

3.3. Soil management in the lowlands

The traditional method of cultivation in the forested lowlands is also shifting cultivation, using similar techniques to those in the highland margins. Gardens are made by cutting the undergrowth, felling the smaller trees and pollarding* the larger ones. The vegetation debris is dried and burnt, with the crops planted in a mixture of ashes and soil. The ash from the burnt vegetation fertilises the soil by increasing bases and phosphorus, and reducing acidity. Soil fertility declines during the cropping period due to nutrient uptake by the crops, and losses by leaching and erosion. Organic matter decomposes in the soil, and as many of the available nutrients are associated with organic matter, like nitrogen and phosphorus, there is a reduction in their availability. The rate of soil fertility decline under cultivation is likely to be as fast if not faster than in the highland margins. Soil fertility is replenished traditionally by abandoning the garden and encouraging the return of the forest, which gradually builds up nutrients and organic matter.

The physical condition of the soil also deteriorates during cultivation, with a breakdown in soil structure and an increase in bulk density. This reduces the rate at which water can enter the soil, thus increasing runoff and making the soil more prone to erosion. In shifting cultivation erosion is controlled by keeping the ground covered with vegetation, leaving tree roots in the ground, and by minimal tillage of the soil.

4. Conclusions

The soil pattern of the Purari catchment is greatly influenced by the distribution of volcanic materials, particularly volcanic ash. Volcanic ash soils are the most important soil type in the catchment occupying the largest area and supporting a large and diverse agricultural population living mainly in the highland valleys. Traditional methods of soil management have evolved over thousands of years to suit local environmental conditions, and influence the way in which particular soil types respond to cultivation. Studies of the changes in soil properties with cultivation in different environments suggest

---

*Pollarding is the removal of the lower branches of trees which would shade the garden.

that the major causes of deteriorating soil conditions in the catchment are the erosion of topsoil, declining soil organic matter levels, and the leaching of soluble plant nutrients.

Soil degradation has been a problem in the catchment for thousands of years, although evidence suggests that this has accelerated, particularly in the period since European contact (Oldfield et al. 1980). Rapid population growth and limited cultivable land, particularly in the highlands, are causing increased pressure on land resources. Gardens are being fallowed for periods too short for the soil to regain its fertility. Steeper, more marginal land has also been cultivated, resulting in accelerated erosion.

One possible solution to the problem of population pressure in the highlands is to move people from areas of dense population to relatively empty areas, such as the highland margins. Government officials and politicians in Simbu Province have suggested that up to 40 000 people could be resettled from the densely populated highland valleys to the Karimui and Bomai areas in the south of the province. Soil fertility investigations suggest that such a move would result in ecological disaster due to the fragility of the ecosystem and a rapid soil fertility decline (Wood 1979).

Another way of solving the problem is to encourage the use of more effective and appropriate soil conservation practices. Soil conservation in the catchment is based almost entirely on traditional methods. Techniques range from comparatively simple shifting cultivation through to highly elaborate forms of soil management where permanent gardens are maintained by regular tillage, composting, water and erosion control. If the land is to be cultivated more intensively, techniques need to be adapted to meet the changing conditions and thus preserve the ecological balance between human activities and the environment. In areas with high populations, the traditional soil conservation methods are no longer effective in controlling soil erosion and declining soil fertility. Wood and Humphreys (1982) have recommended that research is urgently required for the evaluation of ameliorative measures such as the reafforestation of grasslands, contour cultivation, crop rotations, leguminous fallow crops, and composting methods. They suggested that where possible, conservation techniques already used in New Guinea should be encouraged.

Acknowledgements

The helpful comments and suggestions by Dr C.F. Pain and Mr G.S. Humphreys are gratefully appreciated.

References

Allen, B.J., R.M. Bourke, L.J. Clarke, B. Cogill, C.F. Pain and A.W. Wood, 1980. Child malnutrition and agriculture on the Nembi Plateau, Southern Highlands, Papua New Guinea. Social Sci. and Med. 14D: 127 – 132.

Allen, B.J. et al., in Press. The Nembi Plateau. Dept. of Geography, University of Papua New Guinea. Occasional Paper.

Barrie, J.W., 1956. Population land investigation in the Chimbu subdistrict. P.N.G. Agric. J. 11: 45 – 51.

Bleeker, P., 1969. Soils of the Kerema-Vailala area. In: B.P. Ruxton et al. (eds.), Lands of the Kerema-Vailala Area, Territory of Papua New Guinea, pp. 77 – 94. C.S.I.R.O. Land Research Series, No. 23. C.S.I.R.O., Melbourne.

Bleeker, P., 1974. Soils. In: E. Ford (ed.), Papua New Guinea Resource Atlas, pp. 10 – 11. Jacaranda Press, Milton, Qld., Australia.

Bleeker, P. and P.A. Healy, 1980. Analytical data of Papua New Guinea soils. C.S.I.R.O. Div. of Land Use. Research Technical Paper, No. 40. C.S.I.R.O., Melbourne.

Bleeker, P. and J.G. Speight, 1978. Soil – landform relationships at two localities in Papua New Guinea. Geoderma 21: 183 – 198.

Blong, R.J. and G.S. Humphreys, 1982. Erosion of road batters in Chim Shale, Papua New Guinea. Inst. of Engineers Australia. Civil Eng. Trans. CE24: 62 – 68.

Blong, R.J. and C.F. Pain, 1978. Slope stability and tephra mantles in the Papua New Guinea Highlands. Geotechnique 28: 206 – 210.

Bourke, R.M. and B.J. Allen, 1979. Subsistence agriculture and child malnutrition in the Okapa area, Eastern Highlands Province. Mimeo. Office of Environment and Conservation, Waigani, Papua New Guinea.

Bowers, N., 1968. The ascending grasslands: an anthropological study of ecological succession in a high mountain valley of New Guinea. Ph.D. Thesis. Columbia Univ., New York.

Brookfield, H.C., 1962. Local study and comparative methods: an example from Central New Guinea. Ann. Assoc. Amer. Geographers 52: 242 – 254.

Brookfield, H.C., 1964. The ecology of highland settlement: some suggestions. In: J.B. Watson (ed.), New Guinea: The Central Highlands, pp. 20 – 38. Amer. Anthropologist, Special Publ., 66.

Brookfield, H.C., 1966. The Chimbu: a highland people from New Guinea. In: S.R. Eyre and G.R.J. Jones (eds.), Geography as Human Ecology, pp. 174 – 198. Arnold, London.

Brookfield, H.C. and P. Brown, 1963. Struggle for land: agriculture and group territories among the Chimbu of the New Guinea Highlands. Oxford Univ. Press, Melbourne.

Brookfield, H.C. and D. Hart, 1971. Melanesia: a geographical interpretation of an island world. Methuen, London.

Clarke, W.C. and J.M. Street, 1967. Soil fertility and cultivation practices in New Guinea. J. Trop. Geography 24: 7 – 11.

Drover, D.P., 1973. Chemical and physical properties of surface peats in the Wahgi Valley, Western Highlands. Sci. In New Guinea 1: 8 – 10.

FAO-UNESCO, 1974. Soil map of the world, Vol. 1: Legend. UNESCO, Paris.

FAO-UNESCO, 1976. Soil map of the world, Vol. 10: Australasia, 2 map sheets. UNESCO, Paris.

Golson, J., 1981. New Guinea agricultural history: a case study. In: D. Denoon and C. Snowden (eds.),A History of Agriculture in Papua New Guinea: a time to plant and a time to uproot, pp. 55 – 64. Institute of Papua New Guinea Studies, Port Moresby.

Haantjens, H.A., 1969. Detailed soil descriptions and analytical data from the Goroka-Mount Hagen area, New Guinea. C.S.I.R.O. Div. Land Research Technical Memorandum, 69/5, C.S.I.R.O., Melbourne.

Haantjens, H.A., 1970a. Soils. In: R.G. Ward and D.A.M. Lea (eds.), An Atlas of Papua New Guinea, pp. 40 – 41. University of Papua New Guinea.

Haantjens, H.A., 1970b. Soils of the Goroka-Mount Hagen area. In: H.A. Haantjens et al. (eds.), Lands of the Goroka-Mount Hagen Area, Territory of Papua New Guinea, pp. 80 – 103. C.S.I.R.O. Land Research Series, No. 27. C.S.I.R.O., Melbourne.

Haantjens, H.A., J.R. McAlpine, E. Reiner, R.G. Robbins and J.C. Saunders, 1970. Lands of the Goroka-Mount Hagen area, Territory of Papua New Guinea. C.S.I.R.O. Land Research Series, No. 27. C.S.I.R.O., Melbourne.

Haantjens, H.A., J.J. Reynders, W.L.P.J. Mouthaan and F.A. Van Baren, 1967. Major soil groups of New Guinea and their distribution. Royal Tropical Inst. Dept. Agric. Res., Communication No. 55. Amsterdam.

Haantjens, H.A. and G.K. Rutherford, 1964. Soil zonality and parent rock in a very wet tropical mountain region. Trans. 8th Intern. Congress of Soil Science (Bucharest) 5: 493 – 500.

Howlett, D.R., 1962. A decade of change in the Goroka Valley, New Guinea: land use and development in the 1950s. Ph.D. Thesis, A.N.U., Canberra.

Humphreys, G.S., 1981. Soil loss estimates from northern Simbu Province, P.N.G. Unpublished Report, April, 1981. Dept. of Primary Industry, Port Moresby, Papua New Guinea.

Löffler, E., 1972. Pleistocene glaciation in Papua and New Guinea. Z. Geomorphol. 16: 32 – 58.

Oldfield, F., P.G. Appleby and R. Thompson, 1980. Palaeoecological studies of three lakes in the Highlands of Papua New Guinea, I. The chronology of sedimentation. J. Ecol. 68: 457 – 478.

Ollier, C.D., D.P. Drover and M. Godelier, 1971. Soil knowledge amongst the Baruya of Wonenara, New Guinea. Oceania 42: 33 – 41.

Paglau, M., 1982. Conservation of soil, water and forest in the upper Simbu Valley, In: L. Morauta, J. Pernetta and W. Heaney (eds.), Traditional Conservation in Papua New Guinea: Implications for Today, pp. 115 – 119. Institute of Applied Social and Economic Research. Monograph, 16. Boroko, Papua New Guinea.

Pain, C.F., 1975. The Kaugel Diamiction – a Late Quaternary mudflow deposit in the Kaugel Valley, Papua New Guinea. Z. Geomorphol. 19: 430 – 442.

Pain, C.F., 1982. Enga soils: reconnaissance soil map and explanatory notes. Enga Yaaka Lasemana, suppl. Vol. 1. P.N.G. National Planning Office, Waigani, Papua New Guinea.

Pain, C.F. and R.J. Blong, 1976. Late Quaternary tephras around Mt. Hagen and Mt. Giluwe, Papua New Guinea. In: R.W. Johnson (ed.), Volcanism in Australasia, pp. 239 – 251. Elsevier, Amsterdam.

Pain, C.F. and R.J. Blong, 1979. The distribution of tephras in the Papua New Guinea Highlands. Search 10: 228 – 230.

Pain, C.F. and A.W. Wood, 1976. Tephra beds and soils in the Nondugl-Chuave areas, Western Highlands and Chimbu Provinces. Sci. in New Guinea 4: 153 – 164.

Parfitt, R.L., 1975. Clay minerals in recent volcanic ash soils from Papua New Guinea. In: R.P. Suggate and M.M. Creswell (eds.), Quaternary Studies, pp. 241 – 245. Royal Soc. New Zealand, Wellington.

Parfitt, R.L. and B. Mavo, 1975. Phosphate fixation in some Papua New Guinea soils. Sci. in New Guinea 3: 179 – 190.

Perry, R.A., M.J. Bik, E.A. Fitzpatrick, H.A. Haantjens, J.R. McAlpine, R. Pullen, R.G. Robbins, G.K. Rutherford and J.C. Saunders, 1965. General report on lands of the Wabag-Tari area, Territory of Papua New Guinea, 1960 – 61. C.S.I.R.O. Land Research Series, No. 15. C.S.I.R.O., Melbourne.

Powell, J.M., A. Kulunga, R. Moge, C. Pono, F. Zimike and J. Golson, 1975. Agricultural traditions of the Mount Hagen area. Dept. of Geography, University of Papua New Guinea, Occasional Paper, 12.

Rutherford, G.K., 1962. The yellow brown soils of the highlands of New Guinea. Trans. Intern. Soc. Soil Sci., New Zealand, pp. 434 – 439.

Rutherford, G.K., 1964a. The tropical alpine soils of Mt. Giluwe, Australian New Guinea. Canadian Geographer 8: 27 – 33.

Rutherford, G.K., 1964b. Observations on the origin of a cutan in the yellow brown soils of the Highlands of New Guinea. In: A. Jongerus (ed.), Soil Micromorphology, pp. 237 – 240. Elsevier, Amsterdam.

Rutherford, G.K., 1968. Observations on a succession of soils on Mt. Giluwe. Ann. Assoc. Amer. Geographers 58: 304 – 312.

Rutherford, G.K. and H.A. Haantjens, 1965. Soils of the Wabag-Tari area. In: R.A. Perry et al. (eds.), General Report on Lands of the Wabag-Tari Area, Territory of Papua and New Guinea, 1960 – 61, pp. 85 – 99. C.S.I.R.O. Land Research Series, No. 15. C.S.I.R.O., Melbourne.

Ruxton, B.P., P. Bleeker, B.J. Leach, J.R. McAlpine, K. Paijmans and R. Pullen, 1969. Lands of the Kerema-Vailala area, Territory of Papua and New Guinea, C.S.I.R.O. Land Research Series, No. 23. C.S.I.R.O., Melbourne.

Sorenson, E.R., 1976. The edge of the forest: land, childhood and change in a New Guinea protoagricultural society. Smithsonian Inst., Washington, D.C.

U.S.D.A., 1975. Soil taxonomy: a basic system of soil classification for making and interpreting soil surveys. U.S. Dept. of Agric. Agriculture Handbook, No. 436.

Waddell, E.W., 1972. The mound builders: agricultural practices, environment and society in the Central Highlands of New Guinea. Univ. of Washington Press, Seattle and London.

Wood, A.W., 1979. The effects of shifting cultivation on soil properties: an example from the Karimui and Bomai Plateaux, Simbu Province. P.N.G. Agric. J. 30: 1 – 9.

Wood, A.W., 1980. Why is our ground going bad? Soil deterioration in the Southern Highlands and methods for its control. Southern Highlands Public Seminar No. 1, Mendi, Nov. 1980, unpublished.

Wood, A.W., 1982a. The soils of New Guinea. In: J.L. Gressitt (ed.), Biogeography and Ecology of New Guinea, pp. 73 – 83. Dr W. Junk, The Hague.

Wood, A.W., 1982b. Soils. In: D. King and S.R. Ranck (eds.), Papua New Guinea Atlas: A Nation in Transition, pp. 90 – 91. Robert Brown and Associates and University of Papua New Guinea, Port Moresby.

Wood, A.W., in press. The soils of the Nembi Plateau. In: B.J. Allen (ed.), The Nembi Plateau. Dept. of Geography, Univ. of Papua New Guinea. Occasional Paper.

Wood, A.W. and G.S. Humphreys, 1982. Traditional soil conservation in Papua New Guinea. In: L. Morauta, J. Pernetta and W. Heaney (eds.), Traditional Conservation in Papua New Guinea: Implications for Today, pp. 93 – 114. Institute of Applied Social and Economic Research. Monograph, 16. Boroko, Papua New Guinea.

Wood, A.W. and C.F. Pain, 1977. Soils and land use in southern Simbu Province. Unpublished report to Simbu Provincial Government.

6. Clay mineralogy of selected soils and sediments of the Purari River basin

G. Irion and T. Petr

1. Introduction

River-transported sediments mostly derive from erosion of poorly consolidated rocks and from the denudation of soil horizons. In cold climates, where soil horizons are missing or are poorly developed, usually only the erosion products of rocks or sedimentary deposits form river-transported sediments, whereas in rivers of humid tropics the suspended load includes minerals formed in deep soil profiles during intensive tropical weathering (Irion, 1976). Although situated in the tropical belt, the catchment of the Purari River extends altitudinally through a number of climatic zones.

The heterogeneity of the geological formations found throughout P.N.G. (Bain et al. 1972), and consequently in the Purari catchment, is responsible for the occurrence of a wide range of weathering products. Pain (1978) conveniently divided the landforms of the central highlands, the source of most rivers which eventually form the Purari, into four groups: erosional, volcanic, karst and depositional landscapes. Rocks and soils of these have been blanketed by layers of volcanic ash (tephra) (Pain and Blong 1979), some reaching a thickness of more than 15 m.

The scarcity of mature weathering is a major reason for the rarity of 'typical tropical' soils (Haantjens and Bleeker 1970). While volcanic and depositional materials are more weathered than young and unstable landforms, the average estimated percentage area with mature weathering in areas investigated by the Division of Land Research in the 1960s was only 12% (Haantjens 1970). On the other hand, tephras and tephra-derived materials are strongly weathered (Pain 1978).

Haantjens and Rutherford (1964) were the first to investigate the mineralogy of humid, brown clay soils and volcanic ash, clastic sediments, and limestones in altitudes between 1200 and 2700 m of the P.N.G. highlands.

Petr, T. (ed.) The Purari – Tropical environment of a high rainfall river basin
© *1983, Dr W. Junk Publishers, The Hague / Boston / Lancaster*
ISBN-13: 978-94-009-7265-0

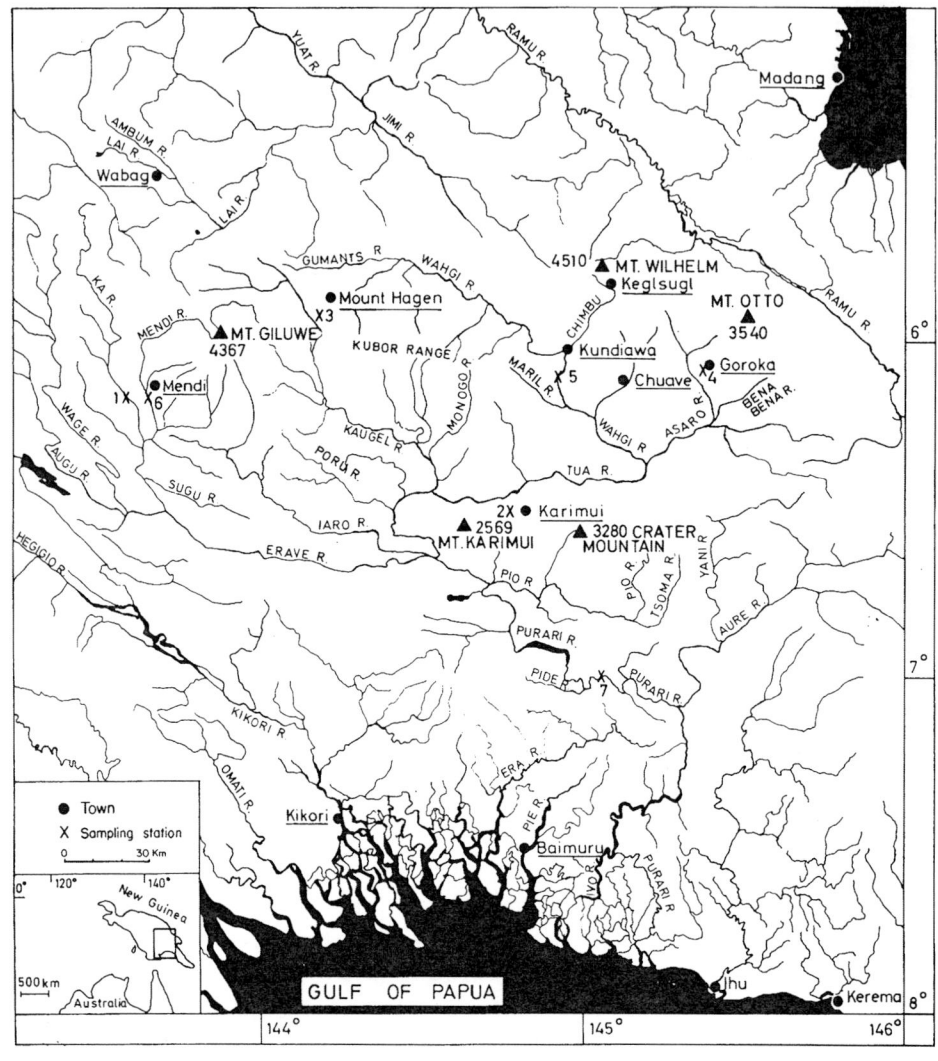

Fig. 1. Soil sampling stations.

They found a dominance of gibbsite in ash soils, of kaolinite and illite in clastic sediments and of kaolinite, gibbsite and montmorillonite in limestones. Parfitt (1975) found in volcanic ash soils in the highlands allophane, gibbsite, halloysite, imogolite and goethite.

The present study concentrated on investigating clays as the most active components in geochemical and mineralogical processes taking place in soils.

88

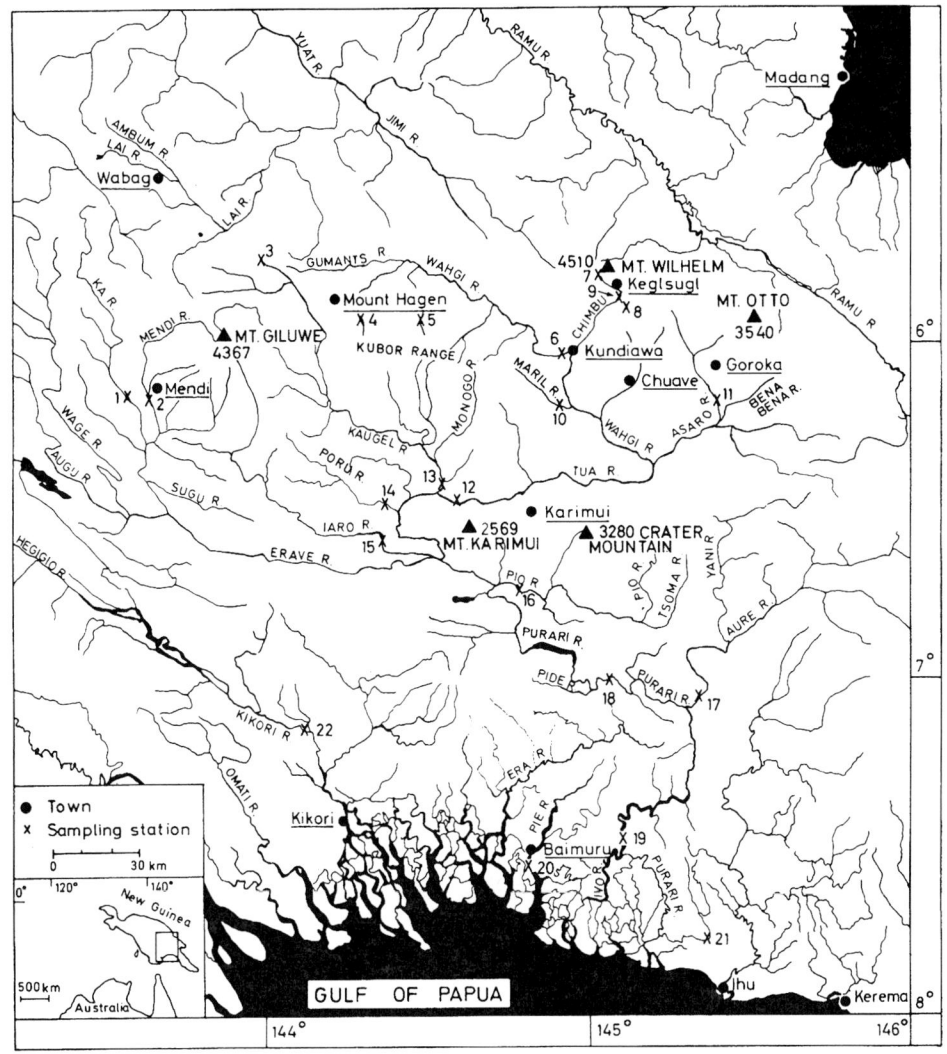

Fig. 2. Sediment sampling stations.

It also aimed at improving our knowledge of the relationships between clay mineralogy of soil profiles and that of river-transported sediments of the Purari catchment.

Fig. 3. Soil collection from a roadcut near Mt. Giluwe.

2. Collection of samples and their analysis

In October and November, 1978, samples of a number of soil profiles and river-transported sediments were collected from the Purari River catchment (Figs. 1 and 2). In the highlands soil profiles were usually collected from road cuts (Fig. 3), while those in the lowlands were collected from steep river banks. River sediments were collected by hand or with a grab.

Each sample was separated into four size categories: smaller than 2, 2 – 6.3, 6.3 – 20, and larger than 20 microns, by using the Atterberg (1912) method of sedimentation in demineralized water. The clay fraction of some samples was subdivided by centrifuging into three fractions: less than 0.2, 0.2 – 0.6, and 0.6 – 2 μm. The coarse silt (20 – 63 μm) and the sand fractions were separated by sieving after applying the Atterberg method, or, if the original sample was coarse grained, by sieving about 50 g of untreated sample.

As iron oxide seriously interferes with the copper radiation method for analysis of minerals, the dithionate-citrate-bicarbonate method (Mehra and Jackson 1960) was used. In order to saturate the clay minerals with Mg^{++} the samples were treated with 1 N Mg-acetate solution and then washed with demineralized water. The same procedure was carried out with K-acetate to get K-saturated samples. For X-ray diffraction analysis of the clay fraction,

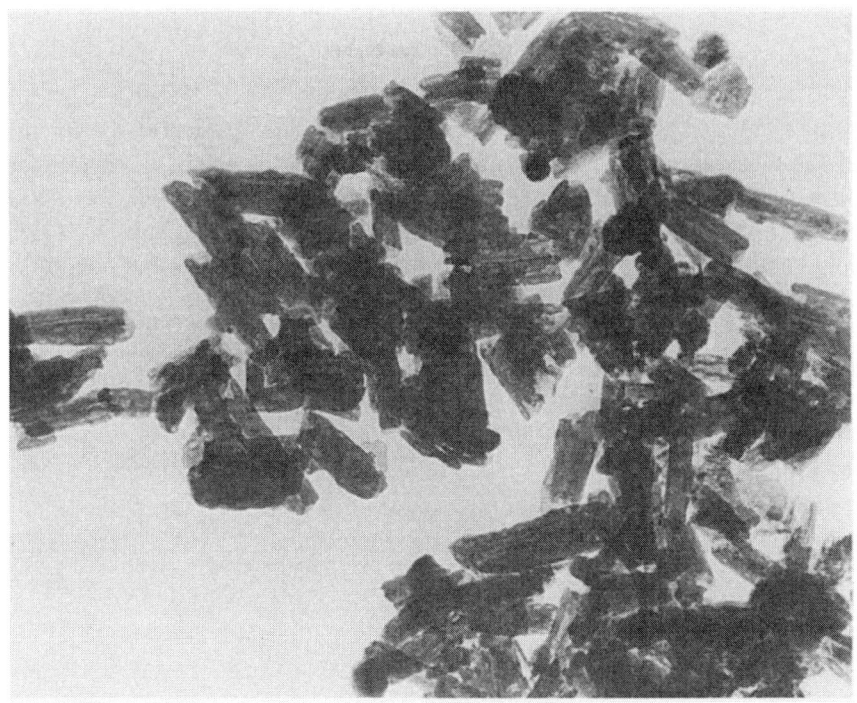

Fig. 4. Electron microscope photograph of halloysite from Purari soils.

smear slides were prepared. The coarser fractions were analysed as powder specimens. For X-ray investigations a Philips commercial standard vertical diffractometer with a linear recorder was used. Halloysite was photographed by using electron microscopy (Fig. 4).

3. Mineralogy of soil profiles

Seven profiles are reproduced here as typical for the area investigated (Fig. 1).

3.1. Highland profiles

3.1.1. Profile formed on limestone. In the southwest of the Purari catchment, especially in the Erave River drainage, limestone is the dominant rock. It is strongly weathered here and forms kegelkarst. The profile near Lai River, west of Mendi, represents a complete weathering sequence (Table 1).

The deeper horizons of this profile are dominated by low-charged mont-morillonite, which is known to form during weathering processes both in tropical and temperate zones. The absence of mica in the nonsoluble fraction of the limestone confirms that the montmorillonite does not originate from mica, and this is also confirmed by the low charge of this mineral. In a closed geochemical system the presence of quartz usually excludes formation of gibb-site (Garrels and Christ 1965). In the present profile the good drainage of the soil seems not to allow the concentration of SiO_2 to reach a level high enough for it to form kaolinite, and therefore only gibbsite is formed. The presence of gibbsite also indicates the beginning of laterisation, as described on limestones by Valeton (1962). Al-chlorite is a relatively stable mineral which usually forms from the triphormic clay minerals chlorite or mica by replacement of the brucite sheet in their lattices by gibbsite. Gibbsite is found in a lower con-centration than Al-chlorite possibly because it is used up in the formation of this mineral. Jackson (1963) calls such a phenomenon the antigibbsite effect.

3.1.2. Profiles formed on volcanic ash. In these profiles the unweathered rock was not reached. The clay fraction of these profiles is dominated by halloysite, a mineral of the kaolinite group, or by aluminium chlorite (Tables 2 – 5). Halloysite has been identified as an intermediate weathering product of volcanic ash (tephras) (Parfitt 1975). Bleeker (1972), who investigated eight weathering profiles at Safia in southeastern Papua New Guinea, found halloysite to be the dominant mineral of the topsoil. Pain and Blong (1976), describing Wanabuga tephra at Keltiga in the Nebilyer River catchment, noted halloysite there. Halloysite also dominates three of the samples in the present study. The widespread distribution of tephras (Pain and Blong 1979) in the highlands of the Purari catchment seems to be a major factor behind the widespread presence of halloysite in soils and in river-transported sediments.

Table. 1. Limestone soil profile above the Lai River near Mendi (Fig. 1, x1; Fig. 5).

Depth (cm)	Al-chlorite	Gibbsite	MM	Quartz
Topsoils	xx	x	x	xx
20	xx	x	xxx	x
50	xx	x	xxx	x
100	unweathered limestone which contains 95% calcite, 2.7% quartz and some montmorillonite (MM)			

x = low content, xx = medium content, xxx = high content.

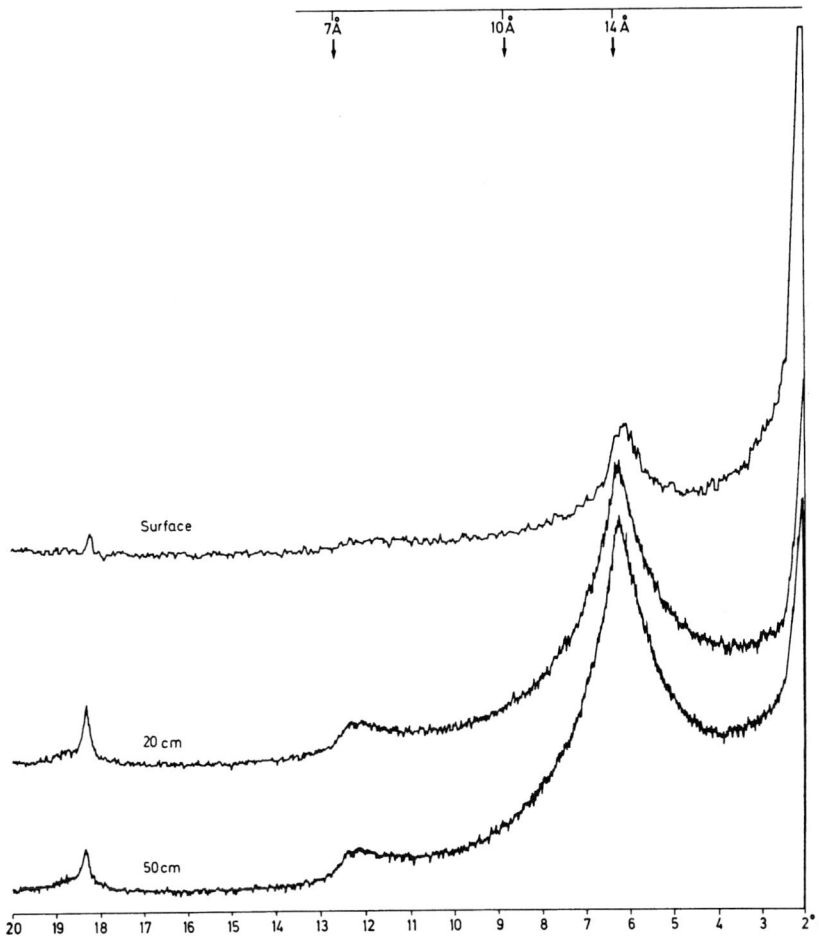

Fig. 5.. X-ray diffractogrammes of the limestone soil profile clay fraction (Lai River catchment, near Mendi). The peaks at 14Å for 20 and 50 cm depths belong predominantly to montmorillonite; in the topsoil it is Al-chlorite. The basal spacing (001) in montmorillonite will change when treated with potassium ions, while that of the Al-chlorite does not. Gibbsite shows the main peak at 18.3°2θ.

Table 2. Soil profile at Karimui, Tua River catchment (Fig. 1, X2).

Depth (cm)	Gibbsite	Halloysite
Topsoil	xxx	xx
200	–	xxxx
300	–	xxxx

Parfitt (1975) suggested that weathering of glass and feldspar from tephras leads to formation of allophane, halloysite and gibbsite. The aluminium chlorite in the present study probably originated from mica and/or chlorite, which, however, were not identified in the samples.

It is suggested here that the source of Al-chlorite in the studied profiles is either an X-ray amorphous material (glass in the Karimui profile), or normal chlorite, or 2:1 minerals (in the Tagoba profile), but the possibility of its origin through weathering of volcanic ash cannot be excluded. More investigations certainly appear essential for further clarification.

The Tagoba and Goroka profiles also contain cristobalite, which has been suggested by Bleeker (1972) to originate from ultrabasic rocks. Most resear-

Table 3. Soil profile at Tagoba, Nebilyer River catchment (Fig. 1, X3).

Depth (cm)	Al-chlorite	Gibbsite
Topsoil	xxxx	–
20	xxx	xx
60	xxx	xx
100	xx	xxx
150	xx	xx
200	xxx	xx (basaltic hornblende and cristobalite are also present)

Table 4. Soil profile in Goroka, Asaro River catchment (Fig. 1, X4; Fig. 6).

Depth (cm)	Halloysite	Note
Topsoil	xxxx	
30	xxxx	
150	xxxx	
300	xxxx	
500	xxxx	(cristobalite is present in the coarser fraction)

Table 5. Soil profile at Maromaul, Wahgi River catchment (Fig. 1, X5).

Depth (cm)	Al-chlorite	Halloysite	Mica
Topsoil	x	xxx	x
30	x	xxx	xx
60	–	xxx	xxx
300	–	xxxx	x

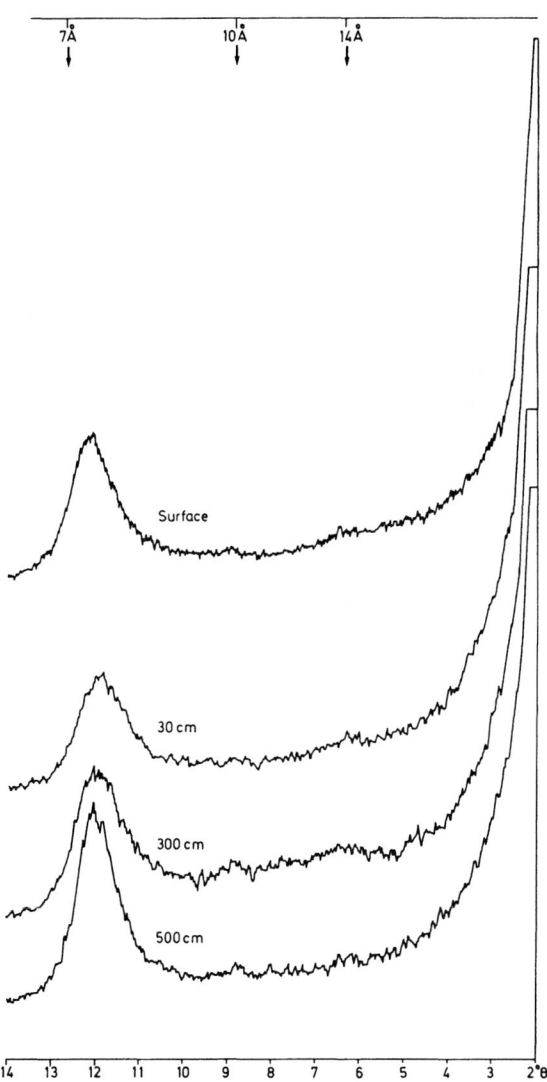

Fig. 6. X-ray diffractogrammes of the volcanic ash soil profile clay fraction (Asaro River catchment, at Goroka). The peaks at 7Å represent halloysite heated to 80°C. Haantjens and Bleeker (1970) consider halloysite as the commonest mineral in Papua New Guinea.

chers of soil profiles suggest that cristobalite cannot form during the soil formation process. However, Eswaran and Coninck (1971) believe that they found evidence for the formation of cristobalite in soils of El Salvador and

95

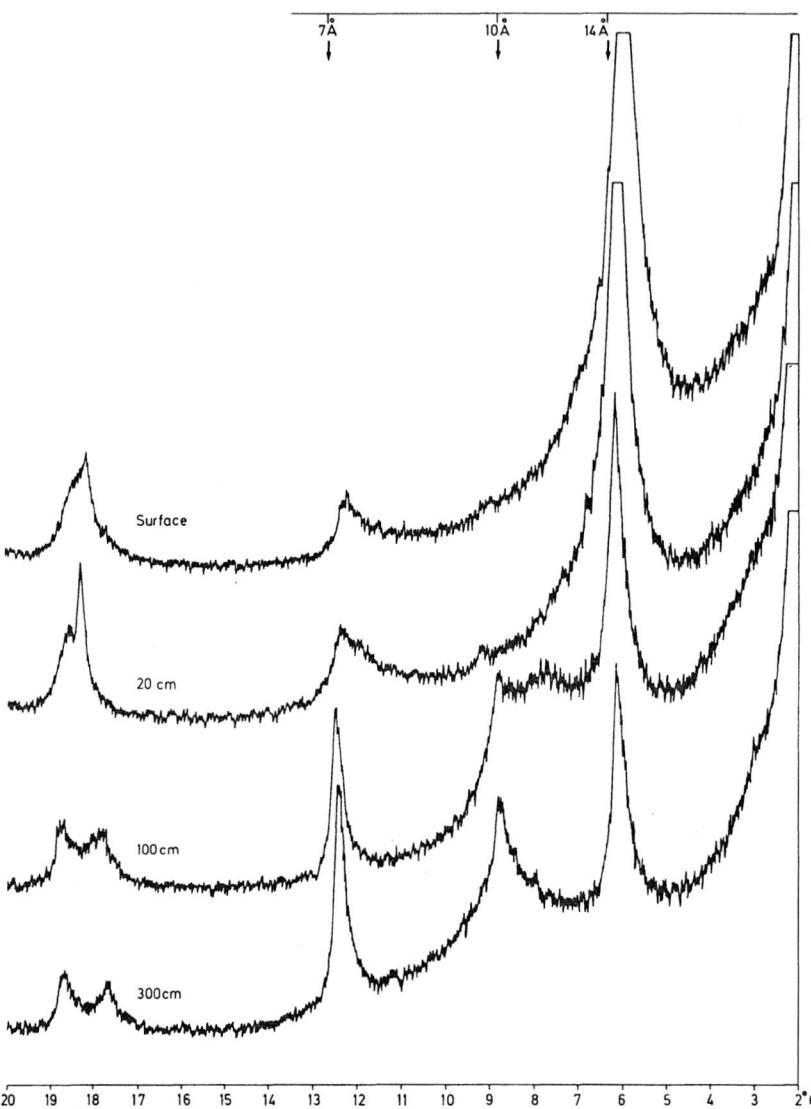

Fig. 7. X-ray diffractogrammes of the shale oil profile clay fraction (Erave River catchment at Mendi). The profile starts at a depth of 300 cm with a relatively high content of chlorite and the mixed layer mineral rectorite. There is an increase in rectorite up to 100 cm depth, but in shallower horizons, gibbsite and Al-chlorite form. Al-chlorite dominates the topsoil. Such profiles with similar sequence of minerals resulting from weathering processes are also known from the Amazon lowlands (Irion 1976), where soil formation has taken place since at least the late Tertiary.

Madagascar. Another possibility is that cristobalite forms in volcanic tuff or lava rich in silica during their dissolution. Frequently this mineral forms in opals, and it is possible that such an origin took place in the investigated profile. Parfitt (1975) found opal in the clay fraction from Hoskins in New Britain but has not reported it for his highlands profile.

3.1.3. Profile formed on shale. Clay shales outcrop in a number of areas of Papua New Guinea. They have erosion rates which are amongst the highest recorded anywhere in the world (Blong and Humphreys 1982). An outcrop of shale (Table 6) was studied on the Southern Highlands highway, where large mudslides have disrupted the traffic several times. The original unweathered rock probably contained mica and chlorite. During tropical weathering the chlorite has dissolved, whereas mica has been transformed into illite, which in turn has changed into the mixed-layered mineral rectorite, to finally form Al-chlorite (Fig. 7).

3.2. Lowland and delta profiles

A profile of old Purari River sediments at Wabo was analysed to show weathering of alluvial sediments (Table 7). Even at 3 m depth the horizon is

Table 6. Shale profile at Mendi, Erave River catchment (Fig. 1, X6; Fig. 7).

Depth (cm)	Chlorite	Illite	Rectorite	Al-chlorite	Gibbsite
Topsoil	–	–	–	xxxx	–
20	–	–	–	xxx	xx
100	xx	(x)	xxx	xx	–
200	xxx	x	xx	–	–
300	xxx	xx	xx	–	–

(x) = very low content.

Table 7. Old river terrace on the Purari River at Wabo (Fig. 1, x7; Fig. 8).

Depth (cm)	Al-chlorite (+)	Kaolinite	Mixed-layer and/or montmorillonite
Topsoil	xxx	xx	–
200	xx	x	xxx
300	xx	xx	xx

(+) = includes chlorite, increasing in quantity with depth.

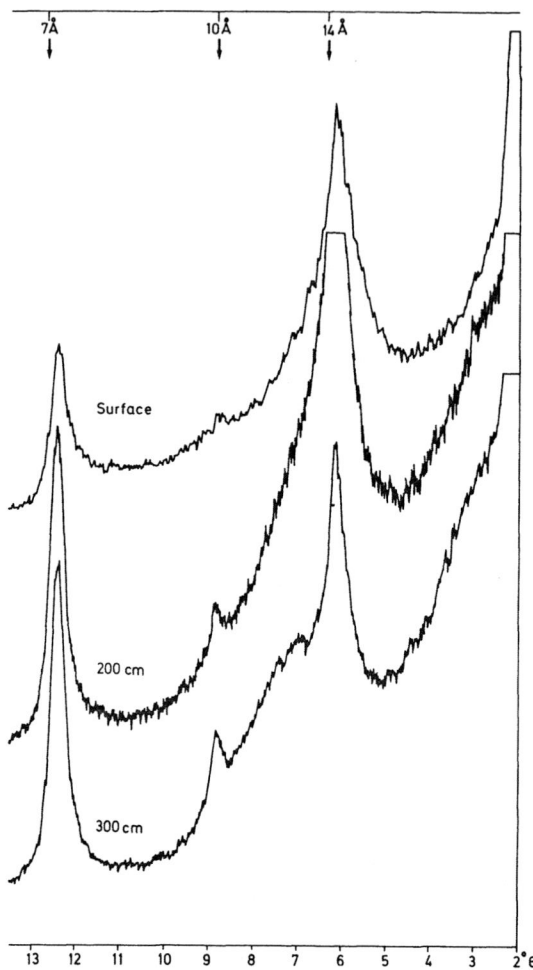

Fig. 8. X-ray diffractogrammes of a river terrace soil profile clay fraction at Wabo on the Purari River. Montmorillonite and mixed layer minerals are common at 200 and 300 cm depth. At the surface, Al-chlorite and chlorite with a peak at about 14Å is the commonest mineral (see Table 7).

strongly weathered. The originally present illite was dissolved or transformed into mixed-layer minerals. A fireclay type of kaolinite is present at all depths. The topsoil has a higher Al-chlorite content than deeper layers, suggesting strong weathering in surface layers. The high rate of weathering is due to good drainage, high temperature and relatively high rainfall in the area.

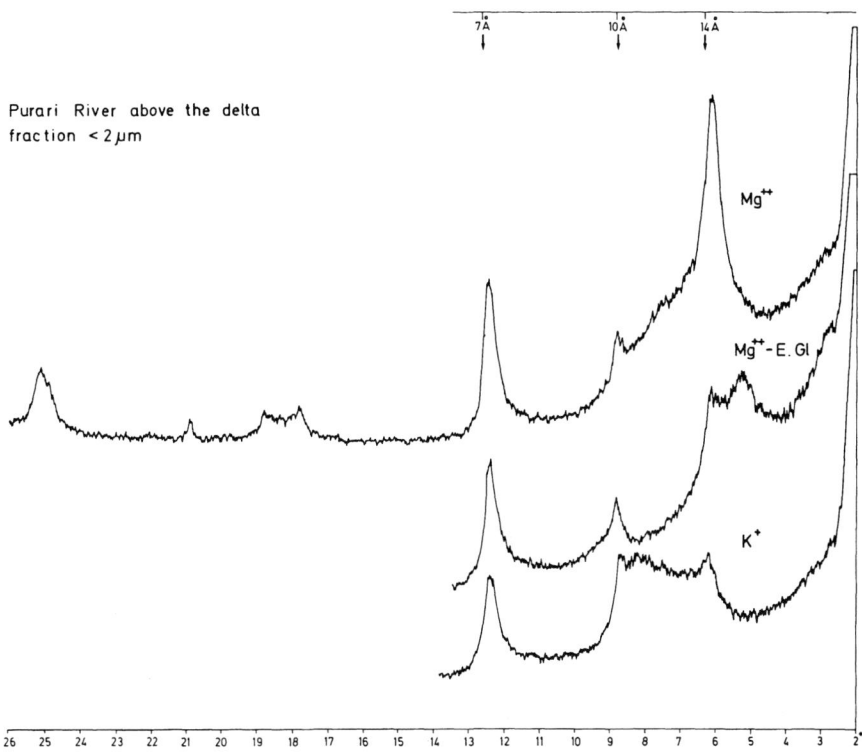

Fig. 9. X-ray diffractogrammes of the Purari sediment clay fraction shortly above delta. After treatment with ethylene glycol the high content of chlorites shows up. The peak of 25°2θ with a clear maximum at 3.52Å shows that the sediments contain about the same amount of kaolinite as chlorite. When treated with K^+, montmorillonite has the basal spacing between 14 and 10Å without showing a clear peak, therefore one can conclude that montmorillonite here is a mixture of montmorillonites with high and low charges. Highly charged montmorillonite forms predominantly from mica, while the low charged forms from volcanic glass and also for example on limestones and in some soils with a high ionic content in pore water (Irion 1976).

4. Mineralogy of river sediments (for sampling stations see Fig. 2)

4.1. River sediments in the Purari catchment above the delta (Table 8) (Fig. 9)

To enable a comparison of mineralogy and geochemistry, sediments were sorted into four fractions. The clay fraction, which is considered as the most suitable for such comparison as it contains the genetically most interesting minerals, is the only one discussed here.

99

Table 8. Clays of the < 2 μm fraction of river sediments.

	MM	C – AlC	I + M	K + H	G
1. Lai at Mendi	xx	x	(x)	x	–
2. Mendi River at Mendi	xx	x	x	xx	–
3. Ambuya River	xx	xx	–	xx	–
4. Kunan River	x	xx	x	x	–
5. Tuman River	x	xx	x	xx	(x)
6. Wahgi River	xx	xx	(x)	xx	–
7. Lake Aunde	xxx	–	–	xx	–
8. Tem River	xx	–	–	xx	–
9. Chimbu River	xx	xx	x	xx	–
10. Maril River	xx	x	(x)	xx	–
11. Asaro River	xx	xx	–	xx	–
12. Tua River	x	xx	x	xx	–
13. Kaugel River	xx	xx	–	x	–
14. Poru River	xxx	–	x	xx	–
15. Iaro River	xxx	–	(x)	x	–
16. Pio River	xxx	x	(x)	x	–
17. Aure River	xx	x	(x)	x	–
18. Purari River at Wabo	xx	x	x	xx	(x)
19. Purari River at Pawaia	xx	x	x	xx	(x)
20. Pie Estuary at Baimuru	xx	x	x	xx	(x)
21. Heperi River	xxxx	(x)	(x)	x	–
22. Kikori River	xxx	(x)	(x)	x	–

xxxx = very high content; xxx = high content; xx = medium content; x = low content; and (x) = very low content.
MM = montmorillonite, C – AlC = chlorite including Al-chlorite, I + M = elite and mica, K + H = kaolinite and halloysite, G = gibbsite. Numbering corresponds to sampling stations on Fig. 2.

Pickup (this volume) has estimated that clays form 35% of suspended sediments at Wabo. Most minerals of the clay fraction are weathering products. In the Purari catchment rivers and the Purari, montmorillonite and halloysite appear to be the most common clay minerals, with the former one found in all investigated river sediments (Fig. 2). Although halloysite was detected in most samples, its identity was somehow obscured as it has the same diffraction peak as kaolinite. Gibbsite, the common mineral in soils, was absent in sediments, or when present, in very low concentrations.

Primary chlorite and Al-chlorite were not separated due to methodological difficulties. However, the high Al-chlorite concentration in topsoils of the catchment leads us to assume that the 14 t mineral found in sediments is mostly Al-chlorite.

Some sediments have illite in low quantities.

The 2 to 125 micron sediment fraction (silt and fine sand) of the Purari on entering the lowlands consists mainly of quartz, feldspar, and heavy minerals such as magnetite, limonite, leucoxene, pyroxene, basalt, hornblende and titanite. In this coarse sediment fraction only limonite was identified as a product of weathering processes. The mineral composition of the coarse fraction is a less suitable indicator of ongoing weathering processes than the clay fraction, and therefore, little attention was paid to it in the present study.

4.2. River sediments of the Pie River (deltaic environment)

The sediment profile at Baimuru on the Pie River (Table 9), bordering the Purari delta in the west, shows no intensive weathering, and the minerals are evenly distributed throughout the profile. From the composition of clay minerals and the lack of vertical gradient in their distribution, one can conclude that the sediments in Baimuru are relatively young. Pickup and Warner (in press) suggest that the period of major deposition in the Gulf of Papua probably ended about 5000 years ago when the sea reached what is more or less its present level. As the levees of the present Baimuru floodplain are only about 1 metre above the king tide level, it is most probable that the area was fully submerged during the last 5000 years when the world ocean was higher than at present, suggesting that the levees are younger than 5000 years.

Table 9. Sediment profile at the Baimuru Gulf Hotel, Pie River.

Depth (cm)	Montmorillonite	Kaolinite	Illite	Chlorite
Topsoil	xx	x	(x)	x
100	xx	x	(x)	(x)
200	xx	x	(x)	x

5. Discussion

As shown by Pain and Blong (1976, 1979) great parts of the Papua New Guinea highlands are covered by tephra layers. The clay mineral halloysite, dominant in well-weathered tephras, is well distributed both in soils and in most river sediments of the Purari catchment. In sediments of the Aure River, in whose catchment tephra is not common (Pain, this volume), the content of halloysite is correspondingly lower. Gibbsite, present in the topsoils of some tephras and in the weathering profile on limestone, showed up very seldom and in barely detectable quantities in some river sediments of the Purari River (for example at Wabo), and in one of the highland rivers (Tuman). Tephras contribute gibbsite only from the topsoil as it is in a low concentration or absent at greater depths, e.g. in landslides, which expose and bring into suspension deeper horizons of weathering profiles, gibbsite is absent. Limestone is another possible source of gibbsite. Limestone deposits cover a large area in the western part of the Purari catchment, drained by the Erave and some of its tributaries. However, mineralogical analysis of a limestone profile near Mendi shows that the content of gibbsite in these soils is not very high.

Montmorillonite is present in the high altitude soils formed on the crystalline basement of Mt. Wilhelm (Irion, unpublished data). From there, montmorillonite enters sediments of the glacial Lake Aunde (alt. 3530 m) (Table 8). Limestones and fine-grained sediments of rivers, lakes, and water-logged mudstones are other important sources of montmorillonite. All these are common in the Purari catchment.

Al-chlorite is common in the highland topsoils of both limestone and volcanic origin. Chlorite, which weathers easily, can be expected to be in a lower concentration in topsoils, but increases with depth (see Table 6). When deeper soils undergo more rapid erosion as a result of disturbance by excavations, landslides, or in some soils, by intensive agriculture, more chlorite can be expected to enter the streams with the surface water drainage and to be carried and deposited with sediments. Most of the Purari catchment river sediments probably contain chlorite, which, in the present analysis, however, was not separated from Al-chlorite. Both Al-chlorite and chlorite are absent from the highest altitude sample of Lake Aunde, because at low temperatures they form only in very low quantities. They are also absent or in low concentrations in river sediments originating predominantly from limestone and marine clastic sedimentary rocks (e.g. Kikori, Poru, Iaro, Aure rivers). The reason for this is unknown.

When comparing the Purari with other tropical rivers both inside and outside Papua New Guinea, a striking feature in their clay mineralogy of sediments is that most of them are dominated by kaolinite/halloysite or montmorillonite (Table 10). For a better understanding of the situation, smaller rivers representing subcatchments are also listed for some of the major basins.

Studies of the Amazon (Irion 1976, 1978) show that montmorillonite formed on a large scale in water-logged, fine-grained sediments of tropical lowlands. The clay fraction of rivers draining areas with this character is formed nearly exclusively by montmorillonite (Fig. 10). In contrast to this, kaolinite formed from coarsely-grained, well-drained, and quartz-rich sediments or well-weathered igneous rocks. Among the rivers draining such areas are Rio Negro, Orinoco and Zaire. Among the smaller rivers, June River (Papua New Guinea) of the Fly River catchment would appear to have a catchment lithology similar to the other kaolinite-rich rivers.

Examples of montmorillonite-rich river sediments are those of the Rio Purús in the Amazon basin, and the Yellow River, a tributary of the Sepik River. The Yellow River and the Purús, draining deeply weathered Cretaceous to Tertiary sediments, and the Sepik River tributaries, draining the relatively stable Pacific plate with fine-grained Tertiary to Pleistocene sediments, also produce large amounts of montmorillonite as a result of intensive tropical weathering. Another source of montmorillonite is the weathering of limestone. Iaro River in the Purari watershed has predominantly a limestone catchment, and consequently its sediments carry a lot of montmorillonite. However, after merging with other rivers and eventually forming the Purari, the latter's mineralogy is dominated by halloysite, with less montmorillonite and chlorite. This is dictated by the overwhelming presence of volcanic rocks, especially tephras (volcanic ash) in its catchment. Halloysite dominance is clearly seen in the Wahgi River sediments as a typical example of a river pre-

Table 10. Mineral content of some tropical rivers.

	Clay fraction				Silt and sand	
	K(H)	MM	C	IM	Q	F
Purari R.	xx(mostly halloysite)	xx	x	x	xxx	xx
Wahgi R.	xx(mostly halloysite)	xx	xx	+	xx	xx
Iaro R.	x	xxx	−	+		
Fly R.	xx	xxx	x	x	xxxx	x
June R.	xxxxx	+	+	x	xxxxx	−
Sepik R.	−	xxx	xx	xx	xxx	xx
Yellow R.	x	xxxx	xx	x	xxx	xx
April R.	−	x[a]	xxx	xxx	xxx	xx
Zaire R.	xxxxx	+	+	x	xxx	+
Orinoco	xxx	x	+	xx	xxx	x(+mica)
Amazon	xx	xx	x	xx	xxx	xx
Rio Negro	xxxxx	−	−	−	xxxxx	x
Rio Purús	+	xxxxx	−	+	xxxx	x

xxxx = extremely high content, xxxx = very high content, xxx = high content, xx = medium content, x = low content, + = very low content, − = absent.
K(H) = kaolinite (halloysite), MM = montmorillonite, C = chlorite, IM = illite and mica, Q = quartz, F = feldspar.
[a] Mixed-layer mineral.

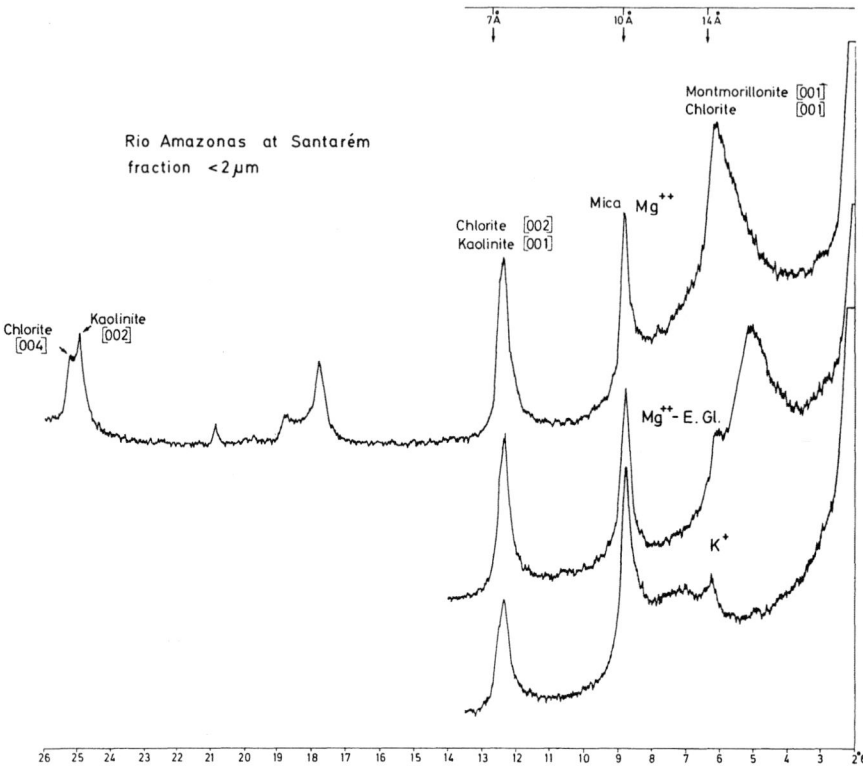

Fig. 10. X-ray diffractogrammes of clay fraction in Amazon sediments at Santarém, approximately 600 km upstream from the sea. In spite of a certain similarity of the Amazon and Purari sediments the former has higher illite and kaolinite content, while the latter has a higher chlorite content.

dominantly draining tephra-rich areas (Pain and Blong 1979).

Although the present study has not quantified changes in the individually suspended clay minerals with increasing salinity, it is known from experimental studies summarized by Burton (1976) that kaolinite and illite flocculate completely during early mixing, by a salinity of about 4‰, while montmorillonite is more slowly flocculated over a wide salinity range. Porenga (1966) established that there is a differential sedimentation of particles transported by the Niger River into the sea. During this transport the kaolinite and montmorillonite separate from each other (if in flocs) and then sediment separately: with increasing distance from the coast the montmorillonite component in the sediment increases.

Gibbs (1977) allowed the $< 2\mu$ fraction of the Amazon River suspended material to flocculate in seawater and found that there was no change in composition with time. He identified as the probable reason for the lack of significant differential flocculation in the natural samples the fact that the natural particles have metallic and organic coatings, while the material used in experiments was lacking them. These coatings effectively eliminate the differential focculating properties of kaolinite, illite and montmorillonite. However, Gibbs also found that among these three clays montmorillonite is by far the dominant clay far away from the Amazon mouth. Gibbs concluded that it is the physical sorting by size rather than other properties of the clays which makes the smallest of the three clays, i.e. montmorillonite, to be carried over the longest distance. If this is valid for the Purari River system, then the relatively high concentration of montmorillonite reaching the Purari Delta will be carried out into the sea for a considerable distance.

Clay particles, especially in flocs together with organic matter, form substrate for bacteria. Paerl and Kellar (1980) found a striking increase in bacterial numbers in the brackish reach of the Purari Delta as compared with that of the Purari freshwater. The brackish water suspended material was clearly preferred by bacteria as evidenced by scanning electron microscopy. After being deposited on mud banks, the bacteria, together with other microorganisms, are utilized by larger benthic organisms. The intensive tidal flow in the delta brings the flocs easily back into suspension, encouraging further bacterial colonization. Studies by Frusher (1980) have shown that penaeid prawns, under conditions of 5‰ or higher salinity, frequent mud banks of the Purari – Kikori channel system. Thus, the presence of flocs appears to be of importance for maintaining the relatively high productivity of estuaries. In open, more saline waters of the Purari plume, clay particles, together with adsorbed nutrients, particulate organic matter, dissolved nutrients and dissolved organic matter will combine in determining the level of productivity of the coastal waters. Present equilibrium in this environment will change if dams are constructed on rivers entering the Gulf of Papua, and the presence of dams could, through changes in the input of suspended clay minerals, alter the aquatic productivity of estuaries and coastal seawaters. Further investigations of clay minerals in this area would help to gain a better understanding of factors determining the present aquatic productivity of the Gulf of Papua.

References

Atterberg, A., 1912. Die mechanische Bodenanalyse und die Klassifikation der Mineralboden Schwedens. Int. Mitt. f. Bodenkunde 2: 312–342.

Bain, J.H.C., H.L. Davies, P.D. Hohnen, R.J. Ryburn, I.E. Smith, R. Grainger, R.J. Tingey and M.R. Moffat, 1972. Geology of Papua New Guinea, 1:1 000 000 map. Bureau of Mineral Resources Australia, Canberra.

Bleeker, P., 1972. The mineralogy of eight latosolic and related soils from Papua and New Guinea, Geoderma 8: 191 – 205.

Blong, R.J. and G.S. Humpreys, 1982. Erosion of road batters in Chim shale Papua New Guinea. Inst. of Engineers Australia. Civil Eng. Trans. CE24: 62 – 68.

Burton, J.D., 1976. Basic properties and processes in estuarine chemistry. In: J.D. Burton and P.S. Liss (eds.), Estuarine Chemistry, pp. 1 – 36. Academic Press.

Eswaran, H. and F. Coninck, 1971. Clay mineral formations and transformations in basaltic soils in tropical environments. Pedologie 21: 181 – 210.

Frusher, S.D., 1980. The inshore prawn resource and its relation to the Purari Delta region. In: D. Gwyther (ed.), Purari River (Wabo) Hydroelectric Scheme Environmental Studies, Vol. 15: Possible effects of the Purari Hydroelectric Scheme on subsistence and commercial crustacean fisheries in the Gulf of Papua, Workshop 12 Dec. 1979, pp. 11 – 17. Office of Environment and Conservation, Waigani, Papua New Guinea.

Garrels, R. and C. Christ, 1965. Solution, minerals and equilibria. Harper and Row, New York.

Gibbs, R.J., 1977. Clay mineral segregation in the marine environment. J. Sediment. Petrology 47: 237 – 243.

Haantjens, H.A., 1970. Geologic and geomorphic history of the Goroka-Mount Hagen area. C.S.I.R.O. Land Res. Ser., No. 27, pp. 19 – 23. C.S.I.R.O., Melbourne.

Haantjens, H.A. and P. Bleeker, 1970. Tropical weathering in the Territory of Papua and New Guinea. Aust. J. Soil Res. 8: 157 – 177.

Haantjens, H.A. and G.K. Rutherford, 1964. Soil zonality and parent rock in a very wet tropical mountain region. Trans. 8th Int. Cong. Soil Sci. Bucharest, 5: 493 – 500.

Irion, G., 1976. Mineralogisch-geochemische Untersuchungen an der pelitischen Fraktion amazonischer Oberböden und Sedimente. Biogeographica 7: 7 – 25.

Irion, G. 1978. Soil infertility in the Amazon rain forest. Naturwiss. 65: 515 – 519.

Jackson, M.L., 1963. Interlayering of expansible layer silicates in soils by chemical weathering. Clays and Clay Minerals 11: 29 – 46.

Mehra, O.P. and M.L. Jackson, 1960. Iron oxide removal from soils and clays by dithionate-citrate system buffered with sodium bicarbonate. Clays and Clay Minerals 7: 319 – 327.

Paerl, H.W. and P.E. Kellar, 1980. Some aspects of the microbial ecology of the Purari River, Papua New Guinea. In: T. Petr (ed.), Purari River (Wabo) Hydroelectric Scheme Environmental Studies, Vol. 11: Aquatic ecology of the Purari River catchment, pp. 25 – 39. Office of Environment and Conservation, Waigani, Papua New Guinea.

Pain, C.F., 1978. Landform inheritance in the central highlands of Papua New Guinea. In: J.L. Davies and M.A.J. Williams (eds.), Landform Evolution in Australia, pp. 48 – 69. Australian National University Press, Canberra.

Pain, C.F., 1983. Geology and geomorphology of the Purari River catchment. This volume, Chapter I, 3.

Pain, C.F. and R.J. Blong, 1976. Late Quaternary tephras around Mount Hagen and Mount Giluwe, Papua New Guinea. In: R.W. Johnson (ed.), Volcanism in Australasia, pp. 239 – 251. Elsevier, Amsterdam.

Pain, C.F. and R.J. Blong, 1979. The distribution of tephras in the Papua New Guinea highlands. Search 10: 228 – 230.

Parfitt, R.L. 1975. Clay minerals in recent volcanic ash soils from Papua New Guinea. In: R.P.

106

Suggate and M.M. Cresswell (eds.), Quaternary Studies, pp. 241 – 245. The Royal Soc. N. Zealand, Wellington.

Pickup, G., 1983. Sedimentation processes in the Purari River upstream of the delta. This volume, Chapter I, 11.

Pickup, G. and R.F. Warner, in press. Geomorphology of tropical rivers. 1. Geomorphic history, hydrology and sediment transport in the Fly and Purari, Papua New Guinea.

Porenga, D.H., 1966. Clay minerals in recent sediments of the Niger delta. In: Clays and Clay Minerals. Proc. 16th Nat. Conf. on Clays and Clay Minerals, pp. 221 – 223. Pergamon Press, Oxford.

Valeton, I., 1962. Petrographie und Genese von Bauxitlagerstättèn. Geol. Rdsch. 52: 448 – 474.

7. Geochemistry of soils and sediments of the Purari River basin

T. Petr and G. Irion

1. Introduction

As a follow-up of mineralogical studies (Irion and Petr, this volume), soils and sediments of the Purari catchment were also analysed for their elemental composition.

As already mentioned by other contributors to this volume (e.g. Pain), the lithology of the relatively small Purari catchment is very diverse. However, two dominant geological formations seem to determine the chemical composition of soils and sediments: volcanic rocks and tephras, on one side, and limestone – widespread especially in the western part of the catchment – on the other.

The major purpose of this study was to assess the geochemistry of soils in typical weathering profiles and in river-deposited sediments. The Purari data, apart from providing baseline information on this system which as yet is largely unmodified by human activities, also are used as comparison with some other tropical catchments elsewhere in the world.

The following elements have been determined in the present study: Zn, Cd, Ti, Cu, Mn, Cr, Hg, Pb, Co, Mo, Ni, K, Mg, and Ca.

2. Methods

Soil samples were collected from road cuts, quarries, or from river banks. Throughout the Purari catchment freshly deposited river sediments were also collected. Clays for analysis were separated from the samples using the Atterberg (1912) method. Bulk samples were used in analysis of freshly deposited, usually brackish water, deltaic sediments consisting of very fine muds. Heavy metals, potassium, magnesium and calcium were determined by

Petr, T. (ed.) The Purari – Tropical environment of a high rainfall river basin
© 1983, Dr W. Junk Publishers, The Hague / Boston / Lancaster
ISBN-13: 978-94-009-7265-0

atomic spectrophotometry. Details of methods are given in Petr (1980) and Irion and Petr (1980).

3. Soil profiles

Table 1 represents two major types of soils of the Purari catchment: those formed on limestone, and those formed on volcanic ash (tephra). The topsoil on limestone (Table 1A), with a high content of montmorillonite (Irion and Petr, this volume), shows intensive weathering, as a result of which the manganese content is low, but chromium high. The relatively high zinc content may be explained by the limited mobility of this element, which originated most probably from limestone.

The chemical composition of the second soil (Table 1B) suggests its volcanic origin. The soil, probably formed on a tephra layer, has low concentrations of potassium and lead, with the exception of the topsoil, and high concentrations of titanium and copper. These elements have their origin in basic rocks and volcanic ash. The low content of manganese indicates that this profile has

Table 1. Concentrations of Zn, Cd, Ti, Cu, Mn, Cr, Hg, Pb, and K in clay fraction ($< 2\,\mu m$) of selected soils of the Purari and Kikori River catchments. Values in ppm.

Depth (cm)	Zn	Cd	Ti	Cu	Mn	Cr	Hg	Pb	K
A. Soil on limestone above Lai River near Mendi									
Topsoil	224	0.1	6900	175	153	225	0.5	6	5200
B. Soil profile at Tagoba (Nebilyer River catchment), volcanic ash									
Topsoil	n.d.	0.2	5300	170	720	16	0.4	20	3800
20	n.d.	0.1	6500	190	290	33	0.6	4	6700
60	53	<0.1	11800	200	130	43	0.7	4	5600
100	67	<0.1	8400	170	230	110	0.9	6	6600
150	63	n.d.	10200	155	1180	250	0.9	10	2800
C. Soil profile at Karimui (Tua River catchment), volcanic ash									
Topsoil	48	0.1	1000	160	260	67	0.3	17	4000
200	138	0.2	12900	195	3600	28	0.4	40	3300
300	120	0.2	7800	130	2200	18	0.1	15	3500
D. Soil profile of an old river terrace on the Purari River at Wabo									
Topsoil	128	<0.1	5900	40	76	102	0.1	7	10700
200	133	<0.1	6200	64	530	155	<0.1	3	14400
300	122	<0.1	5900	71	380	n.d.	<0.1	2	17500
E. Soil on volcanic ash at Kikori (outside the Purari catchment)									
Topsoil	90	0.1	9700	58	76	153	0.3	8	3000

n.d. = not determined.

Table 2. Concentrations of heavy metals and potassium, and selected element ratios, for the clay fraction of 25 Purari soils as compared with basalt, intermediate igneous rock and granite.

	Concentrations									Ratios		
	Zn	Cd	Ti	Cu	Mn	Cr	Hg	Pb	K	Ti/Pb	Cr/Pb	Ti/K
Soils of Purari catchment	115	0.20	6900	115	770	90	0.3	9	5400	740	9.7	1.3
Basalt[a]	105	0.20	13800	87	15500	170	0.1	6	8300	2300	28.0	1.7
Intermediate igneous rock[a]	130	0.13	3500	5	850	2	0.0	12	48000	292	0.2	0.07
Granite[a]	45	0.13	2500	80	540	13	0.1	17	34000	147	0.8	0.07

[a] Turekian and Wedepohl, 1961.

111

been weathering over a long time period. The geochemistry of the soil collected from basic volcanic ash near the Karimui extinct volcano is similar (Table 1C). The lead concentration is high for this sample, but the very low potassium and high titanium are typical of such basic volcanic ashes. In both volcanic soils (1B and 1C) halloysite is the dominant clay mineral (Irion and Petr, this volume).

The Wabo soil profile (Table 1D) is taken from an old sediment of the Purari River. It is well weathered, as indicated for example by the low manganese content in the topsoil. At this location the Purari sediments are a mixture of sediments derived from the whole catchment. These sediments, rich in illite in their clay fraction (Irion and Petr, this volume), contain less basic components, and the illite is responsible for a high potassium content. The less basic character is also in good correspondence with the somewhat lower titanium content. The long weathering has also resulted in a very low content (mean value 3.9 ppm) of lead. In comparison with this old sediment deposit, the fresh river sediment at Wabo has a lead content of 9.5 ppm.

The Kikori topsoil (Table 1E) outside the Purari catchment is high in titanium and low in lead, indicating its volcanic origin. The manganese content (76 ppm) is far below the average value of unweathered intermediate igneous rocks and basalt (Table 2), suggesting that the soil is heavily weathered.

The average elemental content of heavy metals and potassium in the Purari catchment soils is somewhere in between basalt and intermediate igneous rocks (Table 2). This is also evident from the ratios for Ti/Pb, Cr/Pb, and Ti/K. Manganese content is very low due to the relatively high mobility of manganese during weathering. In spite of Turekian and Wedepohl's (1961) data referring to rock material, while the Purari data refer only to clay fraction, the concentrations are not greatly dissimilar.

4. River and lake sediments (Fig. 1)

Of the 16 samples collected throughout the catchment, one originates from Lake Aunde at 3530 m altitude. The Pie sample was collected from the brackish environment of an estuary bordering the Purari Delta in the west. Sample No. 17 comes from the Kikori River further west. All the remaining samples originate from Purari catchment rivers or the Purari itself.

The highlands' river Asaro has markedly high concentrations of lead (99 ppm), cadmium (1.16 ppm) and copper (216 ppm) (Table 3). Such concentrations are known to occur in polluted rivers, but in the Asaro they probably indicate the presence of metamorphic rocks rich in copper and lead. In contrast

Table 3. Heavy metals, magnesium and calcium in the clay fraction of river sediment. Purari catchment and the Kikori River. Values in ppm (see Fig. 1 for stations).

Station	No.	Zn	Cd	Ti	Cu	Mn	Cr	Hg	Pb	K	Mg	Ca
Lai	1	131	0.11	4900	60	1100	120	0.3	5.7	15700	14400	56900
Kunan	2	148	0.21	6100	100	1400	40	0.1	12.8	14100	8400	10300
Tuman	3	142	0.22	5900	120	1900	60	0.2	12.2	8800	23800	17000
Wahgi	4	156	0.05	7100	110	910	110	0.2	3.2	13600	11600	9900
Lake Aunde	5	50	n.d.	800	190	120	n.d.	n.d.	1.0	9100	1000	4400
Tem	6	206	0.28	5700	150	2900	60	0.7	9.5	10500	9600	12200
Maril	7	129	0.05	6700	170	2400	120	0.2	13.1	11900	18700	10700
Asaro	8	164	1.16	3800	216	2300	60	0.3	99.0	4300	7600	20800
Kaugel	9	120	0.10	6200	120	1100	100	0.2	8.2	7800	8900	9900
Poru	10	127	0.02	5000	30	500	100	0.4	1.3	18700	14500	30000
Pio	11	160	0.41	4700	100	1600	120	0.1	15.2	16100	14600	39100
Aure	12	130	0.11	6600	120	1200	140	0.2	10.0	10100	14900	21700
Purari at Wabo	13	132	0.10	5500	90	1000	110	0.1	9.5	11300	12500	29200
Purari at Pawaia	14	130	n.d.	6100	80	830	120	n.d.	8.0	16100	16000	35200
Pie at Baimuru	15	n.d.	n.d.	5600	90	910	100	n.d.	n.d.	16800	22100	10300
Heperi	16	130	n.d.	5200	80	990	150	n.d.	8.0	19500	9100	59600
Kikori	17	164	0.28	6400	120	1100	120	0.2	8.4	19500	9100	59600
Mean[a]		137	0.24	5369	114	1323	94	0.25	14.4	12775	12981	23575

n.d. = not determined.

[a] Mean calculated for stations 1 – 16.

113

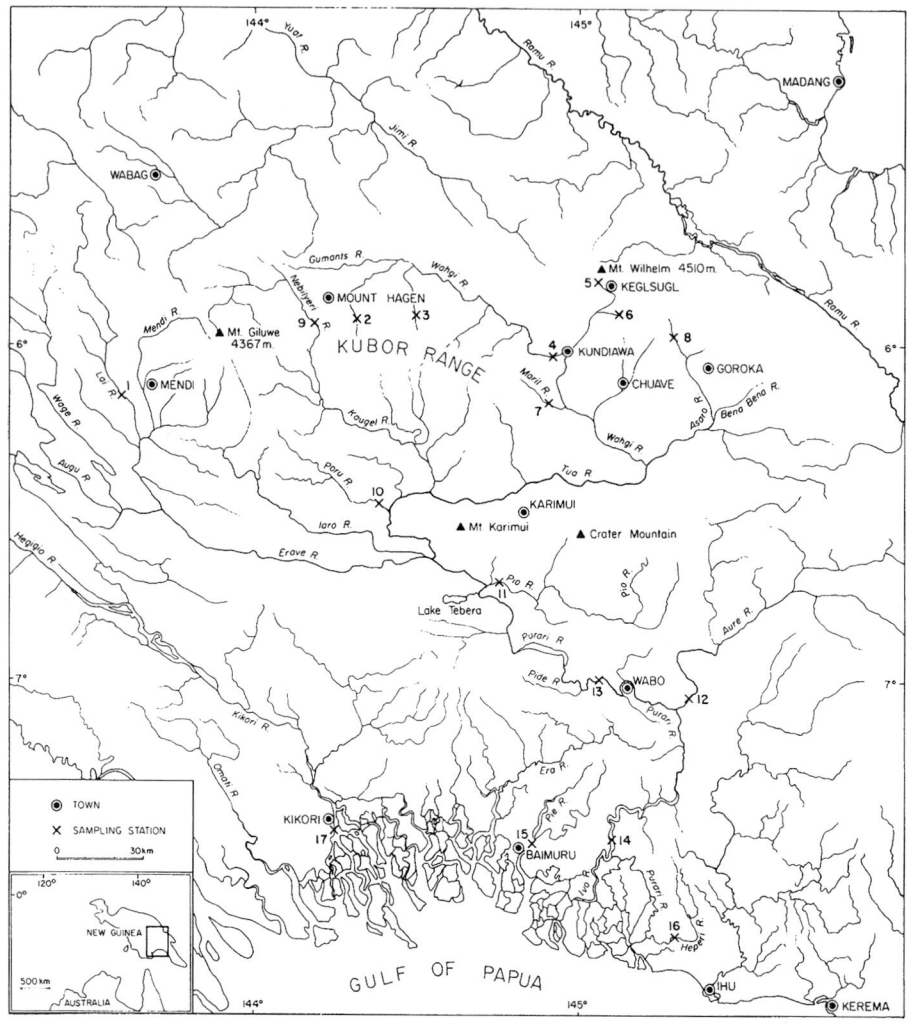

Fig. 1. Purari catchment: river sediment sampling stations.

with this river, the clay fraction of the sediments from the high mountain Lake Aunde is poor in almost all metals except copper. High calcium values in sediments of the Lai, Heperi, Pio and Kikori rivers is a reflection of extensive limestone and/or clastic marine sedimentary rocks in their catchments, and correspond to high levels of dissolved calcium in water (Petr, this volume).

Sediments in the main distributaries of the delta and the more active inter-connecting channels have grain sizes between 200 and 400 μm (Fig. 2). In the

114

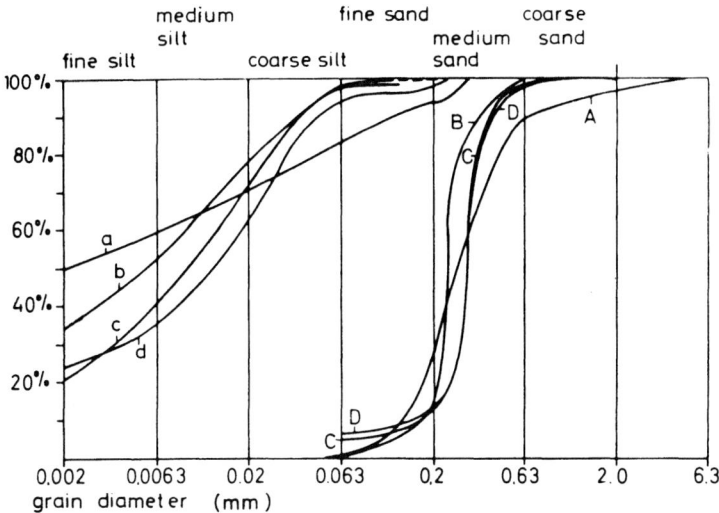

Fig. 2. Comparison of grain size distribution of sediments from tidal channels and from the Purari (sands A-D) with those from neighbouring mangrove forests (clay-silt a-d). A/a Wame River; B/b between Wame and Ivo River; C/c in the mouth of the Purari; D/d on the Muroa River.

tidal Pie/Port Romilly River, adjacent to the Purari in the west, clay content ranges between 30 and 50% (Irion and Petr 1980) (Figs. 3, 4). As expected, the clay fraction shows higher metal (Zn, Cu, Mn) concentrations than bulk samples (Table 4).

The average chemical composition of the clay fraction of freshly deposited sediments, calculated for all catchment rivers (and Lake Aunde), including the Purari above the delta, is close to that of the sediment sample taken from the Purari at Pawaia, i.e. just above the head of the delta. The rather low manganese concentration (830 ppm) in the Purari entering the delta could be explained by dissolution of manganese during sediment transport. The high calcium content in sediments above the delta may signify the high input of sediments through rivers having predominantly limestone drainage such as Lai, Poru and Pio. There is little difference between elemental composition of the Purari and Pie brackish water sediments, with the exception of calcium, clearly showing that input of calcium-rich sediments into the Pie estuary is considerably smaller than that passing through the Purari dominated delta. The Pie, a largely tidal river, receives only small tributaries (i.e. Eia Creek), draining swamps which are poor in cations (Petr, this volume). The brackish water character of the Pie also suggests that calcium dissolution is more rapid

115

Fig. 3.

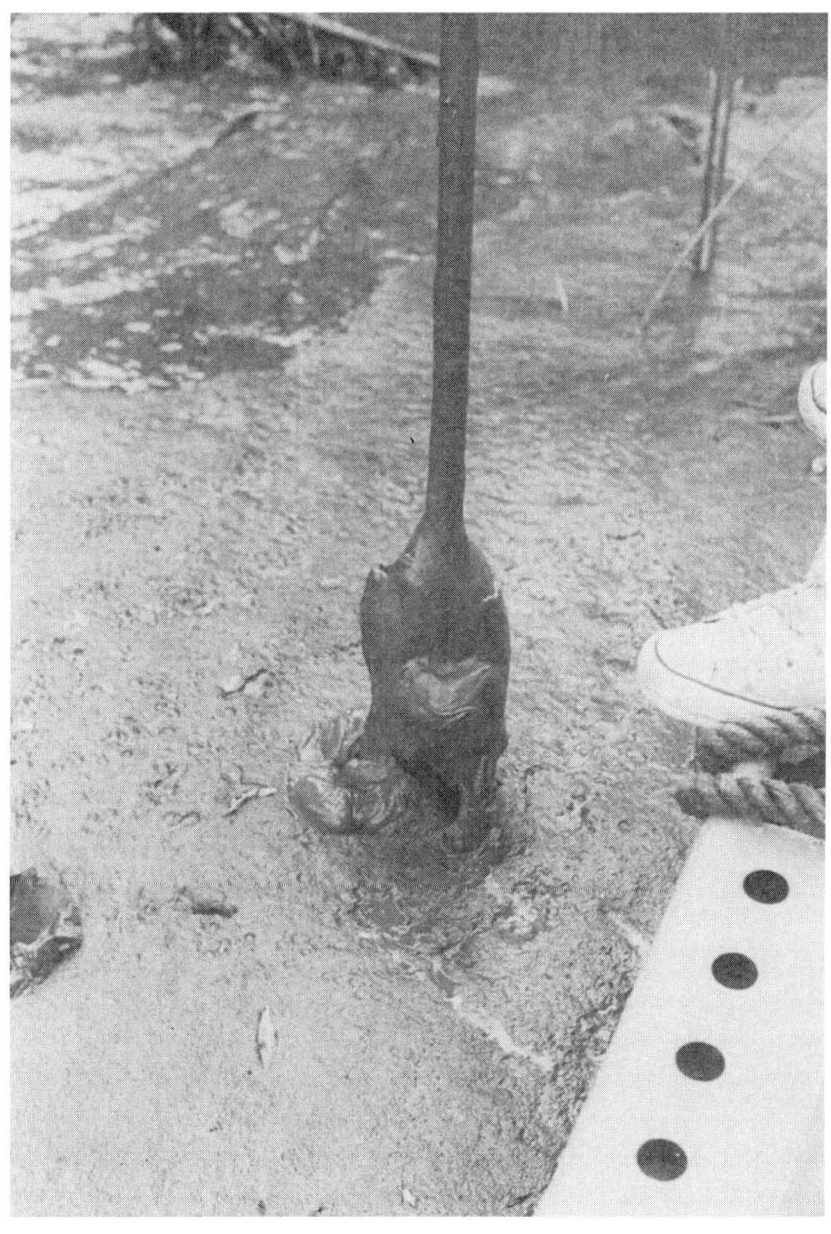

Fig. 3 and 4. The typical consistency of tidal muds of the brackish Pie River at Baimuru.

117

Table 4. Geochemistry of the brackish reach of the Purari and Pie River. Values in ppm.

	Zn	Ti	Cu	Mn	Cr	K	Mg	Ca	Co	Hg	Mo	Ni	Pb
Clay fraction													
Purari brackish reach (n=6)	140	5800	82	967	111	13833	17400	18200	n.d.	n.d.	n.d.	n.d.	n.d.
Pie brackish reach (n=3)	140	6067	89	1063	101	15833	18633	7267	n.d.	n.d.	n.d.	n.d.	n.d.
Total surficial sediment (bulk samples)													
Purari brackish reach (n=7)	82	n.d.	41	523	n.d.	n.d.	n.d.	n.d.	15	0.06	8	35	10
Pie brackish reach (n=2)	94	n.d.	49	700	n.d.	n.d.	n.d.	n.d.	18	0.10	4	40	14

n.d. = not determined.

there than in the largely freshwater Purari Delta. On contact with saline water, calcium and magnesium in sediments are partially replaced by sodium. However, since most of the magnesium is fixed in the lattice, while calcium is attached to the surface of the crystalline structure of clays, much more calcium than magnesium is lost.

5. Discussion

The mean chemical composition of topsoils is in good agreement with that of river sediments, perhaps with the exception of manganese and potassium (Table 5). As already mentioned above, manganese is a relatively mobile element and therefore it easily migrates into deeper soil layers. Much potassium is lost from soils during the weathering process. This is also seen from the clay mineralogy profiles, with higher concentrations of illite (and hence potassium) in deeper layers than in the topsoil (Irion and Petr, this volume).

There is some difference between the heavy metal content of the freshwater Purari sediment clay and that of the brackish water of the Pie, with some metals in a lower concentration in saline environment. Due to the limited number of samples it is difficult to say whether these differences are of significance.

Table 6 compares river sediment chemical composition of the Purari with

Table 5. Elemental composition in clay fraction of topsoils and sediments. Values in ppm.

	Zn	Cd	Ti	Cu	Mn	Cr	Hg	Pb	K
Topsoil (*n* = 25)	115	0.20	6900	115	770	90	0.30	9.3	5400
Purari catchment sediments (*n* = 16)	137	0.24	5369	114	1323	94	0.25	14.4	12775
Purari Delta (brackish) sediments (*n* = 6)	140	n.d.	5800	82	967	111	0.06	10.0	13833
Pie River (brackish) sediments (*n* = 3)	140	n.d.	6067	89	1063	101	0.10	14.0	15833

n.d. = not determined.

119

the two largest PNG rivers, the Fly and the Sepik, and with the Amazon. All drain mountain areas with complex catchment lithologies. In the Amazon, all elemental concentrations except Zn, Ti, K and Pb are lower than those in the Purari. PNG major rivers are broadly similar in their elemental composition, with some exceptions: the Fly has considerably lower Mg and Cu concentrations than the Purari, while it is higher in K; the Sepik is very low in Ca, but other elements have values very close to those in Purari sediments. The Amazon has only about 30% the Ca that the Purari has, and in this respect is close to the Sepik; it also has only about 50% of the Purari concentrations of Cu and Mn, but a considerably higher K concentration, and three times higher Pb concentration. The relatively low Pb concentration in the Purari sediments is probably mostly due to the widespread presence of volcanic ash and volcanic rocks in its catchment. The calcium concentration in the Purari is high, as limestone and marine clastic sediments are common. One may conclude that in spite of the integrated character of large river sediments, some chemical components give them a specific character.

Table 6. Comparison of heavy metals, potassium, magnesium, and calcium in the clay fraction of river sediments of some tropical rivers, and the world average. Values in ppm.

	Zn	Ti	Cu	Mn	Cr	Pb	K	Mg	Ca
Purari River	137	5369	114	1323	94	14	12775	12981	23575
Wahgi River	150	7100	110	910	110	3	13600	11600	9900
Pio River	160	4650	100	1600	115	15	16000	14600	39200
Heperi River	140	5200	80	1000	150	8	12500	17300	6800
Fly River									
upper drainage area	150	n.d.	50	900	85	15	17600	8300	24400
June River	140	n.d.	40	270	65	8	6300	2000	1170
Sepik River	140	n.d.	80	780	165	13	14900	16200	7300
Yellow River	140	n.d.	85	970	230	6	15100	26100	11400
April River	130	n.d.	130	1440	340	11	14400	26600	9000
Zaire River[a]	300	4900	100	1200	210	220	9100	7200	5700
Rio Orinoco	190	n.d.	45	630	105	n.d.	16300	5200	3400
Rio Amazonas	145	6200	65	670	85	45	20000	10200	8000
Rio Negro	65	n.d.	35	60	30	45	6500	2100	150
Rio Purús	150	4600	60	930	40	30	17000	10000	7500
World river average of suspended matter[b]	95[c]	5800	100	1050	100	20[c]	21000	11000	24500

[a] Sholkovitz, Martin and Van Grieken (1978); clay fraction was not separated.
[b] Martin and Meybeck (1979).
[c] Turekian and Wedepohl (1961) values for shale.
n.d. = not determined.

The smaller rivers reflect the catchment lithology and weathering processes through their sediment elemental composition much better than large rivers. For example, the Wahgi River sediments are high in Ti and low in Pb and Ca, indicating a catchment rich in volcanic rocks and volcanic ash, and poor in limestones. The Pio River catchment, rich in marine clastic sedimentary rocks and limestone, has high Ca and Pb concentrations, but a low Ti concentration. Rivers of tropical lowlands like the Heperi, June, Yellow, Zaire, Orinoco, Rio Negro and Rio Purús, usually have at least one element in noticeably low concentration. Calcium, being a very mobile element, is low in all lowland river sediments listed in Table 6, with the exception of the Yellow River, which drains some Quarternary sediments probably rich in calcium carbonate. Some of these rivers are known to arise in deeply weathered areas. The June River is one of them and has low concentrations of manganese, potassium, magnesium and calcium in its sediments. The sediments of Rio Negro in South America have a similar composition. Both rivers' sediments contain mainly kaolinite, a mineral formed during intensive weathering by losing all major and minor elements except Si, Al, 0 and H. Montmorillonite-rich rivers like the Wahgi, Pio and Heperi of the Purari catchment, Rio Purús and Yellow River have a relatively high content of magnesium and calcium, since these elements are part of the montmorillonite itself.

The elemental composition of the Purari sediments can be compared with that of the world river average suspended matter as calculated by Martin and Meybeck (1979), and for some elements in shales (Table 6). The lower K values correspond to the lower values of this element in weathered topsoils (see above).

In conclusion it can be said that elemental values for the Purari River sediments clay fraction, as represented by the sample from the lowest Purari station situated above the delta, and by the calculated mean for the catchment, are close to those of the world average for unpolluted sediments. The present results, although only preliminary and based on a limited number of samples, should represent a reference point for further investigations.

Acknowledgements

The Bougainville Copper Analytical Laboratory in Panguna, Bougainville Island, analysed some of the sediment samples. Comments on some parts of this manuscript by Mr J. Van Der Linden, chief chemist of the above laboratory, are most appreciated.

References

Atterberg, A., 1912. Die mechanische Bodenanalyse und die Klassifikation der Mineralböden Schwedens. Int. Mitt. f. Bodenkunde 2: 312–342.

Irion, G. and T. Petr, 1980. Geochemistry of the Purari catchment with special reference to clay mineralogy. In: T. Petr (ed.), Purari River (Wabo) Hydroelectric Scheme Environmental Studies, Vol. 18. Office of Environment and Conservation, Waigani, Papua New Guinea.

Irion, G. and T. Petr, 1983. Clay mineralogy of selected soils and sediments of the Purari River basin. This volume, Chapter I, 6.

Martin, J.-M. and M. Meybeck, 1979. Elemental mass-balance of material carried by major world rivers. Mar. Chem. 7: 173–206.

Martin, J.-M. and R. Van Grieken, 1978. Trace element composition of Zaire suspended sediments. Netherlands J. Sea Res. 12: 414–420.

Meybeck, M., 1981. Pathways of major elements from land to ocean through rivers. In: J.-M. Martin, J.D. Burton and D. Eisma (eds.), River Inputs to Ocean Systems. Proc. of a SCOR/ACMRR/ECOR/IAHS/UNESCO/CMG/IABO/IAPSO Review and Workshop, Rome, 26–30 March 1979, pp. 18–30. UNEP & UNESCO.

Petr, T., 1980. A preliminary note on baseline levels of nine metals, including mercury, in freshly deposited sediments of the lower Purari in Papua New Guinea. In: T. Petr (ed.), Purari River (Wabo) Hydroelectric Scheme Environmental Studies, Vol. 11: Aquatic ecology of the Purari River catchment, pp. 59–68. Office of Environment and Conservation, Waigani, Papua New Guinea.

Petr, T., 1983. Limnology of the Purari basin. Part 1. The catchment above the delta. This Volume, Chapter I, 9.

Sholkovitz, E., R. Van Grieken and D. Eisma, 1978. The major-element composition of suspended matter in the Zaire River and Estuary. Netherlands J. Sea Res. 12: 407–413.

Turekian, K.K. and K.H. Wedepohl, 1961. Distribution of the elements in some major units of the earth's crust. Bull. Geol. Soc. Amer. 72: 175–192.

8. The hydrology of the Purari and its environmental implications

G. Pickup and V.H. Chewings

1. Introduction

The Purari is Papua New Guinea's third largest river after the Fly and the Sepik. Its catchment area is 26 300 km^2 at Wabo and about 33 670 km^2 at the coast so it is small by the standards of major world rivers such as the Amazon with a catchment area of 5 776 000 km^2 and the Nile which drains 2 978 000 km^2. Much of the Purari basin is covered by undisturbed humid tropical rain forest which is maintained by frequent and often heavy rainfall although there are some areas of grassland in the highlands.

Hydrological records for the Purari basin are generally poor although there are a few exceptions. Of 47 gauging stations listed in the Inventory of Papua New Guinea Water Resources Data (Bureau of Water Resources 1978), less than half have stage-discharge rating tables and most of these are in the highlands, well upstream of the wettest areas of the basin. Many of these gauging stations operated for only a few years and are on very small catchments rather than major tributaries. Even where useful data are available, it is quite normal for 20 – 30% of records to be missing and only a few stations have information for a period of more than ten years.

The best hydrological records for the Purari catchment are for the Purari itself at Wabo and for the Aure, the only major tributary downstream from Wabo. A dam has been proposed for the Aure as well as the Purari. These records were collected by the Commonwealth Department of Works before independence and the Bureau of Water Resources since then. Snowy Mountains Engineering Corporation and Nippon Koei also undertook supplementary data collection during dam feasibility studies. In spite of this activity, runoff data are still inadequate to describe fully the behaviour of the system so two models are derived and used to extend the record. The first is a stochastic model for the Purari and the second is a transfer function model

Petr, T. (ed.) The Purari – Tropical environment of a high rainfall river basin
© *1983, Dr W. Junk Publishers, The Hague / Boston / Lancaster*
ISBN-13: 978-94-009-7265-0

relating the Purari and the Aure. These models are then applied to potential environmental problems.

2. Rainfall and runoff patterns in the Purari basin

The mean annual runoff of the Purari is 2617 mm compared with an average annual rainfall input of 3500 mm. Because the basin is so large, there are substantial variations in the amount and time distributions of both rainfall and runoff. Most of these can be attributed to seasonal climatic factors and topographic effects. Some variations in runoff are due to differences in geology, slope and other catchment characteristics but these cannot be separated out because there are usually not enough recording stations to determine catchment rainfall input with sufficient accuracy.

The factors affecting rainfall distribution are described in detail in the chapter on the climate of the Purari (Evesson, this volume), so only brief comments are made here.

Annual rainfall increases inland from about 3500 mm at the coast to a maximum of 7000 – 8000 mm or more along the edge of what Löffler (1977) has called the southern fold mountains. After this, it decreases inland and many of the highland valleys have a rainfall of only 2000 – 2500 mm with perhaps twice that amount on high mountains of the interior.

The seasonal distribution of rainfall also changes inland. On the coast and inland as far as the edge of the southern fold mountains, about 70% of the annual rainfall falls during the southeasterly season between May and October. Further inland, this trend reverses and the highlands receive most of their rainfall in the northwesterly season between December and March. Between these areas are a number of stations such as Kagua and Erave which represent a transition zone in which rainfall is evenly distributed throughout the year (for map of rainfall in the Purari basin, see Evesson, this volume).

The amount and seasonal variation of runoff reflects the rainfall pattern. Little is known about the coastal area but runoffs equivalent to a rainfall of 2000 – 4000 mm/year or more occur in catchments of the southern fold mountains. Further inland, in the highlands, runoff declines to about 1000 – 2000 mm/year. Runoff coefficients also vary from one area to another with values of 40 – 60% being common in the highlands and 70 – 85% in the southern fold mountains. Data from temperate forested areas leads Carter (1980) to suggest that the runoff coefficients for the southern fold mountain catchments are too high and indicate that some of the runoff present may be derived by leakage from adjacent basins. While some leakage may occur,

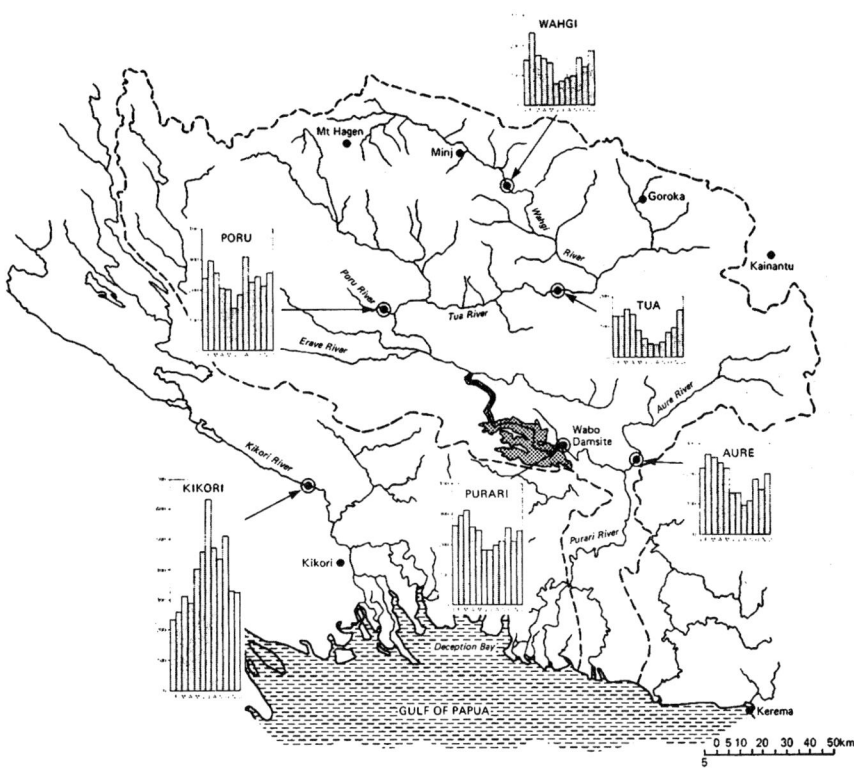

Fig. 1. Seasonal distribution of runoff (in mm) for selected basins draining into the Gulf of Papua. Data from Bureau of Water Resources records, SMEC (1977) and Carter (1980).

similar values may be calculated for high rainfall, non-karst basins on the headwaters of the Fly River in western Papua. This suggests that high runoff coefficients are the norm in areas of heavy rainfall and occur because evaporation losses are limited due to long periods of cloud cover and the fact that the catchment is almost permanently saturated which promotes runoff.

Seasonal variations in runoff are illustrated in Fig. 1. As with rainfall, there are two basic patterns. Firstly, rivers draining the highlands have high runoff during the northwesterly season and low flows at the height of the southeasterly season from June to September, when they are sheltered from prevailing winds. Secondly, rivers draining the southern fold mountains, such as the Kikori, experience their highest flows in the May to October period when they are exposed to air moving in from the southeast. Other systems such as the Purari and the Poru have a composite pattern because they drain catchments exposed to both northwesterly and southeasterly airflows. This reduces

125

seasonal variations but, on average, there is still a tendency for the larger monthly flows to occur in the January – March period during the northwesterly season.

3. Floods and low flows

Floods may occur at any time of year on the Purari, although the largest ones seem to coincide with the southeasterly season from June to September when heavy rain falls on the southern fold mountains. The annual maximum floods in the period 1961 – 1978 vary between 6290 m³/sec and 10450 m³/sec, and the annual series flood frequency curve is presented in Fig. 2. The slope of this curve is very gentle, indicating that variability is low. The size of the floods is relatively large given that they come from such a large catchment. For example, the maximum flow on the Purari is equivalent to a discharge of 0.4 m³/sec/km². This is similar to the amount of runoff which might be expected for a flood of similar frequency on a catchment one third the size of the Purari in the humid tropical area of northern Australia (see, for example, Ash 1978).

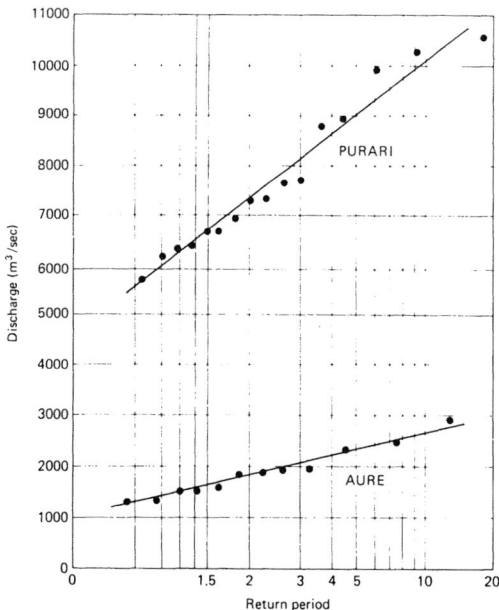

Fig. 2. Annual series flood frequency curves for the Purari and Aure Rivers. Data from Carter (1980) and SMEC (1977).

126

The Aure, like the Purari, is subject to flooding at any time of year but the flood regime is dominated by the northwesterly season with 50% of recorded annual maxima occurring in February and March. Like the Purari, the slope of the flood frequency curve is very gentle and the maximum discharges for each year in the period 1966 – 1978 lie in the range of 1341 – 2887 m³/sec. The largest flow represents a discharge of 0.66 m³/sec which is similar to values occurring during major tropical cyclones in northern Australia on similar sized catchments.

The regularity of flooding may be expressed by the ratio of the ten-year flood to the two-year flood. On the Purari and the Aure this value is less than 1.5, whereas ratios of 3 or more occur over much of Australia. Both climatic and topographic factors contribute to this low variability. The catchment of the Purari has a reliable seasonal climate in which flood-producing storms occur every year. At the same time, the length of the wet season is such that much of the catchment remains close to saturation so losses due to infiltration are small and do not vary greatly from storm to storm. This is in contrast to the tropical cyclone belt where huge rainfalls may occur over short periods and large areas on the irregular occasions when a tropical cyclone penetrates a landmass producing high flood variability. The topography of the Purari catchment also has some effect and results in a large number of localised, intense, short duration thunderstorms instead of general rain over the whole catchment. This means that runoff occurs from a number of scattered areas rather than the whole catchment and the impact of heavy rain in one area is offset by smaller amounts of rain elsewhere.

Like floods, very low flows may occur at any time of year on both rivers, but they tend to be concentrated in a particular season. On the Purari, most low flows occur in the southeasterly season which is also the time when the largest floods may be expected. The lowest recorded daily discharge is 487 m³/sec compared with a mean flow of 2360 m³/sec. This occurred in September 1972, when much of Papua New Guinea was experiencing drought (see SMEC 1972). Most of the 1972 records are missing but flows of less than 600 m³/sec persisted for at least 20 consecutive days, indicating a sustained lack of rainfall by Papua New Guinea standards.

Low flows on the Aure also tend to occur in the southeasterly season when daily values as low as 44 m³/sec compared with a mean of 307 m³/sec have been experienced. Sustained low flows are particularly common in August and the lowest average flow for that month is 66 m³/sec. These values occurred in 1972.

The seasonal distribution of low flow conditions illustrates the important role played by the high rainfall area around Wabo in maintaining flows. This area receives much of its rain during the southeasterly season when rainfall in the highlands is low. If an abnormal or irregular southeasterly season occurs, groundwater discharge and localised runoff from storms in the upper catchments are sufficient to maintain only very low discharges. This occurred during the 1972 drought when unusually stable conditions persisted through the second half of the year, seriously inhibiting cloud formation. Although some droughts have occurred in the northwesterly season, no major period of sustained low flow has yet been recorded.

4. Statistical characteristics of Purari and Aure River flows

While information on seasonal variations and extreme conditions like floods and droughts is useful, it does not give a full picture of river behaviour. It is also of very limited value in constructing the models which are necessary to design a dam and to assess environmental impact.

A more detailed story is provided by the data in Tables 1 and 2 which list a set of descriptive statistics for the Purari and Aure for both daily and monthly mean flows. These statistics are presented for both raw data and log transformed values, since both can be used to develop hydrologic models for long-term prediction of river behaviour. The discussion which follows is mainly based on statistics for the raw values.

The daily flow statistics for the Purari show two aspects of seasonal variation: the means change from month to month as discussed earlier; and, although the data are positively skewed throughout the year, skewness increases by a factor of $2-3$ in the southeasterly season. This reflects the presence of major floods which produce large positive deviations from the mean, which is relatively low because of limited runoff from the highlands, where rainfall is relatively small at this time of year. Other parameters are less affected by season. Flow variability, as expressed by the standard deviations, changes little from month to month and the lag one serial correlation is also fairly constant.

Daily flow statistics for the Aure are somewhat different to those of the Purari. Systematic seasonal variations occur in not only the mean but also the standard deviation and skewness. In the southeasterly season, the standard deviation ranges from $44-72$ which is one half to one quarter of what may be expected for the rest of the year. This indicates that fewer high flows occur and there are long periods of baseflow recession in the data.

Skewness remains high throughout the year but undergoes an increase of about 50% in May and remains high through to December. This suggests that

128

Table 1. Flow statistics for the Purari River.

Daily discharge statistics (m³/sec)

Month	Raw values				Log_e values			
	Mean	Standard deviation	Skewness coefficient	Lag one serial correl.	Mean	Standard deviation	Skewness coefficient	Lag one serial correl.
Jan.	2603	832	0.732	0.804	7.814	0.319	−0.084	0.828
Feb.	3118	1170	0.706	0.837	7.976	0.376	−0.062	0.857
Mar.	3107	1216	0.585	0.854	7.962	0.408	−0.313	0.884
Apr.	2748	1129	0.868	0.831	7.839	0.400	0.1132	0.863
May	2491	919	1.140	0.796	7.756	0.360	−0.054	0.836
Jun.	2026	1003	1.599	0.825	7.513	0.434	0.506	0.837
Jul.	2024	1198	1.753	0.815	7.464	0.543	−0.002	0.838
Aug.	2156	1202	1.229	0.869	7.524	0.566	−0.190	0.900
Sep.	2112	1239	1.532	0.866	7.490	0.592	−0.233	0.884
Oct.	2594	1177	0.559	0.875	7.741	0.529	−0.915	0.916
Nov.	2119	1084	0.883	0.884	7.530	0.515	−0.036	0.911
Dec.	2529	1111	0.630	0.847	7.736	0.459	−0.224	0.883

Monthly discharge statistics (m³/sec)

Month	Raw values				Log_e values			
	Mean	Standard deviation	Skewness coefficient	Lag one serial correl.	Mean	Standard deviation	Skewness coefficient	Lag one serial correl.
Jan.	2647	483	0.242	−0.341	7.831	0.184	0.019	−0.330
Feb.	3081	847	0.016	0.045	7.955	0.297	−0.549	0.012
Mar.	3208	815	0.355	0.130	7.993	0.286	−0.371	0.183
Apr.	2755	697	0.329	0.212	7.841	0.255	0.124	0.199
May	2432	611	−0.004	−0.144	7.730	0.265	−0.575	−0.156
Jun.	2026	609	0.999	−0.069	7.513	0.270	0.608	−0.078
Jul.	2038	851	0.525	0.161	7.475	0.422	−0.065	0.242
Aug.	2151	1028	0.676	0.115	7.497	0.514	−0.312	−0.067
Sep.	2204	900	0.525	0.406	7.542	0.461	−0.471	0.154
Oct.	2601	836	−0.305	−0.054	7.744	0.408	−1.051	−0.103
Nov.	2173	856	0.606	−0.419	7.550	0.406	0.126	−0.496
Dec.	2532	762	0.227	0.055	7.739	0.324	−0.114	0.119

Table 2. Flow statistics for the Aure River.

Daily discharge statistics (m³/sec)

Month	Raw values				Log$_e$ values			
	Mean	Standard deviation	Skewness coefficient	Lag one serial. correl.	Mean	Standard deviation	Skewness coefficient	Lag one serial correl.
Jan.	336	151	1.284	0.716	5.724	0.435	−0.079	0.782
Feb.	449	272	1.662	0.771	5.958	0.532	0.449	0.818
Mar.	388	196	1.253	0.747	5.848	0.471	0.218	0.816
Apr.	383	188	1.331	0.619	5.840	0.459	0.219	0.731
May	335	173	2.079	0.687	5.709	0.452	0.263	0.753
Jun.	199	101	2.152	0.704	5.192	0.433	0.482	0.784
Jul.	201	139	1.955	0.663	5.103	0.639	0.043	0.814
Aug.	150	92	1.700	0.757	4.853	0.548	0.352	0.867
Sep.	155	113	2.044	0.791	4.834	0.628	0.394	0.851
Oct.	298	199	1.583	0.839	5.488	0.671	−0.316	0.882
Nov.	241	162	2.083	0.777	5.312	0.574	0.397	0.860
Dec.	318	190	1.802	0.653	5.600	·0.580	−0.088	0.818

Monthly discharge statistics (m³/sec)

Month	Raw values				Log$_e$ values			
	Mean	Standard deviation	Skewness coefficient	Lag one serial. correl.	Mean	Standard deviation	Skewness coefficient	Lag one serial correl.
Jan.	329	107	−0.039	−0.619	5.962	0.354	−0.582	−0.598
Feb.	460	195	0.661	−0.160	5.976	0.391	0.334	−0.132
Mar.	404	98	−0.093	0.063	5.905	0.261	0.565	0.039
Apr.	394	106	0.706	−0.061	5.876	0.266	0.271	−0.050
May	312	107	−0.637	0.172	5.606	0.404	−1.169	0.061
Jun.	188	59	0.808	−0.137	5.137	0.300	−0.294	−0.138
Jul.	189	72	−0.225	0.025	5.055	0.455	−0.516	−0.049
Aug.	142	44	−0.368	0.166	4.822	0.336	−0.614	0.025
Sep.	142	57	0.712	0.834	4.758	0.362	−0.212	0.623
Oct.	238	111	0.398	0.499	5.271	0.525	−0.147	0.400
Nov.	288	215	1.881	−0.125	5.405	0.586	1.207	−0.096
Dec.	319	132	0.488	−0.187	5.604	0.438	0.039	−0.198

although the southeasterly season has long periods of baseflow, there are still sufficient high flows to give a markedly asymmetric flow frequency distribution. Serial correlations do not seem to vary with season but they are lower than those on the Purari as might be expected with a smaller catchment.

Monthly flow data statistics remove or reduce many of the patterns in the daily discharges. While the means continue to show seasonal variation on both the Purari and Aure, variance is reduced and skewness also tends to be reduced or to disappear completely. Serial correlations become highly variable but without any significant trend from month to month.

5. Flow data analyses and modelling techniques

The techniques involved in using hydrologic data for engineering design and environmental impact assessment are essentially the same. Studies of safe yield for power generation, dam overflow and storage/inflow ratio all use simple storage-yield models of the type described in the chapter on sediment transport (Pickup, this volume). A series of inflows is passed through the models which simulate the behaviour of the storage under various conditions and for various dam designs. The performance of the structure is then assessed to see whether it meets requirements or if it can be improved and if necessary, the design is modified. A major part of this process involves the examination of extreme or unusual inflow conditions likely to put the system under stress.

The accuracy achieved in modelling studies of this type depends on the quality of the input data. In most cases these data are inadequate and the Purari Hydroelectric Scheme is far from unusual in having to rely on less than 20 years of information for the design of a structure costing hundreds of millions of dollars and with a life of perhaps 500 years. Given this problem, it is necessary to extract the maximum possible useful information from the data available.

The first step in extracting such information relies on the fact that rainfall records have usually been kept for a longer period than flow records. Rainfall and runoff are related so it may be possible to use a catchment model which simulates the main hydrologic processes involved and predicts runoff from observed rainfall and evaporation. SMEC-NK (1977) have done this for monthly flows on the Purari and the Aure. Cross and Higgins (undated) have estimated weekly flows on the Purari and Pickup (1976) has developed a model for predicting daily flows in highland tributaries of the Purari.

These models can only be used when rainfall data are available so it has only been possible to fill in gaps in the runoff record and extend it back in time for a short period. The next step is to examine this record and to set up a stochastic model whose parameters are derived from statistical analysis of existing data. The stochastic model is then used to generate a long period of flow which is known as a synthetic series. The assumption behind this operation is that existing records are a sample from a time series which conforms with a particular probability distribution. If the parameters of this distribution can be estimated from the sample the synthetic series will provide an indication of the range of flow conditions likely to be encountered in the future.

The stochastic model used in this chapter to generate flows for the Purari is the autoregressive moving average ($ARMA$) process. This model is described in detail by Box and Jenkins (1970) and has the following basic form:

$$n_t = a_6 + \theta_1 a_{t-1} + \theta_2 a_{t-2} + \ldots + \theta_q a_{t-q}$$
$$- \varphi_1 n_{t-1} - \varphi_2 n_{t-2} - \ldots - \varphi_p n_{t-p}$$

in which n is the output, a is assumed to be white noise with a mean of zero, φ are the autoregressive parameters and θ are the moving average parameters. The model was applied using the CAPTAIN package of programs on the C.S.I.R.O. computing system (Freeman 1981). CAPTAIN is based on parameter estimation procedures developed by Young, Shellswell and Neethling (1971).

The first stage of model selection is to examine the autocorrelation function (ACT) and partial autocorrelation function ($PACF$) of the data. The autocorrelation function describes the serial correlations between values in the series at progressively increasing lags, thus the lag one autocorrelation is the correlation between x_1 and x_2, x_2 and x_3, x_3 and x_4 etc. The partial autocorrelation function at lag k describes the serial correlation remaining in the data after the correlation at lag $k-1$ has been removed.

The first two graphs in Fig. 3 show the ACF and $PACF$ of monthly mean discharges for the Purari. According to the guidelines suggested by Freeman (1981), the damped fluctuations in the ACF suggest an autoregressive (AF) process and the $PACF$, which cuts off sharply at lag one indicates that a first order model is appropriate. The general form of this model may be written as:

$$n_t = \varphi_1 n_{t-1} + a_t$$

where n is the parameter value, φ is the partial autocorrelation coefficient and a_t is an error or noise term.

Another feature of the ACF is a significant correlation at lag 12. This in-

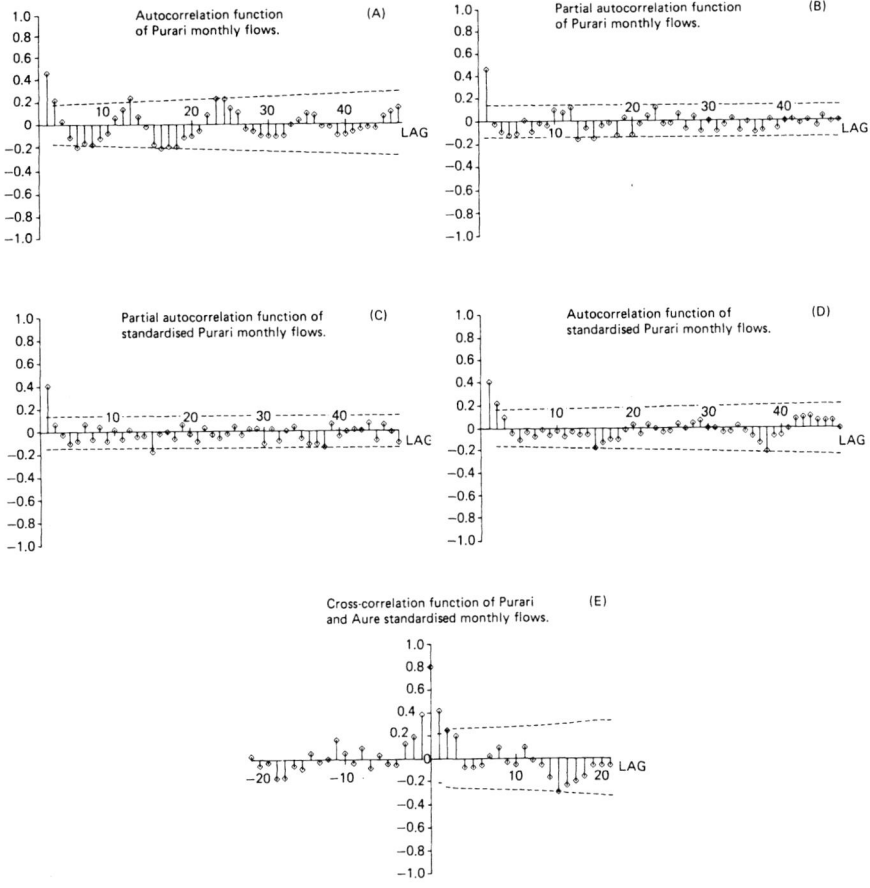

Fig. 3. Autocorrelation, partial autocorrelation and cross-correlation functions for Purari and Aure River monthly mean discharges.

dicates the presence of a seasonal component in the data which should be removed since the modelling process requires stationarity in the data. Two approaches were tried in removing the seasonal component. The first was seasonal differencing* and the second was to standardize the data by subtracting the mean of all the values for each month from the individual observa-

---

* Differencing is a technique advocated by Box and Jenkins (1970) for removal of trends. It consists of calculating the difference between values in the series with a one month lag (i.e. $x'_t = x_t - x_{t-1}$). The differenced series is then used in the analysis. Seasonal differencing is the same process but with a 12-month lag to remove seasonal trends.

tions and dividing by the standard deviation. Seasonal differencing was found to enhance rather than remove the lag 12 autocorrelation so it is not a useful approach. The standardized data with the monthly mean subtracted shows no seasonality, confirms the presence of a lag 1 AR process (Fig. 3) and was used in all subsequent analyses. This results in models with the form:

$$Q'_t = \varphi_1 Q'_{t-1} + a_t$$

where:

$$Q'_t = [\, Q_t - (\sum_{t=1}^{n} Q_t)/n \,] \,/\sigma_t$$

where Q_t is the mean discharge in month t and σ_t is the standard deviation.

While this type of model may be used to produce a synthetic flow series for the Purari, it is not suitable for the Aure. This is because if we are to model river behaviour below the dam during the filling period, it is necessary to preserve the correlation between flows in both the Purari and the Aure. Two independent lag 1 AR seasonal models would only preserve the correlation between overall means for each month, which is not adequate because, as the cross-correlation function (XCF) of Fig. 3 shows, deviations from the overall mean in the Purari are related to similar deviations in the Aure. It is therefore necessary to construct what Box and Jenkins (1970) refer to as a transfer function model relating the two sets of data.

Transfer function modelling was carried out using the CAPTAIN package on the C.S.I.R.O. computer system. A description of this package is presented by Freeman (1981) in which the basic form of the transfer function model is presented as:

$$y_t = x_t + n_t$$

in which

$$x_t = \omega_0 u_{t-d} + \omega_1 u_{t-d-1} + \ldots + \omega_r u_{t-d-r}$$
$$- \delta_1 x_{t-1} - \delta_2 x_{t-2} - \ldots - \delta_s x_{t-s}$$

where u is the input series, y is the measured output series, x is the model output and n is the difference between the model output series and the measured output series and is sometimes referred to as the noise series. The ω terms are known as moving average (MA) parameters and the δ terms are autoregressive (AR) parameters.

The noise series, n, may be random or it may be serially correlated. If serial

134

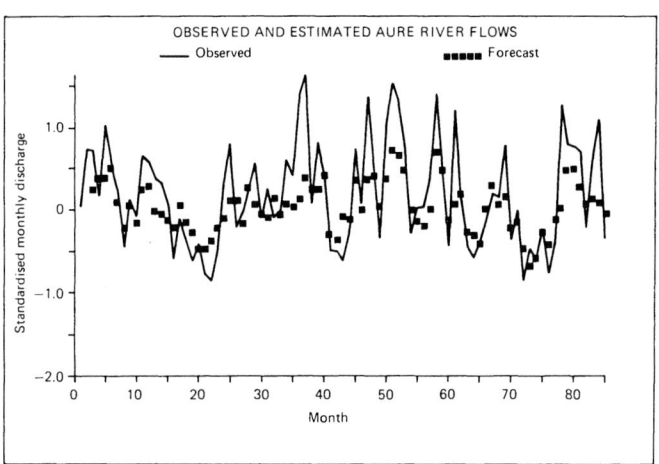

Fig. 4. Standardised monthly mean discharges for the Aure River predicted from Purari River flows using the transfer function model.

correlation is present, the noise series may be modelled using a stochastic model of the type described earlier.

Selection of a transfer function model is essentially a process of trial and error, although the *XCF* may provide some guidelines. In this case, a range of models was tried out, the most satisfactory of which was a lag 0, two parameter moving average type. The associated noise series shows lag 1 autocorrelation so a first order autoregressive process was fitted. The final structure of the model describing the joint behaviour of the Purari and the Aure may be stated as:

$$QA'_t = 0.4334 \, QP'_t + 0.0608 \, QP'_{t-1} + 0.1473 \, QA'_{t-1} + a_t$$

where QA' and QP' are the standardized monthly flow data for the Aure and Purari Rivers, respectively. The error series, a, has a mean of 0.1489 and a standard deviation of 0.5819, allowing it to be generated by random number techniques. The quality of fit is shown in Fig. 4.

The stochastic model for generating a series of flows for the Purari is:

$$QP'_t = 0.472 \, QP'_{t-1} + a_t$$

in which the noise series has a mean of 0.0091 and a standard deviation of 0.5278.

135

6. Model application to environmental impact assessment

The flow characteristics of the Purari have important environmental implications. The size and variability of dam inflows directly influence the storage/inflow ratio. This ratio may affect dissolved oxygen content, BOD, the degree of stratification and the speed of water movement through the reservoir. Inflows therefore influence such important environmental factors as weed growth, fish population behaviour and the degree of eutrophication. The storage – inflow ratio may be estimated using the stochastic model of Purari flows and the storage behaviour model described in the chapter on sedimentation, but residence time cannot be assessed without information on reservoir circulation. At present this information is not available so we can go no further.

The environmental effects of flows in the Purari extend below the dam, especially during the reservoir filling period. The bed of the river is below sea level as far upstream as Bevan Rapids which is about 65 river km inland from the delta mouth so there is potential for a major saltwater intrusion in the period between dam closure and the commencement of overflow.

The risk of saltwater intrusion depends on the tides and on two hydrologic factors. Firstly, there is the length of the reservoir filling period which determines how long the lower river is likely to be vulnerable. Secondly, there is the effect of the Aure River which enters the Purari downstream of the dam and supplements the flow. If filling occurs during periods of high flow on both the Purari and the Aure, the risk of saltwater intrusion will be small. On the other

Table 3. Filling periods for Wabo Dam.

Starting month	Maximum	Percentile					Minimum
		10%	20%	50%	70%	90%	
Jan.	3.10	2.24	2.19	2.10	2.04	1.54	1.45
Feb.	3.13	2.15	2.11	2.01	1.48	1.41	1.28
Mar.	3.07	2.22	2.15	2.07	2.00	1.40	1.26
Apr.	4.04	3.06	2.29	2.20	2.14	2.07	1.43
May	4.19	3.19	3.12	2.33	2.25	2.18	2.07
Jun.	4.31	3.31	3.20	3.03	2.31	2.23	2.05
Jul.	4.16	3.27	3.19	2.39	2.27	2.14	1.48
Aug.	4.11	3.16	3.09	2.31	2.23	2.11	1.47
Sep.	4.20	3.14	3.06	2.24	2.16	2.07	1.45
Oct.	4.20	3.07	2.39	2.26	2.18	2.06	1.40
Nov.	4.00	3.06	2.40	2.27	2.20	2.10	1.37
Dec.	3.13	2.33	2.27	2.19	2.14	2.08	1.44

Table 4. Mean monthly flows on the Aure from a synthetic 200-year record.

% Time flow ≤ specified value	Flow (m³/sec)											
	Jan.	Feb.	Mar.	Apr.	May	Jun.	Jul.	Aug.	Sep.	Oct.	Nov.	Dec.
5	232	295	327	307	206	129	113	102	83	159	94	208
10	260	327	342	333	234	149	134	111	100	181	144	223
20	291	400	372	358	264	165	153	123	123	199	209	270
50	354	508	429	414	314	201	198	149	153	261	309	331

hand, if the filling period coincides with low flows on the Aure, a serious environmental problem could result.

In an attempt to estimate the risk of saltwater intrusion while Wabo Dam is filling, the stochastic model developed in the previous section has been used to generate a synthetic 200-year record for the Purari. This record has been used to determine how long the filling period is likely to be and what variation may be expected. These results are summarised in Table 3.

The length of the filling period varies from 1.26 to 4.31 months which indicates how regular the hydrologic regime of the Purari is. Some months are slightly more variable than others. For example, the greatest range of filling periods occurs through the southeasterly season and in the transition period which follows it during October and November. The shortest and most reliable filling period is between December and March. This would be the best time to begin filling the dam because, no matter what conditions occur, it exposes the delta to the risk of saltwater intrusion for 25 − 30% less time than at any other period.

The other hydrologic factor which determines exposure to saltwater intrusion is the flow which may be expected from the Aure River during the filling period. This flow has been estimated using the synthetic 200-year record for the Purari as input to the transfer function model described in the previous section. This model simulates seasonal effects, the joint behaviour of the Purari and the Aure, and the stochastic characteristics of flows on the Aure as far as they are known.

A selection of model results is presented in Table 4. These show that the lowest flows on the Aure may be expected in the June to November period. The risk is not the same in every month of this period because some months have more variable flows than others. for example, low flows can be consistently expected in August with a range of $102 - 149$ m^3/sec between the fifth and fiftieth percentile while October and November have ranges of $159 - 261$ m^3/sec and $94 - 309$ m^3/sec. These months present a smaller risk of low flows than August but it is still a significant one.

The best period for filling Wabo Dam from the point of view of flows on the Aure is from December to April. During these months, it is unlikely that the mean monthly discharge will fall below 200 m^3/sec and values of $300 - 500$ m^3/sec are fairly normal. This is still less than the minimum recorded mean monthly discharge of the Purari (665 m^3/sec) but is probably enough to protect much of the lower river. It also coincides with the months during which the filling period of Wabo Dam is at a minimum.

References

Ash, R.A., 1978. Assessment of flood risk and flood estimation practices. In: G. Pickup (ed.), Natural Hazards Management in North Australia, pp. 93 – 108. Australian National University, Canberra.

Box, G.E.P. and G.M. Jenkins, 1970. Time series analysis – forecasting and control. Holden-Day, San Francisco.

Bureau of Water Resources, 1978. Inventory of Papua New Guinea water resources data. Dept. Minerals and Energy, Port Moresby, Papua New Guinea.

Carter, J.H., 1980. Discharge data for the Purari River and some of its tributaries. In: A.B. Viner (ed.), Purari River (Wabo) Hydroelectric Scheme Environmental Studies, Vol. 16, pp. 17 – 23. Office of Environment and Conservation, Waigani, Papua New Guinea.

Cross, R.W. and R.J. Higgins, undated. Report on hydrology of the Purari River system. Commonwealth Dept. of Works, Papua New Guinea.

Evesson, D.T., 1980. The climate of the Purari River catchment above Wabo. In: A.B. Viner (ed.), Purari River (Wabo) Hydroelectric Scheme Environmental Studies, Vol. 16, pp. 1 – 16. Office of Environment and Conservation, Waigani, Papua New Guinea.

Freeman, T.G., 1981. Introduction to the use of CAPTAIN for time series analysis. C.S.I.R.O. Div. Computing Research. Computing Note 43. Australia.

Löffler, E., 1977. Geomorphology of Papua New Guinea. A.N.U. Press, Canberra.

Pickup, G., 1976. A self-calibrating model for the simulation of daily runoff from rainfall for humid tropical drainage basins. Dept. of Geography, Univ. of Papua New Guinea. Occasional Paper 14.

Pickup, G., 1983. Sedimentation processes in the Purari River upstream of the delta. This volume, Chapter I, 11.

Snowy Mountains Engineering Corporation (SMEC), 1972. An investigation of the severity of the 1972 drought in the Fly River basin. Kennecott Pacific Pty. Ltd.

Snowy Mountains Engineering Corporation – Nippon Koei (SMEC-NK), 1977. Purari River Wabo Power Project feasibility report, Vol. 6.

Young, P.C., S.W. Shellswell and C.G. Neethling, 1971. A recursive approach to time series analysis. Dept. of Engineering, Univ. of Cambridge. Report No. CUED/B-Control/TR 16.

9. Limnology of the Purari basin
Part 1. The catchment above the delta

T. Petr

1. Introduction

The drainage basin of the Purari lies between latitudes 5°35' and 7°56'S, and
longitudes 143°22' and 146°03'E. The catchment is very rugged and mountainous, with the top of the catchment draining the highest mountain in Papua
New Guinea, Mt. Wilhelm (4510 m above sea level). The river basin occupies
the greatest part of the agricultural and relatively densely populated highland
provinces: the Western Highlands, Enga, Eastern Highlands, Southern
Highlands and Simbu, as well as the little populated mid-altitude and lowland
tropical forest of the Gulf Province in the south. Pickup and Warner (in press)
describe the major morphological zones for this river:

> Some of the major tributaries rise in high mountains and then pass through intramontane basins
> before falling steeply to the coast. This means that there is usually a gorge, valley and alluvial
> plain section within the mountains above the main gorge. Once the river leaves the basin, the mor-
> phological zones are repeated.

The rainfall of the Purari catchment is variable, ranging from 1936 mm in the
highlands at Goroka to over 8500 mm at Wabo (Evesson 1980). The principal
Purari basin catchment rivers above Wabo are the Asaro, Wahgi, Kaugel,
Iaro, Pio and Erave. After the junction of the Asaro and Wahgi, the river is
known as Tua; downstream from the junction of the Tua and Erave, the river
is known as the Purari (Fig. 1). The total length of the main stem is about 630
km. The catchment area above Wabo, the proposed dam site in the foothills,
is 26300 km², while the total catchment is 33670 km². The only major tributary
below Wabo is the Aure, with a catchment of 4400 km². Some 40 km from the
sea, the Purari forms its delta by dividing into three major distributaries: the
Purari, the Ivo – Urika, and the Wame – Varoi. These, in turn, further branch
out on their approach to the sea. The Ivo arm carries by far the greatest por-

Petr, T. (ed.) The Purari – Tropical environment of a high rainfall river basin
© *1983, Dr W. Junk Publishers, The Hague / Boston / Lancaster*
ISBN-13: 978-94-009-7265-0

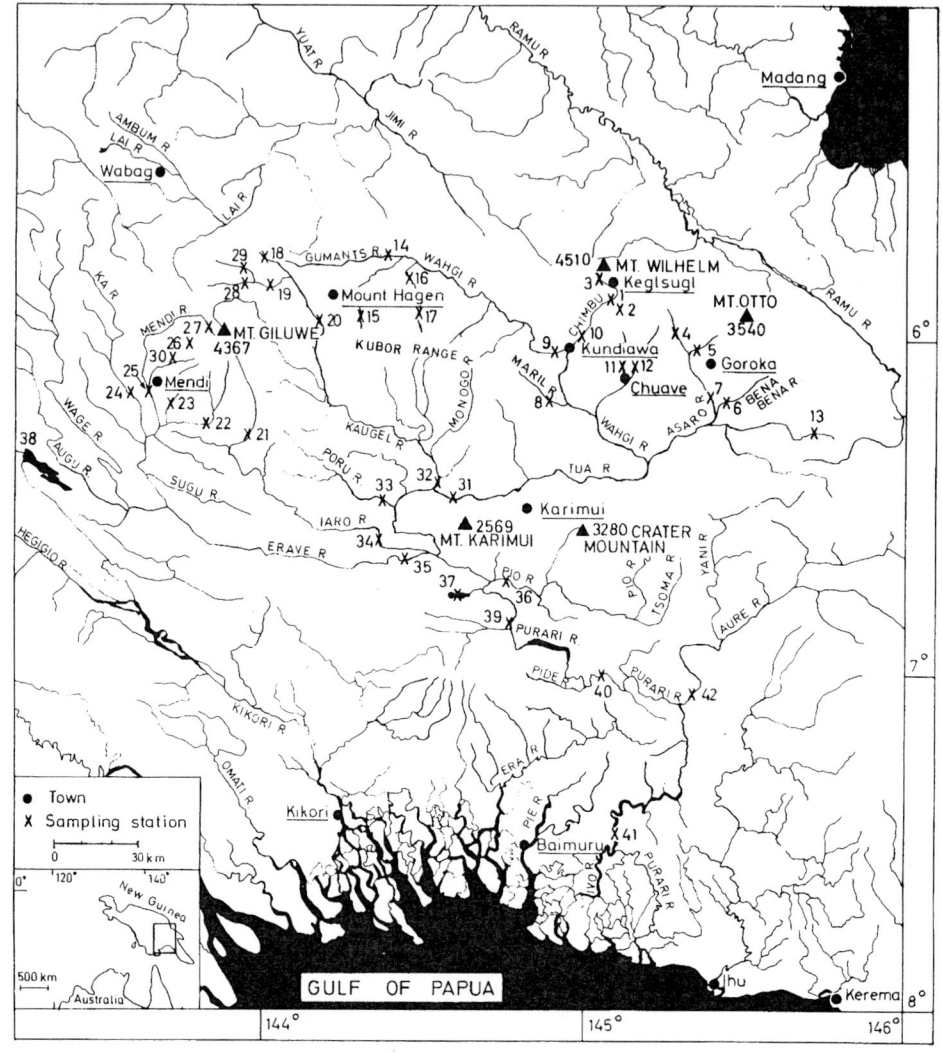

Fig. 1. Purari River basin with sampling stations.

tion of the water (Thom and Wright, this volume).

The mean annual discharge of the Purari at Wabo is 2360 m³ per second, and that of the Aure is 307 m³ per second (Pickup, this volume), corresponding to a total annual water mass transport of 74.4 km³ and 9.7 km³ respectively. The total mean annual volume of water discharged by the Purari into the Gulf of Papua is thus 84.1 km³. This high discharge reflects the high rainfall, with the highest precipation falling in the foothills.

The geological formations and soil types of the basin are complex, and are described in more detail in this volume in chapters by Pain and Wood. In the high mountains of the Mt. Hagen area (Bismarck and Kubor ranges), and in the Mt. Wilhelm massif drained mainly by the tributaries of the Kaugel River, the dominant rocks are granodiorite, low grade schist, and in the Kubor Range also gabbro and phyllite. At low altitudes, greywacke, shale, and some tuff and lava are present. The Erave River drains mainly polygonal karst, with islands of granite and granodiorite. The upper course of the Asaro River has a geology similar to that of the Kubor Range, while in the lower course there is mainly greywacke, shale, tuff, lava and agglomerate (Haantjens 1970, Löffler 1974). Large areas of the catchment are covered in volcanic ash (Pain, this volume). In the Purari valley the dominant rocks are Cretaceous greywackes and mudstone (Ruxton 1969).

The Aure catchment lithology is dominated by marine clastic sedimentary rocks, with some limestone.

The vegetation is very diverse, and is assessed by Paijmans (this volume) in more detail. In the highest altitudes above 4000 m, trees and tree ferns are usually absent, and grassland predominates. High mountain grasslands above the upper limit of agriculture at about 2850 m alternate with montane forest, with the lower montane/higher montane forest boundary at about 3000 m a.s.l. Intensive agriculture is carried out on deforested and grassland areas between 2600 and 1500 m. Below that altitude spreads largely uninterrupted tropical humid forest. Freshwater swamp vegetation, which in the brackish reach adjoins the mangroves, is present on the alluvium.

While geographically the Purari catchment is situated in a tropical belt, altitudinally it transgresses several climatic zones: tropical, subtropical, temperate and cool temperate/alpine zones, with corresponding ranges of temperatures. Mt. Wilhelm frequently has snowfalls, but the snow melts during the daytime. Air temperature gradually increases towards sea level, with a corresponding increase in water temperature.

As a matter of convenience, the area investigated is subdivided into the following zones: glacial lakes (4000 – 3500 m), streams and ponds of high mountains and highlands (3500 – 1300 m), streams and lakes of the humid tropical forest (1300 – 20 m), and the Purari Delta, which is dealt with in Part 2.

2. Sampling and methods

No regular sampling programme was possible due to access difficulties created by weather conditions and restricted availability of transport and assistance. Thus, sampling was mostly of expeditionary character. On a few occasions, intensive sampling at one station was carried out over a period of several days, which then made it possible to follow the dynamics of the physico-chemical environment during varying discharge. Synoptic sampling over much of the area was done only once, and concentrated on the upper part of the catchment. Such sampling required hiring carriers and use of four-wheel drive vehicles and helicopters. Major catchment rivers with gauging stations were visited more frequently by helicopter. More frequent sampling was possible on the Purari River downstream of Hathor Gorge, and in the delta. These stations are accessible most of the time by boat.

Water samples were collected from the surface in PVC bottles. When possible, these were kept in a refrigerator at temperatures of $+4$ to $+10°C$, otherwise at ambient air temperature. At the end of each expedition they were air freighted to Port Moresby and delivered for analysis to the Mines Analytical Laboratory of the Department of Minerals and Energy. Field measurements included water temperature, and on some occasions Secchi disc transparency, and in the delta, salinity using Goldberg T/C (temperature compensated) refractometer, with an accuracy of reading $±0.25‰$.

Bicarbonate was determined by titration to pH 4.5 with 0.01 M HCl. Sodium, potassium, calcium and magnesium were determined by atomic absorption spectrophotometry (Techtron Model 1200). Chloride was measured by silver nitrate titration with potassium chromate indicator. Sulphate concentration was estimated as the difference between the equivalent sum of cations and that of bicarbonate and chloride ions. Dissolved silica was determined by the molybdate method.

3. Glacial lakes (4000 – 3500 m a.s.l.) (Fig. 2)

The glacial lakes are situated in the Bismarck Range at an altitude between 3530 and 3920 m and are the highest lakes of Papua New Guinea. They were formed by Pleistocene glaciers which receded by 12 000 B.P. (Löffler 1972). The lakes receive runoff from rocky and grassed slopes extending above 3500 m and draining gabbro, and the associated ultrabasic rocks pyroxenite and hornblendite (Dow and Dekker 1964), with some apatite present. The soils are predominantly acidic (Hnatiuk et al. 1976). The weather pattern throughout

144

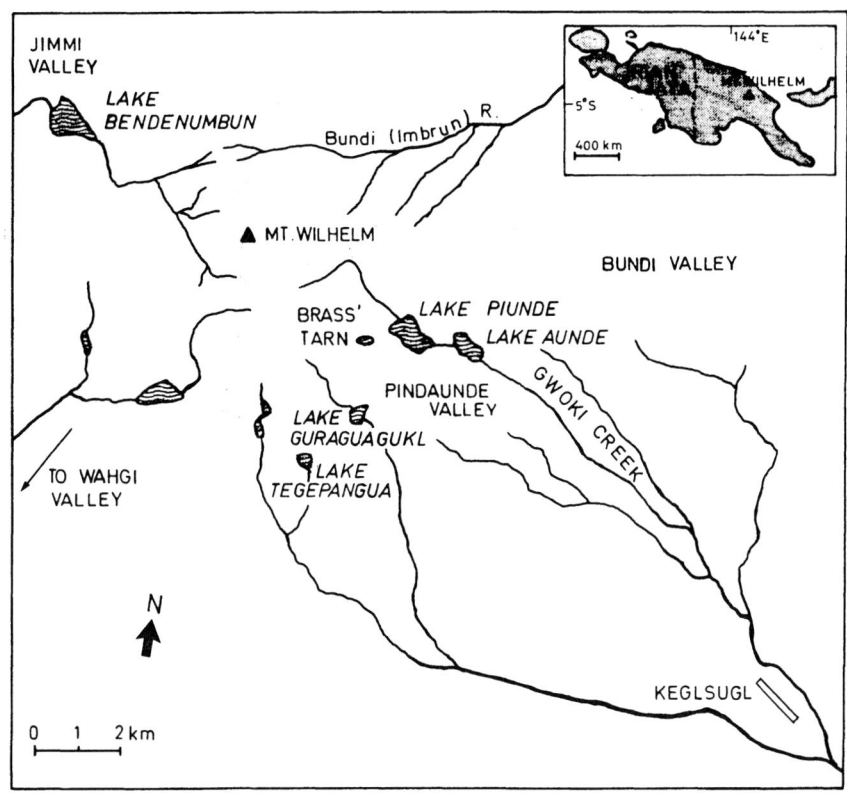

Fig. 2. Glacial lakes, Mt. Wilhelm area.

the year reflects the regional wet and dry season cycle. Hnatiuk et al. (1976) give annual rainfall on the ridge of Mt. Wilhelm as 4380 mm, with frequent snowfall above 4000 m. At Lake Aunde* (3530 m), the precipitation is ca 3450 mm; at Keglsugl (2510 m), 2284 mm; and at Kundiawa (1500 m), 2184 mm. All lakes drain southeast (Fig. 2). Brass' Tarn (3920 m) (Fig. 3), and lakes Piunde* (3630 m) and Aunde (3530 m) (Fig. 4) drain into Pindaunde Valley (Fig. 5) through Gwoki Creek, while Guraguagukl (3750 m) and Tegepangua (3800 m) drain into a neighbouring valley south of Pindaunde; their discharge joins Gwoki Creek further downstream. Other glacial lakes are situated west of the investigated area, but access to them is difficult and they have never been investigated.

Glacial lakes on Mt. Wilhelm were visited by a number of scientists, but

* On the PNG 1:100 000 topographic sheet Bundi the names are reversed.

145

Fig. 3. Brass' Tarn (3900 m a.s.l.), a small glacial lake on Mt. Wilhelm.

Fig. 4. Lake Piunde (left bottom) and Lake Aunde, on Mt. Wilhelm. Note the research hut, built by the Australian National University in 1966.

Fig. 5. Tree fern *Cyathea atrox* in Pindaunde Valley, Kombuglomambuno moraine. (Photo G. Irion.)

limnological collections and measurements were first obtained only in late 1950s and 1960s (Brass, Balgooy, Illies, Löffler). Brass' (1964) expedition collected the first information on some chemical parameters of Lake Aunde. Löffler (1973) used the material to support a theory of circumtropical distribution of zooplankton in high mountain lakes. Illies (1969) collected adult caddisflies. Other samples provided material for description of new species of Ostracoda (McKenzie 1971), and for assessing the lakes' phyto-

147

plankton (Thomasson 1967). More information was gathered since the Australian National University field station at Lake Aunde was established in 1966. This information covered especially the terrestrial vegetation and climate, but also included some data on Lake Aunde temperature.

All five glacial lakes visited during the present survey are characterised by low turbidity, as reflected by the high water transparency ranging up to 10 m as measured by Löffler (1973) for Lake Aunde. Thomasson recorded 9.5°C for Lake Guraguagukl, and 10°C for Lake Tegepangua. During the present observations the range for lakes Piunde and Aunde was between 10.8°C and 14.4°C, the high value temperature being above that recorded by Löffler (1973), Thomasson (1976) and Hnatiuk et al. (1967). The last authors give for Lake Aunde a temperature range of 8.9°C to 12.8°C, with a mean of 10.7°C for 121 morning recordings. The exceptionally high temperature recorded in the present survey was measured on a sunny day in late afternoon in shallow water near the shore.

The pH values between 6.50 and 7.50 (Table 1) are higher than would be expected in a runoff from predominantly acidic soils, with some peat. As the pH was not measured in the field but on stored samples, there is a possibility that under natural conditions it might be lower. However, the few data obtained by Thomasson (1967) and Löffler (1973) are within the present range. Brass' Tarn, circa 90 m in diameter, is the highest and smallest bog lake, with a thick growth of submerged *Scirpus*, *Callitriche* and *Spirogyra*. It is surrounded by an inner ring consisting of *Carex gaudichaudiana* fen and an outer one of *Carpha alpina* fen (Wade and McVean 1969). Underlying these is semi-liquid, brick red peat, which is probably responsible for the low water pH. The relatively high pH of Tegepangua and Guraguagukl lakes, which are the second and third highest in altitude, may reflect a difference in soils: these lakes are bordered by high mountain slopes, unlike Brass' Tarn. They also have the highest water conductivity (43 and 42 μS/cm respectively) among the five lakes, and apart from one high value for Lake Aunde, they have also the highest concentrations of total cations, anions and dissolved silica (Table 1). Elsewhere in New Guinea, samples collected by Peterson (1976) from seven lakes and two ponds situated between 3747 and 4260 m in Puntjak Jaya, Irian Jaya (Indonesia), gave a surprisingly high mean conductivity value of 122 μS/cm (range 109 – 152 μS/cm). Lake Discovery, situated at the lowest altitude of this range, had the highest conductivity.

With the exception of pH, a rainwater sample (Table 2) collected at Aunde had a conductivity and water chemistry very similar to that of Brass' Tarn, indicating that runoff water receives little in the form of ions from rocks and soils at this high altitude; the submerged aquatic plants of Brass' Tarn

Table 1. Glacial lakes of Papua New Guinea.

Name	Brass' Tarn[a]		Tegepangua	Guraguagukl	Piunde[a]		Aunde[a]	
Altitude (m)	3920		3800	3750	3630		3530	
Approximate size (m)	20 × 40[b]		100 × 200[b]	200 × 200[b]	1300 m long[a]		950 m long[a]	
Maximum depth (m)	3[a]		10[b]	20[b]	54[c]		22[c]	33[c]
Date sampled	8.11.78	1.4.69	9.11.78	9.11.78	6.11.78	1.4.69	7.11.78	30.3.69
Time	07.00	n.d.	n.d.	n.d.	16.00	n.d.	13.00	n.d.
Temperature °C	n.d.	n.d.	n.d.	n.d.	14.4	n.d.	10.8	n.d.
Turbidity J.T.U.	2	n.d.	2	2	5	n.d.	2	n.d.
pH	6.60	6.50	7.50	7.20	6.70	6.80	7.20	6.50
Conductivity μS/cm	13	11	43	32	18	17	22	16
TSS	nil	n.d.	nil	nil	6	n.d.	nil	n.d.
Total hardness $CaCO_3$	5	n.d.	20	20	9	n.d.	20	n.d.
Alkalinity $CaCO_3$	10	n.d.	23	25	19	n.d.	21	n.d.
Ca	2	2.4	6	6	2	2.0	6	2.4
Mg	0.4	1.9	1	1	1	1.5	1	1.9
Na	1	1	1	1	1	0.2	1	0.2
K	0.1	0.1	0.4	0.5	0.1	0.1	0.1	0.1
Total cations me/1	0.13	0.17	0.46	0.46	0.20	0.24	0.41	0.18
HCO_3	12	n.d.	28	31	23	n.d.	26	n.d.
Cl	nil	1.8	nil	nil	nil	n.d.	nil	n.d.
SO_4	2	n.d.	3	2	2	n.d.	2	n.d.
Total anions me/1	0.13	n.d.	0.51	0.45	0.24	n.d.	0.47	n.d.
SiO_2	8	5	13	13	7	4	7	4

[a] Löffler (1973).
[b] Thomasson (1967).
[c] Brass (1964).

Values in mg/1 where not otherwise stated.
n.d. = not determined.

149

Table 2. Rainwater.

Name	Aunde	Wabo	Muro
Altitude (m)	3530	61	3
Date sampled	7.11.78	11.2.76	19.3.78
Turbidity J.T.U.	2	1	5
pH	7.20	6.70	6.30
Conductivity μS/cm	16	2.5	9
TSS	nil	nil	2
Total hardness $CaCO_3$	7	4	3
Alkalinity $CaCO_3$	10	4	5
Ca	3	1	0.6
Mg	nil	0.3	0.3
Na	0.3	nil	0.7
K	nil	nil	0.03
Total cations me/l	0.15	0.08	0.08
HCO_3	12	5	7
Cl	nil	nil	nil
SO_4	2.0	0.5	nil
Total anions me/l	0.13	0.09	0.11
SiO_2	1	nil	0.4

Values in mg/l where not otherwise stated.

possibly also remove some dissolved ions. The pH range of 6.5 to 7.5 for the high mountain lakes and the rainwater probably reflects a series of factors influencing the pH. Apart from the type of soil, there is also the possibility of the impact of distant vegetation burning, and the timing of the rainwater sampling. It has been pointed out that combustion of vegetation leading to accumulation of nitrogen oxides in the atmosphere, may result in drastically increased hydrogen ion and nitrate loading rates (Lewis 1981). While Lewis (1981) finds low pH to be most likely at the onset of the rainy season, Tricart (1965) gives a pH of 7.85 for the beginning of tropical thunderstorms. It is clear that a number of factors are involved, and for PNG conditions more studies are necessary.

Two other rainwater samples were collected from the catchment, one at Wabo, and one at Muro (in the east of the Purari Delta). Their conductivities of 2.5 and 9 μS/cm respectively reflect the very low ionic content. The rainwater at Wabo was collected in the middle of a long-lasting, extremely heavy rainstorm, one of many in this area, which has an average annual rainfall of ca 8500 mm. In the undisturbed environment of all three stations, with an extremely low population density, little input of ions into the atmosphere should be expected, with the exception of distant vegetation burning. It is interesting

150

to note that the station with the lowest annual rainfall and highest altitude had the highest conductivity, while that with the highest rainfall had the lowest conductivity, and that of Muro was somewhere in the middle. The Wabo conductivity of 2.5 μS/cm is identical with that of meltwater from the Carstensz Glacier of Puntjak Jaya (Bayly 1976). The relatively high content of sodium in Muro rainwater can be explained by the proximity of this station to the coast.

Nutrient concentrations in the Mt. Wilhelm lakes are very low. Löffler (1973) found 0.3 μg/l of total phosphorus in the surface water of Aunde, but no traces of nitrates.

4. Streams and ponds of high mountains and highlands (3500 – 1300 m a.s.l.)

There is very little difference in the water chemistry of streams draining four major mountain massifs of the Purari catchment. Conductivity correlates inversely with altitude (Table 3). If limestone drainage streams are excepted, Mt. Giluwe streams have the lowest conductivity, calcium and pH values. Mt. Giluwe (4368 m) drains a similar type of lithology as Mt. Hagen, i.e., basic to intermediate lavas and pyroclastic rocks, lahars, fanglomerate and lacustrine deposits. Much of this area is covered in volcanic ash. The low concentration of chemicals could partially reflect the highest man elevation, with the lowest mean temperature contributing to a slower rate of weathering than at lower altitudes. The peaty soils which overlie lava, scoria and morainic material at high altitude (Bic 1967), apart from slowing down their weathering, probably also remove some solutes from water percolating downstream. However, the pH values close to neutral do not indicate the presence of extensive low pH bogs in the upper catchment of Mt. Giluwe. If the weathering process were to

Table 3. Water chemistry of streams draining major mountain massifs of the Purari catchment.

Mountain massif	Mt. Wilhelm, Crater Mt.	Mt. Otto, Mt. Karimui	Kubor Range	Mt. Hagen	Mt. Giluwe	
Number of streams	9	2[a]	2	4	8	1[a]
Mean altitude (m)	1764	1575	1650	2130	2141	1400
Conductivity μS/cm	61	130	62	60	43	170
pH	7.19	7.65	7.10	7.20	7.00	7.60
Ca mg/l	7.4	18.0	6.5	5.3	4.8	33.0
HCO_3 mg/l	38.6	83.0	39.0	38.5	29.0	108.0
K mg/l	0.9	1.0	1.0	1.1	0.8	1.0
SiO_2 mg/l	23.4	20.5	17.0	23.0	16.3	13.0

[a] Limestone drainage.

Table 4. Streams and ponds in the altitudinal range 3500–1300 m a.s.l.

Station	1	2	3	4	5	6	7	8	9	10
Altitude (m)	2050	2050	2600	1600	1600	1370	1360	1750	1550	1550
Date sampled	9.11	9.11	10.11	11.11	11.11	11.11	11.11	12.11	12.11	12.11
Time sampled	16.00	16.00	11.30	11.00	12.40	13.30	14.00	13.30	15.15	16.30
Temperature °C	16.7	19.6	15.6	18.8	21.3	23.5	24.9	19.0	22.8	20.5
Turbidity J.T.U.	5	5	2	5	5	20	25	350	250	200
pH	7.40	7.10	6.90	7.20	7.40	7.40	7.40	7.10	7.10	7.80
Conductivity μS/cm	62	62	44	49	57	70	69	49	75	126
TSS	15	32	4	10	7	27	18	400	257	680
Total hardness $CaCO_3$	29	27	19	17	22	25	26	20	30	58
Alkalinity $CaCO_3$	38	37	25	27	27	33	35	25	38	67
Ca	9	8	6	5	6	6	7	5	8	18
Mg	2	2	1	1	2	2	2	2	3	3
Na	4	4	3	4	4	5	4	3	3	4
K	0.8	0.8	0.4	0.9	0.4	0.7	0.9	0.3	0.7	1
Total cations me/l	0.81	0.71	0.50	0.52	0.60	0.74	0.73	0.52	0.77	1.36
HCO_3	27	45	31	33	33	40	42	31	47	82
Cl	0	0	0	0	0	0	0	0	0	0
SO_4	2	3	2	3	3	4	3	4	4	4
Total anions me/l	0.82	0.78	0.54	0.60	0.61	0.75	0.76	0.58	0.85	1.42
SiO_2	29	23	20	27	22	21	26	16	18	21

List of stations: 1. Kuragamba River at Gembogl; 2. Tem River at Gembogl; 3. Goenigl River at Keglsugl; 4. Asaro R. at Kongi Village; 5. Manyata R.; 6. Bena Bena R.; 7. Asaro R. at Bena Bena R. confluence; 8. Maril R.; 9. Wahgi R. at Kundiawa; 10. Chimbu R. at Kundiawa. All stations sampled in 1978.

Values in mg/l where not otherwise stated.

Table 4. (continued). Streams and ponds in the altitudinal range 3500 – 1300 m a.s.l.

Station	11	12	13	14	15	16	17	18	19	20
Altitude (m)	1600	1650	1600	1960	1640	1600	1540	2700	1600	2200
Date sampled	12.11	18.15	13.11	16.11	16.11	16.11	16.11	17.11	17.11	18.11
Time sampled	18.00	18.15	15.00	11.30	14.00	14.30	16.00	17.00	18.00	11.00
Temperature °C	20.8	18.3	22.4	19.6	19.3	22.6	22.9	12.6	17.7	18.3
Turbidity J.T.U.	20	5000	45	2	25	5	45	4	5	250
pH	7.50	6.70	7.20	7.30	7.00	7.30	7.30	7.30	7.30	7.10
Conductivity $\mu S/cm$	135	69	67	84	52	50	72	25	74	62
TSS	22	6861	108	2	42	2	83	0	6	131
Total hardness $CaCO_3$	60	27	41	34	20	19	28	9	27	26
Alkalinity $CaCO_3$	69	31	48	44	29	27	27	15	38	35
Ca	18	7	13	6	5	6	7	2	7	6
Mg	4	2	2	5	2	2	3	1	2	2
Na	4	2	4	4	4	3	4	1	4	3
K	1	2	1	2	0.6	0.3	0.8	0.4	1	1
Total cations me/l	1.38	0.68	1.03	0.87	0.55	0.53	0.73	0.25	0.75	0.66
HCO_3	84	38	59	54	35	33	33	19	47	43
Cl	0	0	0	0	0	0	0	0	0	0
SO_4	5	5	3	2	4	3	4	3	2	2
Total anions me/l	1.48	0.71	1.02	0.92	0.66	0.60	0.62	0.36	0.71	0.64
SiO_2	20	19	24	33	21	17	22	16	24	22

List of stations: 11. Mangiro Creek; 12. Mairi Creek at Kenangi; 13. Dunantina River; 14. Gumants River; 15. Kunan River at Mt. Hagen; 16. Tuman River; 17. Wahgi River at Banz; 18. Nebilyer River at 2700 m; 19. Nebilyer River at Tagoga Village; 20. Kaugel River east of Giluwe. All stations sampled in 1978.

Table 4. (continued). Streams and ponds in the altitudinal range 3500 – 1300 m a.s.l.

Station	21	22	23	24	25	26	27	28	29	30
Altitude (m)	2250	1780	1900	1400	1740	2200	2800	2260	2260	2100
Date sampled	18.11	18.11	18.11	18.11	18.11	19.11	19.11	19.11	19.11	19.11
Time sampled	12.30	13.15	14.30	17.30	18.30	11.00	12.00	14.00	14.00	10.00
Temperature °C	14.7	18.1	17.7	20.8	19.7	14.6	12.3	17.8	19.8	23.8
Turbidity J.T.U.	5	5	5	35	5	2	2	2	30	4
pH	6.90	7.10	7.15	7.60	7.30	6.90	6.80	6.80	7.00	6.50
Conductivity μS/cm	20	47	58	170	88	15	12	46	58	14
TSS	0	0	0	44	15	0	0	5	21	4
Total hardness $CaCO_3$	8	20	23	90	36	7	5	18	22	5
Alkalinity $CaCO_3$	10	25	31	88	48	10	6	25	29	8
Ca	2	6	7	33	10	2	1	4	6	2
Mg	1	1	2	2	3	1	1	2	2	0.3
Na	1	2	2	2	3	1	1	3	3	1
K	0.2	1	1	1	1	0.3	0.2	1	1	1
Total cations me/l	0.31	0.59	0.67	1.91	1.01	0.17	0.16	0.48	0.62	0.19
HCO_3	12	31	38	108	59	12	7	30	34	9
Cl	0	0	0	0	0	0	0	0	0	0
SO_4	6	4	2	3	2	3	2	2	2	2
Total anions me/l	0.31	0.58	0.65	1.84	1.01	0.15	0.16	0.45	0.62	0.19
SiO_2	10	17	20	13	23	9	8	21	19	4

List of stations: 21. Iaro (Yelo) River at 2300 m; 22. Ankuro River; 23. Anga (Angu) River; 24. Lai River above confluence with Angurra River; 25. Mendi River at Mendi; 26. Kumi Creek, western Giluwe at 2200 m; 27. Mt. Giluwe Stream at 2800 m; 28. Kaugel River at Tambul; 29. Ambuga River at Tambul; 30. village pond at Karil at 2100 m. All stations sampled in 1978.

154

reach the underlying mafic lava rich in potassium (Blake and Löffler 1971), this should be reflected in a higher potassium content in drainage water, which it is not.

Water chemical composition of streams of Mt. Wilhelm, Mt. Glendako, Mt. Otto and the Kubor Range differs very little. The widespread presence of volcanic ash (tephras) in the upper basin of the Purari (Pain and Blong 1979; Pain, this volume), because of its function as a long-term stabilizer of various lithologies, and as a contributor of some trace elements, such as for example titanium (Petr and Irion, this volume), and of nutrients, must be considered in the context of the present catchment water chemistry and that of transported and deposited sediments. This subject still requires considerable study.

Limestone drainage, as represented by that of Lai River bypassing Mt. Giluwe in the west, and that of largely marine clastic sedimentary rocks (Chimbu River (Fig. 6), Mangiro Creek, situated south and southeast of Mt. Wilhelm) (Table 4) results in higher pH, conductivity, calcium and bicarbonate values, but also in 16% lower dissolved silica values than those measured in noncalcareous drainage rivers. The total dissolved ionic content, as reflected in conductivity, is two to four times higher in limestone drainage rivers; calcium, three to 7 times higher; and bicarbonates, three times higher. The highest total suspended sediment content of 6861 ppm recorded for the Purari basin was that in Mangiro Creek, where the sample was collected during an intensive rainstorm. Mangiro Creek drains part of Simbu Province, which is very densely populated, with much of the original vegetation stripped, and agriculture/gardening activities extending over the steepest slopes.

5. Streams and lakes of the humid tropical forest (1300 – 20 m)

5.1. Purari tributary rivers above Hathor Gorge (640 – 180 m) (Table 5) (Figs. 7, 8)

Using helicopter transport, six rivers of the catchment were irregularly sampled. The sampling stations were all situated at gauging stations, positioned in uninhabited valleys of the high rainfall forest zone.

The mean water temperature of 23.1°C (Table 10) is only slightly higher than that for water in the altitudinal range 1500 – 1000 m. The rapid passage of water from highlands does not seem to provide sufficient time for water to warm up. This is even more so when all rivers join into the Purari, and the huge volume of water rapidly advances towards the sea.

With the exception of the Poru and Pio rivers, the highland reach of these

155

Fig. 6. Chimbu River at Kundiawa.

Fig. 7. Kaugel River gauging station.

156

Fig. 8. Confluence of the Tua (left) with the Erave. The Tua carries a high suspended sediment load.

Fig. 9. Purari River in Hathor Gorge.

Table 5. Purari tributary rivers above Hathor Gorge.

River	Tua	Kaugel	Poru	Iaro	Erave	Pio
Station number	31	32	33	34	35	36
Altitude (m)	500	518	490	640	600	180
Number of samples	3	3	4	4	4	4
Mean discharge m^3/sec	n.d.	364	63	n.d.	n.d.	n.d.
Temperature °C	24.8	21.9	22.6	22.2	22.6	24.6
Turbidity J.T.U.	312	88	31	99	30	60
pH	7.63	7.51	7.76	7.83	7.80	7.68
Conductivity ɥS/cm	95	68	116	132	185	115
TSS	412	108	41	187	45	81
Total hardness CaCO$_3$	36	29	53	70	91	50
Alkalinity CaCO$_3$	43	32	58	93	95	53
Ca	10.3	11.3	18.3	23.9	31.0	16.5
Mg	2.7	2.3	2.3	2.8	3.3	2.5
Na	4.3	2.8	2.3	2.8	4.0	4.0
K	0.9	0.7	0.6	0.7	0.7	1.0
Total cations me/1	0.93	0.88	1.17	1.89	1.95	1.16
HCO$_3$	52	51	70	113	116	65
Cl	0.3	0.3	0.8	0.5	nil	nil
SO$_4$	3.3	3.8	4.3	4.0	4.3	4.5
Total anions me/1	0.93	.92	1.23	1.94	1.97	1.17
SiO$_2$	19	19	13	12	13	16

Where not otherwise stated, values in mg/l; n.d. = not determined.

rivers has a medium to dense population, and this is reflected in some physico-chemical parameters. The Tua has the highest mean of total suspended sediments, as a result of human interference with the soils in its upper catchment, especially in the Wahgi and Asaro valleys. The Poru, which is the shortest and most heavily forested, with lowest human density, has the lowest TSS. The human impact on water quality is also evident from the presence of the highest number of coliform bacteria in the Tua as compared with elsewhere (Petr, this volume). Due to the complex lithology of some catchments, such as Pio, the conductivity may change considerably and be dominated temporarily by inputs from subcatchments when these receive localized heavy rainstorms. This gives a wide range of 86 to 145 ɥS/cm for this river.

Water chemistry of these rivers is rock dominated. As with higher altitude streams, elevated bicarbonate values suggest a widespread presence of calcareous rocks (Erave, Iaro, Poru rivers). Amongst the six rivers, the higher bicarbonate corresponds to a lower dissolved silica, a higher conductivity, and a higher pH.

5.2. Lake Tebera (Table 6)

This is the only large low altitude natural lake of the Purari basin. It is accessible by helicopter, or walking a number of days through the rugged terrain of tropical rain forest. Situated in the Mamesu syncline at an altitude of 586 m, and surrounded by limestone karst, the morphometry of the lake is still to be determined. The lake is approximately 6 – 7 km long and up to 1.5 km wide. It drains east, joining within a short distance of some 5 km and a drop of some 400 m the Purari below its confluence with the Pio. Conn (1979a) described briefly the floating plant islands of this lake, which are formed by *Leersia hexandra*, numerous sedges, grasses, *Ludwigia adscendens*, *Limnophila indica* and *Polygonum*. *Azolla pinnata*, *Spirodela oligorrhiza* and *Utricularia* have also been identified. The smaller of these 'sudd' islands are moved by wind. The shoreline of Lake Tebera is formed in many places by *Phragmites karka*, and Conn (1979a) has suggested that this reed prefers to sit on the ground for

Table 6. Mid-altitude lakes in karst area.

	Lake Tebera (St. 37)			Lake Kutubu[a] (St. 38) (outside the Purari)
	Mid-lake	Shore (west)	Outflow	
Altitude (m)	586	586	586	808
Date sampled	18.4.78	18.4.78	18.4.78	Nov. 1975
Time	11.30	15.00	15.00	n.d.
Turbidity J.T.U.	2	5	18	2
Temperature	n.d.	n.d.	n.d.	n.d.
pH	7.55	7.65	7.30	8.00
Conductivity μS/cm	110	134	182	170
TSS	0.5	18.5	60.5	9
Total hardness $CaCO_3$	48	67	91	94
Alkalinity $CaCO_3$	58	70	95	90
Ca	16	22	33	33
Mg	2	3	2	3
Na	2	2	3	1
K	0.5	0.5	0.8	0.3
Total cations me/l	1.02	1.39	1.99	1.92
HCO_3	71	86	116	110
Cl	nil	nil	nil	nil
SO_4	4	3	10	2
Total anions me/l	1.65	1.48	2.11	1.83
SiO_2	17	5	6	n.d.

[a] Collected and analysed by Geological Survey of the Department of Minerals and Energy. Where not otherwise stated, values in mg/l; n.d. = not determined.

159

ease of obtaining nutrients rather than float over deep water. The aquatic flora of the lake shows a very close similarity in species composition, zonation and succession to that of Lake Kutubu, draining into the Kikori west of the Purari catchment (Conn 1979b).

The lake water (only surface water samples were collected) has a chemistry similar to that of rivers draining calcareous deposits, and is dominated by calcium bicarbonate. Conductivity increases in the sequence: mid-lake − shore − outflow, with corresponding trends in calcium, bicarbonates, sodium and potassium ions. In the outflow, the concentration of sulphates is approximately 2.5 times higher than in the mid-lake. Dissolved silica has a reversed trend, with highest concentration found in the mid-lake and lowest in the outflow.

The lake water conductivity (182 and 134 μS/cm for mid-lake and shoreline samples respectively) is near to that of Lake Kutubu (153 μS/cm, Bayly et al. 1970). Lake Kutubu, which is positioned in a similar geomorphological situation 145 km to the west, had surface water temperature of 24°C, and it has been suggested that Lake Kutubu is oligomictic. Conn (1979b) recorded Secchi disc water transparency of 6.5 to 8 m, which he found to coincide with the maximum depth in which rooted plants grow in Lake Kutubu. Lake Tebera's high water transparency should allow a reasonably high algal productivity, subject to availability of nutrients. Lake Kutubu plankton has a large quantity of *Botryococcus* (Bayly et al. 1970), and a great diversity of desmids, dominated in samples collected by Conn (1979b) by *Staurastrum*, with a second most common being *Cosmarium*. Amongst other identified algae are *Anthrodesmus*, *Spirogyra*, and *Microcystis*. The dominant form of zooplankton in Lake Kutubu is *Calamoecia ultima*, with *Mesocyclops* sp., an unidentified cladoceran, and rotifer *Keratella* also being identified by Bayly et al. (1970).

Lake Tebera is known to dry out periodically. The floating islands then settle down, apparently overgrowing the whole lake area. However, the lake has a fish population of unknown species composition, and this is being fished by the only Pawaia family inhabiting (in 1978) the shores of this lake.

6. The Purari and the Aure

Past the confluence with the Poru, the Purari passes over a series of rapids through Hathor Gorge (Figs. 9, 10). Afterwards, the river emerges into the foothill zone at an approximate altitude of 120 m a.s.l. (Fig. 11) and continues its rapid advance towards Wabo, taking in small tributaries on each side. Be-

Fig. 10. Hathor Gorge, with outcrops of Darai limestone.

Fig. 11. Purari River upstream of Wabo.

Fig. 12. Purari River armoured bed, above Wabo.

Fig. 13. The Purari at Uraru upstream of Wabo, at low water level. Two Pawaia in the right foreground.

162

tween Wabo and Pio-Purari confluence, an area of 260 to 290 km² and delimited by the 150 m contour line, is to be flooded in the future if a dam is constructed at Wabo. At present, the river flows rapidly over heavily armoured beds (Figs. 12, 13), with low rapids navigable by a dinghy with a powerful outboard motor. In places, the width of the valley allows the river to braid, with fairly stable islands. At present there is almost no human habitation in this area, most of the Pawaia people preferring nowadays to live at Wabo with its school, medical post and small tradestore (see Toft, this volume). Only small tributaries join the Purari in this reach. About 40 km downstream from Wabo, the Purari is joined by the Aure River.

6.1. Small tributaries between Hathor Gorge and Wabo

These carry only a small volume of water, at a rate usually not exceeding a few m³/sec. Some of them are known for their sudden surge of water carrying stones, the result of a heavy thunderstorm somewhere in their catchment. But usually the streams have very clear water (Fig. 14). This is a high precipitation zone of some 8500 mm per year, known for extremely heavy downpours. In some places, the dense tropical forest is scarred by landslides.

Among these streams, Moa Creek has the highest conductivity, total hardness and calcium levels, indicating a limestone drainage (Table 7). The highest surface water temperature was measured in a calm section of Pide River, the largest and most sluggish of these small tributaries above Wabo. Wabo Creek, entering the Purari downstream of the proposed dam area, has a lower conductivity and total anions and cations content than any other stream in this area.

6.2. Purari River above the delta, and Aure River

Three stations were irregularly sampled on the Purari: at the lower end of Hathor Gorge, at Wabo camp (Fig. 15) just above the confluence of the Purari with Wabo Creek, and about 1 km upstream of the delta. The Aure was sampled either 300 to 500 m upstream from its confluence with the Purari, or at the gauging station further upstream.

Hydrographic records for Wabo were used for the Hathor station, as the difference in discharge between Hathor Gorge and Wabo is negligible.

The Purari transport of total suspended sediments in the surface water ranged from 29 to 778 mg/l, with a mean of 254 mg/l (Table 8). The Purari

Table 7. Small tributaries between Hathor Gorge and Wabo, and Wabo Creek.

Station	Wai-i Creek		Moa Creek		Gipi River	Pide River		Wabo Creek			Mean
Date sampled	16.6.78	24.8.79	16.6.78	24.8.79	24.8.79	16.6.78	24.8.79	11.2.76	17.6.78	23.8.79	
Time	nd..	12.50	n.d.	14.00	14.35	n.d.	15.15	n.d.	n.d.	16.45	
Temperature °C	n.d.	24.9	n.d.	23.2	25.7	n.d.	26.3	n.d.	n.d.	24.1	24.8
Turbidity J.T.U.	15	5	15	5	5	15	5	5	15	5	9
pH	7.40	7.30	7.00	7.40	7.30	7.20	7.40	7.50	7.60	7.10	7.40
Conductivity $\mu S/cm$	108	200	274	235	192	94	165	94	87	125	157
TSS	1	10	2	4	5	18	4	327	11	7	39
Total hardness $CaCO_3$	54	70	152	94	67	50	61	49	40	45	68
Alkalinity $CaCO_3$	54	76	141	96	73	56	61	44	42	47	44
Ca	16	22	56	36	22	16	19	18	14	16	23.5
Mg	4	4	3	1	3	2	3	1	1	1	2.3
Na	4	6	4	2	6	4	5	2	3	4	4.0
K	1.0	1.5	0.8	0.3	0.8	0.8	0.9	0.2	0.5	0.4	0.7
Total cations me/l	1.25	1.70	3.25	1.97	1.63	1.19	1.40	1.11	0.93	1.07	1.6
HCO_3	66	93	172	103	79	68	74	54	51	57	82
Cl	1	nil	nil	nil	nil	1	nil	nil	1	nil	0.3
SO_4	5	5	5	2	5	4	3	6	4	3	4.2
Total anions me/l	1.19	1.63	2.92	1.80	1.45	1.22	1.32	1.03	0.95	1.00	1.45
SiO_2	23	21	17	7	11	10	13	9	8	11	13

Where not otherwise stated, values in mg/l; n.d. = not determined.

Fig. 14. Moa Creek at Uraru, a small tributary of the Purari between Hathor Gorge and Wabo.

Fig. 15. Purari River at Wabo camp. Airstrip and camp. The proposed dam site was cleared of the high forest in 1974 – 75. The river flows from the top left corner.

Table 8. Lowland reach of the Purari and Aure rivers above the delta.

Station Station number	Hathor Gorge 39		Wabo 40		Above confluence 41		Weighted mean of means 23	Aure 42	
Altitude (m) Number of samples	90[e] 3		40 16		25[e] 4			31 7	
	Mean	Range	Mean	Range	Mean	Range	Weighted mean of means	Mean	Range
Discharge m^3/sec	2809[a]	n.d.	1608	828 – 4015[c]	3341[d]	2499 – 4183	1930	171	58 – 388
Temperature °C	23.3[b]	n.d.	24.6	23.8 – 24.9	25.9[b]	n.d.	n.d.	23.8	23.0 – 24.2
Turbidity J.T.U.	30	15 – 50	102	20 – 200	433	150 – 1000	150	23	2 – 80
pH	7.70	7.40 – 7.89	7.64	7.40 – 8.00	7.70	7.20 – 8.10	7.66	7.61	7.30 – 8.25
Conductivity μS/cm	154	134 – 180	131	108 – 158	142	120 – 170	136	153	115 – 265
TSS	221	34 – 510	205	29 – 759	473	246 – 778	254	110	5 – 504
Total hardness CaCO$_3$	70	67 – 75	51	50 – 77	61	53 – 72	55	70	53 – 98
Alkalinity CaCO$_3$	70	66 – 74	65	52 – 75	65	63 – 67	65	71	50 – 98
Ca	22.7	21 – 25	20.4	16 – 26	20.3	18 – 24	20.7	23.1	16 – 26
Mg	3.3	3 – 4	2.7	2 – 4	2.5	2 – 3	2.8	3.1	2 – 4
Na	3.3	2 – 4	3.0	2 – 4	3.0	2 – 4	3.0	4.0	3 – 6
K	1.3	1.0 – 2.0	0.9	0.5 – 2.0	0.9	0.8 – 1.0	1.0	0.7	0.4 – 1.0
Total cations me/l	1.49	1.25 – 1.64	1.42	1.16 – 1.72	1.38	1.26 – 1.58	1.42	1.61	1.40 – 2.17
HCO$_3$	51	66 – 86	79	63 – 92	78	76 – 81	69	87	61 – 120
Cl	0.3	nil – 1.0	0.8	nil – 2.0	1.3	nil – 3.0	0.8	0.4	nil – 2.0
SO$_4$	4.6	3 – 6	3.8	1 – 6	2.3	nil – 3.0	3.6	4.3	2 – 6
Total anions me/l	1.42	1.19 – 1.59	1.41	1.17 – 1.62	1.38	1.27 – 1.59	1.41	1.55	1.20 – 2.01
SiO$_2$	18	15 – 23	14	9 – 24	15	10 – 19	15	17	9 – 24

[a] Wabo discharge applied for $n=1$: 11.2.1976.
[b] $n=1$.
[c] Available only for periods 6 – 10.7.1978 and 17 – 22.8.1978 ($n=11$).
[d] $n=2$.
[e] Approximate altitudes.
Where not otherwise stated, values in mg/1; n.d. = not determined.

can thus be considered as a turbid river. The total suspended sediment transport is 40.1 × 10⁶m³ per year, of which the clay fraction is 34.7%, silt 55.8%, and sand 9.5% of the total (Pickup, this volume). The clay fraction of river sediments is dominated by montmorillonite and chlorite, with a medium content of kaolinite/halloysite, gibbsite, and usually a low concentration of illite and mica (Irion and Petr 1979, this volume). Elemental analysis of sediments is discussed in more detail by Petr and Irion (this volume).

The dilution impact on water chemistry of a rapid rise in water volume transported by the Purari was followed on one occasion. Within three days the discharge increased from 777 to 4015 m³/sec (Table 9), the river reached its

Table 9. Purari at Wabo (17 – 22.8. 1978).

Date	17.8	18.8	19.8	20.8	21.8	22.8
Mean discharge m³/sec	977.20	827.86	777.27	1196.04	3490.68	4015.18
Time	18.00	08.45	08.00	07.00	07.00	07.00
Temperature °C	24.8	24.3	24.7	24.8	24.7	23.8
Turbidity J.T.U.	20	50	45	150	200	150
pH	7.45	7.60	7.65	7.40	7.55	7.80
Conductivity μS/cm	141	141	142	116	124	109
TSS	29	56	31	256	150	196
Total hardness $CaCO_3$	64	64	64	50	57	53
Alkalinity $CaCO_3$	66	75	66	52	56	52
Ca	20	20	20	16	19	18
Mg	3	3	3	2	2	2
Na	4	3	4	3	3	2
K	0.9	0.9	0.9	0.6	0.6	0.6
Total cations me/l	1.46	1.44	1.46	1.16	1.29	1.17
HCO_3	81	81	81	63	68	63
Cl	1	1	1	2	2	1
SO_4	4	4	4	4	5	4
Total anions me/l	1.43	1.44	1.44	1.17	1.27	1.14
SiO_2	17	16	17	12	12	9
PO_4-P	n.d.	2.7	1.46	1.2	1.6	1.46
NH_4-N	n.d.	11	8	29	60	86
NO_3-N	n.d.	61.5	25	29.4	28.5	26.4
Total dissolved P	n.d.	35.0	36.8	39.8	39.8	26.2
Total dissolved N	n.d.	104	112	103	78.3	110
P adsorbed	n.d.	82	78	92	60	164
P organic	n.d.	4.1	6.3	20.4	21.7	21.1
P particulate	n.d.	6.4	4.6	4.9	7.0	4.5
P ads./orthophosphate	n.d.	30	53	77	38	112

Values for nutrients in microgrammes per litre; elsewhere, where not otherwise stated, values in mg/1; n.d. = not determined.

levies (Fig. 16), and a large number of forest trees were carried downstream. The concentration of suspended solids increased with rising water level, but lower values were recorded at the peak discharge than during the rise. Pickup (this volume), who sampled at considerably longer intervals, found for his range of 730 to 6489 m^3/sec that the sediment transport correlates well with discharge. In our case the water conductivity was lowered by one third during floods, and the concentration of bicarbonates by one fifth to one quarter, as compared with pre-flood conditions. The concentration of dissolved silica was also reduced by one third to one half of original values, suggesting that there is little buffering in the system, which would maintain concentrations of dissolved silica through changes in discharge as suggested by Edwards and Liss (1973). Turvey (1975) found very high concentrations of silica in the leaf litter of the tropical rain forest, and one would expect this litter to enter the Purari, especially during long, heavy rainfalls. However, no increase has been noticed, and Turvey's observations are probably valid only for the small streams which he investigated. Purari floods also result in lower concentrations of calcium, magnesium and sodium. A slight increase in chlorides at a higher discharge might suggest a weak precipitation influence as a result of the proximity of the ocean.

Viner (1979, 1982) investigated the impact of flood situation on nutrient chemistry, and some of his data on phosphorus and nitrogen are listed in Table 9. Dissolved nutrients did not change conspicuously with changes in discharge, except ammonia nitrogen which seems to increase with discharge. Nitrate concentration decreased on one occasion, and on the other, increased with an increase in discharge. For dissolved orthophosphate, total dissolved phosphorus and total dissolved nitrogen, no correlation with changes in discharge were found. Viner placed emphasis upon that portion of the nutrients associated with suspended particles. There was a general increase of adsorbed P and N with an increase in particulate material. Clays have at least an order of magnitude more phosphorus than silt. In spite of the dissolved orthophosphate concentrations being generally low, the nutrient loading into any zone of the river is largely due to high discharge rates. Viner calculated the annual nutrient loading of inorganic nitrogen at Wabo to be 2000 to 11000 tonnes, and that of inorganic phosphorus, 102 – 462 tonnes. Between about one and two thirds of this phosphorus is particle borne.

The Aure River (Figs. 17, 18) is the last large tributary to the Purari above the delta. The Aure basin is dominated by marine clastic sedimentary rocks interbedded by limestone, with some bioclastic limestone, and this is reflected in higher values of calcium and bicarbonates than measured in Purari at Wabo. The Aure also has higher concentrations of Mg, Na, Cl and

168

Fig. 16. The Purari at Wabo with dugouts. The river is in flood with an approximate discharge of 4000 m³/sec.

Fig. 17. The Aure River gauging station.

169

Fig. 18. Confluence of the Aure (right) with the silt-laden Purari.

SiO_2. The higher values for the Aure may be biased by sampling there mostly during very low discharges, with only one discharge above the mean (the mean for the sampling period being 171 m^3/sec, as compared with the annual mean discharge of 307 m^3/sec given by Pickup, this volume). Samples for the Purari were for a mean sampling period discharge of 1930 m^3/sec, which, although lower than the annual mean discharge of 2360 m^3/sec, included five readings above the mean (Table 8).

7. Discussion

The Purari River, with its tributaries, is a short fast-flowing river. It can be compared with the upper reaches of the Amazon River to which it seems to have a very similar hydrological character (Irion, personal communication).

The amount of suspended sediments in the lower Purari River, with a mean value of 254 mg/l for this study, is close to the wet season mixed-type environment of Amazon River tributaries, and thus it is in between the montane environment with the highest concentrations, and the tropical environment with the lowest concentrations of suspended sediments as defined by Gibbs (1967). In spite of the intensive greyish brown coloration of the Purari River, the concentration of sediments is considerably lower than, for example, that of monsoonal Burmese rivers, where in the Rangoon River values range from 3.2 to 7.4 g/l, and in the Moulmein arm of the Salween River the range is 1.0 – 5.0 g/l (Schnitzer 1974). There, the rainfall, which is largely concentrated in a wet season, results in intensive erosion in the catchment area and the erosion is further aggravated by the intensive agricultural activity.

The Purari water chemistry places this river amongst the 98% of surface waters which, according to Meybeck (1980), are dominated by calcium and bicarbonate ions. This type corresponds to the rock-dominated type as defined by Gibbs (1970). Meybeck (1980) has concluded that the influence of rock composition on the contents of major elements in river water is predominant in more than 95% of all rivers.

Limestone drainage streams are conspicuous with their high calcium and bicarbonate contents, higher pH and conductivity, and lower dissolved silica than the other streams.

The ionic dominance in the Purari, which integrates water from all its tributaries, is $Ca > Mg > Na > K$ and $HCO_3 > SO_4 > Cl$. The same dominance is valid for all its tributaries outside the influence of seawater. There is a longitudinal gradient, with dissolved concentrations of a number of major ions and conductivity increasing downstream. The conductivity rises from the initial 27.0 – 27.6 μS/cm at the altitude between 2500 – 4000 m, to 133 μS/cm at 10 – 100 m (Table 10). Water temperature rises from a mean of 11.4°C for glacial lakes (3500 – 4000 m) to 24.7°C at the lowest altitudes outside the delta, and 24.9°C at Mapaio in the delta (see Petr, Part 2, this volume). In the lower Sepik in the north of Papua New Guinea, the mean temperature at a similar altitude is 29.0°C (Mitchell, Petr and Viner 1980). The rather cool water in the Purari appears to result from the fast altitudinal drop and advance of the large Purari water mass through its lower course before reaching the sea.

171

Table 10. Altitudinal gradients of temperature, conductivity and dissolved silica content in aquatic environment of the Purari River basin.

Altitude (m)	Number of stations	Number of samples	°C		Conductivity $\mu S/cm$		SiO$_2$ mg/l	
			Range	Mean	Range	Mean	Range	Mean
4000 – 3500[a]	5	5	9.1 – 14.4	11.4	13 – 43	27.6	7 – 13	9.6
3000 – 2500	3	3	12.3 – 15.6	13.5	12 – 44	27.0	8 – 20	14.7
2500 – 2000	7	7	14.7 – 19.8	17.3	15 – 62	54.0	9 – 29	19.0
2000 – 1500	16	16	17.7 – 22.8	20.1	49 – 135	72.0	16 – 33	21.4
1500 – 1000	3	3	20.8 – 24.9	23.0	69 – 170	103.0	21 – 26	23.5
1000 – 100	6	23	21.0 – 24.6	23.1	86 – 195	160[d]	12 – 19	15.5
					63 – 137	93[e]		
100 – 20[b]	1	23	23.8 – 26.5	24.7	108 – 180	136	9 – 24	15.0
10[c]	1	10	24.2 – 26.5	25.0	120 – 150	136	12 – 16	13.8

[a] Glacial lakes.
[b] Purari River.
[c] Purari delta – freshwater reach.
[d] Limestone drainage dominates.
[e] Other drainage dominates.

The concentration of dissolved silica seems to depend on altitude and temperature, and also on the lithology of the catchment. Streams of Mt. Giluwe, an extinct volcano with a number of bogs, have the second lowest dissolved silica concentrations (8.9 mg/l) in the Purari catchment, with the absolute minimum being recorded for glacial lakes of Mt. Wilhelm (range of 7 to 13 mg/l). There is a gradual increase in dissoved silica to a mean of 23.5 mg/l for streams between 1000 and 1500 m altitude, followed by a steep decrease to 15.0 mg/l for the Purari at 20 – 100 m, and 13.8 mg/l at 10 m altitude. As there is no possibility that this loss in silica would be due to algal production (the river is too turbid and turbulent), it could be that this is due to adsorption of silica onto clay particles. However, since small tributaries at 150 – 50 m altitude, usually with very low TSS, also have relatively low dissolved silica concentrations (mean 13.0 mg/l, cf. Table 7), there may be another mechanism behind the conspicuously low silica concentrations at low altitudes. The Sepik and Fly river values of dissolved silica, as measured in their lower courses, are close to those of the lower Purari, supporting the suggestion that concentrations of dissolved silica decrease in tropical rivers passing through lowlands.

Inorganic combined nitrogen (NH_4-N + NO_2-N + NO_3-N), like most of the other components, was found by Viner (1979) to increase progressively downstream. Orthophosphate, on the other hand, had a relatively higher concentration in the highlands than downstream, suggesting its adsorption onto silt.

Floods result in a lower dissolved ion transport per litre, as seen in the decrease in conductivity. In August 1978, within three days, the Purari discharge at Wabo increased from 777 m^3/sec to 4015 m^3/sec, which led to a decrease in conductivity from 147 μS/cm in pre-flood condition to 115 μS/cm at full flood. A major change from 17 mg/l to 9 mg/l took place in the concentration of dissolved silica, while bicarbonates decreased by about 25% only. Concentrations of Ca, Mg and Na also dropped, but those of Cl and SO_4 remained stable. There was also a decrease in concentration of total dissolved phosphorus, but not in total dissolved nitrogen (Viner 1979). Pickup (this volume) found a direct progressive correlation between discharge rate and transport of suspended solids. This is of significance for transport of elements and especially nutrients bound with solids.

The average dissolved content (Cd in Table 11) delivered to the lower delta and the sea was calculated from a series of samples collected at Mapaio. The total ionic content is about four times that of the world average, and with the exception of chlorides, all chemical component concentrations are also higher. The Purari also delivers the highest total ionic content in mg/l among

Table 11. Average dissolved content and specific transport of the Purari surface water discharging to the delta.

		SiO_2	Ca^{2+}	Mg^{2+}	Na^+	K^+	Cl^-	SO_4^{2-}	HCO_3^-	Σi
Purari	Cd	13.8	20.6	2.6	3.2	1.0	1.2	2.4	80.7	111.8
	Td	34.5	51.5	6.5	8.0	2.5	3.0	6.0	201.5	279.0
World	Cd	3.9	5.0	1.25	1.9	0.5	2.15	3.1	19.4	33.3
	Td	10.4	13.4	3.35	5.15	1.3	5.75	8.25	52.0	89.2
Sepik[a]	Cd	12.5	15.5	4.0	3.5	0.4	0.0	4.5	73.5	101.4
Fly[b]	Cd	9.0	21.3	1.67	2.33	0.43	0.0	2.67	78.3	106.7
Guyana[c]	Cd	10.9	2.6	1.05	2.55	0.75	3.9	2.0	12.2	25
Philippiness[c]	Cd	30.4	30.9	6.6	10.4	1.7	3.9	13.6	131	198

Cd = average dissolved content in mg per 1; Td = specific transport (in t per km^2 per year); Σi = total ionic content in mg per 1.
[a] Mitchell, Petr and Viner (1980).
[b] Mean for three samples collected by Mr F. Pratt and Dr R.F. Warner in 1978−1979 downstream of the Fly River/Strickland River confluence.
[c] Meybeck (1980).

the PNG major rivers, which include the Fly (approx. discharge 6000 m^3/sec), and the Sepik (approx. discharge 4000 m^3/sec). Comparative values for some other tropical environments taken from Meybeck (1980) indicate that, for example, the Philippine rivers drain mostly a limestone catchment rich in calcium bicarbonate and have higher concentrations of sodium and chloride, suggesting an impact of the sea on inland water chemistry. In the Philippine rivers dissolved silica is also higher than in the Purari. However, Guyana rivers, draining largely lowland tropical humid forest, are poor in ions.

The transport rate for dissolved material is given by Td = qCd, where q is the water runoff, which for the Purari catchment of 33670 km^2 is equal to 79.21 l/s/km^2. The transport rate is about 3.5 times higher than that of the world average river and reflects the fairly intensive weathering of the catchment.

The significance of particle transport was studied in relation to nutrient chemistry by Viner (1979), who found a considerable pool of adsorbed phosphorus, especially on clays. He suggested that such adsorbed phosporus would be in dynamic equilibrium with the phosphorus in solution.

The implications for the water chemistry and aquatic productivity of damming the Purari at Wabo have been considered in connection with the proposal to build a dam there. Much of the suspended sediments, organic debris and detritus carried by the river would be deposited in the reservoir, which would cover some 260 to 290 km^2. Only part of the lightest sediment would spill or pass through the hydroelectric generating system. Viner's (1979) data suggest that the Wabo reservoir fertility would be at the very least mesotrophic, but probably eutrophic. He draws these conclusions from his data an nutrient chemistry and from the expected fast flushing rates resulting in frequent mixing of a large portion of the reservoir. According to him, the high nitrogen to phosphorus ratio would probably be limiting for blue-green algal development. But there would always be a potential risk for explosive growth of some noxious aquatic weeds, such as *Salvinia molesta* and perhaps even *Eichhornia crassipes*, if they were to get into the reservoir.

After damming the Purari at Wabo, the Aure would become the major supplier of sediments to the lower Purari and its delta. Retention of most of the Purari sediments in the reservoir would thus have a major impact on the river between Wabo and Purari/Aure confluence, with intensified channel erosion. Viner suggested that dissolved concentrations of major ions and nutrients below the dam would be similar to those in the river entering the reservoir. Hence major downstream ecological effects of the reservoir would involve the impact of increased water transparency and low sediment concentrations on aquatic communities, with a probable shift in the dominance of major species, or even invasion of the modified environment by new ones. In this respect the impact of the reservoir on the deltaic system and the sea would probably be considerable.

Acknowledgements

The success of the organisation and field work was much dependent on the assistance of a number of Papua New Guinea government officials, both in the central and provincial governments, and in Baimuru. Pacific Helicopters provided transport to the middle altitude rivers. Mrs Jasmin Lucero greatly assisted with the preparatory, field and evaluation parts of the programme. Mr John Bird of Baimuru provided material and moral support whenever necessary. Mines Analytical Laboratory carried out all analyses on numerous water samples. Dr Bayly critically commented on the manuscript. My wife kindly typed this and many other manuscripts for this book. To all of them I am grateful for their assistance.

References

Bayly, I.A.E., 1976. Report on CGE water samples and zooplankton collection. Appendix I to J.A. Peterson, The lakes. In: G.S. Hope, J.A. Peterson, I. Allison and U. Radok (eds.), The

Equatorial Glaciers of New Guinea, pp. 107–110. Balkema, Rotterdam.

Bayly, I.A.E., J. Peterson and V.P. St. John, 1970. Notes on Lake Kutubu, Southern Highlands, of the Territory of Papua and New Guinea. Newsl. Aust. Soc. Limnol. 3: 40–47.

Bik, M.J., 1967. Structural geomorphology and morphoclimatic zonation in the central highlands, Australian New Guinea. In: Landforms Studies from Australia and New Guinea, pp. 26–47. Australian National University Press, Canberra.

Blake, D.H. and E. Löffler, 1971. Volcanic and glacial landforms on Mt. Giluwe, Territory of Papua New Guinea. Geol. Soc. Amer. Bull. 82: 1605–1614.

Brass, L.J., 1964. Results of the Archbold Expedition, No. 86: Summary of the Sixth Archbold Expedition to New Guinea. Amer. Mus. Nat. Hist. Bull. 127: 145–215.

Conn, B., 1979a. *Phragmites karka* and the floating islands at Lake Tebera (Gulf Province), with notes on the distribution of *Salvinia molesta* and *Eichhornia crassipes* in Papua New Guinea. In: T. Petr (ed.), Purari River (Wabo) Hydroelectric Scheme Environmental Studies, Vol. 10: Ecology of the Purari River catchment, pp. 31–36. Office of Environment and Conservation, Waigani, Papua New Guinea.

Conn, B., 1979b. Notes on the aquatic and semi-aquatic flora of Lake Kutubu (Southern Highlands Province), Papua New Guinea. Ibid., pp. 63–90.

Dow, D.B. and F.E. Decker, 1964. The geology of the Bismarck Mts., New Guinea. Bureau Min. Resources, Geol. Geophys. Rep., No. 76. Australia.

Edwards, A.M.C. and P.S. Liss, 1973. Evidence for buffering of dissolved silicon in fresh water. Nature, Lond. 243: 341–342.

Evesson, D., 1980. The climate of the Purari River catchment above Wabo. In: A. Viner (ed.), Purari River (Wabo) Hydroelectric Scheme Environmental Studies, Vol. 16. Office of Environment and Conservation, Waigani, Papua New Guinea.

Gibbs, R.J., 1967. The factors that control the salinity and the composition and concentration of the suspended solids. Bull. Geol. Soc. Amer. 78: 1203–1232.

Gibbs, R.J., 1970. Mechanisms controlling world water chemistry. Science 170: 1088–1090.

Haantjens, H.A., 1970. Geologic and geomorphic history of the Goroka–Mount Hagen area. In: H.A. Haantjens et al. (eds), Lands of the Goroka–Mount Hagen Area, Territory of Papua New Guinea, pp. 19–23. C.S.I.R.O. Land Research Series, No. 27. C.S.I.R.O., Melbourne.

Hnatiuk, R.J., J.M.B. Smith and D.N. McVean, 1976. Mt. Wilhelm Studies 2. The climate of Mt. Wilhelm. Res. School Pacific Studies. Dept. Biogeography and Geomorphology. Publication BG/4. Australian National Univ., Canberra.

Illies, J., 1969. Trichoptera from the high mountain lakes Pinde and Aunde, New Guinea. Pacific Insects 11: 487–493.

Irion, G. and T. Petr, 1979. Geochemistry of the Purari catchment with special reference to clay mineralogy. In: T. Petr (ed.), Purari River (Wabo) Hydroelectric Scheme Environmental Studies, Vol. 18. Office of Environment and Conservation, Waigani, Papua New Guinea.

Irion, G. and T. Petr, 1983. Clay mineralogy of selected soils and sediments of the Purari River basin. This volume, Chapter I, 6.

Lewis, W.M., 1981. Precipitation chemistry and nutrient loading by precipitation in a tropical watershed. Water Resources Res. 17: 169–181.

Löffler, E., 1972. Pleistocene glaciation of Papua and New Guinea. Z. Geomorph., N.F. Suppl. 13: 32–58.

Löffler, H., 1973. Tropical high mountain lakes of New Guinea and their zoogeographical relationship compared with other tropical high mountain lakes. Arctic and Alpine Res. 5, pt. 2: A193–A198.

Löffler, E., 1974. Explanatory notes to the geomorphological map of Papua New Guinea,

C.S.I.R.O. Land Research Series, No. 33. C.S.I.R.O., Melbourne.

McKenzie, K.G., 1971. Ostracoda from Lake Piunde, near Mt. Wilhelm, New Guinea. Zool. Anz. 186: 391 – 403.

Meybeck, M., 1980. Pathways of major elements from land to ocean through rivers. In: J.-M Martin, J.D. Burton and D. Eisma (eds.), River Inputs to Ocean Systems, pp. 18 – 30. UNEP, IOC, SCOR.

Mitchell, D.S., T. Petr and A.B. Viner, 1980. The water-fern *Salvinia molesta* in the Sepik River, Papua New Guinea. Environmental Conservation 7: 115 – 122.

Paijmans, K., 1983. The vegetation of the Purari catchment. This volume, Chapter II, 1.

Pain, C.F., 1983. Geology and geomorphology of the Purari River catchment. This volume, Chapter I, 3.

Pain, C.F. and R.J. Blong, 1979. The distribution of tephras in the Papua New Guinea highlands. Search 10: 228 – 230.

Peterson, J.A., 1976. The lakes. In: G.S. Hope, J.A. Peterson, I. Allison and U. Radok (eds.), The Equatorial Glaciers of New Guinea, pp. 93 – 106. Balkema, Rotterdam.

Petr, T., 1983. Aquatic pollution in the Purari basin. This volume, Chapter II, 6.

Petr, T., 1983. Limnology of the Purari basin. Part 2. The delta. This volume, Chapter I, 10.

Petr, T. and G. Irion, 1983. Geochemistry of soils and sediments of the Purari River basin. This volume, Chapter I. 7.

Pickup, G., 1983. Sedimentation processes in the Purari upstream of the delta. This volume, Chapter I, 11.

Pickup, G. and R.F. Warner, in press. Geomorphology of tropical rivers. 1. Geomorphic history, hydrology and sediment transport in the Fly and Purari, Papua New Guinea.

Ruxton, B.P., 1969. Geology of the Kerema – Vailala area. In: Ruxton et al. (eds.), Lands of the Kerema – Vailala Area, Territory of Papua and New Guinea, pp. 58 – 64. C.S.I.R.O. Land Research Series, No. 23. C.S.I.R.O., Melbourne.

Schnitzer, W.A., 1974. Sediment loads and flow velocities of monsoonal rivers and their effect on shipping, taking Burma as the example. Appl. Sci. Dev. 3: 143 – 153.

Thom, B.G. and L.D. Wright, 1983. Geomorphology of the Purari Delta. This volume, Chapter I, 4.

Thomasson, K., 1967. Phytoplankton from some lakes on Mt. Wilhelm, East New Guinea. Blumea 15: 285 – 296.

Toft, S., 1983. The Pawaia of the Purari River. Social aspects. This volume, Chapter III, 2.

Tricart, J., 1965. Le modèle des régions chaudets, forêts et savanes. SEDES, Paris.

Turvey, N.D., 1975. Water quality in a tropical rain forested catchment. J. Hydrology 27: 111 – 125.

Viner, A.B., 1979. The status and transport of nutrients through the Purari River (Papua New Guinea). In: T. Petr (ed.), Purari River (Wabo) Hydroelectric Scheme Environmental Studies, Vol. 9. Office of Environment and Conservation, Waigani, Papua New Guinea.

Viner, A.B., 1982. A quantitative assessment of the nutrient phosphate transported by particles in a tropical river. Rev. Hydrobiol. trop. 15: 3 – 8.

Wade, L.K. and D.N. McVean, 1969. Mt. Wilhelm Studies 1. The alpine and subalpine vegetation. Res. School Pacific Studies. Dept. Biogeograhy and Geomorphology. Publication BG/1. Australian National Univ., Canberra.

Wood, A.W., 1983. Soil types and traditional soil management in the Purari catchment. This volume, Chapter I, 5.

10. Limnology of the Purari basin
Part 2. The delta

T. Petr

1. Introduction

The Purari Delta takes its origin some 35 km inland of the Gulf of Papua. Three distributaries branch at the head of the delta; in order of water volume, these are: the Ivo – Urika in the centre of the delta, the Purari in the east, and the Wame – Varoi in the west (Fig. 1). The last branch enters into the Pie/Port Romilly tidal river (Fig. 2) bordering the Purari Delta in the west with a side branch (Varoi) meandering south (Figs. 3 – 5). Short tributaries (Heperi River and Muro Creek) enter the Purari distributary from the east, and Eia River enters the Pie River from the northeast.

The delta area is covered by lowland tropical forest, freshwater and brackish swampy vegetation, and mangroves; its plant cover is dealt with in more detail in this book in chapters written by Cragg and Paijmans. Geomorphology of the delta is described by Thom and Wright (this volume), while some aspects of sediment elemental composition are considered by Petr and Irion (this volume).

The limnology of the delta was studied during a number of expeditions between 1976 and 1979.

2. Results

2.1. The delta (Fig. 5)

Above the tidal reach, the limnology of the delta is similar to that of the Purari at Wabo further upstream (see Petr, Part 1, this volume). It is clear that the Purari River water is not greatly modified by the addition of the Aure River water. The major distributary, Ivo – Urika, which carries 60 – 70% of

Petr, T. (ed.) The Purari – Tropical environment of a high rainfall river basin
© *1983, Dr W. Junk Publishers, The Hague / Boston / Lancaster*
ISBN-13: 978-94-009-7265-0

Fig. 1. An aerial view of the effluence of the Purari Delta: the Ivo arm (river flow from the right = north, to the left = south) makes a bend and meanders into the left top corner; the Purari arm (front); the Wame (small patch of water in the right top corner).

Fig. 2. The Port Romilly/Pie River between Baimuru and the Gulf of Papua.

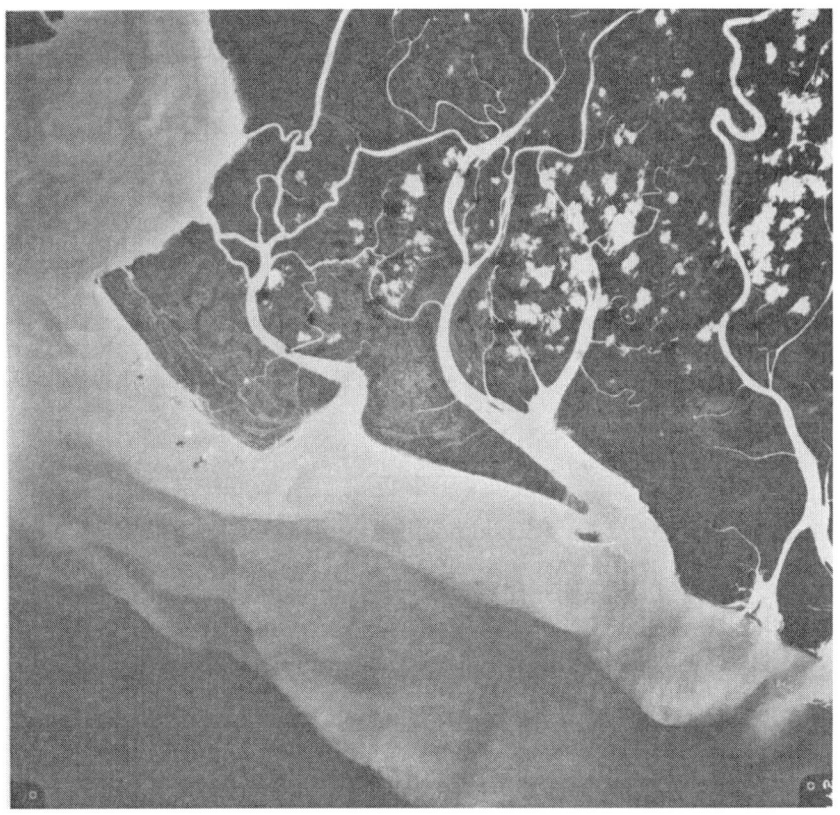

Fig. 3. The Purari Delta – western fringe. Top left: mouth of the Port Romilly/Pie River. Kaimari Island (rectangular). Right of centre: Varoi estuary. Further right: Urika estuary. Note the plume along the coast. (Photographed 20.5.1974) (Courtesy of the National Mapping Bureau, P.N.G.)

the high stage discharge of the Purari (Thom and Wright, this volume), has a fluvial character throughout the delta down to the Urika mouth opening into the Gulf of Papua. There is only a very slight increase in sodium and chloride in the Urika mouth as compared with the Mapaio station (Table 1), indicating slight sea mixing. The Ivo – Urika has the same ionic dominance as the Purari at Wabo, i.e., $Ca > Mg > Na > K$, and $HCO_3 > SO_4 > Cl$.

The chemistry of Varoi estuary west of Urika is much more influenced by seawater and semi-diurnal tides. Total suspended solids (TSS) concentrations are much less than those recorded for Ivo River at Mapaio and Urika estuary, confirming the dilution by seawater input. The minimum conductivity in

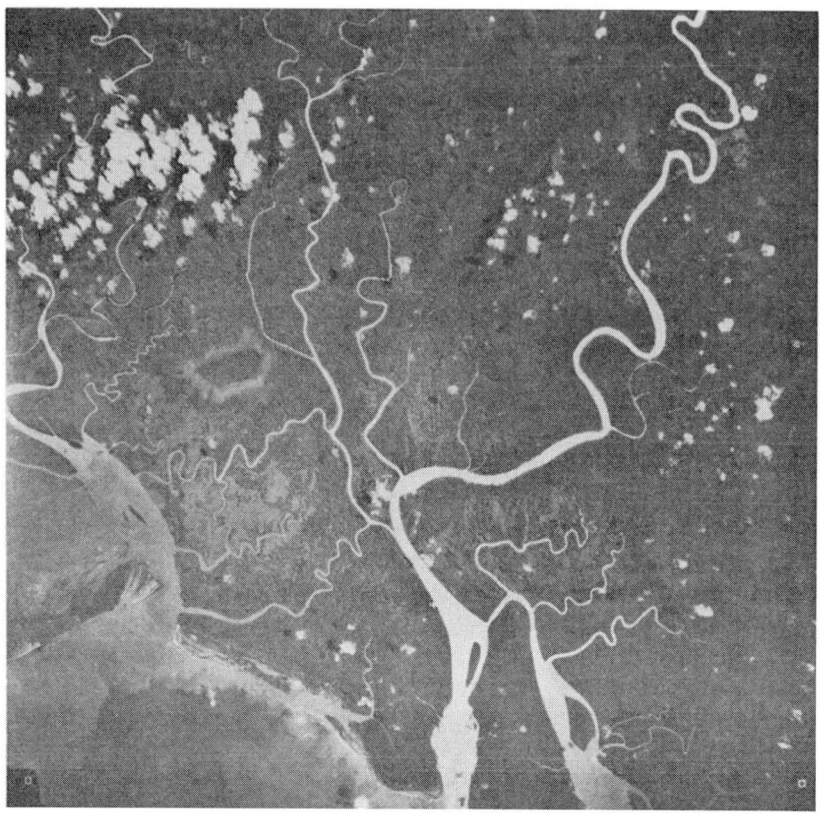

Fig. 4. The Purari Delta – eastern fringe. Left: Panaroa estuary. Center right: Purari distributary with the Aievi arm (left) and Alele arm (right). (Photographed 20.5.1974) (Courtesy of the National Mapping Bureau, P.N.G.).

Varoi mouth is considerably higher than maximum conductivities of the Ivo – Urika distributary. The chemical character of the Panaroa estuary situated east of Urika is similar to that of Varoi. A longitudinal profile of 27.1.1978 shows no seawater admixture to surface water 3 km upstream the Panaroa estuary, although the widespread presence of mangroves there suggests that the salinity of interstitial water is high. In the Panaroa mouth the conductivity of water measured on three different occasions ranged from a fluviatile 135 μS/cm to a brackish 7200 μS/cm, confirming that the estuary is chemically highly dynamic. The presence of freshwater there coincided with the highest TSS, 1277 mg/l, suggesting that the river was in full flood.

Temperature stratification was observed to exist at a number of stations. In

182

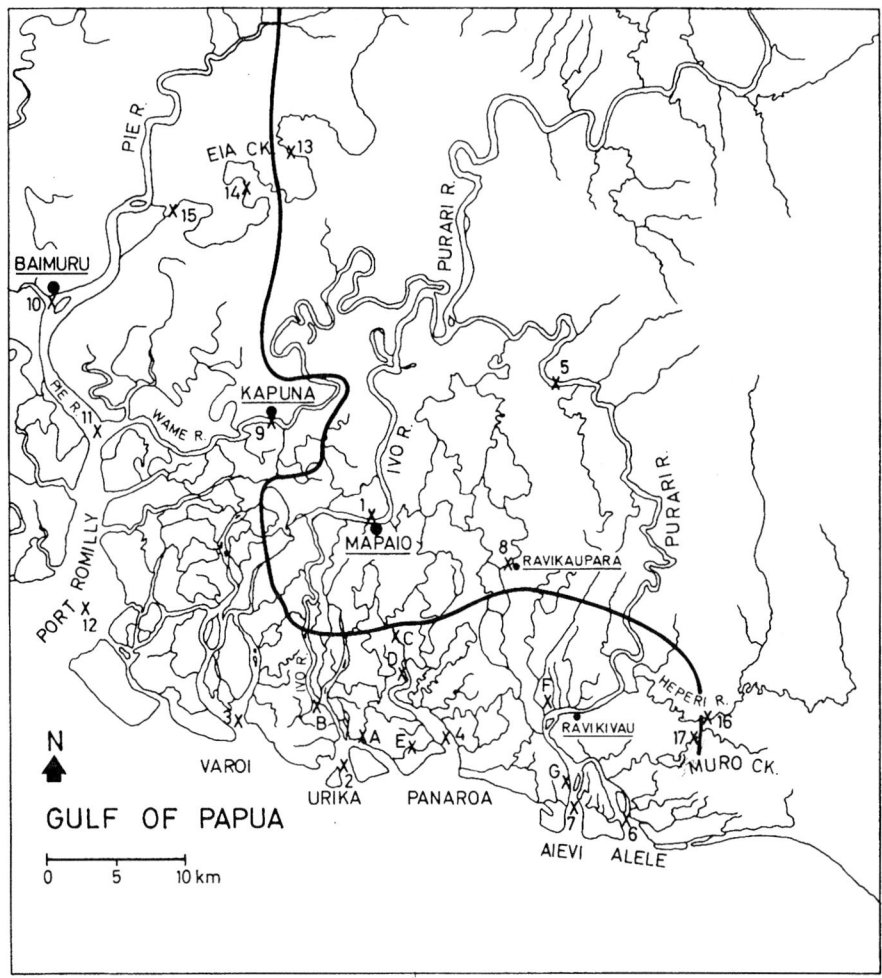

Fig. 5. The Purari Delta. Numbers refer to water sampling stations, letters refer to the interstitial water samples. The line marks the upper limit of the mangrove *Sonneratia lanceolata* distribution, which is taken as a boundary between the brackish and freshwater reaches of the delta.

Port Romilly (6.4.1978) the surface water temperature was 28.7°C, while at 3 and 6 m depth it was 29.3°C. A similar discontinuity was observed on the same day in Varoi estuary (surface: 26.8°C; 3 m: 28.8°C), and in a tidal channel 6 km inland (surface: 27.2°C; 3 to 7.5 m depth: 29.4°C). In Varoi estuary, underlying the freshwater advancing towards the sea, was a strong upstream bottom current, indicating that a salt wedge was creeping upstream as a bot-

Table 1. Purari Delta: Ivo River, Urika, Varoi and Panaroa estuaries.

Station	Ivo River at Mapaio		Urika estuary (mouth)		Varoi estuary (mouth)		Panaroa estuary[a]		
Number of station on map	1		2		3		4		
Number of samples	10		5		4		3		
	Range	Mean	Range	Mean	Range	Mean	1.	2.	3.
Temperature °C	24.2 – 26.5	24.96	25.9 – 26.8	26.35	24.8 – 28.4	26.73	26.8	27.4	30.0
Turbidity J.T.U.	25 – 335	191.3	150 – 250	207.5	7 – 250	79.25	140	39	54
pH	7.4 – 8.1	7.69	7.25 – 7.95	7.62	7.00 – 7.75	7.38	8	7.5	7.5
Conductivity μS/cm	120 – 150	136.2	127 – 150	140.4	2520 – 12000	7667.5	127	7200	12200
TSS	15 – 1188	461.7	147 – 1277	528.8	18 – 101	60.75	294	58	282
Total hardness $CaCO_3$	53 – 74	62.1	45 – 66	56.6	288 – 1643	870.3	62	809	1544
Alkalinity $CaCO_3$	59 – 77	67.0	55 – 74	62.0	66 – 80	73.0	63	63	77
Ca	18 – 24	20.6	14 – 21	16.2	36 – 194	102.0	21	67	144
Mg	2 – 4	2.6	2 – 4	2.8	48 – 281	124.5	2	156	288
Na	2 – 4	3.2	2 – 4	3.6	360 – 2442	1401.3	3	1332	2442
K	0.7 – 2.0	1.0	0.9 – 1.0	0.98	17 – 90	53.75	0.8	50	70
Total cations me/l	1.20 – 1.69	1.42	1.08 – 1.53	1.32	21.74 – 127.83	79.69	1.41	75.39	139.05
HCO_3	72 – 93	80.7	68 – 90	75.6	81 – 97	91.0	77	77	94
Cl	0 – 5	1.2	0 – 4	2.0	694 – 4338	2626.75	3	2194	4002
SO_4	0 – 7	2.4	0 – 8	4.0	74 – 570	308.5	3	273	375
Total anions me/l	1.31 – 1.61	1.54	1.10 – 1.64	1.38	22.45 – 126.00	79.06	1.41	69.3	122.25
SiO_2	12 – 16	13.8	10 – 15	13.0	10 – 14	11.75	12	13	9

[a] Panaroa estuary: 1 = 3 km inland from estuary mouth; 2 = estuary mouth; 3 = sea between Panaroa and Urika, 1 – 2 km offshore. Sampled on 27.1.1978.

Where not otherwise stated, values in mg/l.

tom layer. In a number of measurements the temperature gradient was accompanied by a pH gradient, as observed in the Panaroa estuary (surface: 25.7°C, pH 7.53; 3 m: 26.4°C, pH 8.20). In the Pie/Port Romilly mouth, vertical gradients of temperature and salinity were recorded in the deep channel at flood tide on 29.7.1979. The surface values were 24.6°C and 1.1‰ respectively, and gradually increased to 26.0°C between 9 and 14 m depth, and to 5.2‰ (5 m), 8.1‰ (9 m) and 9.3‰ (14 m). At ebb tide (26.7.1979) a shallow part of the Pie estuary was found well mixed, with temperature from the surface to the bottom uniformly at 26.4°C and salinity of 1.0‰.

A longitudinal gradient extending from the Urika mouth into the Gulf of Papua just outside the plume was recorded on 4.12.1979 (Table 2). The impact of the large input of freshwater through the Purari can thus be traced to a considerable distance offshore, and the river undoubtedly must have a major impact on the biota of coastal waters.

During the major rainy season, which in the central coastal area of the Gulf of Papua results from the impact of southeasterly tradewinds between April and October, the cooling impact of a heavy rainfall was observed in Port Romilly estuary on 27.8.1979 when surface water temperature was 25.8°C, which corresponded to the temperature of rainwater. The heavy rainfall in this month also resulted in a lowest temperature of 26.6°C (26.8.1979) for the surface water at Baimuru, where the range for other observations was 28.2°C to 29.0°C.

The Purari distributary in the east of the delta has in its upper section (some 30 km upstream from its mouth) a chemical composition similar to that at Wabo, or to the Ivo – Urika distributary at Mapaio (Table 3). The Purari distributary discharges into the sea through two outlets, the eastern Alele and western Aievi passages. A larger volume of water appears to enter the Gulf through Alele, although to confirm this requires more hydrological observations than available at present. The conductivity indicates seawater admixture in both passages, although on one (19.3.1978) out of two measurements the water discharged through Aievi was almost of a fluviatile character, with con-

Table 2. Longitudinal gradient Urika mouth – Gulf of Papua.

	Temperature °C	Secchi disc transparency (cm)	Salinity ‰
Urika mouth (outside sandbar)	25.7	7	8.5
Open sea, 10 m inside plume	27.8	30	13
Open sea, 10 m outside plume	31.0	70	16
Open sea, 5 km distant from plume	30.8	95	22.5

Table 3. Purari Delta: Purari distributary, a channel, Wame River.

Station	Purari distributary		Aievi		A channel at Ravikaupara	Wame River at Kapuna	
	30 km upstream Alele						
Number of station on map	5		6	7	8	9	
Number of samples	4		2	2	2	7	
	Range	Mean	Mean	Mean	Mean	Range	Mean
Temperature °C	26.2 – 26.5	26.37	26.1	26.1	25.6	24.7 – 27.3	26.06
Turbidity J.T.U.	50 – 500	223	71.5	85.0	125	5 – 250	83.0
pH	7.20 – 7.50	7.40	7.4	7.5	7.15	7.35 – 8.30	7.70
Conductivity $\mu S/cm$	120 – 138	128	2000	3277	141	124 – 258	155.86
TSS	35 – 665	217	94.5	95	271	13 – 323	126.86
Total hardness $CaCO_3$	60 – 64	61.8	201	392	60.5	61 – 90	71.43
Alkalinity $CaCO_3$	64 – 66	65	68	65	62	60 – 90	73.57
Ca	19 – 21	20	30.5	41	18.5	20 – 29	23.71
Mg	3	3	30.5	70.5	3	2 – 4	3.14
Na	3 – 6	4	336	714.5	7	2 – 10	4.0
K	0.9 – 1.0	0.98	22.0	30.0	1.5	0.7 – 1.6	0.98
Total cations me/l	1.30 – 1.52	1.42	19.23	39.69	1.47	1.32 – 2.26	1.63
HCO_3	78 – 81	79.5	81	84.5	75	72 – 105	83
Cl	0 – 6	1.5	576	1235.5	7	0 – 14	3.57
SO_4	3	3	58	164	6	0 – 4	2.57
Total anions me/l	1.38 – 1.53	1.42	18.7	39.7	1.51	1.29 – 1.91	1.53
SiO_2	14 – 20	17	14.5	14.5	15	12 – 19	14.5

Where not otherwise stated, values in mg/l.

ductivity of 155 μS/cm, and with only slightly elevated concentrations of Cl = 9 mg/l and Na = 8 mg/l. Corresponding values for 24.11.1978 were μS/cm 6400, Na = 1421 mg/l, Cl = 2462 mg/l.

Although the Purari arm is the second major distributary, the volume of water discharged through it is considerably less than that passing through the Ivo – Urika (Thom and Wright, this volume). At low water discharge, as observed for example on 5.12.1979, a section of this distributary upstream of Ravikivau, some 10 to 15 km from the river mouth, was found virtually stagnant. The water was rich in suspended sediments, as indicated by a low Secchi disc transparency of 21 cm. The surface water temperature of 29.8°C was much higher than the mean of the Ivo distributary, suggesting that little freshwater input was reaching this region from upstream. The high water temperature might have resulted from adsorption of heat by suspended particles. The water had a fully riverine character, and isolated groups of the mangrove *Sonneratia lanceolata* along the banks indicated an infrequent upstream saltwater penetration. The water was green with phytoplankton, again suggesting that for some time this reach has not been flushed out. This zone may function as an entrapment zone, usually positioned at the upstream end of the freshwater – saltwater mixing zone of an estuary (Arthur and Ball 1978), and characterized by accumulation of suspended materials at concentrations exceeding those found either upstream or downstream. Such entrapment zones are known elsewhere to be highly productive. In the Purari, its existence is always threatened by the highly eratic river discharges, and in such an environment the entrapment zone has little chance of reaching a climax, as after each flushing out, the system has to completely regenerate, since the previous physico-chemical environment and part of the organisms (especially the plankton) are washed downstream into the sea. More studies are certainly needed to better understand the functioning and importance of the entrapment zone in the Purari deltaic system.

In the eastern part of the delta, saltwater traces (Na = 10 mg/l, Cl = 12 mg/l, μS/cm = 141) were detected in water samples collected from a sluggish channel 20 km inside the delta at Ravikaupara (Fig. 5). This approximately coincides with the upper limit of *Sonneratia lanceolata* and *Pandanus* sp. distribution. These two plants appear to be better indicators of the maximum extent of seawater penetration upstream, and of occasionally brackish water conditions (Fig. 6), than surface water samples from distributaries and channels. While it is possible to map in the Purari Delta the upper boundary of mangroves (Petr and Lucero 1979), mapping upper limits of saltwater penetration from water sample analysis would require the presence of continuous recording stations positioned at different depth levels. Another alternative is mapping interstitial water salinities. Interstitial (pore) water has on

Fig. 6. *Pandanus* sp., a typical plant of the brackish/freshwater boundary.

the average considerably higher salinity than channel water bypassing mangroves (see also below).

Amongst the three distributaries, the Wame River bordering the delta in the north, carries the smallest volume of water. At Kapuna Mission, 15 km east of the Wame/Pie confluence, the river still has some 150 cm tidal difference and a brackish character. There is typical mangrove vegetation there, with *S.lanceolata*, *Rhizophora*, *Bruguiera*, *Nypa* and sago palms common (Figs. 7 – 9). Amongst the surface water samples, only one out of seven had elevated levels of Na, Cl and conductivity (10 mg/l, 14 mg/l, 258 μS/cm respectively). Due to the sluggish water flow and frequent stagnation periods due to tidal backup of inflowing freshwater, the TSS is on the average low, lower than

Fig. 7. Bruguiera (front) and *Rhizophora* (background), typical tree mangroves of the Purari Delta.

Fig. 8. Excoecaria and *Nypa* palm downstream of Kapuna, Wame River.   189

Fig. 9. Sago palms (*Metroxylon sagu*) near Kapuna.

Fig. 10. Sonneratia lanceolata on the Pie River, some 60 km inland. (Photo G. Irion).

190

mean values for the Purari arm, Ivo River (at Mapaio) and Urika estuary.

The Pie River, called in its lower estuarine reach Port Romilly, is a relatively long tidal river. Its mouth is 5 km wide, but its still tidal upper reach some 60 km inland is a narrow creek of 4 m width, closed over head completely by branches of *S. lanceolata* (Fig. 10). A longitudinal profile of water samples collected 19 – 20.10.1978 gives for the upper station a high TSS of 241 mg/l, but a low conductivity of 67 μS/cm, and a pH of 6.50, indicating that the Pie originates in a swamp. In spite of the low conductivity, the sodium and chloride values are above the freshwater average (Table 4). Part of the fine mud of the bottom and river banks is kept in suspension by tidal water movements, and does not seem to settle down during the short stagnation periods.

There is a longitudinal salinity and temperature gradient between the upper Pie and the sea, interrupted by freshwater input through the Wame River (Table 5). In October 1978, 20 km downstream of the upper station (Station 2), the conductivity reached 660 μS/cm; at Baimuru (27 km from the Pie/Port Romilly mouth), 5400 μS/cm; at Pie/Wame confluence, 930 μS/cm; and in the Pie/Port Romilly mouth, 15500 μS/cm. Temperature increased from 26.3°C at the top station, to 29.3°C twenty km downstream. The Wame water entering the Pie had a temperature identical to that of the top Pie station. In November 1979, a similar temperature gradient was observed, with temperature rising from 26.8°C at the top to 30.1°C at Baimuru, this corresponding to a gradual increase in salinity from zero to 7.5‰, and Secchi disc transparency from 7 cm to 80 cm. There is an opposite gradient of TSS, with highest values at the upper stations, gradually decreasing towards the mouth, with the exception of Pie/Wame confluence, where the TSS rises due to the increased input of sediments carried by the Wame. Dissolved silica values are lowest at the top and bottom stations: swamp water, largely determining the water quality of the top stations, is poor in silica. Similarly poor is the seawater, which is probably diluting the silica brought in by the river. An increased algal, specifically diatom, productivity might also be partially responsible for the lowered silica concentrations in the mouth of the Pie River.

At Baimuru (Fig. 11) the water was usually strongly brackish, although on some occasions the conductivity approached freshwater values (Table 4). The reason for this was probably an especially heavy rainfall in the upper Pie River and its tributary Eia Creek, and an increased input of freshwater through the Wame during flood conditions.

Amongst dissolved nutrients, the mean nitrate-nitrogen concentrations of 45.9 μg per litre (Table 6) of the deltaic brackish reach are considerably lower

Table 4. Pie/Port Romilly River: ranges and means.

Station	Pie River at Baimuru		Pie/Wame confluence		Pie/Port Romilly mouth	
Number of station on map	10		11		12	
Number of samples	8		2		6	
	Range	Mean	Range	Mean	Range	Mean
Temperature °C	26.6–29.5	28.83	26.3–27.9[a]	27.38	25.8–30.9	28.20
Turbidity J.T.U.	2–250	84.29	60	60	7–50	24.17
pH	7.00–7.60	7.20	7.35–7.60	7.48	7.30–8.00	7.54
Conductivity μS/cm	136–12100	3663	149–930	539.5	1600–30000	16267
TSS	20–1075	214.75	51–73	62	13–63	32
Total hardness $CaCO_3$	24–1617	665.13	69–100	84.5	736–4245	2289
Alkalinity $CaCO_3$	21–111	64.75	62–77	69.5	76–94	85.2
Ca	3–194	94.38	6–38	22	89–361	272.8
Mg	4–275	135.5	2–3	2.5	125–813	390.8
Na	16–2553	1187.5	18–144	81	1221–8048	3814.2
K	2–100	53.88	1–7	4	93–114	165.16
Total cations me/l	1.23–135.06	68.63	1.39–8.44	4.92	70.38–440.66	212.05
HCO_3	25–136	76.63	28–76	52	93–114	99.5
Cl	23–4496	2120	29–219	124	2248–11180	6380.5
SO_4	8–515	201.12	5–30	17.5	150–1665	741.83
Total anions me/l	1.32–133.76	64.48	1.38–8.05	4.72	71.32–352.15	197.10
SiO_2	8–13	11.31	10–13	11.5	5–14	9.0

[a] $n = 4$.
Where not otherwise stated, values in mg/l.

Table 5. Pie/Port Romilly River longitudinal profile.

Station	1	2	3	3	4	5
Date sampled	20.10.	20.10.	20.10.	19.10.	19.10.	19.10.
Time sampled	09.15	10.15	14.20	14.00	07.40	08.30
Temperature °C	26.3	29.3	29.5	29.2	26.3	27.3
Turbidity J.T.U.	700	600	50	70	60	25
pH	6.50	7.00	7.20	7.10	7.35	7.30
Conductivity $\mu S/cm$	67	660	5400	5400	930	15500
TSS	241	171	20	68	73	21
Total hardness $CaCO_3$	18	91	651	802	100	2821
Alkalinity $CaCO_3$	22	44	64	68	62	78
Ca	4	18	117	167	38	666
Mg	2	11	88	94	2	281
Na	5	97	777	777	144	2553
K	0.5	7	38	40	7	125
Total cations me/l	0.58	6.20	47.78	48.35	8.44	170.63
HCO_3	27	54	83	78	76	95
Cl	5	167	1616	1616	219	5473
SO_4	3	24	30	60	30	705
Total anions me/l	0.66	6.10	47.57	48.11	8.05	170.65
SiO_2	9	15	13	13	13	9

Stations: 1. Upper end of the Pie R. 60 km from the mouth; 2. Pie R. 40 km upstream; 3. Pie R. at Baimuru; 4. Wame/Pie confluence; 5. Pie/Port Romilly mouth. Samples were collected in 1978.
Where not otherwise stated, values in mg/l.

Fig. 11. Baimuru, Port Romilly/Pie River. Research and transport vessel ('river truck') used for surveying the Purari between Hathor Gorge and Gulf of Papua. (Photo G. Irion).

Table 6. Dissolved nutrients (μg per litre) at brackish stations of the Purari Delta (25 – 26.8.1978) (recalculated from Viner 1979).

	Number of stations	Range	Mean
Conductivity μS/cm	7	1150 – 6200	3290
PO_4-P	7	0.1 – 38	14.6
NH_4-N	7	15 – 47	30.5
NO_3-N	7	22 – 92	45.9
Total P dissolved	7	16.2 – 61	42.8
Total N dissolved	3	136 – 205	165.7
Particulate P	2	67 – 96	81.5
Particulate N	2	331 – 403	367.0

than the mean of 220.6 μg/l for the freshwater reach of the Purari (see Part 1 of this study in this volume). Viner (1979) suggested that this may be the result of nitrogen utilization by microorganisms, which reach relatively high concentrations in the brackish reach of the delta (Paerl and Kellar 1980). He also found the phosphate concentrations in the delta low to moderate by standards elsewhere, and only about two thirds of concentrations recorded from the freshwater reach of the Purari, and he expressed the opinion that phosphorus would be a limiting nutrient were it not for the reservoir of adsorbed phosphate on sediments (Viner 1982).

Five samples of zooplankton were collected from the Pie River and Varoi estuary (conductivity 11300 and 30000 μS/cm respectively, corresponding to approximately 6.5 and 18.6‰).These samples were examined by Bayly (1980). The dominant species were *Paracalanus aculeatus*, *Calanopia australica*, *Bestiola similis*, *Acartia baylyi*. Other planktonic species were *Labidocera moretoni*, *Tornatus barbatus*, *Acartia pacifica*, *Dioithona ?oculata*, *Coryaceus* sp., *Tenagomysis* sp., *Centropages* sp., adult Sergestidae, decapod (including brachyuran) larvae, lamellibranch and gastropod veliger larvae, Cirripedia cypris larvae, planktonic isopods and chaetognaths. Bayly pointed out the close similarity between the copepod assemblage found in the Purari and Pie river estuaries and that from Moreton Bay in Queensland (Australia). The only plankton sample from Urika (conductivity 151 μS/cm) had the true freshwater planktonic species *Calamoecia ultima*, *Mesocyclops leuckarti* and *M. hyalinus*, as well as brachyuran zoea larvae.

2.2. Short tributaries of the Pie River and Purari Delta

Eia Creek entering the Pie River from the northeast has a small catchment in the hills reaching a maximum elevation of 300 m a.s.l. These hills are drained by numerous creeks which enter a large swamp at the base of the hills. From this, the Eia Creek emerges and continues meandering through tropical lowland forest which gradually changes into mangrove forest on approach to the Pie River. At a higher water discharge (11.2.1976, Table 7), the water at the top station had a pronounced swamp character, with a low pH, conductivity and silica content. Low water discharge (20.10.1978) allows for a higher concentration of sodium and chloride, conductivity and dissolved silica. The river water has a blackish colour and low turbidity. *Myriophyllum*, *Azolla*, and *Nymphoides* are common aquatic plants at the top station.

There is a gradual increase in conductivity and chemicals determining the ionic concentration on approach to the confluence of the Eia Creek with the

Table 7. Eia Creek longitudinal profiles as compared with Pie at Baimuru (Station 10).

Number of station on map	13	14	15	13	14	15	10
Date sampled	11.2.1976	ditto	ditto	20.10.1978	ditto	ditto	ditto
Temperature °C	n.d.	n.d.	n.d.	27.5	28.3	30.4	29.5
Turbidity J.T.U.	3	2	2	n.d.	n.d.	n.d.	n.d.
pH	6.6	6.8	7.1	6.9	7.0	7.0	7.2
Conductivity μS/cm	68	95	5900	200	231	3600	5400
TSS	137	515	181	7	36	46	20
Total hardness $CaCO_3$	30	42	169	23	64	453	651
Alkalinity $CaCO_3$	30	40	62	34	34	44	68
Ca	9	12	48	20	16	89	117
Mg	2	3	119	4	6	56	88
Na	4	6	1154	11	11	555	777
K	0.6	0.6	52	2	2	23	38
Total cations me/l	0.78	1.11	53.79	0.96	0.77	33.75	47.78
HCO_3	37	49	76	42	42	61	83
Cl	2	4	1615	0	2	1095	1616
SO_4	7	8	298	2	4	78	30
Total anions me/l	0.82	1.10	53.26	0.90	0.75	33.51	47.57
SiO_2	11	14	15	20	18	12	13

Stations: 13. Upper end of navigable reach; 14. middle section; 15. 1 km above the confluence with the Pie R.; 10. Pie River at Baimuru. Where not otherwise stated, values in mg/l; n.d. = not determined.

Pie River. About 1 km upstream from the confluence (Station 15), Eia Creek has a pronounced brackish character and lies in a tidal zone. About 18 km upstream the Eia/Pie confluence (Station 14), *Nypa* palms are still present, indicating a routine penetration of seawater into this area. There is a slight increase in conductivity and sodium and chloride content recorded both for the low and high water discharge.

The usually low discharge through the Eia Creek probably has only a negligible dilution impact on the Pie River, but during periods of extremely high rainfall its impact is substantial, when, in combination with water entering from the upper Pie and from the Wame, it results in almost freshwater surface water conditions at Baimuru (Table 4, Station 10).

Heperi River and Muro Creek are two short, lowland rivers draining the area between the Vailala River and the Purari arm of the delta. The Heperi (approximately 40 km long) drains hills (max. elevation 118 m) with marine and terrestrial clastic sedimentary rocks and pyroclastics, while Muro Creek (approximately 20 km long) originates largely in swamps. This is reflected in their chemistry (Table 8), with the Heperi having a higher ionic content than Muro, and Muro having a lower pH than the Heperi (Table 7). The Heperi

Table 8. Heperi River and Muro Creek (ranges are given as only two samples were collected for each station).

Station	Heperi	Muro
Station number on map	16	17
Temperature °C	26.5 – 27.8	26.1 – 28.5
Turbidity J.T.U.	50 – 54	16 – 50
pH	7.40 – 7.60	6.80 – 7.10
Conductivity μS/cm	190 – 310	104 – 150
TSS	31 – 80	5 – 22
Total hardness $CaCO_3$	67 – 123	34 – 40
Alkalinity $CaCO_3$	85 – 136	38 – 47
Ca	17 – 29	6
Mg	6 – 12	5 – 6
Na	12 – 21	10 – 14
K	2 – 4	2
Total cations me/l	1.95 – 3.50	1.16 – 1.41
HCO_3	104 – 166	44 – 51
Cl	6 – 13	10 – 14
SO_4	13 – 25	10 – 11
Total anions me/l	1.49 – 3.60	1.19 – 1.44
SiO_2	14 – 19	19

Where not otherwise stated, values in mg/l.

Table 9. Interstitial and free water chemistry in the tidal reach of the Purari Delta (December 1979).

Station	A		B		C		D		E		F		G	
	F	I	F	I	F	I	F	I	F	I	F	I	F	I
Turbidity J.T.U.	200	50	150	350	150	450	40	45	40	150	250	300	350	15
pH	6.90	7.00	7.10	7.10	7.40	7.20	7.30	7.30	7.40	7.30	7.80	7.50	7.40	7.20
Conductivity μS/cm	150	3200	140	300	140	430	170	4600	3600	3600	540	1000	3800	7400
TSS	468	248	481	414	238	330	118	83	121	175	308	337	816	64
Total hardness $CaCO_3$	52	447	57	126	58	142	60	689	447	443	61	88	392	732
Alkalinity $CaCO_3$	57	39	57	84	61	69	59	39	73	86	63	59	63	69
Ca	16	56	19	40	19	42	19	111	33	56	10	14	33	67
Mg	3	75	2	6	2	9	3	100	88	75	8	13	75	138
Na	13	666	2	2	4	38	8	999	666	77	71	333	888	1776
K	2	10	0.4	0.5	0.6	0.7	1	10	40	30	3	10	20	50
Total cations me/l	1.67	38.17	1.29	2.56	1.34	4.50	1.58	57.49	38.85	43.51	4.36	16.50	46.97	93.17
HCO_3	65	43	65	93	69	79	67	38	79	96	67	62	67	65
Cl	0	1285	0	9	0	85	0	1927	1178	1286	101	535	1606	2997
SO_4	3	60	3	27	5	18	4	140	140	260	17	20	60	340
Total anions me/l	1.16	38.23	1.16	2.39	1.27	4.09	1.22	57.96	37.48	43.28	4.35	16.57	47.70	92.80

Stations: A – G as marked on Fig. 1.
Where not otherwise stated, values in mg/l.
F = free water; I = interstitial water.

lithology is reflected in higher calcium and bicarbonate concentrations, and a higher pH. The swamp drainage of Muro results in brown coloured water of higher transparency and lower TSS. The nearness of the coast, with rainfall precipitation containing Na and Cl (see Part 1), is probably largely responsible for the slightly elevated values in the Muro Creek, while the higher sulphate value for the Heperi is probably due to seawater input. Silica values of both rivers are similar to those of Eia Creek, in which values decrease towards the Pie.

2.3. Interstitial water

In connection with investigations of sago palm (*Metroxylon sagu*) tolerance to water salinity, interstitial (pore) water was collected from holes drilled with a geological auger next to selected sago palms fringing tidal channels of the delta (Fig. 5). It was established that in the Purari Delta, sago grows in swamp soils with interstitial water salinities of up to 7.5‰ (Petr and Lucero 1979).

The water chemistry of free water in channels bypassing the investigated sago plants is compared with that of interstitial water collected from drill holes (Table 9). Numbers in the table correspond to numbers on the map. In all situations, the ionic concentration of interstitial water was higher than that of the water bypassing the area. The mean free water salinity was only about 40% of that of the interstitial water. It is known that post-depositional modifications in deltaic sediments, associated with changes in redox conditions, mineral dissolution, or organic complexation, can increase the concentration of dissolved solids in the interstitial water (Burton 1978). The conductivity values in Table 9 correspond to salinities of up to 4‰. In the Purari Delta, *Metroxylon sagu* was found to inhabit a brackish water/mangrove environment, growing next to *Nypa* palm, *Rhizophora*, *Bruguiera*, *Heritiera*, *Excoecaria* and other mangrove plants.

3. Discussion

The Purari Delta forms part of a deltaic system extending from the Purari distributary in the east, to Omati River in the west (Fig. 12). Much of this area is covered in mangroves, which penetrate deeper inland wherever the freshwater input is small. Two major rivers influence this deltaic system: the Purari in the east, with its mean discharge of 2360 m³/sec, and the Kikori in the central west, with an approximate 1500 m³/sec discharge.

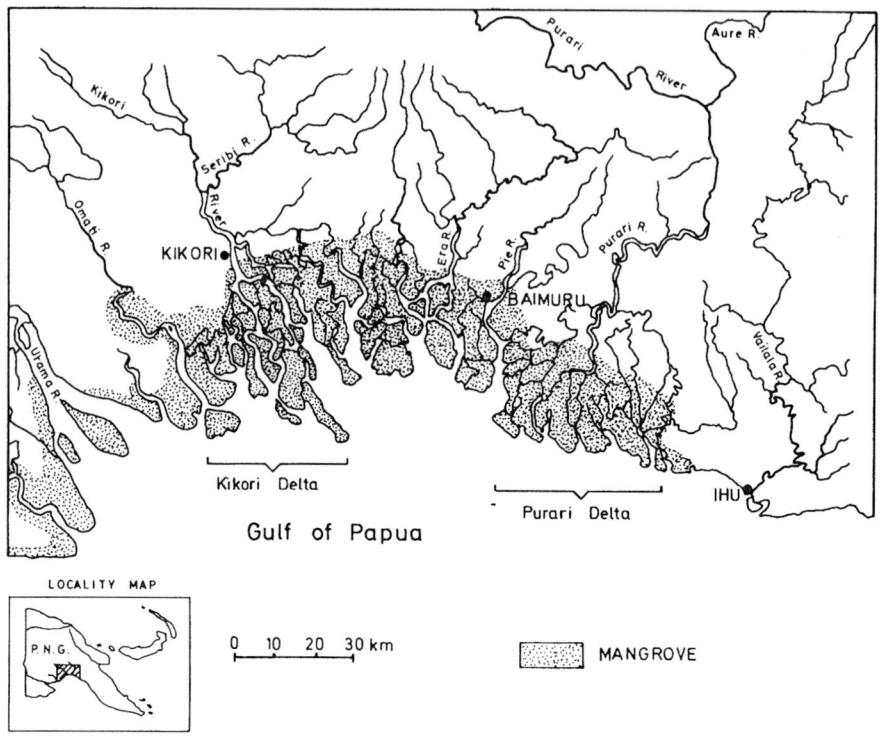

Fig. 12. Kikori – Purari deltaic system, with an approximate area of mangrove distribution.

The maximum (60 km) inland saltwater penetration in the Purari/Pie system, recorded in the Pie River bordering the Purari Delta in the west, is the result of a small input of freshwater into this tidal river which receives the smallest proportion of the Purari discharge through the Wame distributary. As a result, the west of the delta is much more under the influence of the sea than the central and eastern parts. In the Ivo – Urika and Purari distributaries, the upper limit of brackish water is 10 – 12 km from the coast.

Interstitial (pore) water of river banks, being on average 40% more saline than that bypassing in channels and distributaries, clearly determines the upstream extent of mangrove.

Vertical gradients of temperature, pH and salinity show the Purari estuaries to be moderately stratified, with some up-estuary flow in deeper layers. Complete tidal mixing frequently takes place in the distributaries and tidal channels with lower freshwater discharge.

The seasonality in water discharge which is largely determined by southeasterly winds blowing between April and October and bringing heavier rains than during the rest of the year, exerts some influence on the overall pattern of aquatic productivity in estuaries and the Gulf of Papua, as indicated, for example, by prawn breeding and migrations (Frusher and Gwyther, this volume), and by the temporary existence of entrapment zones in the Purari distributary and Pie River. However, unseasonal floods are frequent and affect the deltaic and coastal environments at any time of year.

The outer limit of the estuarine realm (*sensu stricto* Levinton 1982) of the Purari is not known. A longitudinal profile measured on one occasion from the Urika mouth, across the plume's outer limit into the Gulf waters some 8 km offshore, gave a maximum reading of 22‰ at the last station. Frusher (1980), during his survey of juvenile prawn distribution along the Purari Delta, recorded in November 1979 salinities of 30‰ and 28‰ some two kilometres offshore at a station facing Alele Passage. These values were clearly the result of a westward coastal current carrying more saline water from Orokolo Bay, which receives only the Vailala River. At the same time, the discharges through the Aievi and Alele passages were low. Other salinity readings from 3 to 5 km distance along the coast between Alele and Cape Blackwood (the tip of the southernmost island in front of the Kikori River) were less than 25‰, with a mean 18.83‰ ($n = 21$). With the Purari River constituting approximately 20% of the total river discharge into the Gulf of Papua, its impact on the coastal environment cannot be considered as negligible.

Questions have been asked about the possible impact on the deltaic and coastal environments of large scale development projects in the Purari catchment. Amongst the more recent changes is the intensification of agriculture in the highlands, with proposals for resettlement of a large number of people to areas of the upper margin of the lowland tropical forest. Wood (this volume) has pointed out the danger of intensive erosion and loss of soil fertility in forest margins if intensive agriculture is introduced there. The sediment load in rivers would increase, which by itself would probably make little difference with respect to the present deltaic productivity and already high sediment load. If dams were to be constructed on catchment rivers, the silt would get trapped in the reservoirs and the impact of the lower sediment discharge downstream would increase the erosion of the delta (Pickup, this volume). A hydroelectric dam at Wabo on the lower Purari, as planned for the future, would lead to a permanent regulation of the freshwater discharge, resulting in a shift of the present mangrove/freshwater swamp boundaries, and in changes in physico-chemical character of the river water, which would affect the aquatic life. Davies (1979) noted that as a result of the presence of the Kariba,

Kafue and Cahora Bassa dams on the Zambesi/Kafue river system in eastern Central Africa, the whole sea frontage of mangroves exhibited a 1 to 400 m dieback. Other negative developments are known to have taken place in coastal fisheries. Such impacts are associated with changes in water quality, especially in reduction in the input of suspended solids and nutrients. In the Purari, the permanently increased water transparency should have a more profound impact on the biota than losses of particle-bound nutrients in the reservoir. Dissolved nutrients would probably still remain at a similar level as at present (Viner 1979). Their utilization and recycling through plankton would probably be enhanced under clearer water conditions.

The present mangrove input of particulate bound nutrients to estuaries is unknown, but it is believed that due to the large size of the delta it is considerable, in spite of the virtual absence of sea grasses, known to be the major contributors of particulate matter and nutrients elsewhere. After damming the Purari, the continuity in the input of mangrove organic matter into estuaries and the sea would have an ameliorative impact on the system, although not replacing the losses due to deposition of most of the upper catchment input in the Wabo reservoir.

Acknowledgements

All chemical analyses were done in the Mines Analytical Laboratory of the Department of Minerals and Energy, Port Moresby. Field assistance was provided by the Baimuru subprovincial government, Mr John Bird of Baimuru, Mrs Jasmin Lucero and numerous other persons. My wife kindly read and commented on the manuscript and further suggestions and critical comments came from Dr I.A.E. Bayly, Dr R.L. Welcomme and Professor B.G. Thom.

References

Arthur, J.F. and M.D. Ball, 1978. Entrapment of suspended materials in the San Francisco Bay – Delta estuary. U.S. Dept. of the Interior. Bureau of Reclamation, Sacramento, California.

Bayly, I.A.E., 1980. A preliminary report on the zooplankton of the Purari estuary. In: T. Petr (ed.), Purari River (Wabo) Hydroelectric Scheme Environmental Studies, Vol. 11: Aquatic ecology of the Purari River catchment, pp. 7 – 11. Office of Environment and Conservation, Central Government Offices, Waigani, Papua New Guinea.

Burton, J.D., 1978. Chemical processes in estuarine and coastal waters: environmental and analytical aspects. J. Inst. Water Engineers Scientists 32: 31 – 44.

Cragg, S., 1983. The mangrove ecosystem of the Purari Delta. This volume, Chapter II, 5.

Davies, B.R., 1979. Stream regulation in Africa. A review. In: J.W. Ward and J.A. Stanford (eds.), The Ecology of Regulated Streams, pp. 113–142. Plenum Press, New York and London.

Frusher, S.D., 1980. The inshore prawn resource and its relation to the Purari Delta region. In: D. Gwyther (ed.), Purari River (Wabo) Hydroelectric Scheme Environmental Studies, Vol. 15: Possible effects of the Purari hydroelectric scheme on subsistence and commercial crustacean fisheries in the Gulf of Papua, pp. 11–27. Office of Environment and Conservation, Central Government Offices, Waigani, Papua New Guinea.

Frusher, S.D., 1983. The ecology of juvenile penaeid prawns, mangrove crab (*Scylla serrata*) and the giant freshwater prawn (*Macrobrachium rosenbergii*) in the Purari Delta. This volume, Chapter II, 7.

Gwyther, D., 1983. The importance of the Purari River Delta to the prawn trawl fishery of the Gulf of Papua. This volume, Chapter II, 8.

Levinton, J.S., 1982. Marine ecology. Prentice Hall, New Jersey.

Paerl, H.W. and P.E. Kellar, 1980. Some aspects of the microbial ecology of the Purari River, Papua New Guinea. In: T. Petr (ed.), Purari River (Wabo) Hydroelectric Scheme Environmental Studies, Vol. 11: Aquatic ecology of the Purari River catchment, pp. 25–39. Office of Environment and Conservation, Central Government Offices, Waigani, Papua New Guinea.

Paijmans, K., 1983. The vegetation of the Purari catchment. This volume, Chapter III, 1.

Petr, T., 1983. Limnology of the Purari basin. Part 1. The catchment above the delta. This volume, Chapter I, 9.

Petr, T. and G. Irion, 1983. Geochemistry of soils and sediments of the Purari River basin. This volume, Chapter I, 7.

Petr, T. and J. Lucero, 1979. Sago palm (*Metroxylon sagu*) salinity tolerance in the Purari River Delta. In: T. Petr (ed.), Purari River (Wabo) Hydroelectric Scheme Environmental Studies, Vol. 10: Ecology of the Purari River catchment, pp. 101–106. Office of Environment and Conservation, Central Government Offices, Waigani, Papua New Guinea.

Pickup, G., 1983. Sedimentation processes in the Purari River upstream of the delta. This volume, Chapter I, 11.

Thom, B.G. and L.D. Wright, 1983. Geomorphology of the Purari Delta. This volume, Chapter I, 4.

Viner, A.B., 1979. The status and transport of nutrients through the Purari River (Papua New Guinea). In: T. Petr (ed.), Purari River (Wabo) Hydroelectric Scheme Environmental Studies, Vol. 9. Office of Environment and Conservation, Central Government Offices, Waigani, Papua New Guinea.

Viner, A.B., 1982. A quantitative assessment of the nutrient phosphate transported by particles in a tropical river. Rev. Hydrobiol. trop. 15(1): 3–8.

Wood, A.W., 1983. Soil types and traditional soil management in the Purari catchment. This volume, Chapter I, 5.

11. Sedimentation processes in the Purari River upstream of the delta

G. Pickup

1. Introduction

It has been estimated that Wabo Dam and its associated reservoir would trap 92% of the sediment load of the Purari River over a 150-year period (SMEC-NK 1977). The effects of removing such a large proportion of the sediment supply from the river are considerable and the change in sediment supply is the key to environmental impact assessment of the Purari Hydroelectric Scheme.

Extensive reviews of the potential environmental impact of damming the Purari have been presented by Goldman, Hoffman and Allison (1975) who consider that changes in sediment yield will have ecological and socio-economic effects as well as changing the physical environment.

Physical effects include extensive channel erosion in the lower Purari, changes in the sediment balance of the delta, modification of the offshore zone and a possible reduction in sediment supply to other areas of the Gulf of Papua.

Ecological effects may include changes to mangrove forests through changes in delta morphology. Alteration of the estuarine environment could disrupt breeding and nursery grounds for some aquatic organisms, and have an impact on the adults. Further effects are possible through the modification of the sediment-bound nutrient supply and particulate organic matter input into the delta where a number of organisms feed on detrital muds. Socio-economic changes could result from the changes in the physical environment. For example, modification of mud banks could lead to changes in mud crab distribution which would affect both subsistence and commercial fishing.

Because sedimentation processes occupy such an important position in the

Petr, T. (ed.) The Purari – Tropical environment of a high rainfall river basin
© 1983, Dr W. Junk Publishers, The Hague / Boston / Lancaster
ISBN-13: 978-94-009-7265-0

environmental impact assessment, they require detailed studies. This chapter discusses sediment transport in the river under natural conditions, deposition upstream from the dam, erosion below it and the impact of potential developments within the basin.

2. Sediment transport under natural conditions

2.1. Sediment sources

There are few data on sediment transport in the rivers of mainland Papua New Guinea. Only the Fly and the Purari have been studied in any detail and even then the quality and amount of information available makes it possible to draw only tentative conclusions. The discussion which follows is based on experience in the Fly (Pickup, Higgins and Warner 1981) as much as the Purari because the Fly is the only river system in which detailed sampling has been carried out upstream of alluvial plain and piedmont reaches.

The main tributaries of the Purari are the Aure, the Asaro, the Wahgi, the Kaugel, the Iaro and the Erave, all of which have their headquarters in the Papua New Guinea highlands except for the Aure. The highlands are characterised by relatively low local relief, a succession of intramontane plains and broad upland valleys, and by several huge strato-volcanoes whose eruptive products are widespread over the area (Löffler 1977). Much of the area is densely populated and the vegetation cover is mainly anthropogenic grassland. The landscapes of the highlands are of considerable age. They are also fairly stable even though there are high mountains and many of the slopes are mantled by volcanic ashes, most of which are at least 50 000 years old (Pain and Blong 1979). A stable landscape coupled with a relatively low annual rainfall of 2000 – 3000 mm keeps sediment yield small over much of the highlands. Locally, there are areas of rapid erosion such as parts of the Simbu Province where the landscape is underlain by unstable Chim clay-shale formations which produce extensive landsliding (Blong 1981). There are also areas where population pressure has led to cultivation of unfavourable locations, resulting in soil erosion but this contributes only a very small proportion of the total load of the Purari.

Beyond the highlands, the tributaries fall steeply and pass through the southern fold mountains in a series of deep gorges. They meet to form the Purari within the mountain area and then exit through a narrow band of foothills to the Purari Delta after being joined by the Aure River close to the mountain front. The southern fold mountains are made up of the Aure fold

206

area in the east and the Kikori – Lake Kutubu karst area in the west (Löffler 1977). The Aure fold area consists of steep escarpments eroded from over-turned folds in limestone, greywacke, siltstone and mudstone and is drained by the Aure River. The folding is probably Pliocene in age and results in a steep ridge and ravine landscape in which extensive landsliding occurs. This activity plus an annual rainfall which increases from 3000 mm in the southern part of the highlands to 8500 mm at the southern edge of the Aure fold area make this part of the basin the source of most of the sediment load transported by the Purari. The Kikori – Lake Kutubu karst area consists of limestone plateaus and broad ridges separated by narrow corridors formed in clastic sedimentary rocks. Little is known about this area but it has a rainfall of 3000 – 5000 mm a year which is less than the Aure fold area. It probably contributes more sediment than the highlands but not as much as the Aure fold area.

The main processes which add sediment to the river are slopewash, land-sliding and reworking of river bed and bank material. Slopewash occurs quite widely under rain forest in Papua New Guinea (Ruxton 1971; Turvey 1974) and the ground surface is frequently swept clean of surface litter. On steep slopes there may be a well-developed network of rills and gullies in which headcuts are quite common. Slopewash is most important in the water areas of the southern fold mountains on steep claystone and mudstone slopes. It is less significant in the highlands even though much of the original rain forest has been replaced by grassland. The reason for this is the very high infiltration capacity of the volcanic ash soils which mantle many areas. Klaer and Kreiter (1981) report infiltration rates as high as 100 – 1000 cm/hour for soils which are probably developed on volcanic ash suggesting that interflow is likely to be a more important process than overland flow. This would greatly reduce the potential for erosion and could explain why traditional cultivators are able to plant slopes as high as $30° - 40°$ without large scale soil loss.

Landsliding is probably the most significant process which adds sediment to the river. The slides are most common in the folded or steeply dipping claystone and mudstone formations of the southern fold mountains or where limestones are underlain by poorly cemented clastic rocks. Many different types of slide occur (see, for example, Byrne, Ghiyandiwe and James 1978) but many involve the undermining of potentially unstable formations by a river plus a trigger mechanism such as an earthquake. When a landslide does occur, it may add 50% or more of the total material involved to the river almost simultaneously, in some cases producing a debris flow. After that, the landslide scar and the accumulated debris below it continue to erode by slopewash, gullying and secondary slides and it may be many years before the area is recolonised by vegetation and a measure of stability is restored.

Periodic reworking of river bed and bank material is most important in contributing sediment load where there are substantial lag deposits. In the upper gorges of the southern fold mountains, the river bed is frequently swept clean of material or covered by a thin veneer of sediment which is in transit and lag deposits are not well developed. Further downstream, in the lower gorges and the foothills beyond the mountains, old alluvial hills and contemporary floodplain deposits 10 – 30 m in thickness occupy the valleys. The river beds consist of mixed sand and gravel with an armour layer on the surface which limits scour. These alluvial deposits are also common in the highland valleys but they tend to be more complex and include areas of infill in subsiding structural basins, former lake beds and volcanic ashes and lahars as well as river deposits. All this material may be added to the sediment load of the river by lateral cutting in meandering reaches or when flow is great enough to break the armour layer on the bed and expose the gravel sands below.

2.2. Rates of sediment transport

Data on the rate of sediment transport in the Purari system are limited. SMEC-NK (1977) collected 278 gulp samples of suspended load and 18 point integrated samples using three verticals on the Purari at Wabo over a range of flows from 730 m³/sec to 6489 m³/sec. They also collected a few bedload samples. Pickup (1977) sampled the Aure and obtained 14 depth integrated samples using eight verticals over a range of flows from 532 m³/sec, to 1060 m³/sec. These data are presented in the form of sediment rating curves in Fig. 1.

Both rating curves have fairly steep slopes and are well correlated with discharge. This is surprising for a basin the size of the Purari which drains 26 300 km² at Wabo. Normally a much greater scatter would be expected such as that which occurs on the lower Ok Tedi and the upper Fly (Pickup, Higgins and Warner 1981). The reason for this is thought to be that much of the runoff and sediment load of the Purari is derived from sources close to Wabo. The close correlation between sediment concentration on the Aure is partly a result of the limited number of samples but it also reflects the fact that the runoff and sediment load come from nearby.

Calculations using the flow duration – sediment rating curve method indicate that the Purari at Wabo transports 57 million tonnes/year (SMEC-NK 1977) while the Aure contributes 48 million tonnes/year (Pickup 1977). The figure for the Aure is thought to be of the correct order of magnitude but

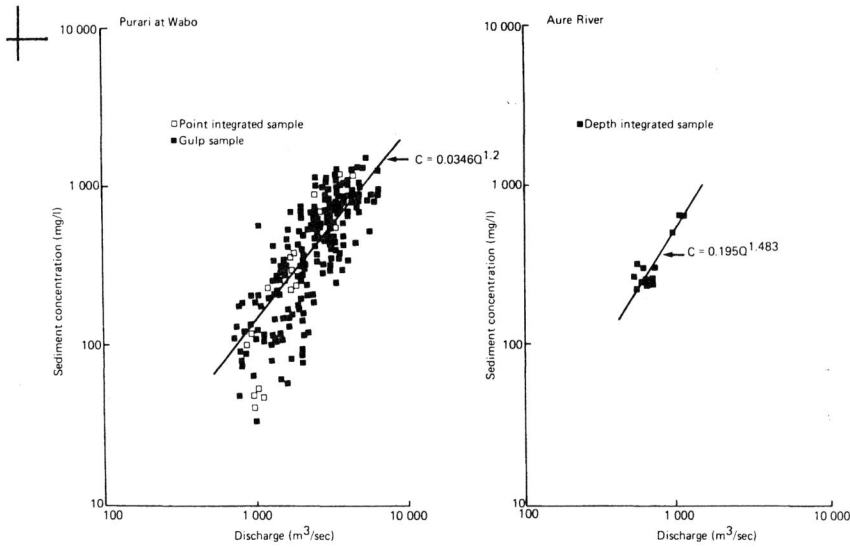

Fig. 1. Sediment rating curves for the Purari River at Wabo and the Aure River.

because it is based on very limited data it cannot be regarded as being precise. The SMEC-NK figures for the Purari include an arbitary 5% added on to allow for bedload. This may be over-generous because calculations using the Meyer-Peter and Muller (1948) bedload equation suggest that at flows of less than 4000 m³/sec a negligible quantity of bedload is transported. After that bedload increases from an amount equivalent to 0.37% of suspended load at 4500 m³/sec to 3.52% at 9000 m³/sec. These calculations do not allow for the effect of bed armouring and subsequent results obtained by Pickup and Warner (in preparation) using the Bagnold (1977) equation in association with an armouring algorithm give an even smaller rate of bedload transport. The Aure has a heavily armoured bed and similar calculations to those for the Purari indicate that the quantity of bedload moved into the Purari is negligible.

3. Estimating deposition above the dam

When a river enters a reservoir, its velocity is reduced and its sediment transport capacity decreases. Coarse particles tend to be deposited in the area affected by backwater from the dam, creating a delta. The finer material remains in suspension, moving slowly downstream until it either settles out or is passed downstream in the outflow.

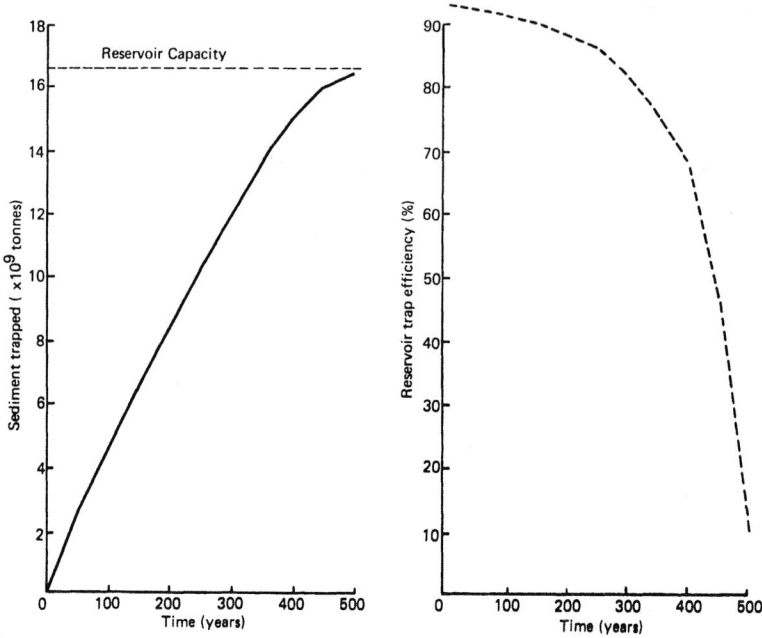

Fig. 2. Changes in trap efficiency and the volume of sediment deposited upstream of Wabo Dam over time.

The amount of sediment passing through the system depends on factors such as the degree of stratification and the water temperature, both of which influence the settling velocity of individual particles within the load. It also depends on the detention storage time and the mode of operation (Gottschalk 1964). If a reservoir has a low storage detention time or is operated in such a way as to keep this time at a minimum, inflowing sediment load will be passed through the system more quickly and will have less time to settle out. Also, if the reservoir is operated in such a way as to evacuate rather than store the sediment-laden water of large floods, this too will reduce the amount of deposition behind the dam.

The ability of a reservoir to trap and retain sediment is known as its trap efficiency, where:

$$\text{trap efficiency} = \frac{\text{sediment inflow} - \text{sediment outflow}}{\text{sediment inflow}}$$

The most important factors governing trap efficiency appear to be the storage

210

detention time and the size distribution of the sediment load itself.

Brune (1953) has derived a series of empirical curves relating trap efficiency to the capacity – inflow ratio of the reservoir. These curves were used by SMEC-NK (1977) in association with the empirical area reduction method (Borland and Miller 1960) and a procedure to calculate the density of deposits (Koelzer and Lara 1958) to provide estimates of the rate of accumulation of sediment within the reservoir. The SMEC-NK (1977) results are plotted in Fig. 2 and show that the storage capacity of the reservoir would be reduced by 15% and 27% after 50 and 100 years and that it would be virtually full after 500 years.

The trap efficiency of the reservoir is also shown in Fig. 2 and varies as the volume of deposited sediment increases. In the first year of operation, SMEC-NK (1977) calculate a value of 93.2% declining to 92.5% after 50 years and 91.5% after 100 years.

4. Estimating sediment transport below the dam

4.1. Modelling techniques

The study of sediment transport downstream from the dam used a large, quasi-analytical computer model developed by W.A. Thomas of the U.S. Army Corps of Engineers (Hydrologic Engineering Centre 1977). The model was designed for the prediction of erosion and deposition and changes in sediment load above and below dams. Originally, it was intended to use HEC-6 to simulate the behaviour of the Purari both upstream and downstream of the dam but finance and the time available for the study prevented this. It was therefore necessary to rely on the SMEC-NK data in providing input to the model.

HEC-6 simplifies river behaviour by treating it as a system within a channel. The horizontal location of the channel banks and the surface of the floodplain are treated as fixed. The behaviour of the moveable bed can then be described by the continuity equation for sediment transport.

$$\frac{\partial G}{\partial x} + W \frac{\partial z}{\partial t} = 0 \tag{1}$$

where:

G = sediment discharge expressed as a volume
x = distance along the channel
W = the width of the moveable bed
z = bed elevation, and
t = time

A finite difference version of equation (1) is used in the model in which the river is divided into a series of reaches. The inflow and outflow of sediment for each grain size present in the model is evaluated for the reach and for a specific time interval and if there is a difference, balance is restored by erosion or deposition. This lowers or raises the bed and changes its grain-size composition.

The basic hydraulic parameters needed to calculate sediment transport capacity in the individual reaches are flow velocity, depth, width and energy slope. All of these are derived from water surface profile calculations using the standard step method.

Once the hydraulic parameters are known, sediment transport may be calculated for each reach given the particle-size distribution of the bed material. Silt and clay particles are transported or deposited depending on whether bed shear stress is greater or less than a critical value which varies with particle size. If bed shear stress is greater than the critical value the sediment is transported. If not, it is deposited at a rate determined as an exponential decay function of river velocity and particle settling velocity. The transport rate for sand and coarser material is calculated using Toffaletti's (1969) modification of the Einstein procedure.

The flow hydrograph is represented by a series of steady flows, each of which occurs for a specified number of days. These flows are entered into the model and the total sediment load, the volume and gradation of deposits, the extent of bed-armouring and the new bed elevation are calculated for a series of points along the river for each time interval. The model also determines sediment outflow from the downstream end of the system for each grain size present.

4.2. Collection of model input data

Because little is known about the behaviour of the Purari downstream of Wabo dam site, environmental impact assessment requires two runs with the model. The first run is necessary to establish baseline conditions in the river and involves the modelling of its sediment balance without the dam. The second run is required to assess the effect of the dam and models the river for conditions after dam construction. Much of the data for these studies was not available and had to be collected or generated using supplementary simulation models.

Fifty-one cross sections were used to represent the channel of the Purari in the simulation exercise (Fig. 3). The number and location of the sections were determined by field inspection and from echo-sounding traverses of the river.

Bed-material samples were obtained from 28 locations in the river to represent the existing sediment characteristics of the river (Fig. 3). Between Wabo dam site and just upstream of the Aure River, the bed samples consist almost entirely of very coarse gravel, 32 – 64 mm in diameter and cobbles 64 – 128 mm in diameter. Other size fractions are present only in small quantities although occasional patches of sand do occur. Starting from cross section 22 which is just upstream from the Purari – Aure confluence, significant quantities of sand begin to appear. This sand is very well sorted and consists almost entirely of material in the medium (0.25 – 0.5 mm) and coarse sand (0.5 – 1.0 mm) categories. Other fractions are present only in very small amounts. Downstream from the Aure, cobbles and very coarse gravel are still present, often in large amounts, as far as cross section 46. Inspections carried out at low flow during reconnaissance field trips to the river indicate that this material is usually present as an armour coat. Beyond cross section 46, the coarse material virtually disappears and the samples consist mainly of coarse and medium sand. Fines are almost totally absent.

The depth of bed-material deposits which determine the possible extent of scour is now known except close to Wabo. Bore holes put down close to the channel centreline at the dam site indicate the presence of 25 – 30 m of gravel or sand and gravel, while at cross section 4, one drill struck bedrock 13.7 m below the surface while a second hole went down 14.0 m without reaching rock (SMEC-NK 1977). Other holes put down between cross sections 9 and 11 found sand and gravel even at depth and one hole went down 14.0 m without striking rock. No data are available for the lower part of the river but a similar, if not greater, depth of sediment could be expected.

Sediment inflow data for modelling the pre-dam construction stage were obtained from the SMEC-NK sampling program and the information collected for the Aure by Pickup (1977). These results have already been described in an earlier section. In both cases more data would be desirable because sediment concentration can vary quite substantially over short periods. This is especially true in seismically unstable areas such as the Aure basin, for as Pain and Bowler (1973) have shown, earthquake-induced landslides can produce very high sediment loads in rivers for short periods. Unfortunately, the time

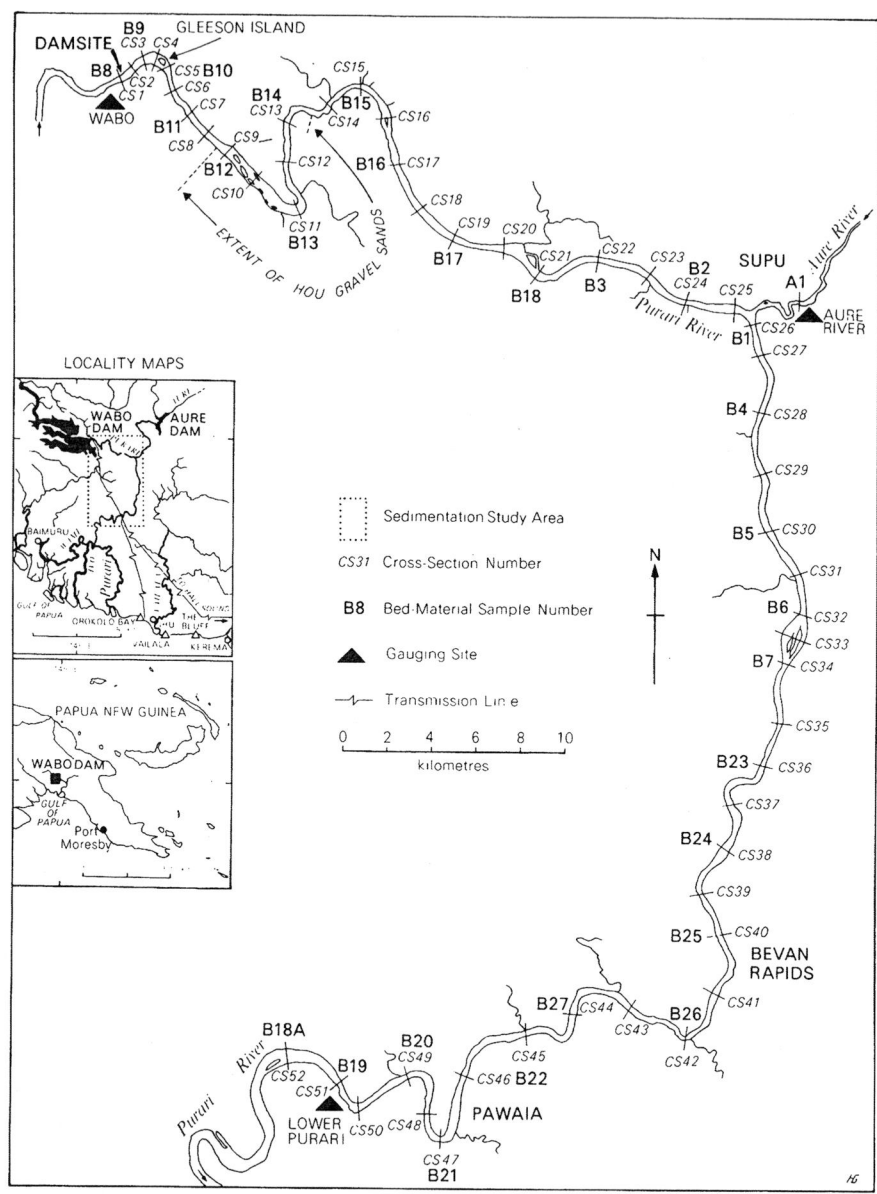

Fig. 3. Location of surveyed cross sections, bed-material samples and gauging sites, lower Purari and Aure Rivers.

limits on the project and the high cost of sediment sampling in remote areas of Papua New Guinea in terms of scarce skilled manpower as well as money prevented a longer period of data collection.

Determination of sediment inflow to the lower Purari after dam construction cannot be done with certainty unless information is available on flow velocity through the reservoir, water temperature, degree of stratification and density currents. A less precise indication can be obtained from the trap efficiency figures of SMEC-NK and for the purpose of modelling, a figure of 93% has been used. The sediment outflow from the dam has been calculated from this as 7% of the normal sediment discharge in the river before dam construction. Only the finest particles will pass through so this load is likely to consist entirely of material in the clay fraction.

4.3. Runoff hydrographs for use in simulation

Modelling of river behaviour before and after the construction of the hydroelectric scheme requires three sets of streamflow data. These are: (i) flows in the Purari at Wabo without the dam; (ii) flows in the Aure River at its junction with the Purari; and (iii) flows in the Purari at Wabo with the hydroelectric scheme in operation. The first two sets of flows were obtained from existing records and from a synthetic record obtained by fitting a probability distribution to existing records and generating a long series of flows using random number techniques. Details of this procedure are presented by Pickup (1977).

Flows in the Purari River downstream of Wabo after dam construction depend on several factors: (i) inflow from upstream; (ii) leakage and evaporation from the reservoir; (iii) the storage-routing effect of the reservoir; and (iv) the water release policy for power generation and the evacuation of floodwater.

Inflow data can be represented using the runoff records already available for the dam site supplemented by SMEC-NK modelling results.

Leakage and evaporation are expected to be negligible (SMEC-NK 1977).

The storage-routing effect of a reservoir may be substantial but this is partly offset by the fact that the dam will be spilling for most of the time. Also, since sediment modelling uses mean daily flows, the attenuation and lag effect of storage will be largely offset by the averaging effect of the time unit. It therefore seems reasonable to neglect the effect of storage.

In the absence of detailed plans for the use of Wabo power, outflows from the dam have been estimated using a water release policy producing a power

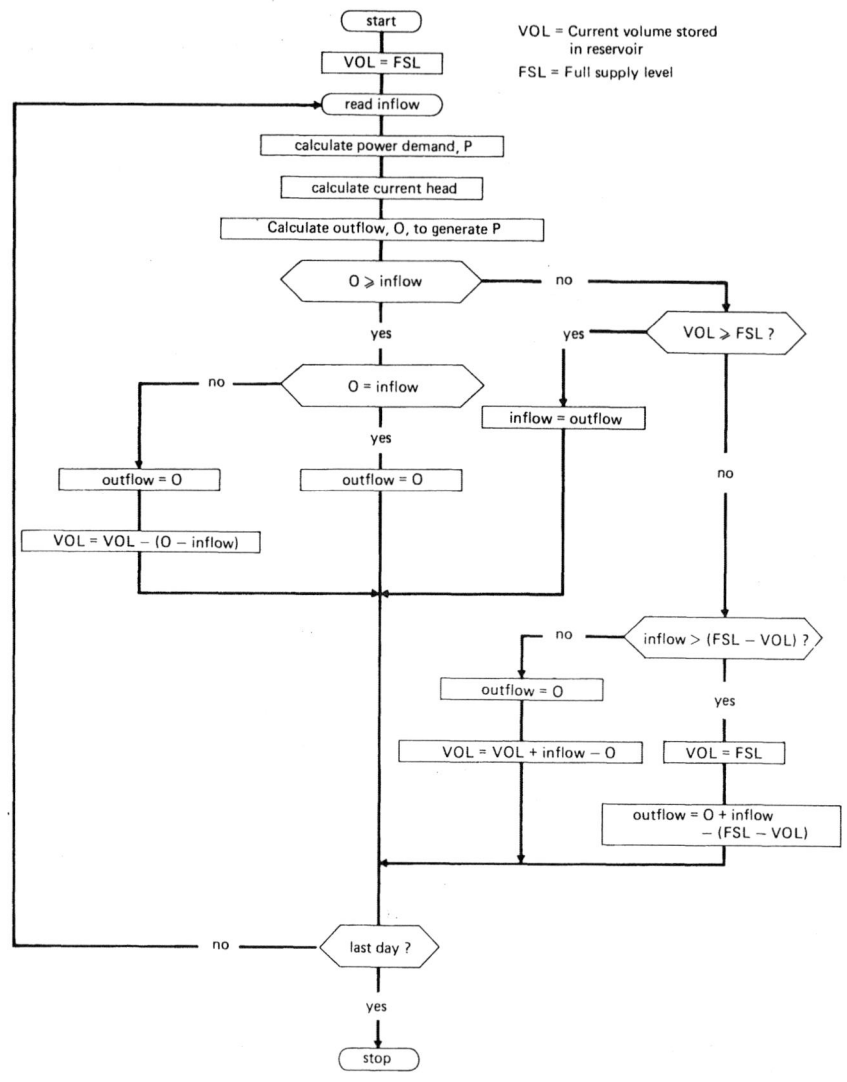

Fig. 4. Simplified flow chart of storage-yield calculations for Wabo Dam under power generation.

output designed to meet firm power demands at the industrial site as specified by the Industrial Bank of Japan (1975). This assumes a power demand increasing at a linear rate from 640 MW at the beginning of the first year of operation to 1735 MW at the beginning of year five. To calculate the water discharge required to generate this power, SMEC-NK (1977) assumed an installed

216

generating capacity of six 360 MW generators and used the relationship:

$$E = 0.08585.e_g.e_t.(H-h_p-h).Q \qquad (2)$$

where:

E = the energy output as the Wabo generator terminals in GWH/
year

e_g = generator efficiency

e_t = turbine efficiency

H = the head of water in the reservoir in m

h_p = the average penstock loss in m

h = the average tailwater level in m, and

Q = the discharge in m³/sec

Above full supply level, H is a function of inflow to the reservoir, whereas below full supply level but above minimum operating level, H is a function of the volume of water stored.

Given empirical equations describing these relationships, outflow from the dam can be determined by a set of storage-yield calculations which are illustrated in Fig. 4. The results of these calculations indicate that for about 80% of the time, reservoir inflows are in excess of the discharge required to generate sufficient power to fulfill demand so the reservoir will be spilling and will have little effect on flow. During dry periods, when inflow is less than the discharge required for power generation, flow in the lower river will be augmented by water released from the reservoir. This flow augmentation will only occur for short periods except where severe drought occurs and there will only be limited drawdown in the reservoir.

5. Results

Before it is possible to assess the impact of the hydroelectric scheme it is necessary to establish baseline conditions in the river. Accordingly the first run with the simulation model was used to provide information on the behaviour of the river without the dam.

The baseline run was carried out using four years of runoff records which are representative of normal conditions in the river. The results of this run may be summarized as follows:

1. Average sediment input to the system:
   - Purari (from upstream of Wabo)
     - 13.90×10^6 m³/year of clay
     - 22.40×10^6 m³/year of silt
     - 3.84×10^6 m³/year of sand
   - Aure
     - 27.07×10^6 m³/year of clay
     - 20.20×10^6 m³/year of silt
     - 1.22×10^6 m³/year of sand

2. Modelled output of the system into the delta:
   - 40.97×10^6 m³/year of clay
   - 42.60×10^6 m³/year of silt
   - 5.54×10^6 m³/year of sand

The baseline run indicates that 88.6 million m³/year of sediment are delivered by the upper Purari and the Aure Rivers. Of this, 55% by volume comes from the Aure. The breakdown of inflowing load by its size composition indicates that 46% by volume is clay, 48% is silt and only 6% is sand. About 2/3 of the clay comes from the Aure but only about ¼ of the sand. Silt is supplied by both rivers in similar quantities.

The sediment outflow into the delta region is very similar to the inflow. The model indicates that all the silt and clay pass through the system, a fact which is verified by the absence of fine sediment in the bed material of the Purari. All of the sand passes through and there is some slight erosion occurring to the extent that the sand outflow is about 9% greater than the sand inflow. The amount involved is very small and is within the limits of model accuracy and sampling error so it is probably safe to say that the lower Purari is in a fairly stable condition.

Although the river appears to be stable, the modelling results suggest that some redistribution of sediment is occurring within the system resulting in local erosion and deposition. This is quite normal, even in a stable river. The distribution and extent of these local changes are illustrated in Fig. 5 which shows that the sandy reach between the Aure junction and cross section 42 has a tendency to scour but most of this sediment is being redeposited in the wider sections of river between cross section 43 and cross section 51. Very little change is occurring in the gravelly section of the river upstream of the Aure, indicating that the bed is fairly stable.

The impact of the Wabo Hydroelectric Scheme on the sediment balance of the lower Purari can be divided into four distinct phases: (1) the construction

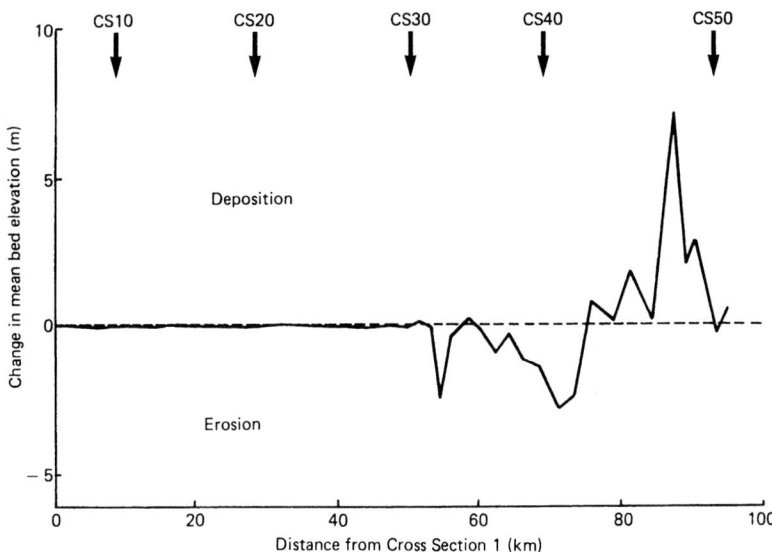

Fig. 5. Local erosion and deposition, baseline run.

period; (2) the reservoir filling period; (3) the transition period as the river adjusts towards a new equilibrium condition imposed by changed sediment input and flow conditions once the dam is operating; and (4) the new equilibrium condition.

No attempt is made to model the effect of the construction period because too many intangibles are involved. It is expected, however, that some deposition would occur in the lower Purari as a result of increased sediment input to the river. Most of this input would be derived from dredging of cofferdam foundations and other river deposits for main dam fill and filter and aggregate material. A qualitative impact assessment of the construction period is presented by Pickup (1977).

The filling period is likely to vary in length depending on rainfall conditions in the Purari catchment. Although the reservoir has a capacity of 16 600 $\times$ 10^6 m³ it would only take 2 – 3 months to fill during an average wet season so the long-term impact on the sediment balance of the lower Purari would be minimal. Consequently, the filling period was not modelled.

The length of the period of transition to a new equilibrium condition is difficult to assess and may involve tens or even hundreds of years. During this period the rate of channel change is not constant but decreases roughly exponentially as the new equilibrium is approached. This means that most of the changes occur early in the transition period and the river may be fairly close to

219

the final adjustment after the first few years. Most of the transition period consists of a long sequence of slow and very minor modifications to channel geometry.

Because the computer time needed would be excessive, the whole of the transition period was not modelled. Instead, a simulation run was carried out to establish what major changes would occur in the early stages, and what conditions would prevail during the long slow progression towards a new equilibrium condition. This run used data for the same period as the baseline run but with water and sediment inflows modified to allow for the effect of Wabo Dam.

The transition period simulation run was carried out for 36 twenty-day periods. The baseline and post-construction sediment input conditions for this run are as follows:

1. Purari (from upstream of Wabo):

baseline condition	post-construction condition
25.905×10^6 m^3 clay	9.66×10^6 m^3 clay
41.752×10^6 m^3 silt	no silt
7.151×10^6 m^3 sand	no sand

2. Aure

baseline condition	post-construction condition
61.983×10^6 m^3 clay	no change
51.858×10^6 m^3 silt	no change
2.753×10^6 m^3 sand	no change

The modelled output of the system to the delta region for the same period is:

baseline condition	post-construction condition
87.887×10^6 m^3 clay	71.641×10^6 m^3 clay
92.695×10^6 m^3 silt	50.944×10^6 m^3 silt
10.093×10^6 m^3 sand	3.691×10^6 m^3 sand

The results indicate that the input of clay to the delta system will be reduced by 18%, the input of silt will be reduced by 45% and the input of sand will be reduced by 63%. These changes are small when compared with other large dams and they clearly indicate that the Aure River will play an important part in maintaining sediment delivery to the delta.

Wabo Dam will not only affect the amount of sediment reaching the lower part of the river. It will also change the size composition of the load. In the baseline run, 46% of the sediment outflow by volume was clay, 48% was silt and 6% was sand. In the transition period run, the outflow is 57% clay, 40%

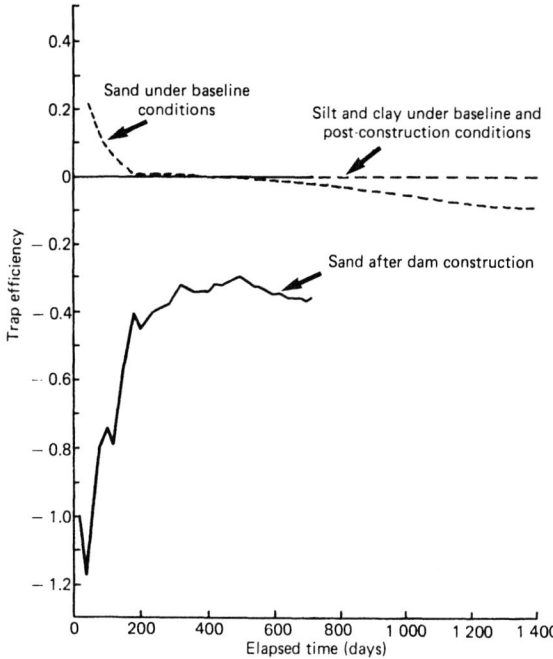

Fig. 6. Trap efficiency values for the lower Purari, baseline and post-dam construction simulation runs.

silt and 2.9% sand. The sediment delivered to the delta will therefore be finer than previously.

The extent to which the river has become adjusted to changed sediment input conditions can be measured by the trap efficiency of the system. A stable channel has a trap efficiency of zero, an eroding channel has a negative value and a silting channel has a positive value.

Trap efficiencies for the lower Purari for the early part of the transition period are plotted in Fig. 6.

The values for clay and silt remain at zero throughout the simulation run, indicating that all the fine material delivered to the river is transported through the system and virtually none is added by erosion of the bed since it is not present in the bed material. The trap efficiency index for sand is about −1.2 in the early part of the transition period but it declines rapidly and after about 300 days, it stabilizes at around −0.35 indicating that outflow of sand is about ⅓ greater than inflow. The average trap efficiency for sand will eventually decline towards zero, equating the outflow of sand into the delta with the input from the Aure which is the only significant source. However, the rate of change will be very slow.

Other changes which may be expected during the transition period are continuing bank erosion, erosion of the bed and armouring of the bed material. Clay and silt output will be maintained at the levels prevailing during the early part of the transition period, but the sand output will gradually decline to an average of about 1.2 million m³/year.

The new equilibrium condition will be attained when, on average, the inflow and outflow of sediment are equal and long-term erosion ceases. It may take a long time for this condition to be attained, but once it is, the following average sediment input and output conditions may be expected.

1. Purari (from upstream of Wabo):
   4.9×10^6 m³/year of clay
   no silt
   no sand
2. Aure:
   27.07×10^6 m³/year of clay
   20.20×10^6 m³/year of silt
   1.22×10^6 m³/year of sand

When compared with baseline conditions, this represents a decrease of 22% in the volume of the clay fraction delivered to the delta, 53% in the silt fraction and 78% in the sand fraction. The amount of sediment passing through Wabo Dam will slowly increase but it is likely to consist almost entirely of clay particles for the first 200 – 300 years of dam life.

The amount of channel erosion which will occur before equilibrium is restored depends very largely on the size composition of the Purari bed material. Because this sediment contains coarse material it will armour quite rapidly as the finer particles are removed by the flow. Once an armour coat of coarse particles is fully developed on the bed, erosion will cease because although the flow has the capacity to transport sediment finer than that in the armour coat, none is exposed. Once this occurs, the river only transports the load delivered from upstream and it is in equilibrium.

To assess the likely extent and distribution of erosion of the river bed, a set of armouring calculations was carried out using the method briefly described in the section on modelling techniques. These produce estimates of the depth of scour required to fully develop an armour layer one grain diameter thick on the bed. The calculations were made for the largest flow used in the simulation run which would be exceeded for less than 0.5% of the time and the results are shown graphically in Fig. 7.

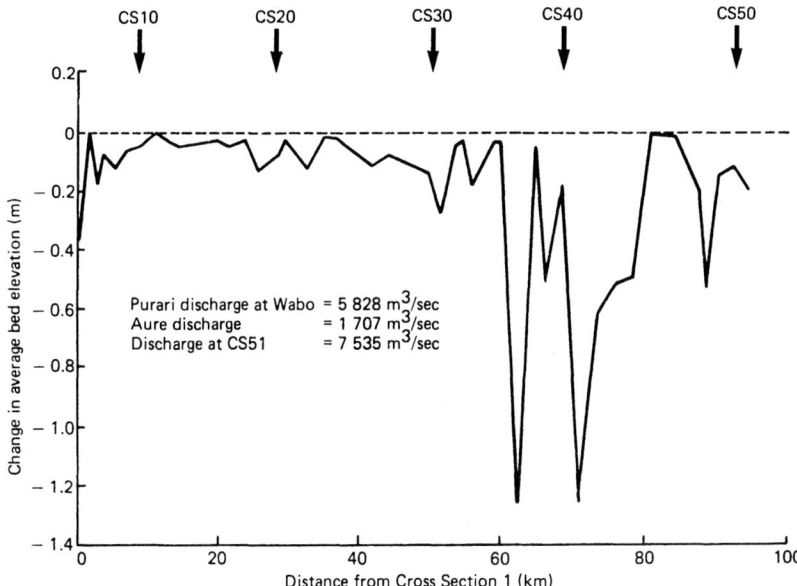

Fig. 7. Depth of scour required for full development of armour coat at maximum discharge used in the simulation runs.

From the armouring calculations, it seems that even before erosion begins, the bed is already armoured to a high degree. Over most of the river, less than 0.2 m of scour is required to fully develop the armour coat although in a few locations in the sandy reach downstream from cross section 37, average scour depths of up to 1.5 m may occur. In general, however, erosion of the bed does not appear to be a problem and will have a minor impact on the system.

6. Potential effects of future development on sedimentation

While many types of development could occur within the Purari basin (Carey 1977), the operations most likely to affect sedimentation processes in the near future are extension of the area under traditional cultivation, logging and large-scale, open-cut mining operations.

Extension of the cultivated area is already occurring, particularly in the highlands where population pressure is on the increase. This is likely to result in soil erosion as new areas are cleared and gardens extend onto progressively steeper areas. Wood (this volume) pointed out the possibility of high erosion if marginal lands of the forest, such as at Karimui, were to be settled and in-

223

tensively cultivated. Whether the additional sediment yield arising from this activity will significantly affect the lower Purari is not known. However, as long as it remains confined to the less wet parts of the basin, the extra material added to the river is likely to be minimal in comparison with that added by natural processes in largely unpopulated areas.

Logging is also unlikely to affect sedimentation in the river above the dam. High transport costs prohibit large operations such as woodchipping in the highlands. The southern fold mountains do not have road access except through the highlands and even if a road was available to the Gulf of Papua, slopes are too steep and the rainfall is too heavy to make the normal type of logging operation found in Papua New Guinea a paying proposition. Some logging may be feasible in the area below the dam but, in relative terms, the contribution of any resultant erosion to the overall load of the river would be small.

A large-scale mining operation could materially affect the sediment balance of the system. This would certainly be the case if free release of tailings was permitted or extensive loss of waste rock occurred. No major operations are at an advanced stage of planning at present but the situation may change. If a mining operation was planned careful monitoring and good sediment control techniques should be obligatory.

Acknowledgements

A version of this paper originally appeared in Earth Surface Processes, Vol. 5. The author is grateful to John Wiley and Sons Ltd. for permission to use copyright material contained in the original paper.

References

Bagnold, R.A., 1977. Bed load transport by natural rivers. Water Resour. Res. 13: 303 – 312.
Blong, R.J., 1981. Stability analysis of Chim Shale mudslides, Papua New Guinea. Int. Assoc. Hydrol. Sci. Publ. 132: 42 – 66.
Borland, W.M. and C.R. Miller, 1960. Distribution of sediment in large reservoirs. Trans. Am. Soc. Civ. Engrs. 125: 166 – 180.
Brune, G.M., 1953. Trap efficiency of reservoirs. Trans. Am. Geophys. Un. 34: 407 – 418.
Byrne, G.M., M.M. Ghiyandiwe and P.M. James, 1978. Ok Tedi landslide study. Geological Survey of Papua New Guinea Report 78/3. Port Moresby, P.N.G.
Carey, W.C., 1977. The Gulf of Papua – a future megapolis. In: J.H. Winslow (ed.), The Melanesian Environment, pp. 309 – 313. A.N.U. Press, Canberra.
Goldman, C.R., R.W. Hoffman and A. Allison, 1975. Environmental studies design, Purari

River development. Report to United Nations Development Program and the Papua New Guinea Government. Ecological Research Associates, Davis, California.

Gottschalk, L.C., 1964. Sedimentation, Part I: Reservoir sedimentation. In: V.T. Chow (ed.), Handbook of Applied Hydrology, pp. 17/1 – 17/34. McGraw Hill, New York.

Hydrologic Engineering Centre, 1977. HEC-6, scour and deposition in rivers and reservoirs, user's manual. U.S. Army Corps of Engineers.

Industrial Bank of Japan, 1975. Report by Purari Industrial Survey Mission. Tokyo.

Klaer, W. and M. Kreiter, 1981. The importance of the humus on the stability of slopes and soils in the highlands of Papua New Guinea. Paper presented at the Symposium on Erosion and Sediment Transport in Pacific Rim Steeplands, Christchurch, New Zealand.

Koelzer, V.A. and J.M. Lara, 1958. Density and compaction rates of deposited sediment. Proc. Am. Soc. Civ. Engrs., J. Hydraul. Div. 84: 1603/1 – 1603/15.

Löffler, E., 1977. Geomorphology of Papua New Guinea. A.N.U. Press, Canberra.

Meyer-Peter, E. and R. Muller, 1948. Formulas for bed-load transport. Int. Assoc. for Hydraulic Structures Res., 2nd Meeting, Stockholm, Sweden, pp. 39 – 64.

Pain, C.F. and R.J. Blong, 1979. The distribution of tephras in the Papua New Guinea highlands. Search 10: 228 – 230.

Pain, C.F. and J.M. Bowler, 1973. Denundation following the November 1970 earthquake at Madang, Papua New Guinea. Z. f. Geomorph. N.F. 18: 92 – 104.

Pickup, G., 1977. Computer simulation of the impact of the Wabo Hydroelectric Scheme on the sediment balance of the lower Purari. In: T. Petr (ed.), Purari River (Wabo) Hydroelectric Scheme Environmental Studies, Vol. 2. Office of Environment and Conservation, Waigani, Papua New Guinea.

Pickup, G., R.J. Higgins and R.F. Warner, 1981. Erosion and sediment yield in Fly River drainage basins, Papua New Guinea. Int. Assoc. Sci. Hydrol. Pub. 132: 438 – 456.

Pickup, G. and R.F. Warner, in preparation. Geomorphology of tropical rivers. II. Channel adjustment to sediment load and discharge in the Fly and lower Purari, Papua New Guinea.

Ruxton, B.P., 1971. Slopewash under mature primary rainforest in northern Papua. In: J.N. Jennings and J.A. Mabbutt (eds.), Landform Studies from Australia and New Guinea, pp. 85 – 94. A.N.U. Press, Canberra.

Snowy Mountains Engineering Corporation – Nippon Koei (SMEC-NK), 1977. Purari River Wabo Power Project feasibility report. 8 volumes.

Toffaletti, W.B., 1969. Definitive computations of sand discharge in rivers. Proc. Am. Soc. Civ. Engrs., J. Hydraul. Div. 95: 225 – 248.

Turvey, N.D., 1974. Nutrient cycling under tropical rain forest in central Papua. Dept. of Geography, Univ. of Papua New Guinea. Occasional Paper, 10.

Wood, A.W., 1983. Soil types and traditional soil management in the Purari catchment. This volume, chapter I, 5.

II. Biological environment

1. The vegetation of the Purari catchment

K. Paijmans

1. Introduction

This account of the vegetation is based partly on field observations, and partly on airphoto interpretation. Most of the field observations were carried out during land resources surveys by plant ecologists and a forest botanist of the Division of Land Use Research, C.S.I.R.O., Canberra. Descriptions of the vegetation in survey areas that cover parts of the Purari catchment are published in Robbins and Pullen (1965), Saunders (1965, 1970), Paijmans (1969), and Robbins (1970). The vegetation of the whole of Papua New Guinea is covered by an annotated vegetation map (Paijmans 1975) and by more detailed descriptions in Paijmans (1976). Comprehensive accounts are given by Smith (1980) of the ecology, and by Hope (1980a, 1980b) of the past and present distribution and composition of New Guinea high-mountain communities.

In order to describe the very diverse vegetation in its relation to altitude and human impact, the Purari catchment has been divided into three physiographic regions: *lowlands*, *uplands*, and *mountains*. The *lowlands* comprise the delta and lower reaches of the Purari River, the hills to the east, and the hilly middle and lower reaches of the Erave and Tua Rivers west and north of Karimui up to 1000 m a.s.l. The *uplands* form a generally narrow zone of hills and low mountains mainly between 1000 and 1500 m. The *mountains*, largely above 1500 m, consist of the ranges and intramontane valleys of the central highlands, where several peaks rise to over 4000 m. Figure 1 outlines the approximate boundaries of the regions. Inclusions of terrain outside the altitudinal limits set are present in each region, but are not shown, due to limitations of scale.

The vegetation types distinguished in the Purari catchment are described mainly on the basis of their structure and physiognomy; characteristic taxa are

Petr, T. (ed.) The Purari – Tropical environment of a high rainfall river basin
© *1983, Dr W. Junk Publishers, The Hague / Boston / Lancaster*
ISBN-13: 978-94-009-7265-0

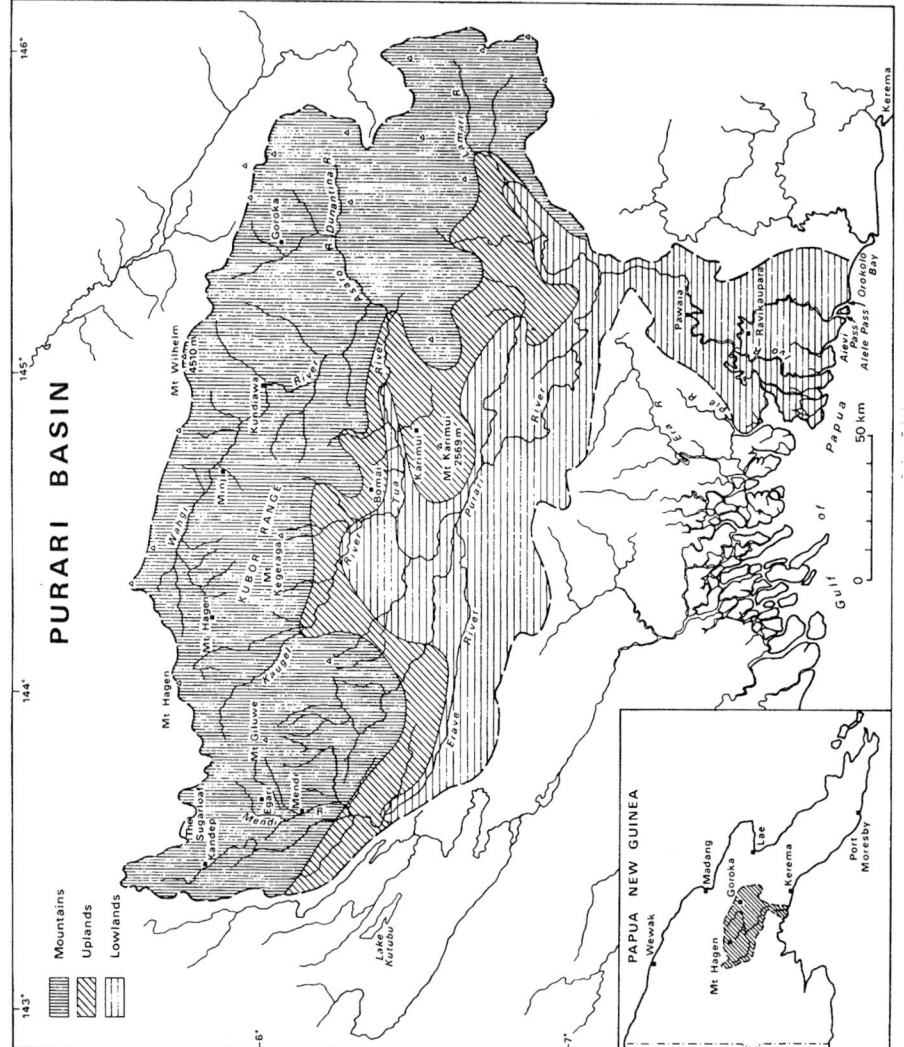

Fig. 1. Boundaries of the Purari catchment and physiographic regions.

Table 1. Vegetation types of the Purari catchment and their environmental relationships.

Vegetation type	Altitude (m)	Location and habitat	Extent
The lowland region	< 1000		
Herbaceous beach vegetation	< 1	First sand ridge behind beach	Very small
Littoral scrub	< 3	Sandy beach ridges	Very small
Littoral woodland and forest	< 3	Inland beach ridges and flats	Very small
(Semi-)aquatic vegetation	< 2	Troughs between beach ridges	Very small
Swamp forest	< 3	Freshwater tidal flats	Small
Swamp woodland	< 4	Freshwater tidal flats	Small
Herbaceous swamp vegetation	< 4	Probably stagnant, non-tidal swamps	Small
Sago palm vegetation	< 20	Freshwater tidal and non-tidal riverine flats	Small
Mangroves	< 2	Saline tidal flats	Small
Nipa palm vegetation	< 2	Brackish tidal flats and creek edges	Small
Sonneratia caseolaris (or *S. lanceolata*)	< 20	Recent mud banks and accreting inner river curves; brackish to near-fresh	Very small
Pandan scrub	< 20	Low river banks; brackish and fresh	Very small
Phragmites karka	< 20	Low river banks; brackish and fresh	Very small
Saccharum robustum	< 20	Low river banks; fresh	Very small
Riverine forest, open	< 20	Ill-drained floodplains	Small
Riverine forest, relatively well-closed	< 25	Levee banks and relatively well-drained plains	Very small
Lowland hill forest	< 1000	Hills throughout	Very large
Small-crowned closed lowland hill forest	< 1000	Limestone in west of region	Medium
Small-crowned open lowland hill forest	< 100	Steep-sided hills in southeast of region	Small
The upland region	1000 – 1500		
Upland hill forest	1000 – 1500	Low mountains throughout	Very large
Upland hill forest, *Castanopsis* dominant	1000 – 1500	Upper slopes and crests	Medium
Small-crowned upland hill, forest, *Nothofagus* dominant	1200 – 1500	Limestone	Small
Garden regrowth and secondary forest	1000 – 1500	Throughout region	Medium
The mountain region	> 1500		
Lower montane forest	1500 – 3000	Throughout zone	Large
Lower montane forest, *Castanopsis* dominant	1500 – 2300	Mainly upper slopes and crests	Small

229

Table 1. Continued.

Vegetation type	Altitude (m)	Location and habitat	Extent
Lower montane forest, *Nothofagus* dominant	1500 – 3000	Mt. Giluwe, Kubor Range, scattered throughout particularly on limestone	Small
Lower montane forest, conifers dominant	2400 – 3000	Throughout zone, particularly upper levels	Small
Montane forest	3000 – 3800	Throughout zone	Small
High-mountain scrub	2000 – 4000	Crests and above tree line	Very small
Swamp forest	> 1800	Fringing or covering shallow depressions in valley floors	Very small
Swamp forest, *Pandanus* dominant	1800 – 2100	Ill-drained sites	Very small
Swamp forest, conifers dominant	> 2200	Ill-drained sites, possibly frost hollows	Very small
Mid-height grassland, lowland species dominant	1500 – 2600	Intramontane valleys	Medium
Mid-height grassland, high-mountain species dominant	2600 – 3800	Above upper limit of agriculture in mosaic with forest	Medium
Low grassland, high-mountain species dominant	> 3800	Above tree line	Small
Phragmites – Leersia swamp grassland	1500 – 2500	Swampy basin sites, seepage footslopes	Very small
Sedge – grass swamp vegetation	1800 – 2500	Swampy basins	Small
Sedge swamp vegetation	2100 – 3500	Swampy depressions	Very small
Sedge – herb swamp vegetation	> 3000	Swampy depressions, wet slopes	Very small
Aquatic vegetation	> 1500	Lakes, deeper parts of swamps	Very small
Garden regrowth	1500 – 2600	Throughout zone	Large
Miscanthus grassland	1500 – 2500	Old garden sites	Small

given for each type. The types are listed in Table 1, together with notes on their relative extent and habitat. Relative extent is estimated in percentages for each physiographic region: very large > 50%, large 25 – 50%, medium 10 – 25%, small 1 – 10% and very small < 1%. Profiles of the main vegetation types are given in Figs. 4, 7, 10, 11. The mangrove and nipa palm communities and the herbaceous vegetation of lakes and mountain swamps are discussed elsewhere in this book.

2. Lowland vegetation

Littoral communities are present on sandy beach ridges and flats fringing the coast. Freshwater swamp vegetation occupies most of the delta and lower reaches of the Purari River, and a variety of riparian communities, reflecting different flooding conditions, occurs on narrow, generally ill-drained and periodically flooded riverine tracts. Dryland forest covers the hills to the north and east of the delta.

2.1. Littoral communities

The vegetation of beach ridges and flats ranges from pioneering herbaceous communities nearest the present beach to tall mixed forest inland. Low sand-binding herbs, grasses and sedges including the pantropical creepers *Ipomoea pes-caprae*, *Canavalia maritima* and *Vigna marina* colonize the first ridge behind the beach from just above high-water mark (Fig. 2). Shrubs and low trees of *Hibiscus tiliaceus* and *Desmodium umbellatum* appear in the inner part of the herbaceous beach vegetation and become denser inland to form a scrub. The epiphytic parasite *Cassytha filiformis* overgrows a variety of host plants, and climbing *Flagellaria indica* in places densely entangles the shrubberies. Elsewhere the pioneering herbaceous vegetation grades into a woodland of wide-crowned low trees including *Barringtonia asiatica* and deciduous *Terminalia catappa*, and the screw palm *Pandanus tectorius*. Forest covers much of the inland beach ridges and flats. Palms are usually plentiful in the shrub and low tree layers of this forest, and the climbing fern *Stenochlaena palustris* covers the tree trunks in many places. Some of the common canopy trees such as *Pterocarpus indicus* and species of *Terminalia* are leafless during the short dry season in November – December.

The littoral zone is usually traversed by more or less permanently swampy depressions aligned parallel to the ridges. These depressions have water plants in their deepest parts, and sago palm (*Metroxylon sagu*), pandans, reeds and other woody and herbaceous swamp plants where the water is just above or at the surface. Nipa palm (*Nypa fruticans*) and mangroves are present where such depressions are within tidal reach.

The largest occurrence of littoral forest and woodland is north of Orokolo Bay. Minor areas of mainly herbaceous and shrubby vegetation are present on sandy strips scattered on both sides of the Aievi and Alele Passages. Elsewhere, much of the original littoral vegetation has been replaced by coconut plantations, which usually have a ground cover of grasses such as

Paspalum conjugatum and, in places, *Imperata cylindrica*.

2.2. Freshwater swamp vegetation

Inland from the tidal saline and brackish belt of mangroves and nipa palm is a zone of tidal freshwater swamps up to 25 km wide. They form the eastern part of a vast area of permanent swamps associated with the Pie, Era, and other rivers outside the Purari catchment, and to the east and north grade into less swampy but still flood-prone country along the Purari River. The swamps are flooded by freshwater backed up by the ocean tide through a maze of creeks and channels. Flooding is daily near the coast, but becomes less frequent inland. The main vegetation types are swamp forest and woodland, and sago palm and herbaceous swamp vegetation.

Freshwater swamp forest, which adjoins the mangroves, has a canopy of small-crowned trees about 25 m high, with emergent trees reaching 30 m. The forest is very mixed generally, but the flat-topped *Campnosperma brevipetiolata* and a *Calophyllum* species with adventitious roots form pure stands in places. In the canopy and lower tree storeys, the stilt-rooted *Myristica hollrungii* and trees of the laurel family are very common, as are other Myristicaceae, together with *Intsia bijuga*, *Sapium indicum*, and species of *Palaquium*, *Syzygium*, *Garcinia* and *Diospyros*. Because of the even, dark-toned, small-crowned appearance from the air and the commonness of stilt-rooted trees, the type is sometimes referred to as 'freshwater mangrove'. Pandans and sago palm form an open storey below the trees, and tall sedges and herbs such as *Mapania macrocephala* and *Hanguana malayana* make up most of the undergrowth (Figs. 3, 4).

Where conditions of drainage and aeration are less favourable, swamp forest merges into lower and more open swamp woodland in which pandans and sago palm are more abundant and the undergrowth is denser. *Nauclea orientalis*, *Elaeocarpus* sp. and *Neuburgia corynocarpa* are common trees in addition to those of swamp forest. Climbers abound and include *Flagellaria indica*, *Nepenthes mirabilis* and *Stenochlaena palustris*.

Near the river where there is a regular influx of freshwater, sago forms almost pure stands which have only widely scattered trees emerging above the dense canopy of sago (Fig. 5). Here its fronds are up to 14 m high, and the flowering, starch-producing stems reach over 20 m. Under dense sago there is no undergrowth, and the palm's numerous tiny pneumatophores form the only live ground cover. The main areas of sago are in the freshwater swamps of the delta, but scattered individuals and groups of the palm also occur on

Fig. 2. Beach vegetation at mouth of Purari River. Coconut plantation in background.

Fig. 3 Interior of tidal freshwater swamp forest, Purari Delta. Tall coarse sedge in foreground. Stilt-rooted *Pandanus* in middle ground. Pinnate leaves of the climbing fern *Stenochlaena* visible on right.

233

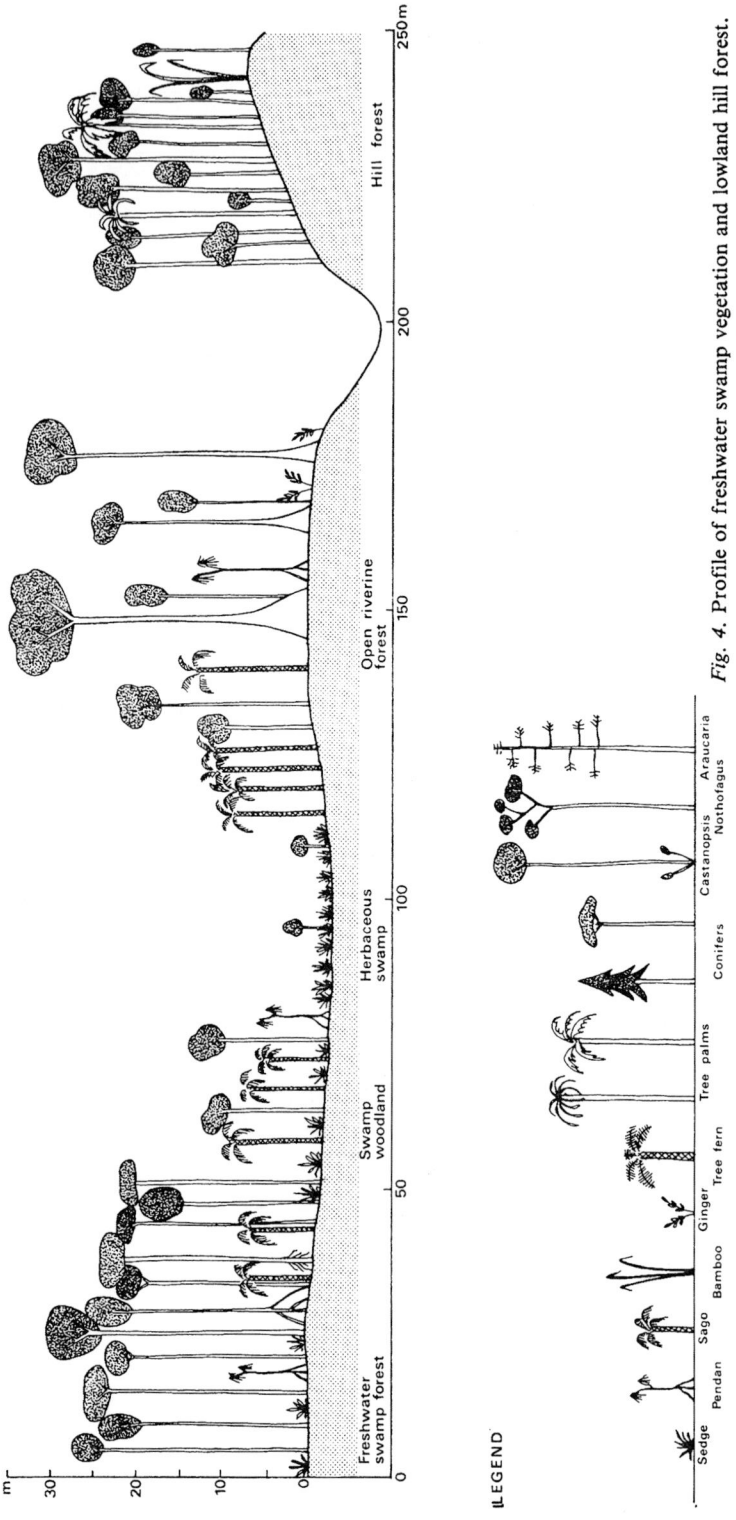

Fig. 4. Profile of freshwater swamp vegetation and lowland hill forest.

slightly brackish sites and in swampy mountain valleys. On the landward side of the mangroves where sago grows together with nipa palm soil water salinities of up to 7.5‰ have been measured (Petr and Lucero 1979). In mountain swamps both natural and planted stands of sago occur to an altitude of about 1000 m (Paijmans 1980).

In the central part of the swamps, between the Ivo and Purari Rivers north of Ravikaupara, swamp woodland grades into vast areas of mainly herbaceous swamp. Here the water appears to be outside tidal reach and more or less stagnant. At the one site visited the vegetation was dominated by *Hanguana malayana*, the tall sedge *Mapania macrocephala*, and the ferns *Cyclosorus* sp. and *Stenochlaena palustris*, which were climbing over the undergrowth and the scattered pandans and low trees.

2.3. Riparian vegetation

Near the mouth of the Purari where flooding is brackish, *Sonneratia lanceolata** colonizes newly formed mud banks and islets. This mangrove persists along the foot of the river banks to where the tidal influence fades out and the water is virtually fresh. It is noteworthy that the riverine variety of *S. lanceolata* has drooping branches and lanceolate leaves, whereas the variety growing in coastal mangroves has stiff erect branches and oval leaves. Further upstream, groups of shrub pandans and *Phragmites karka* grow along the lower banks in brackish to freshwater environment. *Saccharum robustum*, often growing together with the pioneers from further downstream (Fig. 6), is a typical freshwater indicator. Young trees of *Artocarpus incisus* (*A. altilis*) and *Octomeles sumatrana* pioneer on top of levees, and on higher banks are joined by other rain forest trees forming woody communities that approach riverine forest in stature.

Riverine forest that is subject only to short periods of inundation, or is not inundated at all, is present in the environs of Pawaia. The canopy of such forest is 30 – 35 m high generally, but is rather irregular in height and closure, and has many gaps in which only lower trees are present. Scattered trees emerging above the canopy reach 50 m or more. Most emergent and many of the canopy trees have wide crowns, long straight boles, and high and wide buttresses. Common trees in the upper storeys are *Alstonia scholaris*, *Dracontomelon dao* (*D. mangiferum*), *Octomeles sumatrana*, *Pometia pinnata*, and

---

* Some authors name this species *S. caseolaris*.

Fig. 5. Spiny variety of sago palm in tidal freshwater swamp, Purari Delta.

Fig. 6. Stream-bank vegetation, lower course of Purari River. *Saccharum robustum* in centre, *Sonneratia lanceolata* left of centre, *Phragmites karka* right of centre. Swamp forest with scattered sago palm in background.

236

species of *Canarium*, *Dysoxylum* and *Terminalia*. The lower tree strata are usually rather open, while the density of the shrub and tall herb layer varies according to the amount of light penetrating through the canopy. Small tree palms, shrub palms, and young rattan palms are a feature of the shrub layer, and tall gingers, Araceae, and Marantaceae grow densely in places. *Selaginella*, *Elatostema*, tree and rattan seedlings, ferns, and sparse forest grasses and sedges form a patchy low ground cover. Thick woody lianes and climbing fleshy epiphytes and ferns are usually common throughout, but climbing rattan is dense only in gaps.

Where inundation is prolonged, the forest canopy is lower and smaller crowned and has more large gaps. Emergents are more widely spaced, climbing rattan is more common, and sago palm is present in the understorey. Ground observations and airphoto patterns indicate that most forest along the Purari River is of this type.

Damage or destruction by severe floods sets back successional development temporarily, but in the long term the position of the various river-bank communities and the boundary with riverine forest will be shifting southward as the Purari Delta is built out to sea and drainage conditions upstream of the delta improve.

2.4. Lowland hill forest

The forest covering the hills differs from that on alluvium mainly in structure. Owing to generally less favourable conditions of steep slopes and shallow unstable soils, hill forest is not as tall as riverine forest on well-drained alluvium. Only locally does it resemble the latter in stature, as for example on the gentle northern footslopes of Mount Karimui, and the Bomai plateau northwest of Karimui across the Tua River. Compared with well-drained alluvium forest, the canopy of average lowland hill forest is somewhat lower, and trees with a very large girth and large buttresses are less common. The canopy is also less variable in height, closure and crown sizes and, with the exception of *Araucaria*, the tallest trees rarely reach 50 m in height. Other differences are that thick woody lianes, rattan and shrub palms, and fleshy climbers and climbing ferns on tree trunks are less common, but tree ferns are more common. Scrambling bamboo is prominent especially on ridge crests, and tall palms, locally emerging above the forest canopy, are a normal feature.

Most trees present in lowland hill forest also occur in different proportions in alluvium forest and, like the latter, lowland hill forest is very rich in species and very mixed. Frequent canopy trees belong to the genera *Buchanania*, *Celtis*, *Cryptocarya*, *Dysoxylum*, *Elmerrillia*, *Ficus*, *Litsea*, *Pometia* and *Syzygium*. In addition, *Aglaia*, *Canarium*, *Chisocheton*, *Dendrocnide*, *Planchonella* and *Sloanea* are particularly common on the lower slopes on Mount Karimui.

Single trees and groups of *Araucaria hunsteinii* (klinki pine), a species endemic to Papua New Guinea (Enright 1982), are present in both lowland and upland hill forest. The araucarias tend to be most common on ridge crests, and stand out from afar on many of the crests and spurs in the Tua River catchment east and northeast of Karimui, towering some 20–30 m above the associated mixed broad-leaved forest.

Airphoto interpretation indicates that very dense and small-crowned forest southwest of Karimui is associated with limestone, and that an equally small-crowned, but open and landslide-scarred forest occurs on very steep-sided low hills in the southeastern part of the region east of the Purari River.

3. Upland vegetation

The uplands are largely forested, though areas of garden and garden regrowth are present in the northwest and north, and around Karimui. Upland forest is transitional to lower montane forest, a type which is mainly confined to altitudes above 1500 m. The rate of change and the lower limit of lower montane forest appear to be controlled mainly by the frequency and duration of cloud cover close to the ground, factors which generally increase with increasing altitude, but are locally modified by aspect and degree of exposure to the prevailing winds.

3.1. The upland forests

The canopy of average upland hill forest is smaller crowned, more densely closed and more even in height than that of lowland hill forest. Tree density is higher, but the average girth is smaller than in forest at lower altitudes (Fig. 7). Many species characteristic of the lowlands disappear and are replaced by others, notably the oak relative *Castanopsis acuminatissima*. Other frequent canopy trees are *Lithocarpus*, which also belongs to the oak family, and *Canarium*, *Cryptocarya*, *Dillenia*, *Litsea*, and *Syzygium* genera that are also well represented, by different species, in the lowlands. *Nothofagus* or southern beech appears in the upper levels and is also common at lower

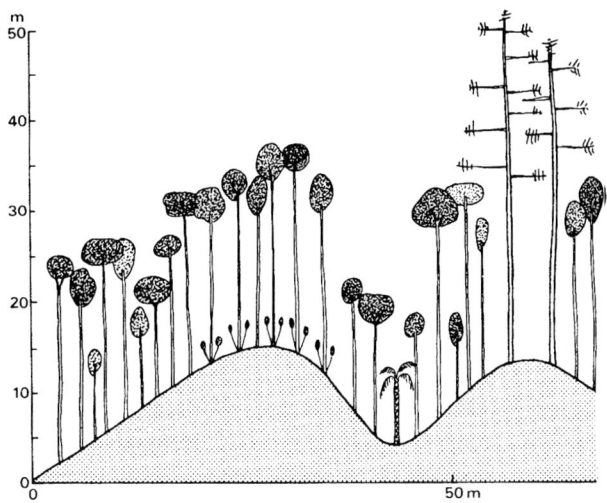

Fig. 7. Vegetation profile of upland hill forest.

altitudes on limestone north of Erave, but generally is more characteristic of the mountain region.

Castanopsis acuminatissima is particularly abundant on ridge crests and upper slopes, where it commonly dominates in the canopy. Such forest has a dense canopy of rounded crowns, an open shrub layer, a very sparse ground cover of herbs, and a thick carpet of fallen leaves. Many *Castanopsis* trees have a ring of coppice shoots around the main trunk. The species appears to regenerate readily, perhaps from rootstock, and often grows densely in young secondary forest. It extends well into lower montane forest to an altitude of about 2300 m.

3.2. Garden regrowth

Abandoned gardens are quickly overgrown by grasses and weeds. *Paspalum conjugatum* is perhaps the most common of the grasses, but others such as *Arthraxon ciliaris*, *Imperata cylindrica*, *Miscanthus floridulus* and *Setaria palmifolia* are usually present and in places co-dominate with *Paspalum*. Besides grasses, bamboo and garden weeds such as *Ageratum conyzoides* and *Urena lobata* are often prominent, and a host of woody regrowth species become established togeth r with or shortly after the herbaceous weeds. They include *Alphitonia incana*, *Castanopsis acuminatissima* and species of *Com-*

239

mersonia, Elmerrillia, Ficus, Macaranga, Rhus, Sloanea and *Trema.*

4. Mountain vegetation

Grassy and low woody regrowth in mosaic with current gardens form the characteristic pattern over most highland valleys and sides (Fig. 8). Generally only pockets of lower montane forest remain on rugged terrain that is less suitable for agriculture, but in places extensive forest tracts persist, as on the flanks of Mounts Giluwe and Hagen, the Sugarloaf, and the Kubor Range.

Around 3000 m a.s.l. lower montane forest gives way to montane forest. Due perhaps to a rather sudden change in climatic conditions from almost daily low cloud-lie for most of the year to more frequent spells of dry sunny weather above about 3000 m, the transition from lower montane to montane forest is more abrupt than that from lowland to lower montane forest.

Extensive stable grasslands akin to those in the lowlands outside the Purari catchment cover the heavily populated valleys of the upper Wahgi, Erave, Mendi and Kaugel Rivers in the western half of the region, and the slightly drier and more seasonal valleys of the upper Asaro and Lamari Rivers in the east. Within the western grasslands, sedge and grass swamp communities occupy ill-drained parts of valley floors and in places cover large areas, as east and west of Kandep and near Egari west of Mount Giluwe. Above about 2600 m, the upper limit of indigenous agriculture, a different grassland composed of high-mountain species occurs in mosaic with forest (Fig. 9), and increases in proportion with increasing altitude. Shrubberies and primary grassland cover much of the area above 3800 m, the approximate present-day limit of tree growth.

4.1. The lower montane forests

The canopy of lower montane forest is generally between 20 and 30 m high, but varies much in height according to growing conditions. It may be 40 m tall on gentle slopes of plateau areas and mountain saddles, and less than 15 m on narrow ridge crests, which have steep slopes and shallow soils, and are swathed in mist more often and for longer periods than the sides of the ridges.

Many tree trunks are low-branched, and bent or leaning, and old trees have thick, crooked and often dead branches. Frequent canopy trees belong to the families Cunoniaceae, Elaeocarpaceae, Fagaceae, Lauraceae and Myrtaceae, the genera *Dryadodaphne, Planchonella* and *Zanthoxylum* and, particularly

240

Fig. 8. Current garden, garden regrowth, and planted groves of *Casuarina* in the Chimbu area at about 2000 m a.s.l.

Fig. 9. Summit area of Mount Giluwe at about 3500 m a.s.l. Grassland and patches of montane forest cover the slopes. Little ponds, some surrounded by sedge swamp, dot the lower parts of the plateau.

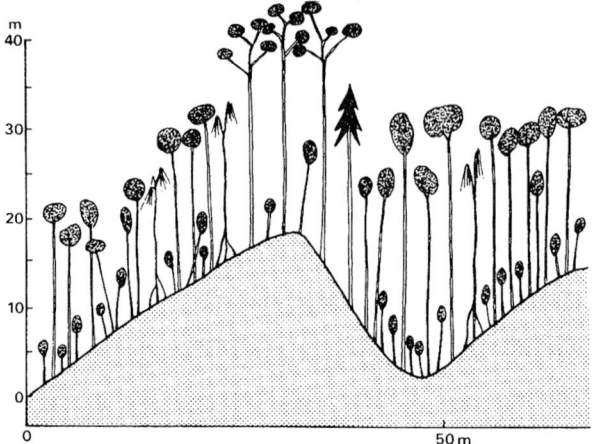

Fig. 10. Vegetation profile of lower montane forest.

in the upper levels, conifers, notably *Podocarpus*. In the lower tree storeys are found *Garcinia, Pittosporum, Polyosma, Quintinia, Saurauia, Sericolea, Symplocos, Tasmannia* (*Drimys*), *Timonius* and Araliaceae. Stilt-rooted pandans are common in places and occasionally reach into the canopy (Fig. 10).

The shrub layer in many places consists of a great number of slender saplings which do not have a high cover, but grow dense enough to hinder passage through the forest. In other places passage is obstructed by tall ferns, or tangles of *Nastus* sp., a thin-stemmed climbing bamboo. Various species of *Cyrtandra, Eurya, Melastoma, Piper, Saurauia* and Rubiaceae are nearly always present. At the highest levels the forest is characterized by tall shrubs in the families Ericaceae, Myrsinaceae and, especially near forest borders with grassland, genera such as *Carpodetus, Coprosma, Olearia* and *Schuurmansia*.

The ground flora of mosses, ferns, herbs and seedlings is also variable in density. In many places mosses almost completely cover the ground and the fallen logs and branches, and in other places ferns, *Begonia, Elatostema* or Zingiberaceae form a dense layer.

Lianes are less common than in lowland forest, and particularly thick woody ones are rare, as are climbing rattan and other palms. Climbing epiphytes include the pandan *Freycinetia*, Gesneriaceae, *Lycopodium* and ferns. Epiphytic mosses become abundant with increasing altitude, as do tree ferns and epiphytic ferns and orchids.

Besides *Castanopsis acuminatissima*, a number of other trees, mainly *Nothofagus* and conifers, tend to form pure stands. Various species of *Nothofagus* are a common component of mixed lower montane forest and,

242

like *C. acuminatissima* and the araucarias, tend to be most frequent on ridge crests and upper slopes, particularly at higher altitudes. However, isolated groves, emerging some 10 – 20 m above the surrounding mixed forest, also commonly occur on side slopes, and in some places *Nothofagus*-dominated forest forms an almost continuous cover over areas in excess of 50 km². Such large areas of beech forest are present on the south side of Mount Giluwe, the northern slopes of the Kubor Range, and on limestone in the southwest of the region. Small areas of stunted beech forest are found on wet sites near Kandep. Regeneration of *Nothofagus* as seedlings and root suckers is usually present both in open forest and under shade, but is densest and most vigorous in patchy openings commonly caused by the death of overmature and possibly even-aged *Nothofagus* trees. The ecology and distribution of *Nothofagus* in New Guinea are discussed in more detail in Ash (1982).

Conifers of the genera *Araucaria, Dacrycarpus, Dacrydium, Papuacedrus, Phyllocladus* and *Podocarpus* dominate in the canopy and emergent tree layers in many localities above about 2400 m. The conifers appear to be well adapted to these altitudes and have a better stem form and reach a larger girth than their broad-leaved associates. Emergent trees of *Papuacedrus papuana* are readily recognized from afar by their narrow conical crowns, and those of *Araucaria cunninghamii* (hoop pine) by their open crowns of long horizontal branches. Patches of conifer-dominated forest are present on the slopes of Mount Giluwe and Mount Hagen, and in the Sugarloaf. Coniferous forest is also found in narrow belts on the sides of intramontane basins between grassland below and mixed forest above. In such situations the more hardy conifers may have started as pioneers where broad-leaved forest was destroyed, or its establishment was prevented, by fire or frost.

Swamp forest in places covers or fringes depressions in valley floors above about 1800 m. The taller trees grow on hummocks separated by pools of water, and form a usually open canopy above a dense layer of small trees and shrubs, and a sparse herbaceous ground cover. Common trees are *Carpodetus, Dacrydium, Pandanus, Podocarpus, Syzygium* and other Myrtaceae. In some places *Nothofagus perryi* or *N. grandis* form an upper storey, and in other places conifers or pandans dominate. Predominance of conifers may have arisen through selective action of extreme frosts. Swamp forest dominated by a tall *Pandanus* with thick and high prop roots occupies ill-drained spots within, or adjacent to forest on dry ground.

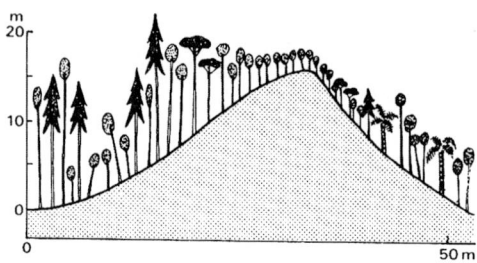

Fig. 11. Vegetation profile of montane forest.

4.2. Montane forest

Montane forest is made up of densely growing trees commonly with thin crooked trunks, and ranging in height from about 18 m at lower levels to 6 m at the tree line. Oaks and southern beeches are absent, and are replaced by trees belonging to the families Ericaceae, Myrsinaceae, Myrtaceae and Rubiaceae. Conifers remain, particularly *Dacrycarpus compactus*. Mature trees of this species have thick and straight stems topped by umbrella-shaped crowns which emerge some 4 – 6 m above the forest canopy (Fig. 11). Tree ferns feature in the shrub layer, but pandans and climbing bamboo are absent, and lianes gradually disappear. Among the most common shrubs are the Rubiaceae *Amaracarpus*, *Coprosma* and *Psychotria*. Mosses, liverworts and filmy ferns mantle the lower branches, tree trunks and exposed roots, and carpet the forest floor. The borders between forest and grassland are marked by a characteristic, fire-seral community of mainly light-demanding shrubs. It forms a buffer between forest and grassland, protecting forest from fire, and enabling forest to recolonize grassland in the absence of fire. Common border shrubs are *Olearia*, *Rhododendron* and *Vaccinium*.

Near the tree line, and on sharp crests, montane forest merges into scrub: a dense growth of shrubs and dwarfed, gnarled trees to about 6 m tall, in which the families Ericaceae, Epacridaceae and Rubiaceae, and the genera *Eurya*, *Olearia*, *Pittosporum*, *Schefflera* and *Tasmannia* are well represented. This type of scrub is most extensive on Mount Wilhelm (Fig. 12), but also occurs throughout the Kubor Range on high ridges, particularly those of Mount Kegeraga.

Fig. 12. Edge of montane scrub on Mount Wilhelm at about 3950 m a.s.l. Tall *Olearia* in centre. Undergrowth of *Coprosma* and other shrubs.

4.3. The grasslands

The relatively stable grasslands of the inhabited highland valleys form usually well-closed communities of tussocky and mainly erect grasses from 1 m to 1.5 m high on favourable sites. They are dominated by species that are common also in the lowlands, but usually have a scattering of grasses that become prominent only at higher altitudes. The main lowland species are *Capillipedium parviflorum*, *Themeda australis*, *Imperata cylindrica* and *Ischaemum*

245

polystachyum, and the mountain grasses are represented by *Dichelachne novoguineensis*.

C. parviflorum is the major dominant in grassland on hill slopes as well as plains, but species composition is mixed and variable, and other grasses may dominate, or co-dominate with *Capillipedium*. These include *Apluda mutica*, *Arthraxon ciliaris*, *Arundinella setosa*, *Eulalia leptostachys*, *E. trispicata*, *Ischaemum barbatum* and *Ophiuros tongcalingii*.

Themeda australis (kangaroo grass) dominates on drier, shallow and gravelly soils of upper hill slopes and crests, usually in association with *Arundinella setosa*. *Imperata cylindrica* takes over after renewed cultivation. This species may remain dominant under conditions of relatively deep soil and regular burning, but is gradually ousted by other grasses on less favourable sites. *Ischaemum polystachyum* typically occupies moist sites in depressions, valley floors, and seepage footslopes, where it is associated with *Paspalum conjugatum*, *P. orbiculare*, and moisture-loving grasses such as *Leersia hexandra*, *Phragmites karka* and species of *Isachne*. In the eastern part of the region, *Themeda intermedia*, with culms growing to a height of 3 m, locally dominates on stream banks and moist river terraces.

Small herbs and sub-shrubs occur throughout the highland valley grasslands. Many belong to the legume family, e.g. *Cassia mimosoides* and species of *Crotalaria*, *Desmodium* and *Indigofera*. Other frequent taxa are *Euphorbia serrulata*, *Osbeckia chinensis*, *Polygala*, *Rhododendron macgregoriae*, the ground orchid *Spathoglottis*, and *Wahlenbergia*. Only a few trees such as *Alphitonia incana* and *Dodonaea viscosa* persist in the open grassland (Henty 1982).

Stream banks are often marked by a woody vegetation which commonly includes species of *Casuarina* and remnants of the original forest, such as oaks and araucarias (Fig. 13). Ferns and tall grasses like *Pennisetum macrostachyum* and *Miscanthus floridulus* often feature in the ground layer.

In contrast to the mainly panicoid grasslands of lower altitudes, the high-mountain grasslands above the upper limit of agriculture at about 2600 m are mainly composed of festucoid grasses (Henty 1982). They grow in tussocks up to 1 m across and from 0.5 to over 1 m high, with smaller tuft grasses, cushion plants, herbs, ferns and mosses covering the spaces between the tussocks. The major dominant grasses are *Chionochloa* (*Danthonia*) *archboldii* and *Deschampsia klossii*. *C. archboldii* occurs mainly on relatively well-drained sites where it forms very large tussocks. The somewhat smaller *D. klossii* tends to be more prominent on wet ground. As in the grasslands at lower altitudes, other grasses dominate or co-dominate locally, and the pattern differs strongly from place to place depending on altitude, soil depth, drainage conditions

246

Fig. 13. Remnant grove of *Araucaria cunninghamii* along Dunantina River at about 1700 m a.s.l. Also shown are stream-bank vegetation and plantings of bamboo and *Casuarina*.

and burning. Common associates of the main grasses are *Agrostis reinwardtii*, *Anthoxanthum horsfieldii* (*A. angustum*), *Arundinella furva* (lower part of the zone), *Deyeuxia brassii*, *Dichelachne novoguineensis*, *Hierochloe redolens*, *Imperata conferta*, *Miscanthus floridulus* and species of *Danthonia* and *Poa*. Single plants and groves of tree ferns, mainly *Cyathea atrox*, are a feature of the high-mountain grasslands. They often grow densest near forest edges and also occupy relatively well-drained stream banks within basin grasslands. Scattered low shrubs such as *Coprosma divergens*, *Hypericum macgregorii*, *Leucopogon suaveolens* and species of *Rhododendron* and *Vaccinium* occur widely and locally grow dense enough to form a heath. Herbs too are common, though leguminous ones are absent; they are most abundant on wet sites and shallow soils, where they grow together with the cushion grass *Danthonia* (*Monostachya*) *oreoboloides* and other low creeping grasses. *Astelia alpina*, *Epilobium*, *Gentiana*, *Lycopodium*, *Potentilla*, *Ranunculus*, various Asteraceae, and the fern *Gleichenia* are among the many herbs that are nearly always present. Sedges are common particularly in wet places, notably *Machaerina rubiginosa* which locally grows in pure swards. Herbaceous swamp communities occupy shallow depressions within the grassed

247

summit plateau areas of Mounts Giluwe and Hagen (Fig. 9), and at higher altitudes are common also in grassland on slopes.

Grasslands above the tree line, i.e. those above about 3800 m, differ in structure from the lower altitude grasslands in that the large bunch grasses are less prominent. In addition to the grasses present in high-mountain grassland below the tree line they contain smaller tussock and sward-forming species of *Danthonia*, *Deyeuxia*, *Festuca* and *Poa* (Walker 1973). Other features are swards of the ferns *Gleichenia vulcanica* and *Polystichum* (*Papuapteris*) *linearis*, and an abundance of low rozette and cushion herbs, mosses and lichens. These become more important with increasing altitude and largely replace the grasses above 4300 m. Tree ferns are rare or absent, and shrubs become lower and sparser with altitude and peter out at about 4400 m.

At higher altitudes aspect plays an increasingly important part in controlling the nature and composition of the vegetation. For example, it has been established (Smith 1980) that on the warmer and less wet east and northeast facing slopes of Mount Wilhelm, tussock grassland and herbs associated with it extend to a higher altitude, and tussocks are larger, than on the damper western slopes. It was also found that shrubs were more numerous, and the forests less widely destroyed by fire and regenerating faster on west-facing slopes. Smith suggests that species distributions are controlled largely by maximum temperatures or some factor related to them, although minimum temperatures cannot be neglected.

4.4. Garden regrowth

Shortly after abandonment, a garden plot becomes overgrown by grasses and weeds. Among the early grasses to become established are *Arthraxon ciliaris*, *Eulalia leptostachys*, *Sacciolepis indica* and *Setaria pallide-fusca*. Pioneer weeds include *Ageratum conyzoides* and other Asteraceae, *Physalis*, *Polygala*, *Sida* and *Verbena bonariensis*. Probably between one and two years after the last harvest, *Miscanthus floridulus* (sword grass) and *Imperata cylindrica* become established, together with species characteristic of stabilized grasslands. If the site is not burned or reworked, sword grass will eventually become dominant and form an almost pure stand 3 m or more high; gradually shrubs and trees begin to appear, and without further interference the community would revert to forest. However, either at this stage or earlier, depending on population pressure and local traditions, the site is commonly cleared again for re-cropping.

Since garden regrowth is affected to various degrees by frequency of burning, length of fallow period, numbers of marauding domesticated pigs, and in places grazing by other livestock, there are many different phases in which *Miscanthus floridulus*, *Imperata cylindrica* and other grasses, together with tree ferns, shrubs, weeds, ferns and sedges occur in different proportions.

Among the most frequent naturally occurring shrubs and trees are *Buddleia asiatica*, *Dodonaea viscosa*, *Ficus*, *Macaranga pleioneura*, *Pittosporum*, *Saurauia*, *Trema* and *Wendlandia*. *Schuurmansia henningsii* and *Timonius belensis* are common particularly near forest borders. Many of the regrowth shrubs and herbs are put to specific uses by the indigenous population (Powell 1976). In addition, garden regrowth contains other useful species which are either self-sown and retained during weeding, or are planted either before or after harvest of the main crop. Among the most conspicuous are groves of *Casuarina* trees planted mainly to provide timber and firewood, *Pandanus conoideus* which produces a rich oily fruit, and *Cordyline* sp. which besides other uses serves as a boundary marker. As planting of *Casuarina* appears to be increasingly popular, the highland valleys are gradually assuming a more woody appearance. A conspicuous but troublesome occupant of old garden sites is *Tritonia crocosmiflora* (*Montbretia*), a probable escape from European gardens which has assumed pest proportions in places.

5. Vegetation history and human impact

During the last glacial episode, which has affected all high mountains in New Guinea, ice reached its maximum extent about 18 – 15 000 years B.P. It has been established (Hope 1980b) that the tree line for the intramontane valleys of the central highlands was then at 2000 – 2250 m, some 1600 m below the present-day limit of tree growth. Pollen records show that in the contemporary montane forests *Nothofagus* was more common than at present, and grassland communities rich in shrubs and tree ferns covered a wide zone above the tree line. As the glaciers began to retreat after 15 000 B.P., the ice-free areas were colonized by herbland, and forest spread upward at the expense of grassland. There is evidence that by about 8500 B.P. climatic conditions were milder than at present, and that the tree line had reached 4000 m altitude. It is also probable that since about 5000 B.P. or somewhat later a deterioration of the climate has caused a slight lowering of the tree line to its present position.

Disturbance of the vegetation by man has partly coincided and interfered with the vegetation shifts brought about by climatic changes after the last glacial era. Archaeological work has established that man has been present in the central highlands for more than 25 000 years. Early agriculture associated with clearance of vegetation has been demonstrated in the upper Wahgi Valley near Mt. Hagen about 9000 years ago, and pollen analytical work at other sites in the same area indicates that clearance of forest occurred at least 5000 years ago (Powell 1976). Until the introduction of sweet potato (*Ipomoea batatas*), perhaps only some 500 years ago, agricultural activities appear to have been restricted mainly to valley floors which were artificially drained, and the adjacent slopes, and staple crops probably included taro (*Colocasia esculenta*) and yam (*Dioscorea* spp.). The arrival of sweet potato, the present staple, which is tolerant of cooler conditions than either taro or yam, has enabled an expansion of agriculture and human settlement to much higher slopes, up to at least 2600 m altitude.

Because of the high population density in the intramontane valleys of the central highlands, the natural vegetation has been more modified here than anywhere else in the Purari catchment. By clearing, cultivation and the use of fire, man has reduced much of the formerly forested land to stabilized grassland, and has converted most of the remaining land below the upper limit of indigenous agriculture into a mosaic of gardens and grassy regrowth. In contrast to populated areas in the uplands region, the proportion of advanced secondary forest is very low, due to the practice of more intensive cultivation and short fallow periods.

Much forest has also been destroyed above the upper limit of agriculture. Although the vegetation here is wet for most of the time, dry spells occur, lasting from a common 2 – 3 days to an occasional fortnight. During such spells, all vegetation, including swamp grasses and even forest, become dry and flammable, and grass fires lit by hunters and travellers may spread into adjoining woody vegetation. As a result of repeated fires, the forest becomes impoverished and is eventually destroyed and replaced by grassland.

Most high-mountain grasslands below the tree line probably occupy formerly forested terrain. Some grassy valleys may be naturally treeless because of low minimum temperatures caused by cold-air drainage, but excessive waterlogging may also play a part, and the area of grassland may have been enlarged at the expense of forest through man-set fires starting from swampy grass patches.

Parts of all high-mountain grasslands are fired each year, (except perhaps on Mt. Wilhelm, where a fire ban has been in force over the last 25 years). Most fires are light and spotty, because of a usual dampness and the presence of many permanently wet and hence less flammable patches. As a result of frequent light firing, the grasslands often consist of a mosaic of short-term successional vegetation stages which relate to the length of time since the last burn. The long-term effect of fire is that only fire-resistant species are able to survive. The natural grasslands above the tree line may be less severely affected by fire because of cooler and wetter conditions, a relative shortage of combustible material, and fewer travellers.

In the uplands and higher parts of the lowlands, gardens, garden regrowth and secondary forest are widespread, particularly in the northwest and around Karimui in the centre. Areas of stable grassland are rare and small.

In the swamps and alluvial plains of the lower Purari, the sparse population, subsisting mainly on a diet of sago and fish, has caused little disturbance to the vegetation, though some has occurred around a few settlements in the tidal freshwater and brackish zones. The river banks are little used for gardening owing to the hazards of flooding, and harvesting of sago in the backswamps causes little damage to the environment. In a narrow broken sandy strip along the coast, much of the original littoral vegetation has either been converted to coconut plantations, or is secondary after gardening.

6. Frost

Ground frosts may occur from about 1500 m upwards, and are common above 2400 m, particularly where cold air drains into more or less enclosed basins. Examples of large 'frost hollows' are the Kandep basin and the upper Mendi and Kaugel valleys. During the dry season the risk of frost is higher because the air is drier and the nocturnal radiation more intense. Above about 2600 m the risk of frost, coupled with generally cold conditions and slow growth of crops, puts a halt to cultivation.

Periodic severe frosts cause extensive damage to both garden crops and regrowth and forest vegetation, occasionally down to altitudes of 1600 m (Brown and Powell 1974). Forest edges are quite often damaged, while larger areas of mixed broad-leaved and beech forests on hill slopes are reported to have been killed occasionally. Once forest has been replaced by grassland, frost retards forest regeneration, and may prevent it altogether.

References

Ash, J., 1982. The *Nothofagus* Blume (Fagaceae) of New Guinea. In: J.L. Gressitt (ed.), Biogeography and Ecology of New Guinea, pp. 355–380. Dr W. Junk, The Hague.

Brown, M. and J.M. Powell, 1974. Frost and drought in the highlands of Papua New Guinea. J. Trop. Geogr. 38: 1–6.

Enright, N.J., 1982. The Araucaria forests of New Guinea. In: J.L. Gressitt (ed.), Biogeography and Ecology of New Guinea, pp. 381–400. Dr W. Junk, The Hague.

Henty, E.E., 1982. Grasslands and grassland succession in New Guinea. In: J.L. Gressitt (ed.), Biogeography and Ecology of New Guinea, pp. 459 – 473, Dr W. Junk, The Hague.

Hope, G.S., 1980a. New Guinea mountain vegetation communities. In: P. van Royen (ed.), Alpine Flora of New Guinea, Vol. 1, pp. 153 – 222. J. Cramer, Vaduz.

Hope, G.S., 1980b. Historical influences on the New Guinea flora. In: P. van Royen (ed.), Alpine Flora of New Guinea, Vol. 1, pp. 223 – 248. J. Cramer, Vaduz.

Paijmans, K., 1969. Vegetation and ecology of the Kerema – Vailala area, pp. 95 – 116. C.S.I.R.O. Land Research Series, No. 23. C.S.I.R.O., Melbourne.

Paijmans, K., 1975. Explanatory notes to the vegetation map of Papua New Guinea. C.S.I.R.O. Land Research Series, No. 35. C.S.I.R.O., Melbourne.

Paijmans, K., 1976. Vegetation. In: K. Paijmans (ed.), New Guinea Vegetation, pp. 23 – 105. Australian National University Press, Canberra.

Paijmans, K., 1980. Sago palm ecology in New Guinea. In: W.R. Stanton and M. Flach (eds.), Sago – the Equatorial Swamp as a Natural Resource, pp. 9 – 12. Martinus Nijhoff Publishers, The Hague.

Petr, T. and J. Lucero, 1979. Sago palm salinity tolerance in the Purari River delta. In: T. Petr (ed.), Purari River (Wabo) Hydroelectric Scheme Environmental Studies, Vol. 10: Ecology of the Purari River catchment, pp. 101 – 106. Office of Environment and Conservation, Waigani, Papua New Guinea.

Powell, J.M., 1976. Ethnobotany. In: K. Paijmans (ed.), New Guinea Vegetation, pp. 106 – 183. Australian National University Press, Canberra.

Robbins, R.G., 1970. Vegetation of the Goroka – Mount Hagen area, pp. 104 – 118. C.S.I.R.O. Land Research Series, No. 27. C.S.I.R.O., Melbourne.

Robbins, R.G. and R. Pullen, 1965. Vegetation of the Wabag – Tari area, pp. 100 – 115. C.S.I.R.O. Land Research Series, No. 15. C.S.I.R.O., Melbourne.

Saunders, J.C., 1965. Forest resources of the Wabag – Tari area, pp. 116 – 131. C.S.I.R.O. Land Research Series, No. 15. C.S.I.R.O., Melbourne.

Saunders, J.C., 1970. Forest resources of the Goroka – Mount Hagen area, pp. 119 – 125. C.S.I.R.O. Land Research series, No. 27. C.S.I.R.O., Melbourne.

Smith, J.M.B., 1980. Ecology of the high mountains of New Guinea. In: P. van Royen (ed.), Alpine Flora of New Guinea, Vol. 1, pp. 111 – 132. J. Cramer, Vaduz.

Walker, D., 1973. Highlands vegetation. Aust. Nat. Hist. 17(12): 410 – 414, 419.

2. The wildlife of the Purari catchment

J.C. Pernetta

1. Introduction

The wildlife of Papua New Guinea is extremely diverse, with an unusually high rate of endemism in many groups, particularly when one considers the relatively recent geological origin of this island land mass. Several factors appear to have contributed to this diversity. The New Guinea land mass is derived from uplift of the Australian continental shelf, which together with part of the northern Australian land mass has moved north and collided with the island arcs running from Sulawesi out into the Pacific. It is thus an area of recent orogenic and volcanic activity. As a consequence the physiography of the country is diverse with the flat plains areas to the south of the central mountain cordillera, the mountain trough area between the cordillera and the northern ranges, and finally the small area of the northern coastal plains.

The recent uplift and formation of the island results in a relatively high rate of erosion and the uplifted areas are generally in a highly dissected state, undergoing active degradation.

The physiography results in a full range of altitudinal zones, from humid tropical lowlands, or coastal savannah in rain shadow areas, to high altitude moss forests and alpine grasslands above the tree line. The mountain cordillera functions as a barrier to dispersal between the northern and the southern coastal plains and foothill forests, allowing regional differentiation in species or species complexes. In addition the high mountains function as habitat islands allowing isolates to develop as separate populations. Thus the fauna is regionally and altitudinally differentiated. Furthermore the location of the island between 2° S and 10° S means that the climate is generally tropical, and terrestrial habitats are dominated by rain forests of various types.

The diverse wildlife has in all areas played a vital role in the subsistence

Petr, T. (ed.) The Purari – Tropical environment of a high rainfall river basin
© *1983, Dr W. Junk Publishers, The Hague / Boston / Lancaster*
ISBN-13: 978-94-009-7265-0

economy of the people and indeed continues to do so (Pernetta and Hill 1981). This role varies from area to area and species to species, with some birds being prized for their decorative plumes in one area and ignored or protected elsewhere (Heaney 1982). The relationship between man and wildlife in New Guinea has not always been one of conservation (for a review of traditional conservation practices in Papua New Guinea, see Morauta et al. 1982). Numerous examples of local extinctions of certain species are cited in Bulmer (1982). Such local extinctions result from two pressures which appear to have operated over many thousands of year; the first is direct over-hunting, which may have caused extinctions such as those of *Protemnodon* and thylacines in New Guinea; the second is habitat modification associated with agricultural activities (Golson 1981; Powell 1982). Habitat modification has resulted in swamp drainage, dating back to 9000 B.P. at Kuk in the Wahgi Valley, the extension of anthropogenic grasslands throughout the highlands as early as 5000 B.P. (Powell 1982) and consequent reduction of wildlife habitats. Such pressures are increasing as a result of changes which are currently underway within the country; monetisation of resources, human population increase, more efficient technology and many others (Haines 1982; Johannes 1982; Pernetta and Hill 1982; Hill et al. 1982; Pernetta and Burgin, this volume).

Despite the fluidity of the current situation, it remains true that wildlife resources continue to play a major role in the subsistence economy as food, by providing decorative items for ceremonial wear and as part of the core of belief and myths which is central to all Papua New Guinea cultural groups (Waiko and Jiregare 1982).

2. The Purari catchment

The headwaters of the Purari River arise in the central cordillera, and upstream of Wabo the river drains some 26 300 km² of mountainous terrain within the Western Highlands, Southern Highlands, Simbu and Eastern Highlands provinces. Some of these areas have the highest human densities in the country (200/km²), where the rate of wildlife depletion is extremely rapid and where pig rearing plays an important role in providing dietary protein to replace the wildlife. Below about 1500 m altitude as the river passes through the Karimui area however, the population density drops markedly and the people are semi-nomadic, hunter-agronomists who rely heavily on wildlife resources for supplies of dietary protein. Lower still in the deltaic area of the Purari basin, sago (*Metroxylon sagu*) replaces taro (*Colocasia esculenta*) in the highland fringes, and sweet potato (*Ipomea batatas*) in the highlands as

254

the major source of dietary energy. Similarly the increased abundance, diversity and size of fish in this area results in these and other aquatic resources reducing the importance of wildlife to the diet (Haines 1977; Liem, this volume). Once on the coast, fish and molluscs with a few species of crustaceans dominate the protein component of subsistence diets, with additional items such as turtle and dugong (Pernetta and Hill, in press).

Within this area the habitats range through the coastal mangrove swamps and their extensive backing of palm and tree swamp which encloses a small area of grass swamp and savannah, to a small belt of lowland alluvial forest followed by the extensive area of lowland hill forest which ranges up to 1000 m altitude and forms the dominant habitat for the middle section of the river. Above 1000 m one passes into lower montane forest zones, whilst above 2800 – 3000 m one encounters the upper montane forest.

In the Ok-speaking area of the Star Mountains to the west, bordering Irian Jaya and outside the Purari River basin, one group of people exploits resources within an altitudinal band from 500 to 3000 m (Hyndman 1979). These people recognise four major ecological zones, within which different mammals, birds and other vertebrates are hunted. These zones correspond quite well with floristically recognised altitudinal zones and the area is in many respects similar to the area under consideration. Unfortunately, apart from Liem's work (this volume), little else has been undertaken in the Purari catchment which is directly concerned with wildlife resource exploitation, so this chapter draws heavily upon the studies of Hyndman (1979) and Pernetta (1982) in the area to the west. Recently Hide has undertaken a study of resource utilisation in the Karimui area as part of the Simbu Land Use Project. This study includes examination of horicultural practices and wildlife exploitation. Although the work is currently in progress, some preliminary results of identifications of hunting trophies made by the present author are used as the basis for the following discussion of mammals.

Data concerning the birds, reptiles and amphibians has been assembled from published sources together with information derived from the collections of the National Museum and the Biology Department Museum of the University of Papua New Guinea, and some personal observations.

3. Wildlife resources

3.1. Amphibians

Frogs are widely eaten throughout the country, although in some areas such as the Star Mountains they form a major part of the diet of women and children only (Hyndman 1979). The ecology of amphibians in this country is poorly understood, although the work of Menzies, and Zweifel and Tyler has contributed to a partial understanding of the distribution and relationships of these animals. Their biogeography is reviewed by Zweifel and Tyler (1982), who list some 184 species in 22 genera, whilst recognising that the total number of New Guinea species may exceed 200. It is clear that altitudinal zonation of frog species occurs throughout the country (Menzies 1975; Zweifel and Tyler 1982), with some species being restricted to high altitudes, others to low altitude zones.

A number of species such as *Litoria angiana* and *Rana* sp. indet. are adapted to torrent breeding in the higher altitude zones, whilst closely related congeneric species such as *Litoria arfakiana* and *Rana grisea* are associated with lower altitudes or swampy areas. Within the frog fauna are found representatives of four families. The introduced cane toad, *Bufo marinus*, found in the Moresby savannah areas and elsewhere, is the single representative of the fifth family, the Bufonidae, and appears not to be eaten anywhere in the country. The dominant family, the Microhylidae, contains approximately 84 species that are burrowing, terrestrial or scansorial and are generally of small size. In contrast the members of the Ranidae are generally larger, terrestrial or arboreal, and the number of species is few, approximately 20. The family Leptodactylidae is also represented by a few species (6), most of which are found in and around the mountain cordillera. The last family, the Hylidae, is represented by some 71 species in two genera, *Nyctimystes* and *Litoria*; frogs in this family are either terrestrial, such as the rocket frog, *Litoria nasuta*, or arboreal such as *Litoria caerulea* and *Litoria infrafrenata*.

The present list of species known from the delta and lower catchment area is small, less than 10% of the total fauna of the country (Pernetta and Burgin 1980), and further work will undoubtedly increase the list of species, particularly in the case of the smaller hylids and microhylids. The presence of only two *Cophixalus* species in the delta region undoubtedly reflects a lack of data rather than an absence of other species in this genus from the area. One might expect a considerable increase in the number of species in this and other genera of small microhylids when further work is undertaken.

Asterophrys turpicula, a burrowing microhylid of large size, might well be expected to occur in the Pawaia speaking area of the catchment, being found in gardens and secondary regrowth areas in the Star Mountains (Pernetta 1982). It is replaced in primary rain forest areas by the microhylid, *Sphenophryne cornuta*, which is also recorded from the upper Purari area.

Litoria bicolor or a member of this species complex could well occur in the unexplored areas of grass swamp behind the mangrove areas of the delta, whilst a variety of species such as *Litoria iris*, *L. darlingtoni* and *L. genimaculata* might be expected in the forested areas above 1400 m. *L. angiana* and a member of the *L. arfakiana* group of species might also be expected to occur in the catchment as they are torrent breeding frogs of wide distribution. *Nyctimystes humeralis*, another widespread species which never appears to occur at high density, should also be present in this area. It would be interesting to know whether *N. papua* or *N. disrupta* occur in the Purari catchment since this is the likely area of overlap or junction between the distributions of these two species.

Species such as *Rana grisea* might be expected to occur in swampy areas bordering rivers in the catchment above 1400 m, whilst the stream breeding frog, *Rana* sp. (Menzies 1975, p. 24), might be expected in the vicinity of small streams within the catchment. *Rana arfaki*, the large river frog, although recorded from the Purari in the vicinity of Poroi (Pernetta and Burgin 1980), does not normally occur above 1000 m altitude.

All the amphibians discussed are edible although their importance to the diet as a whole is difficult to assess. In some places specific 'frogging' expeditions are undertaken, often by women to provide protein for consumption. It is probable that considerable consumption of these animals and smaller reptiles may occur in the garden areas away from the house, thus their importance may be greater for women and children than for men and has certainly been underestimated in many dietary studies to date. In the Ok-speaking areas of Western Province reptiles and amphibians are treated solely as food sources for women and uninitiated males, adult males being permitted to eat only certain game animals (Hyndman 1979).

On the basis of the above one might suggest that the data concerning the distribution of amphibians within the Purari catchment is both inadeqate and sufficiently scanty that the only conclusion which could be drawn is that further intensive survey work is required as a matter of urgency.

3.2. Reptiles

In essence the same comment may be applied to reptiles as has been applied above to amphibians. The basic survey data for the area under consideration is both sketchy and incomplete, allowing only certain broad generalisations to be drawn from the data pertaining to the area. It is almost certain that crocodiles and turtles do not penetrate further into the Purari catchment than the area of Hathor Gorge, which must form a natural barrier to upstream migration of these animals. Although they are important subsistence resources in the deltaic and lower Purari areas, the importance of turtles, in particular the pitted shelled turtle, *Carettachelys insculpta*, is a seasonal one with major harvesting occurring during the period of nesting which runs from October to January. Although nesting for this species is recorded above Wabo (Pernetta and Burgin 1980), it is undoubtedly a more important resource lower in the delta (Liem, this volume).

The larger reptiles of the catchment area include a number of pythons which serve as subsistence food items (Hide, personal communication), and goannas which are eaten and the skins from which are used as drumskins. The most important species are: *Chondropython viridis* and *Varanus indicus*, although species such as *V. prasinus*, *V. salvadori*, *Liasis albertisi* and *L. amethystinus* might be expected to penetrate far into the foothill rain forest. It is possible that the protected species *L. boelini* occurs in the mid-catchment area, replacing the more widely distributed and abundant *L. albertisi*, although the former species is not recorded to date (McDowell 1975).

Amongst the snakes found in the catchment one might expect the following species: *Amphiesma montana*, and a variety of burrowing snakes in the genera *Typhlops* and *Typhlina*, although none are recorded to date (McDowell 1974). Two species of *Stegonotus*, *S. cucullatus* and *S. diehli*, are known from the catchment, of which the former is eaten in the Karimui area(McDowell 1972).

Amongst the endemic snake species known from the catchment area are: *Aspidomorphus muelleri* and *Micropechis ikaheka*, which are widespread; *Toxicocalamus stanleyana* and *T. preussi* from the lowland areas; *T. spilolepidus* and *T. loriae* from the highland area of the catchment (McDowell 1967, 1969, 1970). The altitudinal separation of the *Toxicocalamus* spp. is paralleled in the case of the arboreal species in the genus *Candoia*, with *C. carinata* occurring at lower altitudes, and *C. aspera* replacing it at higher altitudes (McDowell 1979).

Amongst the various skinks known to occur in the area is one conspicuous absentee, *Tiliqua gigas*, the giant blue tongue which is abundant in the lowland forest areas along the south coast from the Trans Fly to Milne Bay. It forms an important food item in the Western Province and is known to occur in the Kikori delta area immediately to the west of the Purari. This species is a forest floor animal feeding on a variety of invertebrates and fallen fruit; it is not recorded above 1000 m altitude but is probably abundant throughout the area below this altitude. The large agamid lizard, *Gonocephalus dilophus* is known from the forest areas below Wabo but probably does not occur above 1000 m altitude where the two congeneric species *G. modestus* and *G. auritus* might be expected to occur. A large increase in the number of *Sphenomorphus* and *Emoia* species can be expected if further survey work is undertaken in the area.

Geckos are recorded from the lower Purari (Pernetta and Burgin 1980) and include such widespread species are *Hemidactylus frenatus* and *Lepidodactylus lugubris* associated with buildings, and *Gehyra mutilata* in the coastal area. *Cyrtodactylus loriae* and *Gecko vittattus* can be expected to occur widely throughout the catchment together with a variety of other species. Again this group of animals is not well collected and their status in the catchment is unknown.

3.3. Birds

Some 670 named forms of 570 currently recognised species of birds are known as residents from the island of New Guinea, more than are known from the continent of Australia (Schodde 1973, Table 1; Pratt 1982). Of these over 350 species are known from the catchment area (Diamond 1972). Coates and Lindgren (1978) record some 262 species of avifauna in the Star Mountains area of the Western Province, in an area of similar range and topography to the middle section of the Purari catchment. Ashford (1979) provides an annotated list of 70 species sighted around the dam site at Wabo. He notes that the majority of these are open environment species which must be recent invaders to the area following the clearing of the camp site. Liem and Haines (1977) record some 89 species as occurring between the dam site and the coast and list 14 species which are of subsistence importance in the mangrove and swamp areas of the delta (see also Liem, this volume). In the Moresby area some 27 species are known to breed or reside in mangrove swamps, of which 6 are apparently restricted to this habitat (Anon., cited in Pratt 1982). It is probable that a detailed study of the avifauna of the Purari mangrove habitat would reveal many more species, due to the greater extent of the habitat in this area.

Schodde (1973, Table 12) provides a list of 17 bird species which are rare, locally endemic, or otherwise noteworthy in the area of the southern scarp of the Kratke Range along the Purari and Kikori rivers between 0 and 3400 m altitude. These include three birds of paradise, two species of *Pachycephala*, the goura pigeon and the megapode *Talegalla fuscirostris*. Diamond (1971, 1972) discusses the anomalous distribution of many lowland species in the Karimui basin which serves as an area of intrusion of lowland species into the highlands section of the catchment. A number of endemic forms and species are described by Diamond (1967).

The New Guinea bird fauna contains numerous species which have their distributions centred on the lowland and hill forests below 500 m, including pigeons of the genera *Ducula*, *Ptilinopus* and *Goura*, lories such as *Lorius* and *Chaleopsitta* spp., kingfishers, *Ceryx* and *Tanysiptera* spp. and many others (Pratt 1982). Above 500 m are found some 190 species of birds which are confined to montane habitats; Diamond (1972) remarks that some 10 of these species 'drop out' from the Purari catchment area of the central cordillera, being present in the mountains of the eastern end of the island and in Irian Jaya (Indonesia) to the west.

Within the catchment the mid-montane grasslands between 600 and 2000 m support an impoverished bird fauna of only some 16 species in comparison with the richer forest habitats. It is generally agreed that these grasslands are largely anthropogenic (Powell 1982) and thus more recent than other habitats. This may well be a causal factor in determining the low number of bird species resident in this habitat which parallels the findings of Dwyer (1982) for mammals. The smaller areas of alpine grassland within the catchment also support a small endemic bird fauna (10 species, Pratt 1982) although this habitat may well have been more extensive during colder climatic phases in the past.

Perhaps the species of greatest subsistence importance for food over the bulk of the catchment area are the megapodes and cassowaries, all of which may function as key species which are the subject of particular subsistence activities. Megapode egg collection is a specialised activity carried out throughout the country wherever these birds are found. *Aepypodius*, *Talegalla* and *Megapodius* species are all present in the catchment, replacing one another altitudinally (for other examples of altitudinal replacement see Diamond 1971, 1972; Pratt 1982). The taking of eggs is regulated only through the practice of 'limited entry'; that is only the traditional owners of the land may collect from known egg mounds. The cassowaries are of importance for food, both as eggs and adults, whilst the plumes are used for adornment both of the person and of artefacts, and the bones are used for knives and arrow points. Over much of their highlands range the cassowaries have declined or disappeared in recent years although they are probably still abundant in the more sparsely populated parts of the catchment. Unlike megapodes these birds are often maintained in a semi-domesticated state in the village although they rarely lay eggs under such conditions.

In addition to these species, nearly all pigeons are hunted for food, whilst birds such as the birds of paradise and Pesquet's parrot are subject to hunting pressure for their plumes.

3.4. Mammals

The mammals of New Guinea are treated in four groups for the purposes of this review: the monotremes, the marsupials, the rodents and the bats.

Only two monotreme species are known from the country, the endemic, long-beaked echidna, *Zaglossus bruijni*, and the short-beaked echidna, *Tachyglossus aculeatus*. The latter is a widespread species of lower altitudes along the south coast, and although not recorded from the Purari area, might be expected to occur in the drier foothills forest and savannah areas. The endemic long-beaked echidna is rare throughout its known range and appears to be an animal of dense, undisturbed forest at higher altitudes. It occurs in the Karimui area and may be widely distributed in the catchment in the more isolated areas. This protected species appears to be susceptible to hunting pressures and its status in the country as a whole is currently uncertain.

Aspects of the altitudinal zonation of marsupials are dealt with by Ziegler (1977) in his general review of the New Guinea marsupial fauna. More recently however the work of Robin Hide in the Karimui has resulted in the collection of over a thousand mammal specimens which are briefly reviewed in the following paragraphs.

Of the thirteen macropods recorded from New Guinea a number are known to occur in the catchment area. Both *Dendrolagus dorianus* and *D. goodfellowi* appear to be relatively common in the highlands' fringe area, although the status of *D. matschie* remains unresolved (Lidicker and Ziegler 1968). The two tree kangaroos, *D. dorianus* and *D. goodfellowi* are declining throughout their highlands' range as a result of over-hunting in the more populated areas (George 1979). Perhaps the most abundant macropod of the forests above 800 m altitude is the lesser forest wallaby, *Dorcopsis vanheurni*, which is replaced in grassland areas at high altitude by *Thylogale bruijni*. Whether *Dorcopsis* occurs in the catchment is unclear although *D. veterum* could be expected to occur in areas below 1000 m.

In the case of phalangers, a number of cuscus species are known to occur within the catchment, the most important and widespread of which is probably the semi-terrestrial cuscus, *Phalanger gymnotis*. The large species *P. maculatus* and the smaller *P. orientalis* occur throughout the bulk of the altitudinal range, but above 1400 m species such as *P. vestitus* and *P. carmelitae* become increasingly important. The taxonomy and systematics of New Guinea *Phalanger* is confused, and the actual number of species present is probably greater than the five currently recognised by Ziegler (1977). Of the ringtails, several species are common, such as the widespread coppery ringtail, *Pseudocheirus cupreus*; up to 2000 m one also finds *P. corinnae*, whilst at higher altitudes one might expect to find *P. forbesi* and *P. mayeri*, a moss forest denizen. *Distoechurus pennatus*, the feather-tailed possum, is common within the altitudinal range of 1400 m to 2000 m, although it also occurs as low down as Wabo. The sugar glider, *Petaurus breviceps*, is also widespread. A single specimen of a very large *Petaurus* species is present in the Karimui collections; the relationships of this to the unnamed species cited by Ziegler (1977) is unknown. *Dactylopsila trivirgata* appears widespread, although not common throughout the Karimui area, whilst *Dactylopsila palpator* occurs only at higher altitudes in moss forest areas.

Of the bandicoots a number are known from the area including *Echymipera rufescens* and its higher altitude replacement *E. kalubu*, as in the Star Mountains area these two species have a broad zone of overlap covering much of the Karimui area and smaller parts of the catchment. *Isoodon macrourus* is a species of lowland savannah and although not recorded from the Gulf Province, may be present in the small area of lowland savannah and alluvial forest backing the swamp areas. The four *Peroryctes* species present in New Guinea all tend to occur commonly at higher altitudes than *I. macrourus* and *E. rufescens*, indeed *P. longicauda* only appears in faunal collections as the frequency of *E. rufescens* declines. Although not currently recorded from the catchment, *P. raffrayanus* might be expected to occur in the dense forest areas where disturbance is minimal.

In the case of the dasyurids the status of many species is difficult to ascertain. These animals appear to occur only at low density and are difficult to trap. Although the people of the Karimui area have a name for the native spotted cat, *Dasyurus albopunctatus*, it appears to have declined in numbers recently. Dwyer (1982) suggests that the recent decline of *D. albopunctatus* over much of its range results from competition with feral domestic cats. Hide has recently collected hunting trophies of feral cats in the Karimui area suggesting that this animal is well established in the area. The only other dasyurid definitely recorded from the catchment area is *Myoictis melas*, although others such as *Neophascogale lorentzi* and *Antechinus* spp. are probably also present.

262

The native rodents of Papua New Guinea total 52 species in 23 genera, with the addition of 6 introduced species, 5 *Rattus* and *Mus musculus*, the house mouse (Menzies and Dennis 1979). The native rodents perhaps serve as better habitat and altitudinal indicators than any other mammalian group, with species in the larger genera *Rattus* and *Melomys* segregating on altitudinal and habitat bases. Dwyer (1982) details changes in the rat fauna of areas subject to varying degrees of disturbance through subsistence activities and shows that both the abundance of individual species and the diversity of species declines with disturbance. He suggests on the basis of these data (Dwyer 1982) that most species are adapted to forest habitats of different types, a conclusion paralleling the findings of Schodde (1977) and Gressitt and Ziegler (1973) for New Guinea faunas generally, and Pernetta and Watling (1978) for Fijian birds.

Numerous examples of altitudinal separation of species in the catchment may be cited; for example, the widespread lowland rat *Uromys caudimaculatus* is found throughout the south coast area up to 1800 m, and above 2000 m it is replaced by the congeneric and larger species *Uromys anak*. Between 1800 and 2000 metres is a zone of overlap between these species. Both species are widely eaten throughout their geographic ranges. The species of *Melomys* are also altitudinally distributed with *M. fellowsi* and *M. rubex* occurring in forest above 2200 metres, *M. levipes* between 2000 and 1500 and *M. platyops* from 1500 down. All the foregoing species are forest animals whilst *M. rufescens* may be encountered in grassland and secondary regrowth areas below 2000 m. Above 1500 m in the Karimui area the most common members of the genus *Rattus* are *R. niobe* and *R. ruber* with the latter species passing into lower altitudes than the former. At lower altitudes one encounters *R. leucopus*.

In addition to the above species the two giant rats, *Hyomys goliath* and *Mallomys rothschildi* both occur in the catchment above 1500 m, where because of their large size and relative abundance they are important subsistence dietary items. The only other two genera presently represented in collections from the Karimui area are *Chiuromys* and *Pogonomys*, both members of Tate's 'Old Papuan Genera' (Menzies and Dennis 1979). Three species are represented: *Pogonomys loriae*, which constructs underground burrows in forests between 1400 and 2000 m altitude; *Chiuromys vates* and *C. lamia*, both of which nest in tree holes over the same altitudinal range.

Of the introduced rodents only *Rattus exulans* occurs in the Karimui area where it is closely associated with houses and gardens. A number of other species can be expected in the catchment area, particularly in habitats above 2000 m which have not been well collected. Species such as *Anisomys imitator, Lorentzimus nouhysi, Macruromys makor, Hydromys habbema,*

Crossomys moncktoni, Parahydromys asper, Neohydromys fuscus, Pseudohydromys murinus, Leptomys elegans, Mayermys ellermani and *Microhydromys richardsoni* are all rare throughout their known ranges which encompass the catchment of the Purari. In general these are either semi-aquatic species or small moss mice whose altitudinal range is generally above 2500 m.

In the case of the bats little or no distributional data is available for the majority of the species; the two species of current subsistence importance are *Pteropus neohibernicus* and *Dobsonia moluccensis*, both large-sized fruit bats which are commonly eaten throughout the mainland areas of New Guinea. A number of other *Pteropus* spp. might be expected in the area together with other Megachiroptera such as *Rousettus* spp.. *Syconycteris crassa* is known from owl pellets collected in the Karimui area. Of some interest is the species *Aproteles bulmerae* first described on the basis of subfossil material by Menzies (1977) from Kiowa rock shelter in the highlands. From an examination of these remains Menzies concluded that the disappearance of this species from the central highlands around 10 000 B.P. resulted from over-hunting. More recently the species has been found to be extant in the Star Mountains area (Hyndman and Menzies 1980), where it is still exploited for food. Whether it still survives in more isolated areas of the catchment is open to question.

Of the Microchiroptera a variety of species in the families Emballonuridae, Rhinolophidae, Molossidae and Vespertilionidae can be expected in the catchment area.

4. Discussion

The current state of knowledge concerning the distribution patterns of terrestrial vertebrates on the mainland of New Guinea is both sketchy and widely scattered throughout the literature although the two volumes edited by the late Dr Gressitt provide a masterful summary of knowledge to date. Individual work of a concentrated nature, in particular locations such as that of Diamond (1972) on birds or the work of various researchers in the vicinity of the Wau Ecology Institute provide detailed insights into some locations. Thus similarly the volume of studies undertaken as part of the Wabo Hydroelectric Scheme Environmental Studies, the Simbu Land Use Project in the Karimui area, and more recently the impact work carried out in association with the Ok Tedi gold and copper mine in the Western Province provide a starting point for building a more comprehensive countrywide picture of patterns of distribution. However is is doubtful whether anything approaching a complete picture of vertebrate distributions in Papua New Guinea will be achieved before the middle of the next century at the rate at which work is currently being undertaken. By this time developments of large-scale cash cropping, mining and other types will have radically altered the natural pattern of distribution.

As a generalisation one can state that the area of the Purari catchment is such that most habitat types and altitudinal zones are represented within it; thus species with known distributions south of the main cordillera can all be expected to occur within the catchment. Already areas of the catchment with their high human population densities show marked depletion in their wildlife stocks such that for many areas animals hunted regularly within living memory have already disappeared.

The direct impact of a hydroelectric scheme such as that proposed for the Wabo site is likely to prove small when the terrestrial vertebrates of the area are concerned, since the area to be inundated is very small in relation to the total catchment. Of far greater impact will be associated indirect effects such as increased hunting pressure resulting from higher population densities, greater cash flow and easier access to remoter areas. The change in the distribution of people will result in increased density in what is now one of the least populated areas of Papua New Guinea. This present low density indeed forms the basis for a number of previous proposals to resettle and develop the area. Any such scheme will have far greater impact upon the wildlife than the formation of a single hydroelectric dam.

References

Ashford, R.W., 1979. A preliminary list of birds for the Purari River between Wabo and Baimuru (Gulf Province, Papua New Guinea.). In: T. Petr (ed.), Purari River (Wabo) Hydroelectric Scheme Environmental Studies, Vol. 10: Ecology of the Purari River catchment, pp. 3 – 8. Office of Environment and Conservation, Waigani. Papua New Guinea.

Bulmer, R., 1982. Traditional conservation practices in Papua New Guinea. In: L. Morauta, J.C. Pernetta and W. Heaney (eds.), Traditional Conservation in Papua New Guinea: Implications for Today, pp. 59 – 78. Institute of Applied Social and Economic Research. Monograph, 16. Boroko, Papua New Guinea.

Coates, B.J. and E. Lindgren, 1978. Ok Tedi birds. Ok Tedi Development Company, Boroko, and Office of Environment and Conservation, Waigani, Papua New Guinea.

Diamond, J.M., 1967. New subspecies and records of birds from the Karimui basin, New Guinea. Amer. Mus. Novitates, No. 2284, 17 pp.

Diamond, J.M., 1971. Birds of the Karimui basin, New Guinea. Nat. Geog. Soc. Research Report. 1965 Projects, pp. 69 – 74.

Diamond, J.M., 1972. Avifauna of the Eastern Highlands of New Guinea. Nuttall Ornithological Club. Publ., No. 12. Cambridge, Mass.

Dwyer, P.D., 1982. Wildlife conservation and tradition in the Highlands of New Guinea. In: L. Morauta, J.C. Pernetta and W. Heaney (eds.), Traditional Conservation in Papua New Guinea: Implications for Today. pp. 173 – 190. Institute of Applied Social and Economic Research. Monograph 16. Boroko, Papua New Guinea.

George, G., 1979. The status of endangered Papua New Guinea mammals. In: M. Tyler (ed.), The Status of Endangered Australian Wildlife. Proc. of the Centenary Symposium of the Royal Zoological Society of South Australia, pp. 93 – 100. Adelaide, Australia.

Gressitt, J.L. and A.C. Ziegler, 1973. The effect on fauna of the loss of forests in New Guinea. In: A.B. Costin and R.H. Groves (eds.), Nature Conservation in the Pacific, pp. 117 – 122. Australian National University Press, Canberra.

Golson, J., 1981. Agriculture in New Guinea: the long view. In: D. Denoon and C. Snowden (eds.), A History of Agriculture in Papua New Guinea, pp. 33 – 42. Institute of Papua New Guinea Studies, Boroko, Papua New Guinea.

Haines, A.K., 1977. The subsistence fishery of the Purari delta. Science in New Guinea 6: 80 – 95.

Haines, A.K., 1982. Traditional concepts and practices and inland fisheries management. In: L. Morauta, J.C. Pernetta and W. Heaney (eds.), Traditional Conservation in Papua New Guinea: Implications for Today, pp. 279 – 292. Institute of Applied Social and Economic Research. Monograph, 16. Boroko, Papua New Guinea.

Heaney, W., 1982. The changing role of bird of paradise plumes in bridewealth in the Wahgi Valley. In: L. Morauta, J.C. Pernetta and W. Heaney (eds.), Traditional Conservation in Papua New Guinea: Implications for Today, pp. 227 – 232. Institute of Applied Social and Economic Research. Monograph, 16. Boroko, Papua New Guinea.

Hill, L., J.C. Pernetta and B. Rongap, 1982. The traditional knowledge base: implications and possibilities for contemporary Papua New Guinea. In: L. Morauta, J.C. Pernetta and W. Heaney (eds.), Traditional Conservation in Papua New Guinea: Implications for Today, pp. 349 – 362. Institute of Applied Social and Economic Research. Monograph, 16. Boroko, Papua New Guinea.

Hyndman, D.C., 1979. Wopkaimin subsistence: cultural ecology in the New Guinea highland fringe. Ph.D. Thesis. Unpublished. University of Queensland.

Hyndman, D.C. and J.I. Menzies, 1980. *Aproteles bulmerae* (Chiroptera; Pteropodidae) of New Guinea is not extinct. J. Mammal. 61: 159 – 160.

Johannes, R.E., 1982. Implications of traditional marine resource use for coastal fisheries development in Papua New Guinea. In: L. Morauta, J.C. Pernetta and W. Heaney (eds.), Traditional Conservation in Papua New Guinea: Implications for Today, pp. 239 – 250. Institute of Applied Social and Economic Research. Monograph, 16. Boroko, Papua New Guinea.

Lidicker, W.Z. and A.C. Ziegler, 1968. Report on a collection of mammals from eastern New Guinea, including species keys for fourteen genera. Univ. Calif. Publ. in Zool. 87: 1 – 60.

Liem, D., 1983. Survey and management of wildlife resources along the Purari River. This volume, Chapter II, 3.

Liem, D. and A.K. Haines, 1977. The ecological significance and economic importance of the mangrove and estuarine communities of the Gulf Province, Papua New Guinea. In: T. Petr (ed.), Purari River (Wabo) Hydroelectric Scheme Environmental Studies, Vol. 3, pp. 1 – 35. Office of Environment and Conservation, Waigani, Papua New Guinea.

McDowell, S.B., 1967. *Aspidomorphus* a genus of New Guinea snakes of the family Elapidae, with notes on related genera. J. Zool. Lond. 151: 497 – 543.

McDowell, S.B., 1969. *Toxicocalamus* a New Guinea genus of snakes of the family Elapidae. J. Zool. Lond. 159: 443 – 511.

McDowell, S.B., 1970. On the status and relationships of the Solomon Island elapid snakes. J. Zool. Lond. 161: 145 – 190.

McDowell, S.B., 1972. The species of *Stegonotus* (Serpentes; Colubridae) in Papua New Guinea. In: Zoologische Mededelingen die 47 Feistbundel L.D. Brongersma, pp. 6 – 24. Rijksmuseum van Natuurlijke Historie, Leiden.

McDowell, S.B., 1974. A catalogue of the snakes of New Guinea and the Solomons with special reference to those in the Bernice P. Bishop Museum, Part I: Scolecophidia. J. Herpetology 8(1): 1 – 57.

McDowell, S.B., 1975. A catalogue of the snakes of New Guinea and the Solomons with special reference to those in the Bernice P. Bishop Museum, Part II: Anilioidea and Pythoninae. J. Herpetology 9(1): 1 – 79.

McDowell, S.B., 1979. A catalogue of the snakes of New Guinea and the Solomons with special reference to those in the Bernice P. Bishop Museum, Part III. Boinae and Acrochordidae. J. Herpetology 13(1): 1 – 92.

Menzies, J.I., 1975. Handbook of common New Guinea frogs. Wau Ecology Institute. Handbook, No. 1. Wau, Papua New Guinea.

Menzies, J.I., 1977. Fossil and subfossil fruit bats from the mountains of New Guinea. Aust. J. Zool. 25: 329 – 336.

Menzies, J.I. and E. Dennis, 1979. Handbook of New Guinea rodents. Wau Ecology Institute. Handbook, No. 6. Wau, Papua New Guinea.

Morauta, L., J. Pernetta and W. Heaney (eds.), 1982. Traditional Conservation in Papua New Guinea: Implications for Today. Institute of Applied Social and Economic Research, Boroko, Papua New Guinea.

Pernetta, J.C., 1982. Ok Tedi Environmental Impact Survey: animal resources of the area. Report to Natural Systems Research Ltd., Melbourne.

Pernetta, J.C. and S. Burgin, 1980. Census of crocodile populations and their exploitation in the Purari area (with an annotated list of the herpetofauna). In: T. Petr (ed.), Purari River (Wabo) Hydroelectric Scheme Environmental Studies, Vol. 14, pp. 1 – 44. Office of Environment and Conservation, Waigani, Papua New Guinea.

Pernetta, J.C. and S. Burgin, 1982. The status and ecology of crocodiles in the Purari. This volume, Chapter II, 11.

Pernetta, J.C. and I. Hill, 1981. Consumer/producer societies in Papua New Guinea: the face of change. In: D. Denoon and C. Snowden (eds.), A History of Agriculture in Papua New Guinea, pp. 283 – 309. Institute of Papua New Guinea Studies, Boroko, Papua New Guinea.

Pernetta, J.C. and L. Hill, 1982. International pressures on internal resources management in Papua New Guinea. In: L. Morauta, J.C. Pernetta and W. Heaney (eds.), Traditional Conservation in Papua New Guinea: Implications for Today, pp. 319 – 332. Institute of Applied Social and Economic Research. Monograph, 16. Boroko, Papua New Guinea.

Pernetta, J.C. and L. Hill, in press. Marine resource use in coastal Papua. J. Societé de Océanistes.

Pernetta, J.C. and D. Watling, 1978. The introduced and native terrestrial vertebrates of Fiji. Pacific Sci. 32: 223 – 244.

Powell, J.M., 1982. Traditional management and conservation of vegetation in Papua New Guinea. In: L. Morauta, J.C. Pernetta and W. Heaney (eds.), Traditional Conservation in Papua New Guinea: Implications for Today, pp. 121 – 134. Institute of Applied Social and Economic Research. Monograph, 16. Boroko, Papua New Guinea.

Pratt, T.K., 1982. Biogeography of birds in New Guinea. In: J.L. Gressitt (ed.), Biogeography of New Guinea, pp. 815 – 836. Monographiae Biologicae, Vol. 42. Dr W. Junk, The Hague.

Schodde, R., 1973. General problems of fauna conservation in relation to the conservation of vegetation in New Guinea. In: B.A. Costin and R.H. Groves (eds.), Nature Conservation in the Pacific, pp. 123 – 144. Australian National University Press, Canberra.

Waiko, J. and K. Jiregare, 1982. Conservation in Papua New Guinea: custom and tradition. In: L. Morauta, J.C. Pernetta and W. Heaney (eds.), Traditional Conservation in Papua New Guinea: Implications for Today, pp. 21 – 38. Institute of Applied Social and Economic Research. Monograph, 16. Boroko, Papua New Guinea.

Ziegler, A.C., 1977. Evolution of New Guinea's marsupial fauna in response to a forested environment. In: B. Stonehouse and D. Gilmore (eds.), The Biology of Marsupials, pp. 117 – 138. Univ. Park Press, Baltimore, U.S.A.

Zweifel, R.G. and M.J. Tyler, 1982. Amphibia of New Guinea. In: J.L. Gressitt (ed.), The Biogeography of New Guinea, pp. 759 – 801. Monographiae Biologicae, Vol. 42. Dr W. Junk, The Hague.

3. Survey and management of wildlife resources along the Purari River

D.S. Liem

1. Introduction

Wildlife plays an important role in the livelihood of the people of most rural areas of Papua New Guinea, not only providing food but also playing an important role in social, cultural and traditional activities of the people (Bulmer 1968; Chowning 1958; Liem 1977; Liem and Haines 1977). Utilization of wildlife is particularly significant in the rural areas, which are not developed and where infringements of modern developments have yet to occur.

At regular intervals, during 1975 – 1977, the Wildlife Division field survey team of the Papua New Guinea Department of Natural Resources, conducted extensive fieldwork in the Purari Delta and the Wabo area in the upper Purari. The objectives of the wildlife survey along the Purari River were to make an inventory of wildlife resources with special emphasis on those utilized by the people, and to evaluate the effects that large-scale developmental projects might have on the existing wildlife resources in the area.

The study areas consisted of two distinct habitats, namely the deltaic mangrove swamp/swamp forest and the lowland/hilly rain forest communities of the Wabo area. The faunal surveys were conducted along the Purari River in a number of transects, by mistnetting, observations, trappings, selective shootings, and spotlighting for nocturnal wildlife. Specimens collected were deposited in the Wildlife Division Museum and in the Papua New Guinea National Museum and Art Gallery in Port Moresby. Wildlife utilized by the people was identified from wildlife brought into the villages, from ornaments and implements made from wildlife products, from wildlife sold in the market, and by interviewing hunters. Once the identity and the local names of the most commonly used wildlife were established, quantitative data on wildlife that were harvested were collected by stationing field workers in villages. From time to time the same villages were visited again for addi-

Petr, T. (ed.) The Purari – Tropical environment of a high rainfall river basin
© *1983, Dr W. Junk Publishers, The Hague / Boston / Lancaster*
ISBN-13: 978-94-009-7265-0

tional data. Quantitative data presented in this paper is the best approximation of the various wildlife harvested from the areas under consideration. Numerous references were consulted, in particular the following: Diamond (1972), Gilliard (1969), Laurie and Hill (1954), Lidicker and Ziegler (1968), Menzies (1973), Menzies and Dennis (1979), Peckover and Filewood (1976), Rand and Gilliard (1967), and others.

Apart from wildlife, vegetation with special emphasis on food plants of wildlife was also studied. Identifications of food plants of wildlife were conducted by feeding observations, stomach content analysis, and interviews with knowledgeable local hunters. Herbarium materials were collected for further identification. Identification of plants was done by consulting publications of Paijmans (1969), Van Royen (1964a, b, c, 1965, 1966), and some others. Some identification was done by comparison of collections with herbarium materials of the University of Papua New Guinea.

2. Results of wildlife and wildlife food plants surveys

There are approximately 50 species of mammals, 250 species of birds, 30 species of reptiles and 12 species of frogs recorded from the areas surveyed. Of these 15 species of mammals, 37 species of birds and 12 species of reptiles have been found to be utilized by the people in the area. In addition to these, one species of mammal, five species of birds and one species of reptile could not be confirmed. These are marked by + in Table 1. Although reported in the literature in the past or in other areas, no frogs were found to be consumed by the people in the areas during the surveys. Some frogs are used as baits for fishing. Numerous other species which are regularly hunted, but have not been identified, include various species of ducks, pigeons, flying foxes, rodents and a possum.

The utilized wildlife comes from a whole spectrum of habitats, from coastal brackish water to arboreal canopy dwellers of hilly, closed forest habitats of the Wabo area. The majority of the wildlife hunted is used for food (about 50% of the total species), and only a small number of species are used for other purposes (Table 2). With the exception of the mammals, the majority of the wildlife harvested is for subsistence use. In most cases wildlife is sold after the subsistence needs of the hunters and their families have been satisfied. Wildlife which was observed to be sold in the markets included bush pig, cassowary, bandicoot, flying fox, wallaby, wildfowl and turtles with their eggs. Two species of crocodiles present in the area are hunted for their skins, which are traded locally and then exported overseas for further processing.

Table 1. List of wildlife species utilized along the Purari River.

Mammals

a. Aquatic – salt and brackish water coastal habitats

Dugong dugon – sea cow	a, d, e
Orcaella brevirostris – Irrawady dolphin	a, d, e

b. *Semi-aquatic – marshy habitats, open vegetation and forest*

Hydromys chrysogaster – common water rat	a, e

c. Terrestrial – open vegetation, forest edge and forest

Dorcopsis hageni – lesser forest wallaby	a, d, e
Echymipera rufescens – rufescent bandicoot	a, d, e
Peroryctes papuensis – Papuan bandicoot	a, d, e
Sus scrofa papuensis – bush pig	a, b, d, e
Thylogale bruijni – dusky wallaby	a, d, e

d. Arboreal – open vegetation and forest edge habitats

Petaurus breviceps – sugar glider	a, b, e
Phalanger maculatus – spotted cuscus	a, b, d, e

e. Arboreal – forest habitats

Dactylopsila trivirgata – common striped possum	a, b, d, e
+ *Dendrolagus* sp. – tree kangaroo	a, b, d, e
Phalanger orientalis – common cuscus	a, b, d, e

f. Aireal/arboreal – forest edge and forest habitats

Pteropus neohibernicus – Bismarck flying fox	a, d, e
Pteropus sp. – flying fox	a, d, e

Birds

a. Water birds – aquatic feeding in open and swamp/marshy habitats

Anas superciliosa – black duck	a, e
Anhinga rufa – Australian darter	a, e
Anseranas semipalmata – magpie goose	a, b, e
+ *Dendrocygna* sp. – tree duck	a, e
Dendrocygna guttata – spotted tree duck	a, e
Pelecanus conspicillatus – Australian pelican	a, e
Phalacrocorax sulcirostris – black cormorant	a, e
Platalea regia – royal spoonbill	a, e
Plegadis falcinellus – glossy ibis	a, e
Xenorhynchus asiaticus – black-necked stork	a, e

b. Forest floor birds – forest floor feeder, roosting on low vegetation

+ *Gallicolumba* sp. – ground dove	a, e
Goura scheepmakeri – Scheepmaker's crown pigeon	a, b, e

c. Terrestrial birds – forest habitats, nest on forest floor, roosting on low vegetation or on forest floor

Casuarius bennetti – dwarf cassowary	a, b, c, d, e
Casuarius casuarius – double wattled cassowary	a, b, c, d, e
Megapodius freycinet – common wildfowl	a, d, e
Talegalla fuscirostris – black-billed brush turkey	a, d, e
+ *Talegalla jobiensis* – collared scrub turkey	a, e

d. Tree/shrub birds – forest edge/open forest habitats near water

Halcyon torororo – lesser yellow-billed kingfisher	b, e

Table 1. Continued.

	Tanysiptera galatea – racket-tailed kingfisher	b, e
	Tanysiptera sylvia – Australian paradise kingfisher	b, e
e.	Tree birds – open vegetation and forest edge habitats	
	+ *Cacatua sanguanea* – little corella cockatoo	b, e
	Cacatua galerita – sulphur-crested cockatoo	a, b, d, e
	Cicinnurus regius – king bird of paradise	b, e
	+ *Diphyllodes magnificus* – magnificent bird of paradise	b, e
	Ducula spilorrhoa – Torres Strait pigeon	a, e
	Pseudeos fuscata – dusky lory	a, e
	Trichoglossus haematodus – rainbow lory	b, e
	Tyto alba – barn owl	b, e
f.	Tree birds – canopy dwellers, closed forest habitats	
	Aceros plicatus – Papuan hornbill	a, b, d, e
	Columba vitiensis – white-throated pigeon	a, e
	Ducula rufigaster – purple-tailed imperial pigeon	a, e
	Ducula zoeae – black-belted fruit pigeon	a, e
	Ducula pinon – black-shouldered fruit pigeon	a, e
	Ducula mulleri – collared fruit pigeon	a, e
	Gymnocorvus tristis – bar-eyed crow	a, e
	Harpyopsis novaeguineae – New Guinea eagle	b, e
	Lorius lory – black-capped lory	b, e
	Manucodia ater – glossy-mantled manucode	b, e
	Probosciger aterrimus – palm cockatoo	b, e
	Psittrichas fulgidus – Pesquet's parrot	b, d, e
	Paradisaea raggiana – Raggiana bird of paradise	b, e
	Seleucides melanoleuca – twelve-wire bird of paradise	b, e
Reptiles		
a.	Aquatic/semi-aquatic – fresh and brackish water swamp/marsh habitats	
	Acrochordus javanicus – Javanese file snake	c, d, e
	Crocodylus novaeguinea – freshwater crocodile	a, b, d, e
	Crocodylus porosus – saltwater crocodile	a, b, d, e
b.	Aquatic/semi-aquatic – coastal saltwater habitats	
	Caretta caretta – hawksbill turtle	a, b, d, e
	Chelonia mydas – green turtle	a, b, d, e
c.	Aquatic/semi-aquatic – river, stream and swamp/marshy habitats	
	Carettachelys insculpta – pitted-shelled turtle	a, e
	Chelodina siebenrocki – Siebenrock's snake-neck turtle	a, e
	Elseya novaeguinea – New Guinea short-neck turtle	a, e
	Emydura subglobosa – red short-neck turtle	a, e
	Pelochelys bibroni – short-shelled turtle	a, e
d.	Arboreal/terrestrial/semi-aquatic – open vegetations and forests	
	Python sp. – python	a, e
	Varanus indicus – common monitor lizard	a, b, c, d, e
	+ *Varanus salvadori* – Salvadori's monitor lizard	a, b, c, e

a = for food; b = pelage, feather, bone, scute, or teeth used for various ornaments; c = skin, bone, tooth or claw used for various implements; d = species entered into the cash economy; e = species used for subsistence; + = utilization of this species has yet to be determined in the area.

Table 2. Percentages of the various wildlife usages along the Purari River.

Wildlife classes	A	B	C	D	E
Mammals	64%	36%	–	79%	21%
Birds	50%	14%	36%	17%	83%
Reptiles	46%	46%	8%	38%	62%

A = for food only; B = for food and other uses; C = for ornament or implement use only; D = for subsistence use, but also sold for cash; E = for subsistence use.

Crocodile meat and eggs are consumed regularly.

In general, large-sized wildlife which is either omnivorous, herbivorous, or frugivorous is hunted as a source of food. Food species which were found to be commonly hunted are listed below in descending order of preference within a class:

Mammals: bush pig, cuscus, wallaby, bandicoot, flying fox, and some others

Birds : cassowary, wildfowl, brush turkey, crown pigeon, hornbill, pelican, spoonbill, ibis, duck, pigeons, and some others

Reptiles : marine turtles (coastal), freshwater turtles, crocodiles, monitor lizard, all including their eggs, and some others

Species harvested and the preferences for wildlife species in the areas under consideration are comparable with other parts of Papua New Guinea. As seen from Tables 1 and 2, only a small proportion of the wildlife hunted were for purposes other than for food. Hunting trips are organized for certain species of wildlife when the need arises, namely for bush pig, cuscus, cassowary, bird of paradise, hornbill, wildfowl, brush turkey, crown pigeon and crocodiles. Most other wildlife is hunted casually during encounters in their daily trips to and from the food gardens.

A number of birds are hunted for their colorful feathers, which are used for body ornaments and adornments to be worn during traditional ceremonial activities. Bird feathers used for this purpose include feathers from Raggiana bird of paradise, Pesquet's parrot, New Guinea eagle, Papuan hornbill, sulphur-crested cockatoo, and various others. Other wildlife-derived ornaments and/or implements are pig teeth, cuscus pelage, and skin of file snake and monitor lizard for kundu drums.

From the above data, it is obvious that wildlife plays a small but significant role in the livelihood of the people of the Purari, and this is similar to other rural areas in Papua New Guinea. To accurately evaluate and manage the wildlife populations in the area, wildlife population dynamics studies are essential. Due to a number of constraints, these were not done in the present survey. An indirect estimate of the status of wildlife was made by recording the quantity of wildlife that is hunted in the study areas. Recording accurate quantitative data of wildlife harvested was difficult, because of the field constraints, semi-nomadic life style of some groups of people in the Wabo area, and the reluctance of some hunters to reveal their harvests. Furthermore, tally of wildlife hunted during extended hunting trips could not be reported with great accuracy.

Thus data presented in Table 3 are just a conservative estimate of the quantity of wildlife harvested in the two study areas, i.e., the deltaic mangrove swamp/swamp forest and the lowland/hilly rain forest of Wabo. The table suggests that wildlife is still hunted in large numbers of primarily for the subsistence use of the people. While water birds such as ducks, pelicans, and spoonbill are harvested in open marshy/swampy habitats of the delta, hornbill, cassowary, crown pigeon and wallaby are harvested from the rain forest areas. Widely distributed species such as bush pig, bandicoot, water rat, and monitor lizard are harvested in both study areas.

Management of wildlife at sustained population levels requires control of hunting, provision of suitable habitats, and the presence of wildlife food plants. Provision of wildlife food plants is an integral part of good resource management planning for any integrated development, which includes the

Table 3. Numbers of wildlife harvested in the Delta and the Wabo study sites.

Wildlife	A	B	C
Bush pig	21	16	6
Common cuscus	50	32	16
Torres Strait pigeon	16	14	4
Wildfowl	2	3	5
Black duck	9	12	2
Pelican	6	3	–
Monitor lizard	4	6	5
Bandicoot	2	6	7
Water rat	2	2	3
Hornbill	1	3	5
Spoonbill	2	4	–
Flying fox	–	14	12
Cassowary	–	1	4
Crown pigeon	–	–	3
Forest wallaby	–	–	3
Possum	–	–	2
Soft-shelled turtle	–	–	2
Short-neck turtle	–	–	5

A = Delta survey in 1976 for 7 villages for a total period of 14 days (Akoma, Kapai, Kinipo, Maipenaru, Mariki, Barea, and Morewan); B = Delta survey in 1977 for 5 villages for a total period of 10 days (Akoma, Kinipo, Mariki, Maipenaru, and Morewan); C = Wabo survey in 1976 and 1977 for 2 villages and Wabo station for a total period of 26 days (Wabo, Wabo station, and Uraru). Wildlife sold at Baimuru were included in the Delta survey count.

Table 4. List of food plants of some wildlife of the upper and lower Purari.

Mammals

Phalanger maculatus – spotted cuscus

Acacia sp.	c
Cocos nucifera	a
Eugenia sp.	a, c
Gnetum gnemon	a, c
Mangifera minor	a, c
Mangifera sp.	a, c
Nauclea sp.	a
Nypa fruticans	a, b
Pandanus sp.	a
Rhizophora mucronata	a
Sonneratia lanceolata	a, b, c
Terminalia microcarpa	a
Xylocarpus granatum	a

Phalanger orientalis – common cuscus

Diospyros sp.	a
Eugenia sp.	a, c
Ficus sp.	a
Gnetum gnemon	a, c
Pandanus sp.	a

Petaurus breviceps – sugar glider

Banksia dentata	b
Cocos nucifera	b, c
Mangifera indicus	c

Pteropus neohibernicus – Bismarck flying fox

Acacia leptocarpa	a
Acacia sp.	a
Artocarpus incisa	a
Artocarpus sp.	a
Carica papaya	a
Ficus sp.	a
Mangifera minor	a
Musa sp.	a
Octomeles sumatrana	a
Psidium sp.	a
Terminalia kaernbachii	a
Terminalia sp.	a
Trema sp.	a

Sus scrofa papuensis – bush pig

Antiaris toxicaria	a
Areca sp.	a
Arenga sp.	a
Artocarpus sp.	a
Cocos nucifera	a
Colocasia esculenta	d

Table 4. Continued

Dioscorea sp.	d
Diospyros sp.	a
Ficus sp.	a
Ipomoea javanica	d
Mangifera minor	a
Mangifera sp.	a
Manihot esculenta	a
Metroxylon sagu	a
Musa sp.	a, d
Pandanus sp.	a
Psychotria sp.	a
Semecarpus australiensis	a
Spondias dulcis	a
Terminalia sp.	a
Birds	
Aceros plicatus – Papuan hornbill	
Arenga sp.	a
Casuarina sp.	a
Chisochiton sp.	a
Diospyros sp.	a
Dracontomelon sp.	a
Eugenia sp.	a
Ficus sp.	a
Gnetum gnemon	a
Mangifera sp.	a
Micromelum minutum	a
Palmae	a
Pometia tomentosa	a
Terminalia kaernbachii	a
Terminalia sp.	a
Cacatua galerita – sulphur-crested cockatoo	
Areca sp.	a
Arenga sp.	a
Cocos nucifera	a
Canarium indicum	a
Elaeocarpus sp.	a
Eugenia sp.	a
Mangifera sp.	a
Manihot esculenta	a
Metroxylon sagu	a
Palmae	a
Spondias dulcis	a
Terminalia catappa	a
Terminalia microcarpa	a
Casuarius bennetti – dwarf cassowary	
Aceratium ledermani	a

276

Table 4. Continued.

Artocarpus sp.	a
Barringtonia calyptocalyx	a
Canarium sp.	a
Cerbera sp.	a
Cyphomandra betacea	a
Ficus sp.	a
Gonocaryum litorale	a
Oenanthe javanica	a
Psychotria sp.	a
Litsea sp.	a
Casuarius casuarius – double-wattled cassowary	
Artocarpus sp.	a
Arenga sp.	a
Aceratium floribunda	a
Elaeocarpus sp.	a
Eugenia sp.	a
Ficus sp.	a
Gnetum gnemon	a
Mangifera minor	a
Mangifera sp.	a
Nauclea sp.	a
Palmae	a
Semecarpus australiensis	a
Spondias dulcis	a
Terminalia catappa	a
Terminalia microcarpa	a
Terminalia sp.	a
Ducula spilorrhoa – Torres Strait pigeon	
Arenga sp.	a
Ficus sp.	a
Myristica sp.	a
Terminalia sp.	a
Ducula rufigaster – purple-tailed imperial pigeon	
Ficus sp.	a
Myristica sp.	a
Terminalia sp.	a
Eclectus roratus – eclectus parrot	
Casuarina sp.	a
Elaeocarpus sp.	a
Ficus sp.	a
Musa sp.	a
Psidium sp.	a
Terminalia sp.	a
Trema orientalis	a
Goura scheepmakeri – Scheepmaker's crown pigeon	
Diospyros sp.	a

Table 4. Continued.

Eugenia sp.	a
Ficus sp.	a
Gnetum gnemon	a
Litsea sp.	a
Mangifera sp.	a
Myristica sp.	a
Pandanus sp.	a
Psychotria sp.	a
Terminalia microcarpa	a
Terminalia sp.	a
Lorius lory − black-capped lory	
Cocos nucifera	b
Eugenia sp.	b
Freycinetia sp.	b
Trema orientalis	a, b
Megapodius freycinet − wildfowl	
Diospyros sp.	a
Barringtonia sp.	a
Eugenia sp.	a
Ficus sp.	a
Gnetum gnemon	a
Myristica sp.	a
Psychotria sp.	a
Terminalia sp.	a
Paradisaea raggiana − Raggiana bird of paradise	
Chisocheton sp.	a
Eugenia sp.	a
Ficus sp.	a
Freycinetia sp.	a
Gnetum gnemon	a
Litsea sp.	a
Pandanus sp.	b
Palmae	a
Terminalia microcarpa	a
Trema orientalis	a
Probosciger aterrimus − palm cockatoo	
Canarium sp.	a
Castanospermum sp.	a
Cocos nucifera	a
Mangifera minor	a
Metroxylon sagu	a
Myristica sp.	a
Palmae	a
Pandanus sp.	a
Parinarium sp.	a
Nauclea sp.	a

Table 4. Continued.

Terminalia kaernbachii	a
Semecarpus australiensis	a
Pseudeos fuscata – dusky lory	
Cocos nucifera	b
Mangifera indica	a
Mangifera sp.	a
Musa sp.	a
Pittosporum ramiflorum	b
Terminalia sp.	a
Psittrichas fulgidus – Pesquet's parrot	
Artocarpus sp.	a
Eugenia sp.	a
Freycinetia sp.	b
Ficus sp.	a
Talegalla fuscirostris – black-billed brush turkey	
Diospyros sp.	a
Elaeocarpus sp.	a
Eugenia sp.	a
Ficus sp.	a
Gnetum gnemon	a
Litsea sp.	a
Mangifera sp.	a
Myristica sp.	a
Palmae	a
Trema orientalis	a
Trichoglossus haematodus – rainbow lory	
Cassia sp.	a
Casuarina equisetifolia	a
Cocos nucifera	a, b
Erythrina sp.	b
Ficus sp.	a
Pittosporum ramiflorum	b
Trema orientalis	b

a = fruit or seed; b = blossom/nectar; c = leaves; d = tuber or root.

management of wildlife. Although our knowledge of food plants of wildlife in the area is far from adequate, some which have been identified are presented in Table 4. Fruits and seeds are the predominant food source for most wildlife identified in the study area. Succulent fruits or those with fleshy pulp are the favourite food source of wildlife. They include *Mangifera* sp., *Ficus* sp., *Diospyros* sp., *Dracontomelon* sp., and some others. Other species of wildlife feed on hard-shelled fruits such as *Canarium* sp. and various palms. Most wildlife feed on fruits directly on trees, but some terrestrial animals such as bush pigs, cassowaries, bandicoots, forest wallabies and crown pigeons feed on fruits after they dropped on the forest floor. Food plants most commonly utilized by wildlife in the area are *Mangifera minor*, *M. indica*, *M.* sp., *Terminalia* sp., *T. microcarpa*, *T. kaernbachii*, *Ficus* sp., *Eugenia* sp., *Cocos nucifera*, *Trema orientalis*, *Gnetum gnemon*, *Artocarpus* sp., *Myristica* sp., *Diospyros* sp., *Pandanus* sp., and various other palms.

3. The outlook for wildlife along the Purari River

The fate of wildlife in the Purari River would follow the same pattern as in other parts of Papua New Guinea if preventive management measures are not implemented. Prior to any large-scale development, in any area hunting pressure is determined by the size of the human population and the number of shotguns. Such hunting pressure generally affects large-sized wildlife such as crown pigeons, hornbills, spoonbills, storks, ibises, cassowaries, wildfowl and brush turkey. At present, the low human population density along the Purari River has exerted only a low hunting pressure on wildlife in the area, as shown by the abundance of various wildlife when compared to areas with larger human population densities.

Developmental projects, such as large-scale forest logging, agriculture, or construction of a large hydroelectric scheme would add more pressure to the existing wildlife populations, not only through the destruction and shrinking of wildlife habitats, and the slow down of the regeneration of suitable habitats, but also through increased hunting pressure on wildlife with the influx of human settlements in the area.

The existence of certain wildlife is beneficial to their habitats. For example, wildlife is closely linked to the regeneration rate and capacity of disturbed habitats. This is so because the way various wildlife consume and digest various fruits and seeds is crucial for the dispersal and the propagation of plant species. If wildlife does not crack open hard-shelled seeds before swallowing, or if these seeds are not ground up in the stomach, it acts as a seed dispersal agent of plants. Included in this beneficial category are cassowaries, pigeons, crown pigeons, hornbills and flying foxes. On the other hand, most parrots, cockatoos and rats, which crack open the hard-shelled seeds and/or grind up the seeds in the stomach are acting as seed predators. They suppress the dispersal and the propagation of seeds and in this way curtail the maintenance of the habitats. This demonstrates the complexity of interrelationship of certain wildlife with their habitats, and the importance of maintaining an ideal balance amongst the various wildlife and their habitats.

Various ways can be employed in the management and the maintenance of sustained population levels of wildlife in the various different situations and various socio-economic and cultural set ups. In Papua New Guinea the conventional management and conservation measures, by introducing hunting bans or setting up conventional conservation areas are not implementable, because they run counter to the existing land tenure system and the traditional life style of the people. Protecting wildlife species through legislation without effective enforcement is insufficient. To ban hunting of wildlife is almost impossible, because the majority of the people live off their land and because the land belongs to the people who live there. Buying off land for conventional conservation purposes would deprive the traditional landowners of their livelihood and in most cases landowners would refuse to part from their ancestral lands. The best approach for the management of wildlife and their habitats in this situation would appear to be to set up wildlife management areas, in which the legal traditional landowners set up their own management rules and enforce them themselves. This management strategy would allow traditional landowners to harvest wildlife from their own land but would prohibit others from hunting there, hence alleviating uncontrolled hunting pressures from outsiders. Appropriate advice from wildlife managers and recognition of these rules by government agencies at the national as well as the local level are essential for the success of this approach. This management strategy has been employed by the Papua New Guinea Wildlife Division for some years with some success.

As in other parts of Papua New Guinea, with the encroachment of large-scale development, in addition to the large influx of human populations and their shotguns, and the increased clearings of habitats of wildlife for food gardens, the outlook of the large-sized, fragile species such as hornbills, palm cockatoos, crown pigeons, cassowaries, wildfowl, brush turkeys and New Guinea eagles is not promising. It is enlightening to note, however, that with proper management and with some enforcements of the rules and laws, resilient wildlife which can live in marginal habitats, such as Raggiana bird of paradise, cuscus, bandicoot, crocodile and certainly bush pigs may have a good chance to maintain their existence, although probably at a lesser population density than the present level.

References

Bulmer, R., 1968. The strategies of hunting in New Guinea. Oceania 38(4): 302–318.
Chowning, A., 1958. Lakalai society. Ph.D. Thesis. University of Pennsylvania.

Diamond, J.M., 1972. Avifauna of the Eastern Highlands of New Guinea. Nuttall Ornithological Club. Publication, No. 12. Cambridge, Mass.

Gilliard, E.T., 1969. Birds of paradise and bowerbirds. Weidenfeld and Nicholson, London.

Laurie, E.M.O. and J.E. Hill, 1954. List of land mammals of New Guinea, Celebes, and adjacent islands. British Museum, London.

Lidicker, W.Z. and R.C. Ziegler, 1968. Report on a collection of mammals from eastern New Guinea. Including species keys for fourteen genera. University Calif. Publ. in Zool. 87: 1 – 60.

Liem, D.S., 1977. Wildlife utilization of the proposed Garu Wildlife Management area. In: J.H. Winslow (ed.), The Melanesian Environment, pp. 285 – 292. Australian National University, Canberra.

Liem, D.S. and A.K. Haines, 1977. The ecological significance and economic importance of the mangrove and estuarine communities of the Gulf Province, Papua New Guinea. In: T. Petr (ed.), Purari River (Wabo) Hydroelectric Scheme Environmental Studies, Vol. 3, pp. 1 – 35. Office of Environment and Conservation, Waigani, Papua New Guinea.

Menzies, J.I., 1973. A key to rats of the genera *Rattus*, *Melomys*, and *Pogonomelomys* on the New Guinea mainland. Dept. of Biology, Univ. Papua New Guinea. Occasional Papers, No. 2.

Menzies, J.I. and E. Dennis, 1979. Handbook of New Guinea rodents. Wau Ecology Institute. Handbook, No. 6. Wau, Papua New Guinea.

Paijmans, K., 1969. Vegetation and ecology of the Kerema – Vailala area. In: B.P. Ruxton et al. (eds.), Lands of the Kerema – Vailala Area, Territory of Papua New Guinea, pp. 95 – 116. C.S.I.R.O. Land Research Series, No. 23. C.S.I.R.O., Melbourne.

Peckover, W.S. and L.W.C. Filewood, 1976. Birds of New Guinea and tropical Australia. A.H. and A.W. Reed, Sydney.

Rand, A.L. and E.T. Gilliard, 1967. Handbook of New Guinea birds. Weidenfeld and Nicholson, London.

Van Royen, P., 1964a. Manual of forest trees of New Guinea, Part 2: Sapindaceae. Papua New Guinea Dept. of Forests, Port Moresby.

Van Royen, P., 1964b. Manual of the forest trees of New Guinea, Part 3: Sterculiaceae. Papua New Guinea Dept. of Forests, Port Moresby.

Van Royen, P., 1964c. Manual of the forest trees of Papua and New Guinea, Part 4: Anacardiaceae. Papua New Guinea Dept. of Forests, Port Moresby.

Van Royen, P., 1965. Manual of the forest trees of Papua and New Guinea, Part 5: Himantandraceae, Part 6: Magnoliaceae, Part 7: Epomatiaceae, Part 8: Dipterocarpaceae. Papua New Guinea Dept. of Forests, Port Moresby.

Van Royen, P., 1966. Manual of the forest trees of Papua and New Guinea, Part 9: Apocynaceae. Papua New Guinea Dept of Forests, Port Moresby.

4. Aquatic and semi-aquatic flora of the Purari River system

B.J. Conn

1. Introduction

The task of presenting an overview of the aquatic and semi-aquatic flora of the Purari River system is an awesome one indeed. The reason is almost entirely because there has been little documentation of this flora for Papua New Guinea. The most extensive publications which describe the vegetation of New Guinea (Clunie 1978; Hope 1980; Johns 1977, 1982; Paijmans 1969, 1971, 1976; Robbins 1968, 1970; Robbins and Pullen 1965; Gressitt 1982; and literature cited therein) emphasize the terrestrial non-aquatic plants (particularly forest species) and only comment briefly on the aquatic and semi-aquatic components. A booklet listing a few common aquatic species was prepared by Conn (1975), but much information was produced by the author and some others for unpublished government reports held by the Department of Primary Industry. However, the aquatic and semi-aquatic flora of New Guinea still has to be largely investigated and documented. Distributional, frequency and altitudinal information for most species is very incomplete or unknown.

This chapter presents an overview of the present situation.

2. The Alpine zone

The watershed of the Purari River system reaches altitudes in excess of 3000 metres (Mt. Wilhelm – 4510 m, Mt. Giluwe – 4367 m, Mt. Hagen – 3795 m). Water-tolerant and semi-aquatic plants form various types of herbaceous bog communities in shallow depressions throughout this Alpine zone (as defined by Van Royen 1980). In flatter areas (particularly above 3500 m) which are subject to inundation, small stands of cushion-forming plants become dominant. Hope (1980) refers to this community as a 'Hard Cushion

Petr, T. (ed.) The Purari – Tropical environment of a high rainfall river basin
© *1983, Dr W. Junk Publishers, The Hague / Boston / Lancaster*
ISBN-13: 978-94-009-7265-0

Bog'. The common species of this community are *Centrolepis philippinensis,* *Ranunculus, Myriophyllum pedunculatum, Gentiana piundensis, Eriocaulon* spp. and *Astelia alpina* (Jessup 1979; Van Royen 1979, plate 81). *Carpha alpina* locally forms pure communities along the edges of tarns that are subject to inundation. Cyperaceous fens ('Open Wet Sedgelands', Hope 1980) occupy peaty areas which are inundated for most of the year by up to ca 20 cm of water. This community is common on several mountains in New Guinea. Wade and McVean (1969) describe three different cyperaceous fens on Mt. Wilhelm (viz. *Carex echinata, C. gaudichaudiana* and *Scirpus crassiusculus* dominated fens, respectively). An open *Scirpus crassiusculus* fen is common on Mt. Giluwe. Similar fen communities occur on Mt. Hagen.

There are few aquatic species in the alpine zone. *Isoëtes stevensii* (Hope 1980, plate 50) grows in shallow (less than 50 cm deep) alpine tarns on Mt. Giluwe at altitudes of ca 3500 m (and also on Mt. Sarawaket, outside the catchment). This aquatic species can apparently withstand dry periods (Croft 1980). *Scirpus crassiusculus* is probably an aquatic species or at least, a partial aquatic. Wade and McVean (1969) record *Callitriche palustris* as occurring in shallow tarns on Mt. Wilhelm, and Robbins and Pullen (1965) recorded it from Mt. Giluwe and Mt. Hagen.

The algal communities of the tarns and lakes have received even less attention than the aquatic vascular flora in general. Thomasson (1967) studied the phytoplankton of the lakes and tarns in and near the Pindaunde Valley of Mt. Wilhelm. He regarded these lakes as ultra-oligotrophic since the plankton proved to be 'quantitatively as well as qualitatively very limited'. The dominant species were *Peridinium pusillum, Strichogloea delicatula,* and (particularly in Lake Aunde) *Tetraspora lacustris.*

3. The Lower Montane zone

The Lower Montane zone (sensu Paijmans 1976) has large areas of gardens and grasslands in the highland provinces of Papua New Guinea, up to altitudes of ca 2700 m. A very small proportion of this area is characterised by swamp vegetation. Large areas of abandoned garden sites and other anthropogenically disturbed areas, mainly between altitudes of 1500 and 2500 m, are covered by *Miscanthus floridulus* dominated grasslands (Fig. 1). This grassland closely parallels that of the high water level of lakes, rivers and swamps (Conn 1979b).

Sedge swampland communities (= 'Sedge – grass swamp', Paijmans 1976) occur above about 1800 m on seepage slopes, small depressions in valley floors and in intermontane basins. This community is relatively extensive at

Fig. 1. Miscanthus floridulus grassland at Lake Onim.

Fig. 2. Broad shoreline community. Mixed grassland merging with cyperaceous zone of open water at Lake Onim.

Lakes Bune, Onim and Papare (near Mt. Giluwe). The most common sedges are *Juncus prismatocarpus, Scirpus mucronatus, Carex* sp., *Eleocharis sphacelata, Fimbristylis salbundia, Gahnia sieberiana, Lipocarpha chinensis, Rhynchospora rugosa* and *Machaerina rubiginosa* (Conn 1979b). There are a number of semi-aquatic herbs and swamp-tolerant herbs and shrubs present. The most common semi-aquatic plants are *Nymphoides hydrocharoides, Villarsia* spp*., *Xyris capensis, Utricularia nivea, Eriocaulon hookerianum* and *Haloragis halconensis*.

Equisetum forms small, locally dominant communities on almost flat areas with surface seepage (e.g. near Lai River, at Kandep) or in the deeper flowing water of small creeks (unnamed creek entering Tongo River, west side of Mt. Giluwe).

With the exception of Lake Egari, the lakes do not have a clearly defined margin. The sedge swampland community is usually gradually replaced by the aquatic community of the lake. Therefore, the shoreline community usually extends over a large area and, by necessity, must be defined such that the fluctuating water level of these lakes is taken into consideration (Fig. 2) (Conn 1979b).

The structure of the shoreline community of the lakes near Mt. Giluwe was described by Conn (1979b) and therefore only the summary of the findings is presented here. The landward zone of the shoreline community is usually dominated by one to three species of the sedge swampland community. At Lake Bune, *Scirpus mucronatus, Rhynchospora rugosa* and *Eleocharis sphacelata* form large stands, often with one of these species completely dominating the community (Conn 1979b, Fig. 2). In general, *Scirpus mucronatus* occurs in water up to 1 m deep, rarely up to 2 m deep, while *Eleocharis sphacelata* occurs in up to 3 m of water (Conn 1979b, Fig. 8). *Leersia hexandra* sometimes forms a floating shelf (Fig. 3) (Conn 1979b), while at Lake Egari and Lake Papare (Fig. 4) (Conn 1979b), it tends to float on the water without forming a shelf. *Paspalum orbiculare* and *Polygonum strigosum* occur in up to 1 m of water. The stems and leaves of the former species tend to float on the water surface. *Hydrocotyle sibthorpioides* and *Nymphoides hydrocharoides* extend from the sedge swampland community to a depth of 3 – 4 m. *Chara* sp. forms moderately dense communities in Lake Sigenamu (Conn 1979b) probably because this shallow lake receives a high level of organic effluent from the nearby village. *Nymphoides indica* occurs at

---

* *Nymphoides hydrocharoides* and many *Villarsia* species appear to survive as aquatic or as semi-aquatic plants, with appropriate changes in habit which are related to the environmental situation. Generic distinctiveness of *Nymphoides* and *Villarsia* is frequently obscure. Cook (1974) regards inflorescence type and habit as being unsatisfactory delimiting characters.

Fig. 3. Floating shelf of *Leersia hexandra*, mainly with *Eleocharis sphacelata* and *Scirpus mucronatus*, Lake Bune.

Fig. 4. Lake Papare with mixed *Miscanthus floridulus* grassland in foreground, Mt. Pebu with *Nothofagus* forest in background.

Lake Egari and near Kerowagi (Western Highlands) in 1 – 2 m of water, rarely in up to 4 m.

The microscopic aquatic flora is very abundant, particularly in those lakes with a high input of domestic waste and sewage. The most common genera represented in the upper 10 cm of water are *Anthrodesmus, Closterium, Spirogyra, Staurastrum* and *Staurodesmus*.

Phragmites karka commonly forms extensive monospecific communities, 3 – 7 m high, on poorly drained to swampy floodplains. This community is common in the Wahgi Valley swamps, Western Highlands, and in small hollows on sandstone or siltstone-derived sediments in the Wabag – Tari area, Southern Highlands (Robbins and Pullen 1965). It is also associated with *Miscanthus floridulus* along river banks and swampy margins (Conn 1977). *Saccharum spontaneum* and *Typha augustifolia* are frequently associated with *Phragmites karka*, particularly in the Wahgi Valley.

Limnophila indica forms extensive communities in shallow swamps and roadside ponds near Kerowagi (Western Highlands). Similarly, the floating *Azolla filiculoides, A. pinnata* and *Lemna* sp. are frequently present in shallow swamps and ponds of the Wahgi Valley.

4. The lowland rain forest zone

The lowland rain forests of Papua New Guinea occupy an area of broken topography with relatively short, steep slopes (= 'The Foothills and Mountains below 1000 m', Paijmans 1976). Therefore, at least in part, the rivers are usually fast flowing, and this is particularly so for the Purari River system. The sandy and rocky banks and beds of streams (below flood level) are characteristically colonized by rheophytic plants (Van Steenis 1952, 1981; Paijmans 1976). The establishment of an aquatic flora, under such harsh conditions, is usually not possible. Occasionally, aquatic or semi-aquatic plants (such as *Phragmites karka*) become established in the relatively still water of the inner shoreline of river bends.

Lake Tebera (Gulf Province - altitude 586 m) is the only large body of water in the immediate vicinity of the proposed Wabo hydroelectric dam site. A brief visit was made to this lake in 1978 (Conn 1979a). The aquatic and semi-aquatic flora of this lake is very similar to that of Lake Kutubu (Southern Highlands) (Conn 1979c), which is in the Kikori River system, adjacent to the Purari catchment to the west. Large floating islands, predominately composed of *Leersia hexandra*, cover extensive areas of the lake's surface (Conn 1979a, Figs. 2 & 3). The subsequent sudd communities which have developed on these floating islands, are commonly made up of a number of sedges (such as *Gahnia* and *Cyperus odoratus*), various grasses (such as *Isachne* and *Ischaemum*), *Ludwigia adscendens*, *Limnophila indica*, *Blechnum* and *Polygonum*. *Azolla pinnata*, *Spirodela oligorrhiza* and *Utricularia* occur in the still water amongst the stems of the various sudd species (Conn 1979a). *Phragmites karka* forms an extensive, almost continuous community as part of the shoreline zone. It occurs in up to ca 2 m of water.

The aquatic and semi-aquatic plant community of this lake is still largely to be investigated. Until our knowledge of the flora of this lake becomes more complete, it seems reasonable to expect a similar flora to that of Lake Kutubu (Conn 1979c), except that Lake Tebera appears to offer fewer niches than does Lake Kutubu.

5. The lowland swamps

In the flatter areas of the southern Gulf Province where the Purari emerges from the hills, at altitudes below 60 m large lowland swamps have developed. The composition of the vegetation of these swamps is largely determined by the amount of drainage and the salinity of the water (ranging from fresh to brackish and saline). The low gradient of the Purari River has resulted in a gradual transition from forest to sago palm, through tidal freshwater swamp forest, to mangrove vegetation on the coast. The aquatic and semi-aquatic flora of the lower Purari River system is incompletely known, largely because access to and movement within these lowland swamps is extremely difficult. Relatively large areas of lowland freshwater swamps form complex mosaics of various vegetation types throughout this region.

Hanguana malayana commonly occurs in small swampy depressions, with a number of sedges and gingers (Zingiberaceae). Occasionally *Metroxylon sagu* and *Pandanus* surround these small depressions on the slightly higher margin. *Leersia hexandra* and various sedges occupy permanently swampy areas of the river plain, including shallow embayments and small islets. *Ischaemum polystachyum*, *Polygonum*, *Ludwigia* and *Ipomoea aquatica* occur on the fringes with their stems and leaves trailing out over the water surface. *Phragmites karka* and *Saccharum robustum* commonly occur in shallow swamps with *Coix lachryma-jobi* frequently occurring on low levees. *Saccharum robustum* appears to be restricted to sites of moving freshwater, while *P. karka* has a high salt tolerance (Paijmans 1969) and so is able to colonise brackish niches.

The aquatic flora of this region is very similar to that of the Moorhead – Kiunga area (Western Province) (Paijmans 1971). *Nymphaea nouchali*

Fig. 5. Cryptocoryne ciliata, a common plant of muddy tidal flats. (Photo T. Petr.)

(Conn, in press), and probably other *Nymphaea* species, *Nymphoides indica, Ceratophyllum demersum, Lemna perpusilla, Pistia stratiotes, Azolla* and *Utricularia* are common aquatic plants throughout the freshwater swamps.

Metroxylon sagu occurs in the uppermost parts of the freshwater tidal flats and on more or less permanently swampy plains, but it also extends well down through the Purari Delta, where it tolerates salinities up to 7.5‰ (Petr and Lucero 1979). *Hanguana,* various sedges and *Phragmites* frequently occur amongst open stands of sago.

Cryptocoryne is a semi-aquatic and aquatic genus which is common in the Western and Gulf Provinces of Papua New Guinea. *C. ciliata* (Fig. 5) is a robust pioneering species common in the tidal muds of the Purari River Delta (Hay 1981). *C. versteegii, C. wendtii* and *C. dewitii* are smaller species which occur along the margin of streams and in muddy areas.

The seaward limit of the freshwater environment is defined by the very extensive *Nypa fruticans* brackish swamp community. *Nypa* forms a transitional zone between the mangrove community and the tidal freshwater swamp forest (Paijmans 1969). There are no aquatic or semi-aquatic species in these latter two communities.

P. karka frequently occurs with *Nypa* along small brackish tidal creeks. The mangrove and *Nypa* communities are discussed by Cragg (this volume).

6. Discussion

The incompleteness of our knowledge of the aquatic and semi-aquatic flora of New Guinea should be emphasized, as there are many areas which are still unknown, or poorly known. In other areas, in-depth studies are necessary. Apart from the pure scientific importance, the necessity of such studies has been highlighted in connection with the rapid spread of some aquatic plants in certain areas of Papua New Guinea. The type of problem which exists at the moment can be illustrated by the two floating aquatic plants, *Eichhornia crassipes* and *Salvinia molesta*, both, however, introduced species.

The occurrence of *S. molesta* and *E. crassipes* in New Guinea was first documented in 1977 (Conn 1977). Mitchell (1979) and Mitchell et al. (1980) discussed the dramatic spread of *S. molesta* and its consequences for the Sepik River environment. As far as it is known these two potentially dangerous aquatic weeds are still absent from the Purari River system.

References

Clunie, N.M.U., 1978. The vegetation. In: J.S. Womersley (ed.), Handbooks of the Flora of Papua New Guinea, Vol. 1, pp. 1 – 11. Melbourne Univ. Press.

Conn, B.J., 1975. Some common aquatic plants of Papua New Guinea. Papua New Guinea Dept. of Forests, Port Moresby.

Conn, B.J., 1977. Notes on aquatic and semi-aquatic flora. In: T. Petr (ed.), Purari River (Wabo) Hydroelectric Scheme Environmental Studies, Vol. 1: Workshop 6 May 1977, pp. 21 – 27. Office of Environment and Conservation, Waigani, Papua New Guinea.

Conn, B.J., 1979a. *Phragmites karka* and the floating islands at Lake Tebera (Gulf Province), with notes on the distribution of *Salvinia molesta* and *Eichhornia crassipes* in Papua New Guinea. In: T. Petr (ed.), Purari River (Wabo) Hydroelectric Scheme Environmental Studies, Vol. 10: Ecology of the Purari River catchment, pp. 31 – 36. Office of Environment and Conservation, Waigani, Papua New Guinea.

Conn, B.J., 1979b. The vegetation of the lakes of Mt. Giluwe area (Southern Highlands Province), Papua New Guinea. In: T. Petr (ed.), Purari River (Wabo) Hydroelectric Scheme Environmental Studies, Vol. 10: Ecology of the Purari River catchment, pp. 37 – 62. Office of Environment and Conservation, Waigani, Papua New Guinea.

Conn, B.J., 1979c. Notes on the aquatic and semi-aquatic flora of Lake Kutubu (Southern Highlands Province), Papua New Guinea. In: T. Petr (ed.), Purari River (Wabo) Hydroelectric Scheme Environmental Studies, Vol. 10: Ecology of the Purari River catchment, pp. 63 – 90.

Office of Environment and Conservation, Waigani, Papua New Guinea.

Conn, B.J., in press. Nymphaeaceae. In: E.E. Henty (ed.), Handbooks of the Flora of Papua New Guinea. Melbourne Univ. Press.

Cook, C.D.K., 1974. Water plants of the world. Dr W. Junk, The Hague.

Cragg, S., 1983. The mangrove ecosystem of the Purari Delta. This volume, Chapter II, 5.

Croft, J.R., 1980. A taxonomic revision of *Isoëtes* (Isoëtaceae) in Papuasia. Blumea 26: 177 – 190.

Gressitt, J.L., 1982. Biogeography and Ecology of New Guinea. 2 vols. Dr W. Junk, The Hague.

Hay, A., 1981. Araceae. In: R.J. Johns and A. Hay (eds.), A Students' Guide to the Monocoty-ledons of Papua New Guinea, Part 1, pp. 31 – 84. Papua New Guinea Forestry College Training Manual, Vol. 13. Forestry College, Bulolo, Papua New Guinea.

Hope, G.S., 1980. New Guinea mountain vegetation communities. In: P. van Royen (ed.), The Alpine Flora of New Guinea, Vol. 1, pp. 153 – 221. J. Cramer, Vaduz.

Jessup, J.P., 1979. Liliaceae. Flora Malesiana ser. 1, 9: 189 – 235.

Johns, R.J., 1977. The vegetation of Papua New Guinea, Parts 1 and 2. Papua New Guinea Forestry College Training Manual, Vol. 10. Forestry College, Bulolo, Papua New Guinea.

Johns, R.J., 1982. Plant zonation. In: J.L. Gressitt (ed.), Biogeography and Ecology of New Guinea, pp. 309 – 330. Dr W. Junk, The Hague.

Mitchell, D.S., 1979. The incidence and management of *Salvinia molesta* in Papua New Guinea. Office of Environment and Conservation, Waigani, Papua New Guinea.

Mitchell, D.S., T. Petr and A.B. Viner, 1980. The water-fern *Salvinia molesta* in the Sepik River, Papua New Guinea. Environmental Conservation 7: 115 – 122.

Paijmans, K., 1969. Vegetation and ecology of the Kerema – Vailala area. In: B.P. Ruxton et al. (eds.), Lands of the Kerema – Vailala Area, Territory of Papua New Guinea, pp. 95 – 116. C.S.I.R.O. Land Research Series, No. 23. C.S.I.R.O., Melbourne.

Paijmans, K., 1971. Vegetation, forest resources, and ecology of the Moorehead – Kiunga area. In: K. Paijmans et al. (eds.), Land Resources of the Moorehead – Kiunga Area, Territory of Papua and New Guinea, pp. 88 – 113. C.S.I.R.O. Land Research Series, No. 29. C.S.I.R.O., Melbourne.

Paijmans, K., 1976. Vegetation. In: K. Paijmans (ed.), New Guinea Vegetation, pp. 23 – 105. C.S.I.R.O. and Australian National Univ. Press, Canberra.

Petr, T. and J. Lucero, 1979. Sago palm (*Metroxylon sagu*) salinity tolerance in the Purari River delta. In: T. Petr (ed.), Purari River (Wabo) Hydroelectric Scheme Environmental Studies, Vol. 10: Ecology of the Purari River catchment, pp. 101 – 106. Office of Environment and Conservation, Waigani, Papua New Guinea.

Robbins, R.G., 1968. Vegetation of the Wewak – lower Sepik area. In: H.A. Haantjens et al. (eds.), Lands of the Wewak – lower Sepik Area, Territory of Papua and New Guinea, pp. 109 – 125. C.S.I.R.O. Land Research Series, No. 22. C.S.I.R.O., Melbourne.

Robbins, R.G., 1970. Vegetation of the Goroka – Mount Hagen area. In: H.A. Haantjens et al. (eds.), Lands of the Goroka – Mount Hagen Area, Territory of Papua and New Guinea, pp. 104 – 118. C.S.I.R.O. Land Research Series, No. 27. C.S.I.R.O., Melbourne.

Robbins, R.G. and R. Pullen, 1965. Vegetation of the Wabag – Tari area. In: J.R. McAlpine et al. (eds.), General Report on Lands of the Wabag – Tari Area, Territory of Papua and New Guinea, pp. 100 – 117. C.S.I.R.O. Land Research Series, No. 15. C.S.I.R.O., Melbourne.

Thomasson, K., 1967. Phytoplankton from some lakes on Mt. Wilhelm, east New Guinea. Blumea 15: 283 – 296.

Van Royen, P., 1979. Liliaceae. In: P. van Royen (ed.), The Alpine Flora of New Guinea, Vol. 2, pp. 37 – 41. J. Cramer, Vaduz.

Van Royen, P., 1980. Definitions and descriptions of the high altitude vegetation types. In: P. van Royen (ed.), The Alpine Flora of New Guinea, Vol. 1, pp. 15 – 27. J. Cramer, Vaduz.

Van Steenis, C.G.G.J., 1952. Rheophytes. Proc. Royal Soc. Queensland 62: 61 – 68.

Van Steenis, C.G.G.J., 1981. Rheophytes of the world. Sijthoff and Noordhoff, The Netherlands.

Wade, L.K. and D.N. McVean, 1969. Mt. Wilhelm studies I: The alpine and subalpine vegetation. Research School Pacific Studies. Dept. of Biogeography and Geomorphology. Publication BG/1. Australian National University, Canberra.

5. The mangrove ecosystem of the Purari Delta

S. Cragg

1. The physical environment of the mangrove ecosystem

1.1. Climate

The mangrove ecosystem of the Purari Delta experiences relatively constant high levels of temperature, humidity and rainfall. Rainfall at Ihu to the east of the delta and Baimuru to the west is fairly evenly distributed throughout the year, with an annual average of 3455 mm at Baimuru and 3098 mm at Ihu (McAlpine et al. 1975). Temperature and humidity data are only available from stations rather more distant from the delta – Kikori to the west and Kerema to the east. At Kerema the mean annual maximum temperature is 29.6°C and the mean annual minimum is 22.7°C; the respective figures for Kikori are 29.9°C and 21.8°C. The average humidity in Kerema is 84% at 09.00 and 74% at 15.00. Humidities are higher at Kikori. The area experiences a monsoonal wind with southeasterlies predominating between May and October and northwesterlies between December and March. Sea breezes tend to occur in the afternoon in the delta (Evesson, this volume).

1.2. Water movement pattern

Patterns of water movement within the delta mangrove ecosystem are determined by a combination of three factors: rate of freshwater input, tidal fluctuations and topography. As each factor varies with time, the patterns of water movement may be expected to show tidal, seasonal, irregular and secular changes. Considerable fluctuations occur in the amount of freshwater input to the head of the delta. Carter (1980) reported that at the Wabo dam site the flow rate of the Purari River ranged from 10 251 m³/s to 425 m³/s and

Petr, T. (ed.) The Purari – Tropical environment of a high rainfall river basin
© 1983, Dr W. Junk Publishers, The Hague / Boston / Lancaster
ISBN-13: 978-94-009-7265-0

in the Aure River it ranged from 2887 m³/s to 44 m³/s. The total amount of freshwater entering the delta is the sum of the flow rates of the Purari and Aure rivers. This water passes into a large number of distributaries of varying importance. Most freshwater flow occurs down the Ivo – Urika, Purari and Wame distributaries (Thom and Wright, this volume). Thom and Wright indicated that during periods of high discharge, the Ivo – Urika distributary carries 60 – 70% of the flow.

Tidal fluctuations are more regular than the weather-dependent fluctuation in freshwater flow rate. At Port Romilly at the mouth of the Pie River, the tidal range is 2.8 m (Australian Dept. of Defence 1980). A similar tidal range is probably at the seaward face of the Purari Delta. Within the delta, however, tidal ranges are likely to vary considerably. Where freshwater flow is strong, very little tidal influence will be evident. On the other hand, estuaries with relatively small freshwater input, for example the Pie estuary, experience reversals of flow due to tidal currents (Thom and Wright, this volume). At Baimuru on the Pie River, the tidal range can be up to 5.5 m (Irion and Petr 1980).

The volume of freshwater discharged by the Purari Delta is sufficient for distinct plumes of water with reduced salinity to be evident extending from the mouths of the delta. These can be easily detected 5 km from shore (Petr, Part 2, this volume), but in the season of southeasterly winds these plumes tend to be trapped close to shore (Thom and Wright, this volume) by marine currents. MacFarlane (1980) indicated that throughout the year there is a general movement of water along the coastline of the delta, moving from east to west. Both surface and seabed drifters released to the east of the delta showed a tendency to move into the shoreline at the west of the delta or into the Pie estuary, presumably due to a combination of open sea and tidal currents.

1.3. Salinity regime

Offshore delta salinities are lower than those of full seawater (Frusher 1980). Frusher measured salinities along much of the Gulf of Papua estuary/delta complex. His data suggest that saline water penetrates further inland in the estuaries to the west of the delta than it does in the delta itself. This is corroborated by the findings of Petr (Part 2, this volume) and Thom and Wright (this volume). Thom and Wright found that there were markedly higher salinity levels within the delta (more than double in some cases) during dry periods than in wet periods. Most salinity measurements made in the delta relate to surface waters. Stratification, with more saline water near the bottom, was observed by Thom and Wright in the Pie River. Saline wedges form in some distributaries (Petr, Part 2, this volume).

1.4. Deposition and erosion

River supply of sediment is considerable: Pickup (1980, this volume) estimates that an average of about 90 million cubic metres of sediment per year reach the head of the delta, consisting of 46% clay, 48% silt and 6% sand. The larger particles travel as river bed load but smaller particles are carried in suspension. Petr (Part 1, this volume) found between 13 and 1277 mg of suspended material per litre at sites within the delta. Higher suspended loads are not uncommon in some tropical rivers (Petr 1976).

The bed load of the river has built the delta, the supply of material being sufficient to more than counteract any sinking of the shoreline (Ruxton 1969). Material for the series of sand bars protecting the seaward face of the delta may also originate from river bed load. Prevailing currents drift the sand in a westerly direction along the shoreline.

Behind the sand bars, the extensive tidal flats receive a portion of the suspended load entering the river. On rising tides, the sediment-laden waters spread slowly over the flats, but waterflow is slowed by mangroves roots and trunks. The resulting low current velocity is sufficient for silt and flocculated clay particles to deposit as a veneer over the flats. This process has been demonstrated experimentally (Spenceley 1977). The sediments in these regions also slowly accumulate peat-like organic material (Ruxton 1969). Once deposited, the tidal-flat sediments are stabilized by ramifying underground root systems (cf. Scoffin 1970).

At the edges of the tidal creeks, the level of the substratum may be raised significantly due to the activity of the mud lobster, *Thalassina anomala* (Löffler 1977). This animal forms crudely conical mounds over one and a half metres high and nearly three metres in diameter. Material for these mounds is brought up from burrows deep beneath the mud surface (Ruxton 1969). Locally these mounds may coalesce to form islands which are sufficiently elevated to allow the growing of food crops (Löffler 1977).

The sediments where the mangroves grow are fine grained (Irion and Petr 1980), soft and poorly consolidated (Ruxton 1969). A change in their colour 2 to 5 cm from the surface indicates the level at which the soil becomes deoxygenated (Viner 1979). Soil pH is generally between 8 and 9 probably due to limestone in the catchment (Viner 1979). Little data is available on soil salinity, but Petr and Lucero (1979) measured salinities of up to 7.5‰ adjacent to sago palms.

Forces of erosion counter deposition processes within the delta. Locally, this dynamic balance favours the forces of erosion so that sediment is actively removed. Erosion is occurring at Baimuru on the Pie River and also at a point some kilometres down the Pie (personal observation). It is also occurring near the mouth of the Varoi River (Thom and Wright, this volume). At Arehava village to the east of the delta, there are coconut stumps in the intertidal zone (Liem and Haines 1977). Floyd (1977) compared aerial photographs of Uramu Island (to the west of the delta) taken in 1955 with those of 1974. From these he calculated that due to erosion, the coastline on that island was receding 20 metres per year.

2. Flora

2.1. Mangrove species of the delta

Mangroves have been defined by Macnae (1968) as woody shrubs and trees growing between the spring-tide high-water mark and a level close to but above mean sea level. The flora of the Purari Delta mangal contains a high proportion of the mangrove species found in New Guinea: at least nineteen of the thirty obligate mangroves listed by Johnstone and Frodin (1982) for New Guinea are represented. Furthermore, the New Guinea mangrove flora is one of the richest in the world (Johnstone and Frodin 1982). Nonetheless, compared with the diversity in the contiguous lowland rain forest, the diversity in the mangal is low. Indeed, the diversity in the seaward part of the mangal is particularly low because there the mangroves tend to occur in virtually pure stands or in mixtures of very few species.

Table 1 gives a list of mangrove species found in the Purari Delta. Species rarely occurring in the mangal, and then only in the boundary region, are excluded. Higher plants with other growth habits are also present; however, with the exception of the palm *Nypa fruticans* and the fern *Acrostichum*, they only form a significant component of the flora on the fringes of the mangal.

2.2. Mangrove distribution and zonation

Figure 1 shows the extent of mangal in the Purari Delta. Compared with the region to the west, the mangal of the delta extends much less far inland. This distribution reflects the difference between the two regions in the extent of tidal influence, with the high freshwater flow rates within the delta curbing the penetration of saline water.

Salinity levels are probably a major factor determining the upstream or downstream limits to the distribution of many plants. Soil – water salinity

Table 1. Trees and shrubs of the mangal of the Purari Delta.

Family	Code[a]	Species
Acanthaceae		*Acanthus ilicifolius* L.
Apocynaceae		*Cerbera floribunda* K. Schum
Barringtoniaceae	Ba	*Barringtonia racemosa* (L) Bl.[b]
Bignoniaceae	Do	*Dolichandrone spathacea* (L.f.) K. Sch.
Bombacaceae	Cs	*Camptostemon schultzii* Mast
Ebenaceae	D	*Diospyros* sp.[b,c]
Euphorbiaceae	Ex	*Excoecaria agallocha* L.
Leguminosae		*Cynometra iripa*[b]
Malvaceae	Hi	*Hibiscus tiliaceus* L.[b]
Meliaceae		*Xylocarpus australasicus* Ridl.
		X. granatum Koenig
	Xm	*X. moluccensis* (Lamk) Roem.[d]
Myrsinaceae	Ac	*Aegiceras corniculatum* (L.) Blanco
Plumbaginaceae		*Aegialitis annualata* R. Br.
Rhizophoraceae	Bc	*Bruguiera cylindrica* (L.) Bl.
	Bg	*B. gymnorhiza* (L.) Lamk
	Bp	*B. parviflora* (Roxb.) Wight & Arn. ex Griff.
	Bs	*B. sexangula* (Lour.) Poir.
		Ceriops decandra (Roxb.) Ding Hou
	Ra	*Rhizophora apiculata* Bl.
	Rm	*R. mucronata* Lamk
	Rs	*R. stylosa* Griff.
Sonneratiaceae	Sa	*Sonneratia alba* Sm.
	Sl	*S. lanceolata* (L.) Engl.[e]
Sterculiaceae	He	*Heritiera littoralis* Ait.
Tiliaceae	Br	*Brownlowia argentata* Kurz.[b]
Verbenaceae	Aa	*Avicennia alba* Bl.[c]
	Ae	*Avicennia eucalyptifolia* Zipp. ex Mold.[g]
	Ao	*Avicennia officinalis* L.

Based mainly on the observations of Liem and Haines (1977), Floyd (1977) and the Australian Institute of Marine Science (unpublished); nomenclature based on Percival and Womersley (1975) and Frodin and Leach (1982).

[a] The code letters are used in Table 2.

[b] These species are not included in Percival and Womersley (1975). They occur in the less frequently inundated parts of the mangal and their range extends inland beyond the limits of tidal and saline influence.

[c] The taxonomic identity of this species is uncertain but it resembles *Diospyros ferrea* (Wild.) Bakh. var. *Geminata* (R. Br.) Bakh. (Duke et al. 1981).

[d] Frodin et al. (1975) state that this species is sometimes confused with *X. australasicus* and that it is not a mangrove. Percival and Womersley (1975) say that *X. moluccensis* sometimes occurs on tidal mud.

[e] The taxonomic status of *S. caseolaris* is under review. (Duke, personal communication).

299

Several varieties are recognised in Australia. The name *S. lanceolata* has been applied by investigators from the Australian Institute of Marine Science, to some of the material ascribed by other authors to *S. caseolaris* including that from the Purari Delta.

f This species was recorded by Floyd (1977). Frodin and Leach (1982) state that this species is not found on the south coast of Papua New Guinea.

g The taxonomic status of this species requires clarification. The names *A. marina* (Forst.f.) Vierh., *A. marina* var. *resinifera* (Forst.) Bakh. and *A. eucalyptifolia* Zipp. ex Miq. have been applied to material from New Guinea. Comparisons suggest that on the south coast of Papua New Guinea these names refer to a single species whose relationship to *Avicennia marina* occurring elsewhere is uncertain (Leach, personal communication).

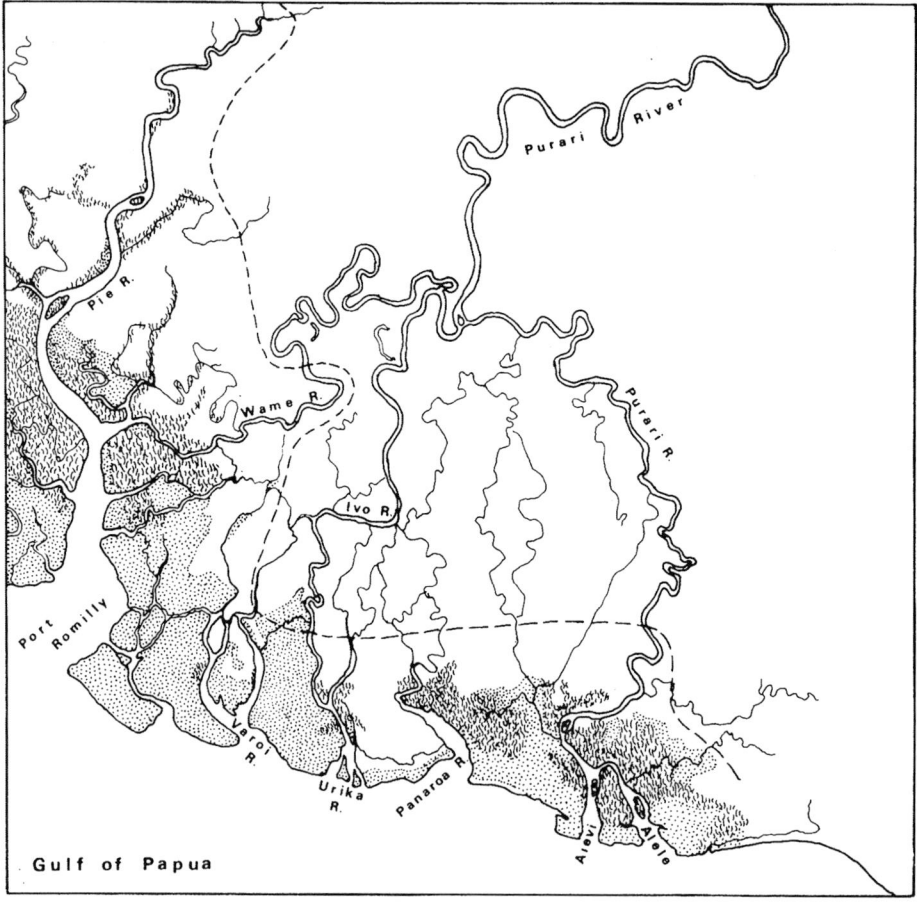

Fig. 1. The distribution of mangroves (dots) and *Nypa fruticans* (lines) in the Purari Delta. Based on P.N.G. Office of Forests (1977 – western part of delta), National Mapping Bureau 1:50 000 series (central part) and Paijmans (1969 – eastern part). The upper limit of *Sonneratia lanceolata* in major creeks and distributaries is shown by a broken line (Petr and Lucero 1979). Scale 1:500 000.

rather than salinity in the adjacent waterways may be the determining factor. Petr and Lucero (1979) found that where sago palms were in saline conditions, soil salinities were generally higher than creek salinities.

Petr and Lucero (1979) mapped the inland limits to the distribution of a species of *Pandanus* tolerant of the brackish water, of the palm *Nypa fruticans* and of the euryhaline mangrove *Sonneratia lanceolata* (= *S. caseolaris*). The limits of the latter is shown in Fig. 1. These authors also investigated the seaward limits to the distribution of a thorn-bearing variety of sago (*Metroxylon sagu*). Their findings indicate that the sago palm is capable of healthy growth in soil salinities as high as 7.5‰ and is quite commonly found in plant associations containing mangroves and the nipa palm. It is not clear whether the sago palm has established naturally under these conditions or whether it has been planted there, as has probably been the case for the coconuts and betel-nut palms which also occur in the same region. *Sonneratia lanceolata* (Figs. 2, 3) and *Nypa fruticans* (Figs. 4, 5) extend much further inland than the major stands of mangrove and nipa palm, but only along river banks. These limits shown in Fig. 1 are corroborated by unpublished findings of investigators from the Australian Institute of Marine Science, who found that *Bruguiera sexangula* and *Cynometra iripa* also occurred well inland. As the upper limits of the *S. lanceolata* distribution were approached, these investigators found that the river salinities dropped to zero. However, saline water is likely to penetrate along the river bed considerably further than the point where surface waters cease to be influenced by sea-derived salts.

The two most common techniques for investigating zonation or the distribution of vegetal associations are aerial photography and ground-level studies of transects. Limited transect studies have been carried out in the mangals of northern Papua New Guinea by Green and Sander (1978) (Manus Island); Cragg (1982) (Umboi Island) and Johns (unpublished) (Labu Lagoon, near Lae). Floyd (1977) provides an illustration of a transect on Uramu Island to the west of Romilly Sound, and Spenceley (1981) presents a summary of a short transect at the mouth of the Varoi distributary of the Purari. These transect studies are too small to provide an insight into the large-scale pattern of distribution of mangroves in P.N.G.

Different zones of vegetation can frequently be distinguished by differences in appearance in aerial photographs. Paijmans and Rollet (1977) and Johnstone and Frodin (1982) have demonstrated the application of this technique to mangrove forests in Papua New Guinea. Paijmans (1970) figures aerial photographs of the Purari Delta showing the seaward margin and the transition area between mangrove and swamp forest. Ground observations were used to corroborate Paijmans' findings from aerial photographs. A map

Fig. 2. Sonneratia lanceolata (= *S. caseolaris*), a common fringing mangrove of the Purari Delta. (Photo G. Irion.)

Fig. 3. Pneumatophores of *Sonneratia lanceolata*. (Photo T. Petr.)

302

Fig. 4. Nypa fruticans in one of the interconnecting channels of the Purari Delta. (Photo T. Petr.)

Fig. 5. Fruits of *Nypa fruticans.* (Photo T. Petr.)

303

of mangrove distribution in the eastern part of the delta (Paijmans 1969) was derived from such corroborated studies. The Office of Forests map (P.N.G. Office of Forests 1977), which not only shows mangrove forest but also subdivides the forest into a variety of vegetal associations, was based almost entirely on aerial photography.

Chapman (1977), Paijmans (1976) and Percival and Womersley (1975) have described mangal zonation in Papua New Guinea, while Floyd (1977), Johnstone and Frodin (1982) and Paijmans and Rollet (1977) provide detailed descriptions of zonation at particular sites. However, the data concerning the distribution of mangroves in the Purari Delta is rather limited. A much simpler model of distribution will be used which is based on observations from the delta and from areas with environments somewhat similar to the delta (see Table 2).

The delta may be considered to have three major types of mangal: fringing, main and transitional. Of course, there are variations within each type and the types grade into one another. The following account is synthesized from the references listed in Table 2 and from unpublished observations. Fringing mangal occurs facing open water, which may be sea, estuary or tidal river. Its character varies according to the nature of the open water. This forest type is generally considered to be a pioneer ecosystem because the species forming it tend to be the first to colonise newly formed mud banks. Within the Purari Delta, *Sonneratia lanceolata* is probably the most important component of the fringing mangal. It ranges from the seaward edge of the delta far inland (see Fig. 1). *S. alba*, *Avicennia eucalyptifolia* and *Aegiceras corniculatum* are significant contributors to the fringing mangal in estuarine conditions (Floyd 1977; Australian Institute of Marine Science, unpublished), while *Pandanus* sp. and *Nypa fruticans* occur alongside but not mixed with *S. lanceolata* in low salinity river reaches. Fringing mangals generally consist of shrubs or small trees which tend to adopt a somewhat recumbent growth form or to develop multiple stems. At the water's edge the canopy may be less than a meter high, but it becomes gradually higher inland. Rarely does this dense closed canopy exceed ten metres in height. There is virtually no undergrowth though there is a dense proliferation of the upward pointing pneumatophores characteristic of *Sonneratia* and *Avicennia*.

The 'main' mangal association is characterized by various species. *Rhizophora stylosa* is generally an important component of mangals both to the east (Johnstone and Frodin 1982) and to the west (Floyd 1977) of the delta; yet only Spenceley (1981) reports its presence within the delta itself, and that record is from the extreme seaward edge. It may be that the lowered salinities of the delta reduce the competitive ability of this species. The main mangal is

Table 2. Characteristic species[a] of mangrove associations on the south coast of mainland Papua New Guinea.

Mangal type	Tidal height[b]	Salinity[c]	Sites[d]				
			1	2	3	4	5
Fringing	Up to HWN	High	As Sc	–	–	–	Ae Sa
		Medium	Aa Ac Sa Sl	Ae Sl	Sl	Sl	–
		Low	Aa Ac Sc	–	–	–	–
		Very low	Sc	–	–	–	–
Main	Around HWN	High	Rs	–	–	Rm	Rs/Ra/Bc/Bg[e]
	Around MHW	Medium	Bp Cs Ra	B Cs R	Bc Bp Bs R	Ra Bg	–
	MHW – HWS	Low	Bp Cs	B Cs Nf R	B Cs R Xm	–	–
		Low	Nf	Nf	Nf	Nf	–
Transitional[f]	HWS – EHWS	Low to very low	Br Ca Ex My	(a) Bg Ca Ex My	(a) Bs Do He Nf Sc Xm	Am Bg Br Bs	–
				(b) Br He Hi Me X	(b) B Br Ca D In My R X		
				(c) Ca Do Me My			
				(d) Ca He In Nf			

a Species letter codes are listed in Table 1; in addition Am = *Amoora cucullata*, B = *Bruguiera* sp., Cu = *Calophyllum* sp., In = *Intsia* sp., Me = *Metroxylon sagu*, My = *Myristica hollrungii*, Nf = *Nypa fruticans*, R = *Rhizophora* sp., X = *Xylocarpus* sp. Frodin (personal communication) has suggested that of the species listed in this table, *Avicennia alba*, may have been misidentified. Dashes indicate that no description is available or that the environment in question is absent at the site.

b Tidal levels represented by: HWN = high water neaps, HWS = high water springs, EHWS = extreme high water springs, MHW = mean high water.

c These descriptions are estimates: High = over 30‰, Medium = over 10‰, Low = over 2‰, Very low = below 2‰.

d Sites are: 1 = Kikori – Port Romilly region, 2 = Kikori to W. Purari, 3 = Purari Delta to Kerema, 4 = Galley Reach, 5 = Hood Lagoon. References: 1: Floyd 1977, 2: P.N.G. Office of Forests 1977, 3: Paijmans 1969, 4: Paijmans and Rollet 1977, 5: Johnstone and Frodin 1982.

e These species occur as a sequence of four distinct zones.

f At sites 2 and 3, more than one type of transitional mangal have been recognised.

Fig. 6. Rhizophora apiculata. (Photo T. Petr.)

Fig. 7. Prop roots of *Rhizophora apiculata.* (Photo T. Petr.)

306

characterized by *R. apiculata* (Figs. 6, 7), plus *Bruguiera parviflora* and *B. sexangula*, individually or in combination. *Camptostemon schulzii, Heritiera littoralis, Rhizophora mucronata* and *Xylocarpus granatum* are also present. The trees generally develop a single stem. Where this association has fairly recently developed, there is a dense closed canopy at about 15 m. Mature trees may exceed 30 m in height and where this occurs the canopy is more open. Trees tend to be taller on the banks of creeks than in the middle of the tidal flats. This forest only has a single main storey with a rather sparse undergrowth of seedlings, saplings, the shrub *Acanthus*, the fern *Acrostichum* and the nipa palm. Otherwise there is only bare mud with knee roots of *Bruguiera* and stilt roots of *Rhizophora*.

Where there is substantial freshwater influence, pure stands of *Nypa fruticans* may develop. These palms form a dense canopy of vertically directed fronds extending about 10 m. There is virtually no undergrowth under this canopy.

It is in the transitional mangal that the greatest variety is encountered – variety of species, species association, plant growth form and forest structure. Due to the lack of detailed information, a generalized description will have to suffice for this ecosystem. In the regions described by Paijmans (1969) there is an uneven open canopy about 20 m high, but Floyd (1977) encountered transition forest up to 30 m high. The principal mangrove species here are *Bruguiera sexangula, Camptostemon schultzii, Dolichandrone spathacea, Diospyros* sp. (see Table 1, Note 3), *Excoecaria agallocha, Heritiera littoralis, Rhizophora apiculata* and *Xylocarpus granatum*. A substantial proportion of the forest is formed, however, by trees which also grow in the neighbouring swamp forest. Among these are *Calophyllum, Intsia bijuga, Myristica hollrungii* and *Amoora cucullata*, which in this transitional region develop root modifications similar to true mangroves (see for example Figs. 6 and 7 in Paijmans and Rollet 1977). Smaller trees of *Barringtonia, Brownlowia, Inocarpus, Hibiscus*, and *Cerbera* form an understorey. There are scattered individuals of the palms *Areca, Arenga, Metroxylon* and *Nypa*. In addition to seedlings and saplings the undergrowth contains the shrub *Acanthus* and the fern *Acrostichum*, but herbs are virtually absent. Unlike the other types of mangrove forest, the trees of the transitional mangal often carry substantial numbers of epiphytes, and climbing plants, notably *Hoya* and *Derris*, make their first substantial contribution to the flora.

The small area at the mouth of the Varoi River exhibits an unusual floral assemblage of mangroves (Spenceley 1981) which does not fit the foregoing account. This unusual flora is probably the product of an environment where the mangroves are particularly highly stressed by wide fluctuations in the physical conditions. The mangroves here appear to be juvenile or stunted.

2.3. Determinants of zonation

Associated with the concept of zonation is the idea of physical variables interacting with the biological potential of the plants to produce the pattern observed. Johnstone and Frodin (1982) considered six factors to be important determinants of mangrove distribution: tidal regime, wave action, drainage, salinity regime, substrate and biota. At a site such as Hood Lagoon (see Fig. 2, Johnstone and Frodin 1982) the zonation pattern is clear-cut: the zone boundaries run parallel to the shoreline. Here tidal regime is probably the major determinant of zonation. In the Purari Delta on the other hand, the inland gradient is too slight to produce such a pattern and the topography is broken up by a maze of streams. A number of environmental factors vary independently or semi-independently across the tidal flats. The net result of such a physical environment is a mosaic of mangrove zones.

Even in the complex physical environment of the delta, tidal regime is possibly the most important factor affecting the distribution of the boundaries of the major mangal types − fringing, main and transitional. Tidal regime affects the dispersal of mangrove propagules (Rabinowitz 1978), which may be adapted for establishment in particular environments (Floyd 1977). However, mangroves may be able to grow outside their usual range if they are transplanted (Rabinowitz 1975) so other factors must also affect their distribution.

Wave action at the seaward margin (Fig. 8) of the delta may select for those species which have a root and stem structure which can resist buffeting from heavy seas. Paijmans (1976) suggests that poor drainage may be the cause of the rather stunted trees which occur on tidal flats away from the tidal creeks, while the better drainage on the creek banks may be the factor which permits better growth and a greater variety of species.

Salinity regime is also an important determinant of zonation in the delta. A few mangrove species such as *S. lanceolata* are tolerant of a wide range of salinities and are probably limited by other factors, yet many, particularly those occurring in the transition zone, are probably limited in their distribution by salinity levels.

The effects of substrate on the distribution of mangroves in the delta are not manifested in observations to date. Most of the delta is composed of fine-grained sediments so this factor is likely to be relatively unimportant.

Fig. 8. Eroding shore, with dying mangroves. (Photo T. Petr.)

2.4. Succession

Floyd (1977) and Johnstone and Frodin (1982) suggested that interspecific competition may be an important factor in the development of zonation. The denseness of the canopy and the close interlocking of root systems in much of the mangal of the delta certainly suggests that there is vigorous competition for space and light. If one mangrove species can out-compete another in an adjacent zone, succession may occur. The idea that mangroves build land by means of such succession is widely believed, though recent authors maintain that changes in sedimentation initiate mangrove colonization.

In the Purari Delta, when sedimentation builds the substratum up to a certain minimum level, propagules from the species such as *S. lanceolata* and *A. eucalyptifolia* become established. These propagules appear to be capable of withstanding longer inundation than those of other mangrove species (Floyd 1977). The roots of these species produce numerous vertical pneumatophores, which by modifying local currents, promote further sedimentation (Spenceley 1977). Once the level of the substratum reaches the next critical level, propagules of *Rhizophora* spp. and *Bruguiera* spp. can become established.

309

Floyd (1977) suggests that these species are more tolerant of shade during early development than the pioneer species so that eventually they shade out the pioneers. Occasional remnants of the pioneer mangal survive (Floyd 1977).

In the case of the Purari Delta, a number of factors are pertinent to the consideration of whether succession is taking place. Firstly, there is the increase in sediment input, resulting in accretion at the delta edge. Secondly, changes in distributary flow may disrupt mangrove distribution. Thirdly, there is the effect of natural catastrophes such as storms, but these are less destructive in the Purari Delta than they are in the regions described by Stoddart (1980) and Lugo (1980). Lightning, however, may have a perceptible effect on the mangal development. Aerial photographs frequently show holes in the mangrove canopy which have been ascribed to lightning strikes (Paijmans and Rollet 1977; White and White 1976). A map of such holes based on photography from 1957, 1972 and 1974 of Galley Reach (P.N.G. Office of Forests 1976) indicates that they are eventually filled by regenerating forest. Holes are formed at a sufficient rate to affect the whole forest within a 2–300-year period (Johns, unpublished). Alternative explanations for such holes are tree death due to the fungus, *Phytophthora* (Bunt, personal communication) or heavy attack by insect herbivores (Johnstone 1981).

If the changes in physical environment due to accretion, erosion and natural disaster were sufficiently widespread, then succession would be arrested and the Purari mangroves would indeed be merely opportunistic colonizers. However, the rate of accretion is probably also affected by the mangroves, areas of marked erosion are fairly localized and natural disasters do not appear to have had a marked effect on the mangal.

3. Fauna

White and White (1976) give a brief account of the fauna of disturbed areas of mangrove forest to the west of the delta. Liem and Haines (1977) and Liem (this volume) provide a list of all the vertebrates, giving details of the niches they occupy. They record only twelve species of mammals – two dolphins, the dugong, four marsupials, three bats and two other placentals. Reptile species have been listed by Pernetta and Burgin (1980), as well as by Liem and Haines (1977) and Pernetta (this volume), who also list bird species. The fish fauna is described by Haines (this volume).

Rather less is known about the invertebrate fauna except for commercially significant crustaceans such as crabs and prawns (Opnai 1980; Gwyther, this volume; Frusher, this volume). Poraituk and Ulijaszek (1981) collected a

310

number of edible bivalves and gastropods. Wood-boring bivalves are described by Cragg (1979). A limited series of collections of zooplankton were described by Cragg (1979) and Bayly (1980). Information about the insects of the mangal is sparse indeed. Some mosquito species are named by Marks (this volume), and some termites and wood-boring beetle larvae were collected by Swift and Cragg from dead and dying wood (unpublished data).

The fauna of the delta mangal can be considered within four broad categories. The first category consists of permanent residents while the other three groups consist mainly of temporary visitors to the ecosystem. Firstly there are those animals which are restricted to niches within the mangal and are specialized for the conditions occurring there. The mudskipper (*Periophthalmus*) is a fish which spends much of its time out of water on mud banks or on mangrove roots. It is capable of respiring out of water and has specialized eyes which function both in and out of water. The mud lobster (*Thalassina anomala*) is also a mangrove specialist which, like the mud crab, *Scylla serrata* and fiddler crabs (*Uca* spp.), can burrow within the anoxic mangrove muds. These muds support little other infauna. Wood-boring bivalves (family Teredinidae) are also specialized for life in the mangal: they bore into dead wood, but also into living trees (Roonwal 1964). The specialist mangrove fauna, like the mangroves themselves, appear to be limited in their distribution by gradients in physical variables. The distribution of teredinids, for example reflects the salinity regime of the delta (Cragg 1979) but also is strongly dependent on the effect of tidal regime on the frequency of inundation (Cragg and Swift, in press).

The second category of fauna consists of those animals which also occur in swamp forest or rain forests. These animals either live in the mangrove canopy, or they move into the mangrove forests as tides recede and retreat again when tides advance. More niches are available for such animals in transitional mangals than in main or fringing mangals. Ground-dwelling animals will of course be disturbed less frequently in transitional zones by tides. The canopy is the main mangrove niche for insects, which are probably the only significant herbivores in this environment (Johnstone 1981). In the less frequently inundated regions of the delta, subterranean termites are common. Their earth-covered runways extend from the substratum up into the canopy (Swift and Cragg, unpublished).

Marine and estuarine animals form the third category of fauna. They move into the mangal on the rising tide penetrating the smaller creeks and moving out over the tidal flats. This category consists mainly of fish. Haines (this volume) describes the fish which undertake such migrations.

The final category consists of fish and crustacea which spend the early part of their lives in the mangrove-dominated environment, within the permanent creeks or over the tidal flats, depending on the state of the tide. Penaeid prawns develop within mangrove swamps where there is a plentiful supply of small food particles before moving to the offshore adult habitat (Frusher, this volume). Liem and Haines (1977) list a considerable number of fish which live within the Purari Delta mangal as juveniles, but eventually migrate either to fresh or saltwater.

The food webs of the delta mangal are mainly detritus-based. Deposit-feeding prawns and fish are numerous, as shown by Haines (this volume) who constructed a food chain for the Purari Delta. Some detritivores may be quite selective – Rose (personal communication) has found that a high percentage of the diet of the turtle *Carettachelys insculpta* consists of fruits of *Sonneratia*. The detritivores are preyed upon by a wide variety of fish. At the top of the food webs are fish-eating birds, crocodiles and man. Due to the temporary status of much of the fauna or to its mobility, food webs initiated within the mangal extend inland and offshore.

4. Ecological interactions of the mangrove ecosystem

4.1. Litter breakdown

Mangrove forests are open ecosystems with substantial inflows and outflows of energy and nutrients. These flows follow various distinctly different pathways whose characteristics are determined by the physical and chemical properties of the various plant tissues and by the location (i.e. aerial, intertidal or subterranean) of these tissues. Wood is less readily moved by currents and is broken down more slowly than leaf litter. A variety of specialized intertidal and terrestrial (aerial) animals burrow into the wood, thus facilitating fungal colonization. Subterranean roots are probably broken down in situ by microbial action alone. This breakdown plus root exudate production may constitute a substantial route for energy flow in the delta mangrove ecosystem.

Mangrove leaves are eaten by insects (Johnstone 1981) and cuscuses (marsupials – Liem and Haines 1977). Johnstone found that herbivory generally removed less than 10% of leaf biomass so that most falls to the substratum. In areas of the mangal subject to regular tidal inundation, much of the leaf litter is removed: accumulations of this litter in shallow waters off the mouth of the delta often clog prawn trawl nets (Frusher, personal communication). Some accumulates as strand-line deposits (Swift and Cragg, unpublished observation). Leaf litter provides food for crabs living in the substratum (Macintosh 1980) and fish (Haines, this volume). Leaves not consumed by detritivores first lose the protection of the cuticle and soluble tannins; then they are colonised by micro-organisms which eventually cause fragmentation of the leaf (Cundell et al. 1979).

A substantial amount of plant tissue breakdown probably occurs in the mangrove soils. Many of the fish and prawns of the saline reaches of the delta feed on these detritus-enriched muds with their microbial flora (Haines, this volume).

Eventually, all mangrove litter is converted into forms readily transported by currents – particulate organic matter (POM) and dissolved organic matter (DOM). Thus, the benefits of mangrove productivity may be felt far out to sea where invertebrates and micro-organisms metabolise this organic matter. Mangrove productivity is also dispersed via food webs extending into neighbouring regions.

4.2. Nutrient fluxes

Nutrients in the mangal are also subject to substantial fluxes. Boto (1982) has identified a number of these: nutrient inputs into mangal take place through rainfall, fresh water, nitrogen fixation, heterotrophic conversion of organic matter to inorganic, seawater, chemical release from soil, human activity, and mobile animals. Nutrients are lost in tidal currents, by leaching of canopy, microbial denitrification, immobilization in soil, leaching of soils by fresh water, and in mobile animals. For Australian conditions, he concludes that knowledge is at present inadequate to determine whether there tends to be a net inflow or loss of nutrients. With the low population of the Purari Delta and lack of industrialization, nutrient input via rainfall and human activity is likely to be insignificant. Nitrogen fixation is probably also unimportant, but the mangrove soils have a considerable potential to supply nutrients and freshwater nutrient input is significant (Viner 1979). Nutrient export in plant litter is clearly significant, but tides and river currents also carry out quantities of particulate organic matter (as demonstrated by Boto and Bunt 1981) and dissolved nutrients both in inorganic form and organically bound. A study of mangrove soils in the Port Moresby region indicates that some nitrogen loss due to denitrifying bacteria occurs (Pearson, unpublished). Where large-scale diurnal movements or migrations of animals occur, as for example is the case with flying foxes, substantial nutrient flow may occur.

5. Utilization of delta mangal resources

5.1. Present subsistence utilization

In Southeast Asia, mangals have been extensively modified by man's activities. In Papua New Guinea, on the other hand, most mangals are virtually unaffected by human exploitation. This is partly due to the less rapid pace of development in P.N.G., and partly to the absence of population pressure on the areas of coastline in P.N.G. where mangroves occur. The population which utilizes the mangal of the Purari Delta is certainly small. Few villages are actually situated within the mangrove forests, but villagers living inland of the mangroves, or on the sand bars at the seaward edge of the delta do intermittantly travel into the mangal in order to harvest the various resources available there.

Sago, the main staple in the diet of the inhabitants of the Purari floodplain, is obtained from the inner fringe of the mangal. However, the mangal itself yields a variety of foodstuffs. In the absence of more palatable alternative, immature fruit and unopened fronds are collected from the nipa palm. Both taste rather salty, but the sap collected from the palm inflorescence is sweet. Mature palm fronds can be used for thatching, though sago palm fronds are deemed to be superior for this purpose (Cragg 1982). The propagules of *Bruguiera* are used as a vegetable elsewhere in P.N.G. and may be so used by villages of the delta.

The mangal ecosystem makes a valuable contribution to the protein component of the diets of villagers living in the southern part of the delta (Ulijaszek and Poraituk, this volume). Prawns and a variety of fish are caught in the numerous tidal creeks. (Stevens 1980). From the tidal flats between the mangrove trees, crabs are collected (Opnai 1980) and molluscs too (Poraituk and Ulijaszek 1981). The subsistence fishing and gathering provide a small entry into the cash economy which has the potential to be increased considerably (Stevens 1980). Terrestrial vertebrates are occasionally hunted in the mangal – goannas (*Varanus* sp.) are not only a source of meat but also a source of skins for traditional kundu drums.

Due to the abundance of rain forest timber in the areas immediately adjacent to the delta villages, villagers probably make only limited use of the excellent mangrove timbers. Elsewhere in P.N.G., *Rhizophora* and *Bruguiera* are prized as first-class fire woods and as strong, straight poles and posts.

5.2. Commercial utilization

Industrial or commercial activity in the delta has to date been extremely limited. Logging has been restricted to rain forest areas, with only minor disturbance occurring in the mangal, due to oil exploration and some mangrove bark and timber harvesting. A small factory, at which tannin was extracted from mangrove bark, was established in the 1950s at the Aird Hills to the west of the delta. The advent of superior leather-tanning chemicals lead to the commercial demise of this operation (Percival and Womersley 1975).

During the exploration of the Gulf of Papua for oil-bearing deposits, drilling rigs and ancillary equipment were installed at various sites within the mangal of the region. The disturbance caused by these operations has had little lasting effect, except in a few small areas where the mud was compacted so that regeneration of the mangrove forest was severely inhibited (White and White 1976).

Commercial exploitation of the wild crocodile population is of minor importance now. Stocks of large crocodiles have probably been severely depleted by hunting carried out before independence (Pernetta and Burgin 1980). The village-based crocodile farms may be currently the most effective way of exploiting the commercial potential of crocodiles.

6. Development options for the delta mangal

It is unlikely that the delta mangal will remain in its virtually pristine state. Quite apart from the potentially far-reaching effects if the dam is constructed at Wabo, a variety of commercial schemes based on the resources of the delta mangal may be initiated. Those who have to decide which schemes are most appropriate for this region are in the fortunate position of being able to draw upon the experience gained (often at considerable cost) in Southeast Asia and elsewhere in the tropics.

6.1. Palm alcohol

Though substantial technical problems remain to be overcome, certain of the trees and palms within the mangal have considerable commercial potential. The sago palm and the nipa palm are both under consideration as possible sources of raw materials from which alcohol could be derived for use as substitute for imported fuels (Holmes and Newcombe 1980; Newcombe et al.

315

1980). In the case of sago, the carbohydrate-rich pith of the mature stem is the raw material; in the case of nipa, it is the syrupy sap of the flower stem. In either case only a small fraction of the vegetable matter produced by the plant would be removed by a commercial operation so that the net drain on the ecosystem need not be excessive. A further advantage is that at least initially the naturally-occurring pure stands of these palms could be exploited. Even if the palms were cultivated plantation style this would not grossly interfere with the palm-based ecosystems, nor would it adversely affect those organisms dependent on the palms for food or shelter.

6.2. Timber and tree-derived products

The trees of the mangal also have undoubtedly economic potential. *Xylocarpus*, marketed under the name 'mangrove cedar', is recognized as a decorative furniture timber and is placed by the Office of Forests of P.N.G. in the second highest category of export timbers. The forests dominated by *Rhizophora apiculata* and *Bruguiera* spp. may also eventually prove to be a valuable resource. These trees tend to predominate where they grow and there are large areas of single species or two species stands – an unusual situation: loggers operating in tropical forests generally have to contend with very mixed forests. *Rhizophora apiculata* and *Bruguiera* spp. usually grow straight with few or no low branches; their timber is dense and strong (Bolza and Kloot 1976), but has no particular decorative value. They are rarely harvested in P.N.G. however, despite the factors in their favour. One barrier to their more widespread utilization in P.N.G. is their inadequate permeability to the standard treatment chemicals used there (Bowers 1977) which means that they cannot be recommended as power line poles, a use for which they would otherwise be eminently suitable. Another major problem is that of harvesting: conventional logging techniques cannot be used in mangrove swamps due to the extremely low load-bearing capacity of the mud, furthermore as the timbers tend to be very dense they cannot usually be floated out when felled. Until supplies of good rain forest timbers dwindle, it is unlikely that logging equipment suitable for mangrove swamps will be introduced into P.N.G.

If the trees of the mangal could be harvested for timber, this would give this resource a fairly high value. Lower value uses of mangrove trees are widespread in the tropics. In some countries, the cost of firewood is sufficient to encourage the harvesting of mangroves for fuel. Rather than burning the wood itself, charcoal may be produced. Mangroves such as *Rhizophora* spp. yield good firewood and excellent charcoal (National Academy of Science

316

1980). In Papua New Guinea for the foreseeable future, the demand for firewood and charcoal can probably be supplied from the waste wood generated by logging and sawmilling operations. The return from exporting firewood or charcoal is at present unlikely to be adequate to support any harvesting operation which was geared to ensuring adequate regeneration of the resource.

Another possible use of the mangroves is as feedstock for a wood-chipping operation. This would also put a relatively low value on the resource. Koeppen and Cohen (1956) showed that timbers taken from the mangal of the Gulf of Papua are suitable for pulping for papermaking. A survey of the mangal near Kikori indicated that the existing economic climate did not yet favour the establishment of a wood-chipping operation based on the mangrove resource (Oji Paper Co. 1975).

Wood-chip harvesting methods used in Southeast Asia appear to offer rapid returns on investment at the expense of the long-term value of the resource (Richardson 1977). The author is unaware of any economically viable method of harvesting mangroves for wood chips which does not entail extensive environmental damage.

Tannins derived from mangrove bark could become a saleable commodity if problems of utilization and market factors were overcome. Tannins extracted from mangrove bark were used in the formulation of an experimental adhesive which was tested in the ply mill at Bulolo, Morobe Province (Hart, personal communication). The adhesive was effective, but could not compete in terms of cost with the adhesive in use for plywood manufacture.

Any method of exploiting the mangal of the Purari Delta should foster adequate regeneration of the forest, unless a completely new use is envisaged for the harvested areas. Harvesting methods in use at the moment frequently cause problems. Some harvesting regimes may keep mangrove forests intact despite regular cropping, however, Gong et al. (1980) have found indications that the productivity of such forests may be declining. The present state of knowledge regarding mangrove silviculture is summarized by Teas (1979).

6.3. Animal resources

The delta mangrove resource should not be considered to consist merely of trees and palms. The mangal harbours a variety of animal resources and supports further stocks of animals which live beyond the mangal. As with the plant resources, a more complete utilization of animal resources depends on the solution of certain technical problems and on favourable economic conditions.

Suitable markets for the mangrove crab exist in the highlands of P.N.G. and Port Moresby, but at present it is difficult to reliably transport adequate quantities of fresh crabs (Haines and Stevens, this volume). Existing crab stocks could sustain a much higher yield than is at present taken (Opnai 1980). The artisanal fishery based on the tidal creeks of the mangal could also be considerably expanded without causing over-fishing (Haines, this volume). The commercial potential of mangrove molluscs is probably fairly limited.

Elsewhere in Southeast Asia, the emphasis has moved away from harvesting fish and shellfish towards farming them. The results of this are not always desirable in the long term. In the Philippines, for example, poorly regulated fish farm construction in many areas has resulted in total loss of the timber resource, declining yields in the farms and complete disruption of the mangal ecosystem. It is unlikely that large-scale fish farming will be attempted in the Purari Delta for some time. Crocodile farming, however, may be expanded.

6.4. Planning utilization

Planning for the most suitable pattern of development for the delta mangal requires adequate maps of the resources present. Existing maps of nipa distribution (Paijmans 1969; P.N.G. Office of Forests 1979) are probably fairly accurate, as stands of the palm have a very characteristic appearance on aerial photographs. It is more difficult to differentiate between the different types of mangrove or even between mangrove and other forms of forest. The Office of Forests map of the mangroves of the Gulf of Papua (P.N.G. Office of Forests 1977) covers the western third of the delta. It differentiates the mangrove forest into a variety of categories, some deemed to have commercial potential and others not. This map was drawn from aerial photographs with little or no confirmation at ground level. The C.S.I.R.O. Land Use series maps (Paijmans 1969) cover about the eastern third of the delta, but little attempt has been made to differentiate between the different types of mangrove forest. Studies are under way in the Departments of Forestry and Surveying at the University of Technology in Lae to develop methods of remote sensing appropriate for surveying mangals in Papua New Guinea. Various types of aerial photography and also Landsat imaging are being tested on areas where the mangal has also been surveyed at ground level. If a suitable remote sensing technique can be developed, it would prove extremely valuable for surveying the delta mangal because ground level surveying on a large scale would be extremely difficult due to the problems of moving about in such areas and consequently would be prohibitively expensive.

As can be seen from the preceding paragraphs a wide variety of options for the development of the Purari mangal could be considered. Some of the options are mutually exclusive, others may complement each other. As is common in considerations regarding development options, a major choice lies between options offering swift returns at a cost to the environment and options yielding lesser returns while requiring a greater investment in terms of cash and expertise. Pressing needs for returns on investment often mean that in developing countries the former sorts of options are selected. However, because of the dynamic instability of the mangrove ecosystem (see the section dealing with the physical environment of the mangal), these options may irreversibly alter the ecosystem. In some cases this may be acceptable, but otherwise, the economic value of the shoreline protection provided by mangroves and of the dependent fisheries should be included in the evaluation of schemes for development.

Where regular harvesting is envisaged, the resource will have to be valued sufficiently high for adequate measures to be taken to ensure regeneration. This means for example that the use of *Rhizophora* and *Bruguiera* for poles and lumber plus the use of *Xylocarpus* for high-grade timber is to be advocated where possible rather than the use of these timbers for wood chips or firewood. The value of the resource can be increased if waste wood can be economically converted into charcoal and if tannin extracted from the bark proves marketable. An integrated timber – charcoal – tannin industry may prove viable where an industry based on a single commodity would not.

7. Possible effects of dam construction

The possible environmental impact of the Purari River (Wabo) hydroelectric scheme has been reviewed by Petr (1979, 1980). He concluded that if the dam was built, a number of environmental changes might affect the mangal ecosystem of the delta.

The predicted changes in sediment load (Pickup and Chewings, this volume) suggest that during the construction period, the delta may actually extend so that new areas will become available for pioneering mangroves. However, the long-term effect of dam construction is more likely to be regression of the coastline of the delta, resulting in compression of mangrove zones or perhaps a slow landward migration of these zones. The data available is insufficient to allow any prediction about the magnitude of this regression (Thom and Wright, this volume).

A fine example of the consequences of reduced sediment input is provided by the Angabunga Delta which lies between the Purari River and Port Moresby (personal observations). Here, due to the river changing its course, sediment input has virtually ceased, and as a consequence, the shoreline of this delta is receding so that now there is no evidence of the fringing zone of mangroves. A tall forest dominated by *Rhizophora apiculata* is exposed to direct wave action which is undercutting the root systems of these trees. Numerous recently fallen trees may be seen along the seaward margin of this delta.

As only a relatively small proportion of clay-sized particles would be retained by the dam, sufficient phosphorus would reach the delta to maintain fertility there (Viner 1979). Dam construction is also unlikely to significantly affect the input of inorganic nitrogen to the delta. Fertility within the mangal is high (Viner 1982), and if water turbidity decreases as projected, increased algal productivity can be anticipated there due to increased light penetration into the water column. These conditions would also make larvae of prawns and fish more visible and thus more vulnerable to predation (Petr 1979, 1980). As the major prawn nursery grounds occur elsewhere along the coast of the Gulf of Papua, this eventuality would not pose a threat to the Gulf prawn fishery (Gwyther, this volume).

References

Australian Department of Defence, 1980. Australian national tide tables 1981. Department of Defence (Navy Office), Canberra.

Bayly, I.A.E., 1980. A preliminary report on the zooplankton of the Purari estuary. In: T. Petr (ed.), Purari River (Wabo) Hydroelectric Scheme Environmental Studies, Vol. 11: Aquatic ecology of the Purari River catchment, pp. 7 – 11. Office of Environment and Conservation, Waigani, Papua New Guinea.

Bolza, E. and N.H. Kloot, 1976. The mechanical properties of 81 New Guinea timbers. C.S.I.R.O. Division of Building Res. Tech. Paper, No. 11. C.S.I.R.O., Australia.

Boto, K.G., 1982. Nutrient and organic fluxes in mangroves. In: B.F. Clough (ed.), Mangrove Ecosystems in Australia: Structure, Function and Management, pp. 239 – 258. Australian National Univ. Press, Canberra.

Boto, K.G. and J.S. Bunt, 1981. Tidal export of particulate organic matter from a northern Australian mangrove system. Estuarine, Coastal and Shelf Sci. 13: 247 – 255.

Bowers, E.A., 1977. Pressure treatment characteristics of 142 commercially important timbers from the south-west Pacific region. C.S.I.R.O. Div. of Building Res. Tech. Paper, No. 13. C.S.I.R.O., Canberra.

Carter, J.H., 1980. Discharge data for the Purari River and some of its tributaries. In: A. Viner (ed.), Purari River (Wabo) Hydroelectric Scheme Environmental Studies, Vol. 16, pp. 17 – 23. Office of Environment and Conservation, Waigani, Papua New Guinea.

Cragg, S.M., 1979. Marine wood borers in the Purari Delta and some adjacent areas. In: T. Petr (ed.), Purari River (Wabo) Hydroelectric Scheme Environmental Studies, Vol. 10: Ecology of the Purari River catchment, pp. 91 – 99. Office of Environment and Conservation, Waigani, Papua New Guinea.

Cragg, S.M., 1982. Coastal resources and the Umboi logging project: An environmental impact study. Office of Environment and Conservation, Waigani, Papua New Guinea.

Cragg, S.M. and M.J. Swift, in press. Wood decomposition in a mangrove community in New Guinea with particular reference to fungi and teredinids.

Cundell, A.M., M.S. Brown, R. Stanford and R. Mitchell, 1979. Microbial degradation of *Rhizophora* mangal leaves immersed in the sea. Est. Coast. Mar. Sci. 9: 281 – 286.

Duke, N.C., J.S. Bunt and W.T. Williams, 1981. Mangrove litter fall in northeastern Australia. Annual totals by component in selected species. Aust. J. Bot 29(5): 547 – 554.

Evesson, D.T., 1983. The climate of the Purari River basin. This volume, Chapter I, 2.

Floyd, A.G., 1977. Ecology of the tidal forests in the Kikori – Romilly Sound area, Gulf of Papua. P.N.G. Office of Forests, Div. of Botany. Ecology Report, No. 4. Lae, Papua New Guinea.

Frodin, D.G., C.R. Huxley and K.W. Kirina, 1975. Mangroves of the Port Moresby region. Univ. of Papua New Guinea. Dept. of Biology. Occas. Pap., No. 3. UPNG, Port Moresby, Papua New Guinea.

Frodin, D.G. and G.J. Leach, 1982. Mangroves of the Port Moresby region, 2nd ed. Univ. of Papua New Guinea. Dept. of Biology. Occas. Pap., No. 3. UPNG, Port Moresby, Papua New Guinea.

Frusher, S.D., 1980. The inshore prawn resource and its relation to the Purari Delta region. In: D. Gwyther (ed.), Purari River (Wabo) Hydroelectric Scheme Environmental Studies, Vol. 15: Possible effects of the Purari Hydroelectric Scheme on subsistence and commercial crustacean fisheries in the Gulf of Papua, Workshop 12 Dec. 1979, pp. 11 – 27. Office of Environment and Conservation, Waigani, Papua New Guinea.

Frusher, S.D., 1983. The ecology of juvenile penaeid prawns, mangrove crab (*Scylla serrata*) and the giant freshwater prawn (*Macrobrachium rosenbergii*) in the Purari Delta. This volume, Chapter II, 7.

Golley, F.B., J.T. McGinnis, R.G. Clements, G.I. Child and M.J. Duever, 1975. Mineral cycling in a tropical moist forest. Univ. of Georgia Press, Athens, Georgia.

Gong, W.K., J.E. Ong, C.H. Wong and G. Dhanarajan, 1980. Productivity of mangrove trees and its significance in a managed mangrove ecosystem in Malaysia. Asian Symp. on Mangrove Environment: Research and Management, Kuala Lumpur.

Green, W. and H. Sander, 1978. Manus Province tuna cannery environmental study. Office of Environment and Conservation, Waigani, Papua New Guinea.

Gwyther, D., 1983. The importance of the Purari River Delta to the prawn trawl fishery of the Gulf of Papua. This volume, Chapter II, 8.

Haines, A.K., 1983. Fish fauna and ecology. This volume, Chapter II, 9.

Haines, A.K. and R.N. Stevens, 1983. Subsistence and commercial fisheries. This volume, Chapter II, 10.

Holmes, E.B. and K. Newcombe, 1980. Potential and proposed development of sago (*Metroxylon* spp.) as a source of power alcohol in Papua New Guinea. In: W.R. Stanton and M. Flach (eds.), Sago: The Equatorial Swamp as a Natural Resource, pp. 164 – 173. Nijhoff, The Hague.

Irion, G. and T. Petr, 1980. Geochemistry of the Purari catchment with special reference to clay mineralogy. In: T. Petr (ed.), Purari River (Wabo) Hydroelectric Scheme Environmental Studies, Vol. 18, pp. 1 – 24. Office of Environment and Conservation, Waigani, Papua New Guinea.

321

Johnstone, I.M., 1981. Consumption of leaves by herbivores in mixed mangrove stands. Biotropica 13: 252 – 259.

Johnstone, I.M. and D.G. Frodin, 1982. Mangroves of the Papuan subregion. In: J.L. Gressitt (ed.), Biogeography and Ecology of New Guinea, pp. 513 – 528. Dr W. Junk, The Hague.

Koeppen, A. von and W.E. Cohen, 1956. Pulping studies on individual and mixed species of a mangrove association. Holzforschung 10(1): 18 – 22.

Liem, D.S., 1983. Survey and management of wildlife resources along the Purari River. This volume, Chapter II, 3.

Liem, D.S. and A.K. Haines, 1977. The ecological significance and economic importance of the mangrove and estuarine communities of the Gulf Province, Papua New Guinea. In: T. Petr (ed.), Purari River (Wabo) Hydroelectric Scheme Environmental Studies, Vol. 3, pp. 1 – 35. Office of Environment and Conservation, Waigani, Papua New Guinea.

Löffler, E., 1977. Geomorphology of Papua New Guinea. Australian National University Press, Canberra.

Lugo, A.E., 1980. Mangrove ecosystems: successional or steady state? Biotropica 12, Suppl: 65 – 72.

MacFarlane, J.W., 1980. Surface and bottom sea currents in the Gulf of Papua and western Coral Sea. P.N.G. Dept. of Primary Industry. Res.-Bull. 27. Dept. of Primary Industry, Port Moresby, Papua New Guinea.

Macintosh, D.J., 1980. Ecology and productivity of Malaysian mangrove crabs (Decapoda: Brachyura). Asian Symp. on Mangrove Environment: Research and Management, Kuala Lumpur.

Macnae, W., 1968. A general account of the fauna and flora of mangrove swamps and forests in the Indo-West Pacific region. Adv. Mar. Biol. 6: 73 – 270.

Marks, E.N., 1983. Mosquitoes of the Purari River lowlands. This volume, chapter III, 7.

McAlpine, J.R., G. Keig and K. Short, 1975. Climatic tables for Papua New Guinea. C.S.I.R.O. Div. Land Use Research, Tech. Pap., No. 37. C.S.I.R.O., Australia.

National Academy of Science, 1980. Firewood crops. N.A.S., Washington.

Newcombe, K., E.B. Holmes and A. Paivoke, 1980. Palm energy: alcohol fuel from the sago and nipa palms of Papua New Guinea – the development plans. P.N.G. Dept. of Minerals and Energy. Report No. 6/80. The Department, Port Moresby, Papua New Guinea.

Oji Paper Company, Ltd./Shin Asahigawa Co., Ltd., 1975. Report on preliminary survey of mangrove forests in the Kikori timber area of Papua New Guinea. Unpublished report.

Opnai, L.J., 1980. The mangrove crab, *Scylla serrata*. In: D. Gwyther (ed.), Purari River (Wabo) Hydroelectric Scheme Environmental Studies, Vol. 15: Possible effects of the Purari Hydroelectric Scheme on subsistence and commercial crustacean fisheries in the Gulf of Papua, Workshop 12 Dec. 1979, pp. 83 – 91. Office of Environment and Conservation, Waigani, Papua New Guinea.

Paijmans, K., 1969. Vegetation and ecology of the Kerema – Vailala area. In: B.P. Ruxton et al. (eds.), Lands of the Kerema – Vailala Area, Territory of Papua and New Guinea, pp. 95 – 116. C.S.I.R.O. Land Research Series, No. 23. C.S.I.R.O., Melbourne.

Paijmans, K., 1970. Land evaluation by air photo interpretation and field sampling in Australian New Guinea. Photogrammetria 26(2/3): 77 – 100.

Paijmans, K., 1976. Vegetation. In: K. Paijmans (ed.), New Guinea Vegetation, pp. 23 – 105. Australian National University Press, Canberra.

Paijmans, K. and B. Rollet, 1977. The mangroves of Galley Reach, Papua New Guinea. Forest Ecol. Manage. 1: 119 – 140.

Papua New Guinea. Office of Forests, 1976. Galley Reach-mangrove vegetation: lightning

322

strikes. Map No. 2676. 1:25,000. Office of Forests, Port Moresby, Papua New Guinea.

Papua New Guinea. Office of Forests, 1977. Forest types: Gulf of Papua. Map No. 2687. 1:150,000. Office of Forests, Port Moresby, Papua New Guinea.

Percival, M. and J.S. Womersley, 1975. Floristics and ecology of the mangrove vegetation of Papua New Guinea. P.N.G. Dept. of Forests. Div. of Botany. Botany Bull., No. 8. Dept. of Forests, Lae, Papua New Guinea.

Pernetta, J.C., 1983. The wildlife of the Purari catchment. This volume, Chapter II, 2.

Pernetta, J.C. and S. Burgin, 1980. Census of crocodile populations and their exploitation in the Purari area (with an annotated checklist of the herpetofauna). In: T. Petr (ed.), Purari River (Wabo) Hydroelectric Scheme Environmental Studies, Vol. 14, pp. 1 – 44. Office of Environment and Conservation, Waigani, Papua New Guinea.

Petr, T., 1976. Some chemical features of two Papuan fresh waters (Papua New Guinea). Aust. J. Mar. Freshwat. Res. 27: 467 – 474.

Petr, T., 1979. Possible environmental impacts on inland waters of two planned major engineering projects in Papua New Guinea. Environmental Conservation 6(4): 281 – 286.

Petr, T., 1980. Purari River hydroelectric development and its ecological impact – an attempt at prognosis. In: J.I. Furtado (ed.), Tropical Ecology and Development. Proc. Vth Intern. Symp. of Tropical Ecology, 16 – 21 April 1979, Kuala Lumpur, pp. 871 – 882. International Society of Tropical Ecology, Kuala Lumpur.

Petr, T., 1983a. Limnology of the Purari basin. Part 1. The catchment above the delta. This volume, Chapter I, 9.

Petr, T., 1983b. Limnology of the Purari basin. Part 2. The delta. This volume, Chapter I, 10.

Petr, T. and J. Lucero, 1979. Sago palm (*Metroxylon sagu*) salinity tolerance in the Purari River Delta. In: T. Petr (ed.), Purari River (Wabo) Hydroelectric Scheme Environmental Studies, Vol. 10: Ecology of the Purari River catchment, pp. 101 – 106. Office of Environment and Conservation, Waigani, Papua New Guinea.

Pickup, G., 1980. Hydrologic and sediment modelling studies in the environmental impact assessment of a major tropical dam project. Earth Surface Processes 5: 61 – 75.

Pickup, G., 1983. Sedimentation processes in the Purari River upstream of the delta. This volume, Chapter I, 11.

Pickup, G. and V.H. Chewings, 1983. The hydrology of the Purari and its environmental implications. This volume, Chapter I, 8.

Poraituk, S. and S. Ulijaszek, 1981. Molluscs in the subsistence diet of some Purari Delta people. In: A. Viner (ed.), Purari River (Wabo) Hydroelectric Scheme Environmental Studies, Vol. 20, pp. 1 – 30. Office of Environment and Conservation, Waigani, Papua New Guinea.

Rabinowitz, D., 1975. Planting experiments in mangrove swamps of Panama. In: G.E Walsh, S.C. Snedaker and H.J. Teas (eds.), Proc. Intern. Symp. Biology and Management of Mangroves, pp. 385 – 393. Inst. of Food and Agric. Sci., Univ. Florida, Gainesville.

Rabinowitz, D., 1978. Dispersal properties of mangrove propagules. Biotropica 10: 47 – 57.

Richardson, D., 1977. A Faustian dilemma. Unasylva 29: 12 – 14.

Roonwall, M.L., 1964. Wood-boring teredinid shipworms (Mollusca) of mangroves in the Sunderbans, West Bengal. In: Scientific Problems of the Humid Tropical Zone Deltas and Their Implications, pp. 277 – 280. Proc. Dacca Symp., UNESCO.

Ruxton, B.P., 1969. Geomorphology of the Kerema – Vailala area. In: B.P. Ruxton et al. (eds.), Lands of the Kerema – Vailala Area, Territory of Papua and New Guinea, pp. 65 – 76. C.S.I.R.O. Land Research Series, No. 23. C.S.I.R.O., Melbourne.

Scoffin, T.P., 1970. The trapping and binding of subtidal carbonate sediments by marine vegetation in Bimini Lagoon, Bahamas. J. Sedimentary Petrology 40: 249 – 273.

323

Spenceley, A.P., 1977. The role of pneumatophores in sedimentary processes. Marine Geol. 24: 31–37.

Spenceley, A.P., 1981. Mangal development at the mouth of Varoi River, Gulf Province. Science in New Guinea 8(3): 192–199.

Stevens, R.N., 1980. The agricultural and fishing development in the Purari Delta in 1978–1979. In: T. Petr (ed.), Purari River (Wabo) Hydroelectric Scheme Environmental Studies, Vol. 13, 1–16. Office of Environment and Conservation, Waigani, Papua New Guinea.

Stoddart, D.R., 1980. Mangroves as successional stages, inner reefs of the northern Great Barrier Reef. J. Biogeography 7(3): 269–284.

Teas, H.J., 1979. Silviculture with saline water. In: A. Hollaender (ed.), The Biosaline Concept, pp. 117–161. Plenum, New York and London.

Thom, B.G. and L.D. Wright, 1983. Geomorphology of the Purari Delta. This volume, Chapter I, 4.

Thom, B.G., L.D. Wright and J.M. Coleman, 1975. Mangrove ecology and deltaic-estuarine geomorphology: Cambridge Gulf-Ord River, Western Australia. J. Ecol. 63: 203–232.

Ulijaszek, S.J. and S.P. Poraituk, 1983. Nutritional status of the people of the Purari Delta. This volume, Chapter III, 8.

Viner, A.B., 1979. The status and transport of nutrients through the Purari River (Papua New Guinea). In: T. Petr (ed.), Purari River (Wabo) Hydroelectric Scheme Environmental Studies, Vol. 9, pp. 1–52. Office of Environment and Conservation, Waigani, Papua New Guinea.

Viner, A.B., 1982. A quantitative assessment of the nutrient phosphate transported by particles in a tropical river. Rev. Hydrobiol. trop. 15(1): 3–8.

White, K.J. and A.E. White, 1976. The effects of industrial development on mangrove forests in the Gulf Province of Papua New Guinea. P.N.G. Office of Forests. Ecol. Progr. Rep., 3. Office of Forests, Papua New Guinea.

6. Aquatic pollution in the Purari basin

T. Petr

1. Introduction

Negative impacts on water quality from human activities in the Purari River basin are still of a very limited nature and can be categorized as follows: (i) increased sediment load due to soil disturbances from intensive agriculture, deforestation and some engineering activities, for example road building; (ii) direct use of stream water for washing and excreta disposal; discharge of sewage into streams; (iii) introduction of pesticides and other harmful components through their use in agriculture, mainly for protection of cash crops.

The virtual absence of mining and processing industries in the catchment allows us to assume that, with the exception of indicators of the above types of interference with water quality, the basic physico-chemical components, both dissolved and particle-bound, will still largely reflect an undisturbed natural environment. Water chemistry of the Purari basin rivers and lakes, and that of deposited sediments is dealt with in other chapters of this book (Petr, this volume a; this volume b; Irion and Petr, this volume; Petr and Irion, this volume). The present chapter considers the current water quality and gives some indication about the possible impact of the planned Wabo reservoir on the self-purification process.

2. Erosion and sediment transport by rivers

In some high rainfall areas of P.N.G., where there is virtually no human pressure on land, extreme sediment loads may result from landslides. After a landslide in January 1977 in the Ok Tedi River catchment in the Western Province, people reported that fish had been killed and disappeared for a number of months from the river (Petr 1979), probably due to a high sediment load.

Petr, T. (ed.) The Purari – Tropical environment of a high rainfall river basin
© *1983, Dr W. Junk Publishers, The Hague / Boston / Lancaster*
ISBN-13: 978-94-009-7265-0

There are no known direct observations from elsewhere in P.N.G. on the impact of natural causes on river water quality, although one can assume that volcanic activity is another relatively common catastrophic cause of elevated sediment inputs into P.N.G. rivers.

Direct documented observations on the impact of deforestation on the sediment transport in Papua New Guinea rivers are unavailable, although commercial large-scale deforestation is taking place in a number of areas outside the Purari basin. Although the Purari itself is rated as a river with high sediment load (Pickup, this volume), its mean load of 205 mg per liter (Petr, this volume a) is still considerably lower than that of many Southeast Asian tropical rivers whose basins have been under intensive human pressure for many years, largely as a result of overpopulation. At present a large proportion of P.N.G. forests is still relatively undisturbed, but the situation is changing rapidly (Lamb 1977). Soil erosion is still on a very limited scale, although areas with the traditional systems of land use have approached the maximum development possible without improvements in technology, or increased labour inputs (Stark 1982). Thus, where new land is not available these systems will not be able to support at an acceptable level further increases in population, and furthermore, such pressure will result in a degree of erosion that causes soil damage or loss (Stark 1982).

In P.N.G. there are a limited number of small catchment rivers where deforestation has led to a drastic increase in sediment transport and in places to the destruction of topsoil. In the area adjacent to the city of Lae, the increasing demand for both firewood and land for subsistence gardens has turned large areas of the heavily wooded hills into eroded wastelands, with severe economic and ecological effects (Newcombe and Pohai 1981). Already prior to disturbance of their catchments, the streams had heavy natural sediment loads due to geologically young topography. Removal of the forests and subsequent intensive gardening of unconsolidated soils led to a rapid erosion of the top soil.

Investigations of sediment transport through the Purari catchment rivers revealed an extremely heavy load of 6861 mg per liter in Mairi Creek at Kenangi in Simbu Province after a heavy rainfall in its catchment (Petr, this volume a). This compares with the maximum value ($n = 173$) of 8493 mg per liter for the Ok Tedi River at Tabubil, Western Province, from an area draining mostly uninhabited forested highlands receiving about 10 000 mm rainfall per year (Petr 1979). In Simbu Province, heavy rainfalls are not infrequent, and one could therefore assume that very high loads are common not only in the Mairi Creek catchment but also elsewhere. This province has the highest density of human population in the highlands of Papua New Guinea, with intensive agriculture evolving a high pressure on soils in areas which have little natural forest left. A further aggravating factor for enhanced erosion is the widespread presence of shale soils in this province (Blong and Humphreys 1982). These soils undergo extremely high erosion when cultivated (Wood, this volume). An ameliorating factor preventing high erosion of the Purari basin highland soils in areas of dense human population is the widespread presence of volcanic ash (Pain, this volume). When cultivated by traditional methods, soils formed on such soils erode minimally even on slopes of $30-35°$ (Wood, this volume). It is the marginal areas with other types of soils which are vulnerable and in which intensification of agriculture should be considered with utmost caution. Wood has suggested that large-scale settlement of such areas should be avoided as the benefits would be short term only.

Amongst the six major rivers draining the highlands of the Purari basin, the Tua appears to have the highest mean sediment transport, i.e. 412 mg/l as compared with the range of $68-185$ mg per liter for the other five rivers. Although this value is only relative, being based on a very limited number of samples (Petr, this volume a), it shows the impact of the high human density on sediment transport. Tua, which forms at the confluence of the Asaro and Wahgi Rivers, receives water from very densely populated highland valleys, with relatively highly disturbed soils. The other rivers (Kaugel, Poru, Erave, Iaro and Pio) have large areas in their catchment covered by undisturbed tropical forest, and as a result have lower erosion, as indicated by their lower sediment load.

The sediment load of the Purari River itself, considered in more detail elsewhere in this book by Pickup, is relatively low when compared with large rivers of the humid tropical environment of Southeast Asia. Only further data collection and evaluation will provide enough information on the impact of human pressure on the Purari basin, both present and future. Any proposals for construction of artificial reservoirs in the basin will have to consider and base their predictions not only on the present sediment transport but also on assessment of any possible deterioration in water quality in the future. At present, the Purari dissolved nutrient content and that of its suspended sediments is considered as being sufficient to maintain a mesotrophic to eutrophic character of the Wabo water reservoir if constructed (Viner 1979). Any further increase above such levels might cause prolonged algal blooms and support considerable quantities of floating macrophytes. Large-scale deforestation in the catchment, followed up by increased sediment input, would become one of the sources of additional nutrient input into such a reservoir.

3. Bacterial densities and faecal pollution

High human population density of the Purari highlands and perhaps an even higher pig density in the same area appear to be the major source of faecal pollution of catchment rivers. There is no estimate of the ratio of pigs to humans for the Purari catchment, but Feachem (1974) estimated that there are two pigs per one person in the Saku Valley in Enga Province, and that the pig population considerably increases the faecal pollution load in the area investigated by him.

Although no regular programme was possible for the whole catchment, expeditional sampling during 1978 provided enough data for drawing some conclusions about the faecal pollution of the Purari basin rivers. All samples were analysed in the Central Public Health Laboratory of the Health Department in Port Moresby, where the total count of bacteria (TCB) (standard plate count) was determined as in Standard Methods (1975); for total coliform count (TC) and faecal coliform count (FC), the multiple-tube fermentation technique was used. For details of methods see Petr (1980). The delay between the collection of samples and their processing in the laboratory could be one reason for the generally high bacterial numbers found in this study.

Percentage faecal in total coliforms was found highest in Tua River (35.4%, $n = 2$) (Table 1), followed by that of the Erave (34.5%, $n = 4$), Pio (24.5%, $n = 3$), Kaugel (12.8%, $n = 3$), Poru (11.1%, $n = 3$), and Iaro (5.6%, $n = 3$).

Table 1. Bacterial densities in highland rivers.

	Tua	Erave	Kaugel	Poru	Pio	Iaro
Altitude (m)	500	600	518	490	180	640
Number of samples	3	4	3	3	3	3
Temperature (°C)	24.7	22.8	22.0	22.2	24.5	22.4
Conductivity μS/cm	95	185	66	106	117	172
Total suspended solids (mg/l)	412	45	111	46	79	170
TCB per 100 ml	5 076 000	2963	17 800	80 200	9500	33 510
TC per 100 ml	13 000 ($n = 2$)	4186	8681	5231	9534	159 014
FC per 100 ml	4600 ($n = 2$)	1444	1112	580	2333	8800
% FC of TC	35.4 ($n = 2$)	34.5	12.8	11.1	24.5	5.6

TCB = total count of bacteria; TC = total coliforms; FC = faecal coliforms.

According to Ford (1974) two of the four areas of highest population density in P.N.G. (with more than 60 persons per km²) are situated in the Purari watershed: these are the Chimbu Valley and the Mt. Hagen area. The Purari catchment also receives the waste water from the major highland towns such as Goroka, Kundiawa, Minj, Mt. Hagen and Mendi. While the high faecal pollution of the Tua comes from the high population density of the Asaro, Wahgi and Tua catchments, that of the Erave River probably comes from the populated areas around Mendi.

In absolute numbers the Poru and Kaugel Rivers have the lowest faecal coliform count among the catchment rivers, probably as a result of their low human (and warm-blooded animal) population density. These rivers have also one of the lowest percentages of faecal coliforms in total coliforms.

In all rivers, higher flow rates, accompanied by higher levels of total suspended solids (TSS), give higher total and faecal coliform counts and usually a higher total count of bacteria. Low water levels with low TSS have low TC, usually low TCB and very low or nil FC (Petr 1980). In highland rivers, the correlations between TSS and TCB ($r = 0.863$), TSS and TC ($r = 0.950$), and TSS and FC ($r = 0.901$) are highly significant ($p = < 0.001$) and support the suggestion of Feachem (1974), based on his study in the Saka Valley of the highlands, who stated that bacterial concentrations tend to rise as turbidity rises.

In the lowlands, as represented by the Purari at Wabo and the Aure River, there is no significant correlation between TSS and TCB, and TSS and TC, but there is a significant positive correlation between TSS and FC ($r = 0.815, p = <$ 0.005). For the whole catchment, TSS does not show a clear relationship with bacteria, suggesting that more factors must be involved, especially in the tidal brackish reach of the Purari Delta.

In the Purari at Wabo, mean bacterial densities (TCB, TC and FC) were lower than in highland rivers (Table 2), suggesting that self-purification of the river waters is taking place between 600 m and 50 m altitude, where hardly any people live. Over a distance of approximately 120 to 180 km, as measured from the Erave and Tua sampling stations along the river to Wabo, there is six-fold drop in mean total bacterial densities, five-fold drop in mean total coliforms, and three-fold drop in faecal coliforms (Table 2). In the temperate climate river Danube in Europe, over a distance of some 200 km, the bacterial count was found to decrease ten-fold (Mucha and Daubner 1969). Feachem (1974) found a rapid die-off of bacteria in larger, briskly flowing and well-aerated rivers of Enga Province, at a higher altitude than rivers in the present study. In spite of the drop, the number of faecal coliforms at Wabo is still relatively high, indicating that faecal coliforms apparently survive the transport better than other bacteria.

Table 2. Bacteriological comparison of individual reaches.

	Mean TCB/ml	Mean TC/100 ml	Mean FC/100 ml	% Faecal of total coliforms	Mean TSS mg/l
Highland rivers	32 604	32 451	2969	9.15	87
	n = 18	*n* = 18	*n* = 18	*n* = 18	*n* = 18
Purari at Wabo	5094	6976	977	14.01	194
	n = 7	*n* = 8	*n* = 8	*n* = 8	*n* = 7
Aure River	12 760	1303	165	12.63	52
	n = 5	*n* = 5	*n* = 5	*n* = 5	*n* = 5
Purari Delta:					
upper reach	23 000	8014	4400	54.90	611
	n = 7	*n* = 7	*n* = 7	*n* = 7	*n* = 6
lower reach	24 714	13 000	6417	49.36	140
	n = 7	*n* = 7	*n* = 7	*n* = 7	*n* = 5
Mean values for the Purari watershed	19 634	17 634	3062	17.36	217

n = number of observations.

When compared with the higher situated reaches, the Purari Delta has a conspicuously high ratio of faecal in total coliforms. In absolute numbers, the delta has about twice the density of coliforms than that found in highland rivers, and 4 to 6 times more than measured in the Purari at Wabo.

Low salinities of the brackish reach appear to enhance bacterial growth, as observed by Paerl and Kellar (1980). They found in the Purari Delta a significant increase in bacterial biomass in response to saltwater intrusion during high tide. When tidal intrusions occurred, particulate matter showed increasing levels of bacterial attachment to minerals and aggregate formation. River-suspended sediments not affected by saline intrusions failed to reveal significant numbers of attached bacteria and algae. Autoradiographs generally confirmed direct microscopic observations. Jensen et al. (1980) suggested that tidal wetlands create an environment suitable for the increased survival or even growth of coliform organisms. They list a number of literature references which suggest that addition of organic substrates and/or inorganic nutrients to enteric bacterial populations in seawater have led to decrease in the die-off rate, and under certain circumstances to the growth of coliform organisms. Sediment resuspension is also known to cause bacterial increases, as coliform organisms tend to adsorb onto and persist in the sediments.

In spite of these studies being very preliminary in nature, they show the differential impact of population densities on water quality of individual catchment rivers, and the capability of the system for self-purification. The level of faecal pollution in the lower Purari is still relatively high if applying Australian water quality standards. However, as any standard reflects the values of the society, the question to be asked is: how much risk is associated with the present pollution levels? Probably little, as people along the Purari prefer drinking rainwater and water from side streams rather that from the Purari itself.

Future studies of faecal pollution problems should be more detailed and diversified. Attention should be paid to pathogenic *Salmonella* which is known to appear almost regularly in association with feacal coliforms when the faecal coliform density per 100 ml is above 1000 organisms (Geldreich and Bordner 1971). Another aspect requiring attention are faecal coliforms and other pathogens in bottom sediments. More data should also be obtained on self-purification processes.

4. Organochlorine residues

A small number of biological samples were collected from the Purari River Delta with the aim of providing a baseline for future assessment of any changes in organochlorine residue levels. Collection and analysis of the material was carried out by Dr Olafson of the Australian Institute of Marine Sciences, Townsville, and the results were evaluated and published by Olafson and Mowbray (1980). The following part of this chapter summarizes their findings.

All samples were collected within the delta of the Purari, and included: surface sediment (Ivo River), clam *Batissa violacea* (from the Wame River near Kairimai village), prawn *Palaemonetes* sp. and crab *Scylla serrata* (from the Wame and Pie Rivers confluence), threadfin salmon *Polydactylus* sp. (from the Wame River), barramundi *Lates calcarifer*, freshwater crocodile *Crocodylus novaeguineae*, and saltwater crocodile *C. porosus* – all from Old Iari village situated on a channel in the center of the Purari Delta.

The concentrations of all residues in animals (and sediment) expressed on a wet weight basis were very low, the highest being 0.41 ng/g lindane in crab and 23.7 ng/g DDT-R in flesh of saltwater crocodiles (Table 3).

The highest residue in animals expressed on lipid weight basis were 36 ng/g lindane in clams and 822 ng/g DDT in flesh of the saltwater crocodiles.

Comparison with selected data from elsewhere and with acceptable maximum residue levels clearly illustrates that residue levels are very low (Table 4).

Table 3. Organochlorine pesticide residues of samples from Purari River Delta, October 1978 (where more than one sample, result is expressed as the geometric mean).

Sample	Tissue	No. of samples	Lindane		Dieldrin		Heptachlor epoxide		DDE		DDD		DDT		DDT-R	
			A	B	A	B	A	B	A	B	A	B	A	B	A	B
Sediment	–	2	1.35	–	0.01	–	–	–	0.01	–	0.13	–	0.16	–	0.39	–
Clam	whole body	4	0.28	36	–	trace	0.02	3	–	trace	0.02	2	0.03	3	0.05	8
Prawn	whole body	3	0.22	9	–	trace	–	–	0.12	6	–	–	–	–	0.12	6
Crab	hepatopancreas	1	0.41	14	1.8	61	–	–	2.0	67	–	–	–	–	2.0	67
Crab	flesh	1	0.13	17	–	–	–	–	1.1	141	–	–	–	–	1.1	141
Threadfin	flesh	2	0.17	34	–	–	–	–	–	–	–	–	–	–	–	–
Threadfin	liver	2	0.06	2	0.07	3	–	–	3.3	53	2.4	39	2.0	7	7.7	99
Barramundi	flesh	1	0.16	31	–	–	–	–	–	–	–	–	–	–	–	–
Barramundi	liver	1	0.31	8	–	–	–	–	0.09	2	–	–	–	–	0.09	2
Freshwater crocodile	flesh	2	0.18	24	–	–	–	–	0.30	41	–	–	–	–	0.30	41
Freshwater crocodile	liver	2	0.40	9	–	–	–	–	3.1	75	0.08	4	–	–	3.2	79
Saltwater crocodile	flesh	2	0.23	8	–	–	–	–	17	592	1.1	37	5.6	193	23.7	822
Saltwater crocodile	liver	2	0.29	9	–	–	–	–	4.9	150	0.67	20	1.7	52	7.3	222

DDT-R = the sum total of all DDT residues and metabolites; A = ng per g wet weight; B = ng per g lipid.

332

Table 4. Comparison of lindane and DDT-R residues found in clams, fish and crocodiles from the Purari Delta with elsewhere. Given as geometric mean. Expressed as ng/g (lindane), μg/g (DDT-R).

Basis		Bivalves (clams or mussels) WW	LW	Fish flesh WW	LW	Fish liver WW	LW	Fish whole body WW	LW	Freshwater crocodile LW	Saltwater crocodile LW
Purari	A	0.28	36	0.17	33	0.10	4	–	–	–	–
	B	0.0005 (clams)	–	–	* (barra)	–	–	–	–	0.04	0.82
Northern Territory (Australia)[a]	A	0.07	7	0.07	0.08 (barra)	0.27	8	–	–	–	–
Great Barrier Reef (Austr.)[b]	B	–	–	–	–	–	–	–	0.08	*	0.10
Namoi River (Australia)[c]	A	1.0	–	–	–	–	–	–	–	–	–
	B	0.17 (mussels)	–	–	15	–	–	–	–	–	–
Lake Nakuru, (Kenya)[d]	A	–	–	–	0.15 (tilapia)	–	–	–	–	–	–
	B	–	–	–	–	–	–	–	–	–	–
Lake Tanganyika (Burundi)[e]	A	–	–	–	–	–	–	50	810	–	–
	B	–	–	–	0.75	–	–	–	–	–	–
World range[f]	A	<10–100	–	–	–	–	–	5–780	–	–	–
	B	*–4000	–	–	*–2700	–	–	–	–	–	–
Maximum residue level[g]	A	–	–	–	–	–	–	–	–	–	–
	B	–	–	–	≐ 15	–	–	1000	≐ 20 000	–	–

A = lindane residue; B = DDT-R residue; WW = wet weight; LW = lipid weight; [a] Best 1973; [b] Olafson 1979; [c] Mowbray 1978; [d] Koeman et al. 1972; [e] Deelstra 1977; [f] Edwards 1973a, b; [g] Australian Dept. Agriculture 1974; [h] in Northern Territory of Australia *Crocodylus johnstoni*; * = means not detected; – = either nil or not determined.

Lindane residues in Purari clams and fish occur in a range of concentrations similar to those in related animals in the Barrier Reef of Australia as reported by Olafson (1979), but these levels in fish are at the extreme lower end of the range of residues that have been reported in both freshwater and marine fish elsewhere in the world as is summarized by Edwards (1973a, 1973b), Deelstra (1977) and Mowbray (1978), and are much lower than the acceptable maximum residue level (Australian Dept. Agriculture 1974).

DDT-R residue levels in clams are lower than in mussels in the Namoi River in Australia (Mowbray 1978) and are much lower than in clams elsewhere in the world as reported by Edwards (1973a, 1973b). DDT-R residues in the barramundi and crocodiles of the Purari are also very low; in the fish the values are much lower than those found in Australian fish in the Namoi River by Mowbray (1978), and elsewhere in the world (Edwards 1973a, 1973b).

For other tropical environments, DDT-R residue levels in tilapia in Lake Nakuru, Kenya, were less than 0.01 ppm wet weight (approximately 0.15 ppm lipid weight (Koeman et al. 1972). In Lake Tanganyika fish sampled close to a cotton area where much DDT was sprayed, levels of approximately 0.05 ppm wet weight (approximately 0.75 ppm lipid weight) were measured (Deelstra 1977). DDT-R residues in the Purari are lower than those reported from Africa.

In the Gulf Province only minimal amounts of pesticides are used. Malarial control spraying with DDT against mosquitoes has not been carried out in the area (Barker-Hudson, personal communication). Use of DDT, lindane, dieldrin and other organochlorine pesticides for agricultural purposes on a very limited scale has occurred in the highland areas drained by the Purari catchment rivers.

One may conclude that pollution of the Purari River Delta with organochlorine residues is minimal with very small residues of lindane, DDT-R, dieldrin and heptachlor epoxide occurring in animal and sediment samples. No PCB's were detected. Use of organochlorines has been negligible in the Gulf Province, and the main source of these chemicals is probably the highlands, where they are known to be used for agricultural purposes. The very limited scale of sampling carried out in the present survey provided only indicator values about the situation. Future surveys should certainly be more detailed.

5. Heavy metals in water and sediments

5.1. Water

Only zinc concentrations in the lower Purari River water were above the detection limit of the analytical methods used. The mean value for zinc is 23 µg/l, which is higher than the 10 µg/l considered as an average natural background (Förstner and Wittmann 1979); this, in turn, is 11 times higher than the mean value for the Fly River (Table 5). At present the only conclusion to be drawn from the results is that the values for the Purari River are probably all well below those indicating even slight pollution, and well below the upper limits for domestic water supply.

5.2. Sediments

During the last three decades sediments have been used for assessing the impact of man-made pollution of aquatic ecosystems. The heavy metals zinc, lead, cadmium, mercury, copper and arsenic are good indicators of mining and industrial expansion.

The results of soil and sediment chemical analyses of the Purari catchment, discussed in more detail elsewhere (Petr and Irion, this volume), show that at present the basin has no heavy metal pollution. The freshly deposited sedi-

Table 5. Heavy metals in the lower Purari River water (values in µg/l in dissolved fraction).

	Purari River ($n = 12$)[a]	Fly River at Kiunga[b]	Mean natural background[c]	Domestic water supply[d]
Cadmium	< 1 – 1	–	0.07	10
Cobalt	< 5	–	0.05	–
Copper	< 5 – 5	1.0	1.8	1000
Lead	< 5 – 10	< 0.5	0.2	50
Manganese	< 5	7.5	5	50
Molybdenum	< 10	–	1	–
Nickel	< 5	< 1	0.3	–
Zinc	< 5 – 60 (mean 23)	0.9	10	5000

[a]   The analysis was performed by the Bougainville Copper Ltd., Analytical Services Department, Panguna.
[b]   Boyden et al. 1978.
[c]   Förstner and Wittmann 1979.
[d]   Train 1979.

ments of the lower Purari, reflecting the total lithology of the catchment, have mean values for individual metals close to median values for uncontaminated sediment clay fractions (Förstner 1977), and to average shale values (Turekian and Wedepohl 1961). The values, as presented in Table 6, can therefore serve as natural background concentrations for any future comparisons.

Manganese, molybdenum, lead and mercury values were significantly higher in predominantly clay material, i.e., in the tidal brackish water reach, than in sediments under the direct influence of the river. In the Purari, mercury, known elsewhere in Papua New Guinea to be concentrated by bioaccummulation (Reynolds and Price 1974; Sorentino 1979; Petr 1978/79; Kyle 1981), has low concentrations in both sediments and fish. A weak bioaccummulative capability has been suggested by Petr (1978/79) for *Salvinia molesta*, an aquatic fern accidentally introduced into the Sepik River backwaters, where it underwent explosive invasion (Mitchell, Petr and Viner 1980). The whole plant contained 0.14 μg/g (dry weight) of total mercury. In Lake Murray in Western Province the predatory fish barramundi (*Lates calcarifer*) was found to have a mean total mercury level of 0.57 μg/g, and freshwater anchovy (*Thryssa scratchleyi*), also predatory, had a range of 0.25 to 0.82 μg/g (Kyle 1981). Such bioaccummulation is of medical significance, as people living at Lake Murray and eating fish regularly, have one of the highest total mercury levels in their hair amongst the people not affected by anthropogenic sources of mercury pollution.

Table 6. Heavy metal concentrations in the recent Purari River sediments (in mg per kg dry weight).

	Purari ($n = 17$)		Lacustrine deposits (Förstner 1977) ($n = 74$)	Range for 90% of the data	Average shale (Turekian and Wedepohl 1961)
	Mean	Range			
Cadmium	<0.1	<0.1	0.35	0.10 – 1.20	0.3
Cobalt	1.7	5 – 22	28	8 – 75	19
Copper	44	22 – 76	43	20 – 80	45
Manganese	620	210 – 1260	750	100 – 1800	850
Mercury	0.06	0.01 – 0.14	0.31	0.15 – 1.20	0.4
Nickel	35	30 – 40	66	30 – 250	68
Lead	12	6 – 14	28	8 – 75	20
Zinc	87	71 – 105	110	45 – 220	95

6. Conclusions

If judged on basis of faecal contamination and suspended sediment concentrations, the present water quality of the Purari River is not very high. However, the water quality is good when heavy metals and pesticide residues are considered. Today, the water quality reflects the human pressure on primary resources, e.g. forest and land, and the agricultural level and intensity, and it shows that the society is still largely on a subsistence and pre-industrial level.

As most water bodies in the world are already polluted, to establish pre-pollution, i.e. natural background levels for indicators of pollution is often difficult. For the Purari River basin the heavy metals in sediments will serve in the future as sensitive indicators of pollution. The present levels can be still considered as natural, determined largely, if not entirely by the lithology of the catchment.

The planned Wabo reservoir on the lower Purari, if ever built, will have a number of functions. Among them will certainly be its ability to purify and clarify water. From this point of view the presence of the reservoir will be beneficial to the downstream areas, which will receive a far better water in the Purari River than at present.

Acknowledgements

Biochemical samples were analysed in the Central Public Health Laboratory of the Health Department in Port Moresby, under the supervision of Dr John McGregor, who was at that time in charge of this laboratory. Organochlorine residues were analysed by Dr R.W. Olafson in the Australian Institute of Marine Sciences, Townsville, Queensland, Australia. Heavy metals in water were analysed by Bougainville Copper Ltd., Analytical Sciences Department, Panguna, Bougainville, and heavy metals in sediment, by Dr G. Irion of the Senckenberg Institut, Wilhelmshaven, Federal Republic of Germany, and by the Mines Analytical Laboratory of the Ministry of Minerals and Energy, Konedobu, Papua New Guinea. I am grateful to all the individuals and laboratories for their input which enabled me to compile this chapter.

References

Australian Department of Agriculture, 1974. Agricultural and veterinary chemicals withholding periods, maximum residue limits and poisons schedules. Pesticide Section, P.B. 158A. Dept. of Agriculture, Canberra.

Best, S.M., 1973. Some organochlorine pesticide residues in wildlife of the Northern Territory, Australia, 1970–71. Aust. J. Biol. Sci. 26: 1161–1170.

Blong, R.J. and G.S. Humphreys, 1982. Erosion of road batters in Chim Shale, Papua New Guinea. Inst. of Engineers Australia. Civil Eng. Trans. CE24: 62–68.

Boyden, C.R., B.E. Brown, K.P. Lamb, R.F. Druckers and S.J. Tuft, 1978. Trace elements in the upper Fly River, Papua New Guinea. Freshwater Biology 8: 189–205.

Deelstra, H., 1977. Organochlorine insecticide levels in various fish species in Lake Tanganyika. Med. Fac. Landbouww. Rijksuniv. Gent 42/2: 869–882.

Edwards, C.A., 1973a. Persistent chemicals in the environment, 2nd ed. C.R.C. Monoscience Series. Butterworth Press, London.

Edwards, C.A., 1973b. Environmental pollution by pesticides. Plenum Press, London.

Feachem, R., 1974. Faecal coliforms and faecal streptococci in streams in the New Guinea highlands. Water Res. 8: 367–374.

Ford, E., 1974. Population. In: E. Ford (ed.), Papua New Guinea Resource Atlas, pp. 32–33. Jacaranda Press, Milton, Qld., Australia.

Förstner, U., 1977. Metal concentrations in recent lacustrine sediments. Arch. Hydrobiologie 80: 172–191.

Förstner, U. and G.T.W. Wittmann, 1979. Metal pollution in the aquatic environment. Springer-Verlag, Berlin.

Geldreich, E.E. and R.H. Bordner, 1971. Fecal contamination of fruits and vegetables during cultivation and processing for market: a review. J. Milk and Food Tech. 34: 184–195.

Irion, G. and T. Petr, 1983. Clay mineralogy of selected soils and sediments of the Purari River basin. This volume, Chapter I, 6.

Jensen, P., A. Rola and J. Tyrawski, 1980. Tidal wetlands and estuarine coliform bacteria. In: P. Hamilton and K.B. MacDonald (eds.), Estuarine and Wetland Processes with Emphasis on Modelling, pp. 385–399. Plenum Press, London.

Koeman, J.H., J.H. Pennings, J.J.M. DeGoeij, P.S. Tjioe, P.M. Olindo and J. Hopcraft, 1972. A preliminary survey of the possible contamination of Lake Nakuru in Kenya with some metals and chlorinated hydrocarbon pesticides. J. appl. Ecol. 9: 411–416.

Kyle, J.H., 1981. Mercury in the people and the fish of the Fly and Strickland River catchments. Office of Environment and Conservation, Waigani, Papua New Guinea.

Lamb, D., 1977. Conservation and management of tropical rain forest: a dilemma of development in Papua New Guinea. Environmental Conservation 4: 121–129.

Mitchell, D.S., T. Petr and A.B. Viner, 1980. The water fern, *Salvinia molesta*, in the Sepik River, Papua New Guinea. Environmental Conservation 7: 115–122.

Mowbray, D.L., 1978. The ecological effects of pesticides on non-target organisms: a study of the environmental impact of pesticides on wildlife in the Namoi River Valley cotton growing area, 1972–1976. Ph.D. Thesis. University of Sydney.

Mucha, V. and L. Daubner, 1969. Die Verwendung neuer hydromikrobiologischer Methoden bei der Erforschung der Donau in der CSSR. Rev. Roum. Biol. Ser. Zool. 14: 139–147.

Newcombe, K. and T. Pohai, 1981. The Lae project: an ecological approach to Third World urbanisation. Ambio 10(2–3): 73–78.

Olafson, R.W., 1979. Effect of agricultural activity on levels of organochlorine pesticides in hard corals, fish and molluscs from the Great Barrier Reef. Unpublished manuscript. Australian Institute of Marine Science, Townsville.

Olafson, R.W. and D. Mowbray, 1980. Organochlorine residues in the Purari River Delta, Gulf

Province, Papua New Guinea. In: T. Petr (ed.), Purari River (Wabo) Hydroelectric Scheme Environmental Studies, Vol. 11: Aquatic ecology of the Purari River catchment, pp. 13–23. Office of Environment and Conservation, Waigani, Papua New Guinea.

Paerl, H.W. and P.E. Kellar, 1980. Some aspects of the microbial ecology of the Purari River, Papua New Guinea. In: T. Petr (ed.), Purari River (Wabo) Hydroelectric Scheme Environmental Studies, Vol. 11: Aquatic ecology of the Purari River catchment, pp. 25–39. Office of Environment and Conservation, Waigani, Papua New Guinea.

Pain, C.F., 1983. Geology and geomorphology of the Purari River catchment. This volume, Chapter I, 3.

Petr, T., 1978/79. Mercury in the Papua New Guinea environment. Science in New Guinea 6: 161–176.

Petr, T., 1979. Possible environmental impact on inland waters of two planned major engineering projects in Papua New Guinea. Environmental Conservation 6: 281–286.

Petr, T., 1980. Bacterial densities and faecal pollution in the Purari River catchment, Papua New Guinea. In: T. Petr (ed.), Purari River (Wabo) Hydroelectric Scheme Environmental Studies, Vol. 11: Aquatic ecology of the Purari River catchment, pp. 41–58. Office of Environment and Conservation, Waigani, Papua New Guinea.

Petr, T., 1983a. Limnology of the Purari basin. Part 1. The catchment above the delta. This volume, Chapter I, 9.

Petr, T., 1983b. Limnology of the Purari basin. Part 2. The delta. This volume, Chapter I, 10.

Petr, T. and G. Irion, 1983. Geochemistry of soils and sediments of the Purari River basin. This volume, Chapter I, 7.

Pickup, G., 1983. Sedimentation processes in the Purari River upstream of the delta. This volume, Chapter I, 11.

Reynolds, L.F. and M.J. Price, 1974. Interim report on the mercury levels found in barramundi caught in Papuan water. Department of Primary Industry, Konedobu, Papua New Guinea.

Sorentino, C., 1979. Mercury in marine and freshwater fish of Papua New Guinea. Aust. J. Mar. Freshwater Res. 30: 617–623.

Standard Methods (1975) for the Examination of Water and Wastewater, 1975. 14th ed. American Public Health Assoc., Washington, D.C.

Stark, J., 1982. The approach to soil conservation in Papua New Guinea adopted by the land Utilization Section, Department of Primary Industry. In: L. Morauta, J. Pernetta and W. Heaney (eds.), Traditional Conservation in Papua New Guinea: Implications for Today, pp. 163–165. Institute of Applied Social and Economic Research. Monograph 16. Boroko, Papua New Guinea.

Train, R.E., 1979. Quality criteria for water. Castle House Publications.

Turekian, K.K. and K.H. Wedepohl, 1961. Distribution of elements in some major units of the earth's crust. Bull. Geol. Soc. Amer. 72: 175–192.

Viner, A.B., 1979. The status and transport of nutrients through the Purari River. In: T. Petr (ed.), Purari River (Wabo) Hydroelectric Scheme Environmental Studies, Vol. 9, pp. 1–52. Office of Environment and Conservation, Waigani, Papua New Guinea.

Wood, A.W., 1983. Soil types and traditional soil management in the Purari catchment. This volume, Chapter I, 5.

7. The ecology of juvenile penaeid prawns, mangrove crab (*Scylla serrata*) and the giant freshwater prawn (*Macrobrachium rosenbergii*) in the Purari Delta

S.D. Frusher

1. Introduction

The inhabitants of the lower regions of the Purari Delta traditionally live on a subsistence basis and obtain most of their animal protein from the fish of the delta. The mangrove crab *Scylla serrata* (Forskal) is the most important crustacean, and large numbers found in some parts of the delta resulted in a commercial venture being started in 1978 (Haines and Stevens, this volume). Carid prawns, particularly the large freshwater prawn *Macrobrachium rosenbergii* (De Man), and juvenile penaeid prawns are also caught, and can form up to 20% by weight of the traditional subsistence catch. The adult penaeid prawns form the basis of a commercial fishery in the Gulf of Papua, with the annual catch varying between 800 and 1100 tonnes tail weight (Gwyther, this volume). This fishery, which incorporates an annual lobster trawl fishery, is a valuable resource to Papua New Guinea, being second only to the tuna fishery. The banana prawn, *Penaeus merguiensis* (De Man) is the dominant species and constitutes 50 – 60% of the total catch. *Metapenaeus demani* (Roux), *M. ensis* (De Haan) and *M. eboracensis* (Dall), collectively known as endeavour prawns, form 15 – 20%; and tiger prawns (*P. monodon* (Fabricus) and *P. semisulcatus* (De Haan)) form 10 – 15% of the catch.

The importance of coastal deltaic zones as nurseries for the early life history of penaeid prawns has been well documented throughout the world, for example *P. indicus* in Madagascar (Le Reste 1973), *P. duorarum* in Florida (Tabb et al. 1962), Penaeidae of India (Mohamed and Rao 1971), Penaeidae of Australia (Walker 1974), and *P. merguiensis* in Australia (Staples 1979). In the Gulf of Papua all the major species exept *P. semisulcatus* use the deltaic zones of the Gulf of Papua as nursery regions. Thus the recruitment and survival of juvenile prawns within the nursery zone is of primary importance for commercial fisheries.

Petr, T. (ed.) The Purari – Tropical environment of a high rainfall river basin
© *1983, Dr W. Junk Publishers, The Hague / Boston / Lancaster*
ISBN-13: 978-94-009-7265-0

2. Purari Delta – physical and chemical environment

The Purari Delta is the easterly portion of a large delta system which feeds the northern Gulf of Papua. The Aird Delta, which interconnects with the Purari Delta in the west, is larger in extent. However, output from the Purari is greater as it drains a substantial portion of the southern slopes of the Papua New Guinea highlands.

The Purari River enters the Gulf of Papua via five main distributaries: the Wame which drains into the Pie River, the Varoi and the Urika arms of the Ivo River, the Panaroa, and the Purari River. The amount of water which flows out of these distributaries is variable and is reflected in the different degrees of saltwater penetration up the river (Fig. 1; see also Thom and Wright, this volume). As saltwater penetrates in the form of a wedge which tapers upstream, the actual degree of penetration is probably greater than indicated by the surface water salinities. Alexander et al. (1932) found that the saltwater wedge can extend even further upstream in the mud of the bottom, and Petr and Lucero (1979) found saline interstitial water around sago palms in areas of the Purari Delta with otherwise fresh water in the bypassing channels.

The highest salinities occur along the banks of the Pie River, the Varoi River and the Alele Passage of the Purari River. The fringing vegetation also reflects the differing salinities. The dominant fringing vegetation also reflects the differing salinities. The dominant fringing vegetation is the mangrove *Sonneratia lanceolata*. Floyd (1977) found the presence of this mangrove to be indicative of accretion in brackish and near freshwater habitats. Petr and Lucero (1979) found that the presence of *S. lanceolata* (also referred to as *S. caseolaris*) marks the extent of saline penetration in the Purari Delta. *Cyperus ferax*, known to be a primary colonizer in near freshwater conditions (Floyd 1977), was only found at the mouth of the Urika River where salinity was zero.

Differences in salinity regimes and in fringing vegetation, together with variable discharge rates and thus substrate composition have created a range of habitats within the coastal fringe of the Purari Delta.

Sea current studies (MacFarlane 1980) have shown a westerly surface and bottom drift from Orokolo Bay (east of the Purari) to Deception Bay, and then northwest into the Aird Delta, throughout the year. Surface drifters released at sea were recovered as far as 40 kilometres inland in the Aird Delta system. Suspended silt and sediment discharged from the Purari Delta would therefore be expected to move towards and into the Aird Delta (cf. Thom and Wright, this volume).

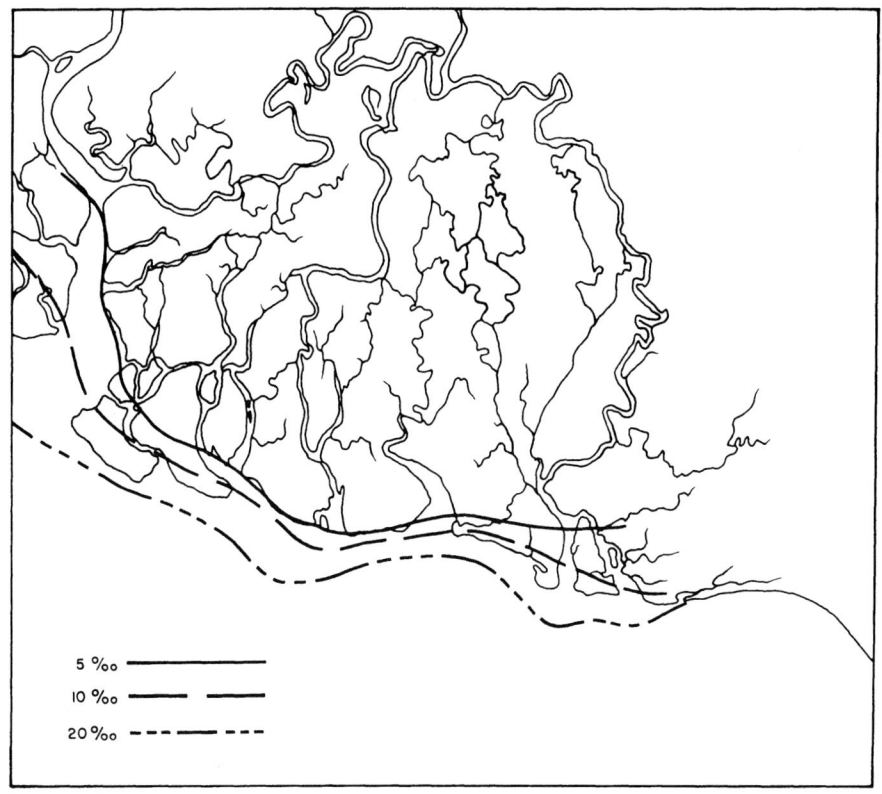

Fig. 1. Approximate isohalines in the Purari Delta during the dry season.

3. Distribution and abundance

3.1. Penaeid prawns

Within the delta regions of the Gulf of Papua *P. merguiensis* is more abundant in the regions of higher salinities. The main nursery region for this species was found to be the more saline Aird Delta where salinities range from 15 – 20 ppt (Frusher 1980). In contrast, *M. demani* is common in the less saline regions and dominated in areas of less than 5 ppt. Within the Purari Delta, *P. merguiensis* was the main species found in Port Romilly, Varoi and Alele Passage, whereas *M. demani* was more abundant in the Panoroa and Kinipo areas and dominated at the Urika River sampling site (Figs. 2 and 3). *M. eboracensis* is found in medium salinity, such as at Kinipo on the Varoi, and

343

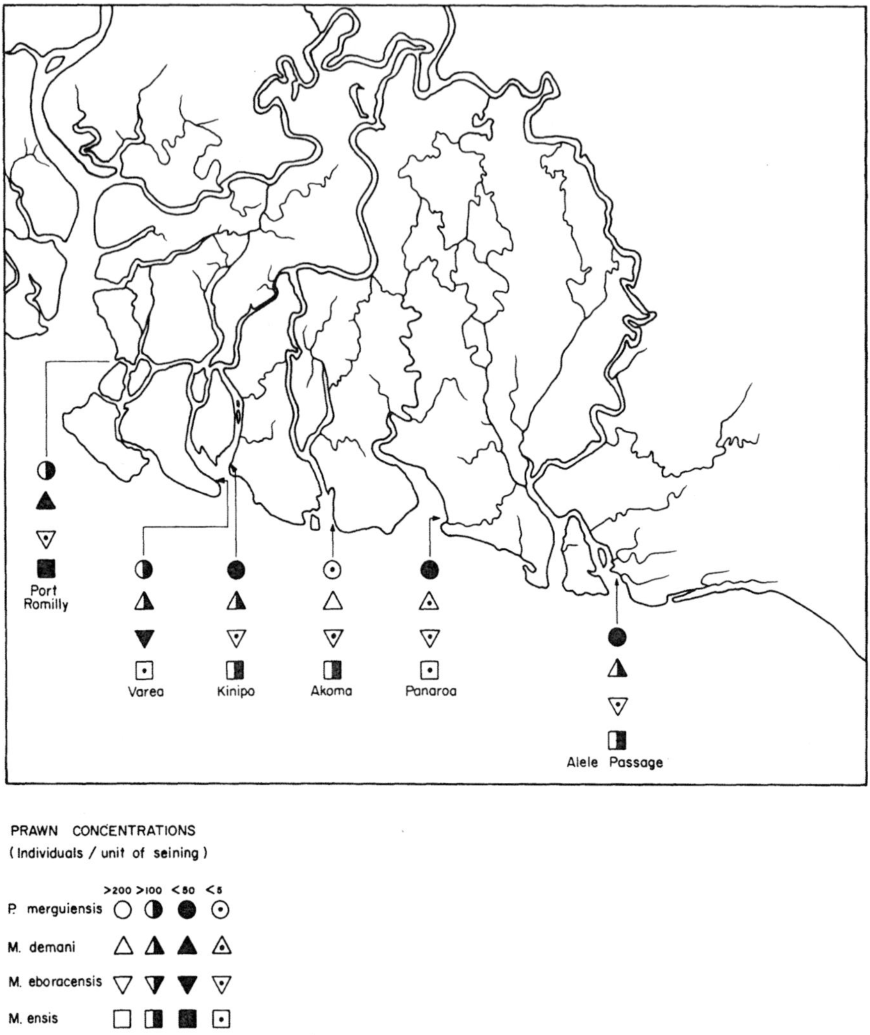

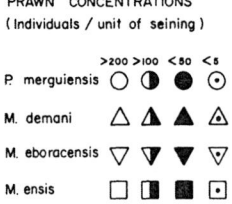

PRAWN CONCENTRATIONS
(Individuals / unit of seining)

	>200	>100	<50	<5
P. merguiensis	○	◑	●	⊙
M. demani	△	▲	▲	◬
M. eboracensis	▽	▼	▼	▿
M. ensis	□	◪	▨	⊡

Fig. 2. Prawn concentrations in the Purari Delta during the wet season (April to October).

in the Alele Passage, where *P. merguiensis* and *M. demani* were in medium to low concentrations. However, there appears to be a definite seasonality of *M. eboracensis* with highest occurrence in August/September. *M. ensis*, the major metapenaeid of the Gulf of Papua prawn fishery (Tenakanai 1980), is only encountered in medium to low concentrations in the less saline regions of the Purari Delta. It is however, possible that this species occurs in regions further upstream where salinity is lower (McPadden 1980). Tenakanai (1980) found

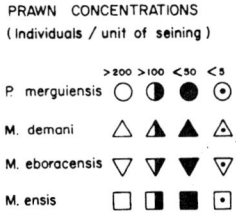

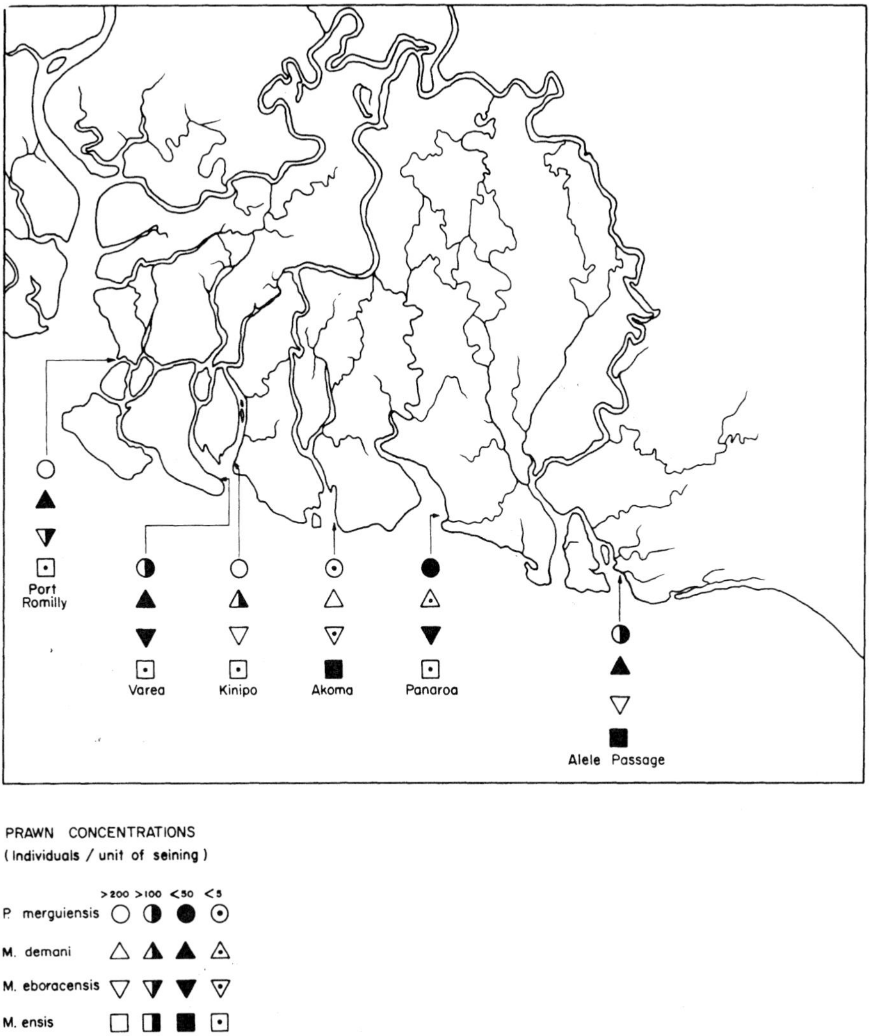

PRAWN CONCENTRATIONS
(Individuals / unit of seining)

	>200	>100	<50	<5
P. merguiensis	○	◑	●	⊙
M. demani	△	▲	▲	△
M. eboracensis	▽	▼	▼	▽
M. ensis	□	◪	■	⊡

Fig. 3. Prawn concentrations in the Purari Delta during the dry season (November to March).

mature females of all metapenaeid species in the Gulf of Papua throughout the year, suggesting both continuous spawning and also recruitment. However, as only 19.5% of *M. ensis* sampled were less than 10 mm carapace length (C.L.), indicating recent post-larval settlement, the coastal regions of the Purari Delta appear to be poor nursery regions for *M. ensis*. In contrast, 98.7% of *M. demani* and 98.2% of *M. eboracensis* sampled were less than 10 mm C.L., from which it can be concluded that the fringe of the delta is a suitable nursery region for *M. demani* and *M. eboracensis*.

Although salinity is used here to delineate nursery habitats, it is only one of a number of parameters which are likely to control the selection of a nursery zone.

3.2. Mangrove crab

The distribution of *S. serrata* within the Purari Delta is limited to a narrow coastal fringe (Opnai 1980), and appears to be restricted to regions of saltwater penetration. Hill (1979a) and Pagcatipunan (1972) found that *S. serrata* could tolerate salinities as low as 2 ppt in South Africa and the Philippines respectively. Macnae (1968) reported that *S. serrata* ascended upstream, well above the influence of saline water but possibly still under the influence of the salt wedge.

The population density of *S. serrata* in the Purari is about twice that of neighbouring regions (i.e. 21.4 crabs/ha in Purari and 10.4 crabs/ha in the Aird Delta), although the average size is smaller. This may be due to crabs leaving the region at an earlier stage in their life history as their osmoregulatory ability shifts to higher salinities (as suggested by Dall (1981) for penaeid prawns). This could explain the observations by Stephenson and Campbell (1960) in Queensland, where they found *S. serrata* to be displaced downstream after periods of large rainfall. The smaller size of the crabs also results from lower salinities which may suppress the growth of *S. serrata*.

3.3. The giant freshwater prawn

The distribution of *Macrobrachium rosenbergii* within Papua New Guinea is not known. Questionnaires have revealed that they are caught by villagers at least 100 km inland from the coast up the Vailala River (east of the Purari River) and in areas greater than 100 km from the coast up the Sepik River in northern P.N.G. Ling (1969) found adults more than 200 km inland in Malaysia.

Although juveniles have been observed throughout the year, freshwater prawns become abundant in village markets of the Gulf Province at the end of the dry season and beginning of the wet (January to June). This may be an artifact of fishing activities as berried females were caught in beach seines throughout the year. Ling (1969) reports that *M. rosenbergii* breeds throughout the year in Malaysia.

346

4. Nutrition

Within brackish water communities, penaeid and carid prawns normally fill the niche of detrital feeders and scavengers. Branford (1981a, 1981b) found a positive relationship between the organic carbon content of the substrate and the distribution and abundance of penaeid prawns in the Red Sea. Ling (1969) describes the common food items for *M. rosenbergii* to be aquatic worms and insects, insect larvae, small molluscs and crustaceans, flesh and offal of fish and other animals, grain, seeds, nuts, fruits, algae and tender leaves and stems of aquatic plants. In an analysis of stomach contents of Australian penaeid prawns, Moriarty (1981) found the meiofauna, especially the foraminifera, to be the important food for prawns in the Gulf of Carpentaria. He suggests that prawns select food with a higher nitrogen content. Aquaria observations have shown that prawns will stop eating detritus and immediately head towards dead fish when placed in a tank.

The importance of prawns in the aquatic food chain is shown by Haines (this volume), who found them to be the major constituent in the diet of many fish species of the Purari Delta.

The importance of the crustacean fauna to the inhabitants of the delta is demonstrated by the large variety of fish capture methods. These are described by Frusher and Subam (1982). Hand traps, seine fences and scoop nets are used to capture small fish, carid and juvenile penaeid prawns.

S. serrata fills a higher tropic niche than prawns, being more carnivorous. Hill (1979b) found the major prey of *S. serrata* in South Africa to be burrowing bivalves and small crabs, preference being given to the latter food item because of its larger mass and high energy content.

5. Life cycles

5.1. Penaeid prawns

All the commercially important penaeids of the Gulf of Papua (except *P. semisulcatus* which uses coastal seagrass flats as nursery areas) have a typical penaeid life cycle. During its first stage, planktonic postlarvae settle in estuaries and deltas. Here the prawns feed and grow until adolescence (18 – 20 mm C.L. for *P. merguiensis* and 12 – 14 mm C.L. for the metapenaeids). Theoceanic stage begins when the prawns emigrate from the nursery areas inshore from where, after a period of growth, they move into deeper water to spawn. *P. merguiensis, M. demani* and *M. eboracensis* spawn in 10 – 20 metres of water, and *M. ensis* and *P. semisulcatus* in 35 – 60 metres. The planktonic larval stages then move shoreward using oceanic currents.

Postlarval *P. merguiensis* initially settle in the headwaters of small creeks and then congregate as juveniles and adolescents at the mouths of these creeks prior to emigrating out to sea. The main observed difference between the headwaters and mouths of these creeks is a stronger current and coarser substrate found at the mouth. Settlement may be associated with the finer substrate, which is essential both for protection (burying in the substrate) and for the presence of food micro-organisms.

Gwyther (1980) and Tenakanai (1980) found mature females of all the commercially important species throughout the year. Frusher (unpublished data) found that recruitment into the nursery areas is continuous throughout the year, but there is a reduction in the size of the nursery areas during the wet season (April to October). Barret and Gillespie (1975) found that in Louisiana increased freshwater output reduces available nursery habitat for *Penaeus aztecus* and limits the consequent yield.

5.2. Mangrove crab

S. serrata uses the mangrove creeks of the Purari Delta as a nursery and the open sea to spawn. The smallest crab found in the Purari has been of 60 mm carapace width (C.W.) (Opnai 1980). As the crabs grow they move towards the mouth of the creeks and estuaries. The crabs range from 60 to 130 mm C.W. Mating takes place just after moulting in either the mangrove swamps or close inshore. In the Gulf of Papua both sexes have been caught in prawn trawls up to depths of five metres. Although ovary development has begun in the crabs found in the mangrove regions of the Purari Delta, no berried specimens have been found. However, berried specimens are caught by prawn trawlers (and discarded as they are of no commercial value) in depths of 7 – 70 m throughout the year. Approximately 70 – 80% of trawled crabs are berried, the rest being in mature or ripe condition except crabs which contain an unidentified external parasite (probably *Sacculina* sp.) which harbours itself between the ventral thorax and the vestigial abdomen in the same space where eggs are carried. The parasite appears to have a similar effect as maturation, in causing the crabs to migrate out to sea. Parasitized individuals of as small as 60 – 80 mm C.W. often occur in the catch. Berried females have been caught 20 – 30 km from the coast, although the depth and limit to which these crabs migrate is not known. Crosiner (1962) found virtually undigested specimens in the stomach of a tiger shark (*Galeocerdo cuvieri*) in the

Mozambique channel about 50 km from the Madagascar coast.

The eggs are carried by the female crab for about 17 days (Arriola 1940) before hatching. The planktonic larvae (zoea) use ocean currents to return to the delta systems.

Only in the outlying island of New Ireland in Papua New Guinea have berried females been found in the mangrove systems (Chapau, personal communication). Although no reason has been found to explain this, it may be associated with the relatively small amount of freshwater runoff.

5.3. The giant freshwater prawn

Macrobrachium rosenbergii uses the mangroves of the Gulf of Papua as a nursery, and juveniles of 1 – 2 cm total length swim upstream into fresh water. Swarms of yet unidentified juvenile carid prawns are seen swimming up the rivers adjacent to the river bank throughout the year. Ling (1969) describes a similar migratory pattern for *M. rosenbergii* in Malaysia, so it could be that juvenile migrations sighted in the Purari concern the same species.

In river fresh water the prawns grow and mature before migrating back to the river mouth. In the Gulf of Papua female prawns with orange eggs have been caught in beach seines at the mouths of rivers. As males have never been caught, it is thought that only females move the final distance. The larvae require brackish water of at least 2 – 4 ppt (Ling 1969). In the Aird Delta of the Gulf of Papua berried females were caught in salinities as high as 15 – 18 ppt. The juveniles usually remain in brackish water for one to two weeks prior to migrating upstream.

6. Environmental modifications and their possible impact on crabs and prawns

Any dam constructed on the Purari will reduce freshwater supply to the delta during the filling period, which may extend over several months. During that time, the water salinity in estuaries will increase, favouring *P. merguiensis* which requires a higher salinity than other Purari penaeid species. *M. demani* and *M. ensis* will be found further upstream at the limits of brackish water. As a result, the overall prawn stocks of the Purari Delta could temporarily increase.

Pickup (1977) has estimated that there would be a considerable decrease in sediment input to the Purari Delta following the construction of the proposed dam. Viner (1979) has suggested that after construction of the dam, the inorganic nitrogen input to the delta would remain the same, and the overall nutrient supply, despite the overall reduction in sediment input, would still be sufficient to maintain the present Purari Delta mangrove system. However, the increased carrying capacity of the river would lead to channel erosion and changes in the sediment balance of the delta, leading possibly to an increase in erosion of some areas covered by mangroves (Pickup, this volume). This erosion would be enhanced if the tolerance of the fringing vegetation (especially C. ferax) was such that it could not withstand the higher salinities which would occur for the 2 – 5 months of filling the dam. Wolanski et al. (1980) has shown that the density of fringing vegetation in tropical Australian creeks determines the physical characteristics of the creeks. An alteration in the delta due to the increased carrying capacity would be from a finer to coarser sediment and some loss in particulate inorganic matter (Petr, personal communication) which would lead to changes in benthic fauna as well as having a detrimental effect on the juveniles of penaeid prawns.

The process of sorting, deposition and removal of particles within the sediment is important in fulfilling different requirements in the form of food and shelter for the various species of penaeid prawns. Branford (1981a, 1981b) found that the particle size of the sediment in the Red Sea influences the abundance and distribution of penaeid prawns. *Scylla serrata*, which exhibits a burrowing habit, would also be expected to have particle-size preferences.

The effect of damming the Purari on regions adjacent to the Purari Delta are difficult to assess. A diminished sediment input might affect both the productivity and substrate composition of the Aird Delta, with a probable effect on the commercial fishery. However, the actual effect on the Aird Delta is very difficult to assess without further complex geomorphological, hydrological, chemical and biological studies.

Tagging studies have shown a movement of *P. merguiensis* from the Aird Delta to Orokolo and Kerema Bays in the east (P.N.G. Fisheries Division 1980, 1982). Gwyther (1980), in analysis of catch statistics, found recruit sized prawns in 10 – 40 cm depth ranges in the inshore waters adjacent to the Purari Delta. The actual requirements for migrating prawns in the form of food and shelter is not known. With the decreased sediment and organic matter input into the inshore region of the Gulf of Papua a change in the substrate composition and associated micro-organisms is expected. Increased light penetration due to the decreased suspended sediments would allow a buildup in the phytoplankton population. Presently plumes from the Purari River extend several kilometres out to sea, depressing water transparency for light. The overall productivity of the substrate on which the penaeids feed might decrease as a consequence of damming of the Purari.

Rich prawning areas of the Gulf of Carpentaria have probably much lower nutrient input than the Gulf of Papua will receive even after damming the Purari. There is much more still to be researched to find out how the system works, and what are the factors determining the prawn productivity of the area.

7. Summary

The delta region of the Gulf of Papua is important as a nursery zone for penaeid prawns, the giant freshwater prawn *Macrobrachium rosenbergii* and the mudcrab *Scylla serrata*. In turn, these crustaceans are important as a protein food source to the subsistence dwellers of the delta region. Adult penaeid prawns form Papua New Guinea's second most valuable fishery.

Within the delta, the salinity appears to be one of the major factors controlling the distribution of the various crustaceans. Alteration of the salinity, substrate or fringing vegetation by the proposed damming of the Purari River would be expected to change the productivity of the crustacean fauna in not only the Purari, but also the adjacent Aird Delta.

Acknowledgements

I wish to thank Mr Allan K. Haines and Dr David Gwyther for their constructive criticism of the manuscript.

References

Alexander, W.B., B.A. Southgate and R. Bassingdale, 1932. The salinity of water retained in the muddy foreshore of an estuary. J. mar. biol. Assoc. U.K. 18: 297 – 298.

Arriola, F.J., 1940. A preliminary study of the life history of *Scylla serrata* (Forskal). Philipp. J. Sci. 73: 437 – 455.

Barret, B.B. and M.C. Gillespie, 1975. Environmental conditions relative to shrimp production in coastal Louisiana. La. Wildl. & Fish. Comm. Tech. Bull., No. 15.

Branford, J.R., 1981a. Sediment and the distribution of penaeid shrimp in the Sudanese Red Sea. Estuarine, Coastal and Shelf Sci. 13: 349 – 354.

Branford, J.R., 1981b. Sediment preferences and morphometric equations for *Penaeus monodon* and *Penaeus indicus* from creeks of the Red Sea. Estuarine, Coastal and Shelf Sci. 13: 473 – 476.

Crosnier, A., 1962. Crustacés décapodes, Grapsidae et Ocypodidae. Faune Madagascar 18: 1 – 143.

Dall, W., 1981. Osmoregulatory ability and juvenile habitat preference in some penaeid prawns. J. exp. mar. Biol. Ecol. 54: 55–64.

Floyd, A.G., 1977. Ecology of the tidal forest in the Kikori/Romilly Sound area of the Gulf of Papua. P.N.G. Office of Forests. Division of Botany. Ecology Report, No. 4. Lae, Papua New Guinea.

Frusher, S.D., 1980. The inshore prawn resource and its relation to the Purari Delta region. In: D. Gwyther (ed.), Purari River (Wabo) Hydroelectric Scheme Environmental Studies, Vol. 15: Possible effects of the Purari Hydroelectric Scheme on subsistence and commercial crustacean fisheries in the Gulf of Papua, Workshop 12 Dec. 1979, pp. 11–27. Office of Environment and Conservation, Waigani, Papua New Guinea.

Frusher, S.D. and S. Subam, 1982. Traditional fishing methods and practices in the northern Gulf of Papua. Harvest 7(4): 150–158.

Gwyther, D., 1980. The Gulf of Papua offshore prawn fishery in relation to the Wabo Hydroelectric Scheme. In: D. Gwyther (ed.), Purari River (Wabo) Hydroelectric Scheme Environmental Studies, Vol. 15: Possible effects of the Purari Hydroelectric Scheme on subsistence and commercial crustacean fisheries in the Gulf of Papua, Workshop 12 Dec. 1979, pp. 29–52. Office of Environment and Conservation, Waigani, Papua New Guinea.

Gwyther, D., 1983. The importance of the Purari River Delta to the prawn trawl fishery of the Gulf of Papua. This volume, Chapter II, 8.

Haines, A.K., 1983. Fish fauna and ecology. This volume, Chapter II, 9;

Haines, A.K. and R.N. Stevens, 1983. Subsistence and commercial fisheries. This volume, Chapter II, 10.

Hill, B.J., 1979a. Biology of the crab *Scylla serrata* (Forskal) in the St. Lucia system. Trans. R. Soc. S. Afr. 44(1): 55–62.

Hill, B.J., 1979b. Aspects of the feeding strategy of the predatory crab *Scylla serrata*. Mar. Biol. 55(3): 209–214.

Le Reste, L., 1973. Etude de la répartition spatiotemporelle des larves et jeunes postlarves de la crevette, *Penaeus indicus*, H. Milne Edwards en baie d'Ambaro (côte nord-ouest de Madagascar). Cah. O.R.S.T.O.M., Ser. Océanogr. 11(2): 179–189.

Le Reste, L., 1980. The relation of rainfall to the production of the penaeid shrimp (*Penaeus duorarum*) in the Casamance estuary (Senegal). In: J. Furtado (ed.), Tropical Ecology and Development. Proc. 5th Intern. Symp. Trop. Ecol., pp. 1169–1173. Intern. Soc. Trop. Ecol., Kuala Lumpur.

Ling, S.W., 1969. The general biology and development of *Macrobrachium rosenbergii* (De Man). FAO Fish. Rep., No. 57, Vol. 3: 589–606.

MacFarlane, J.W., 1980. Surface and bottom sea currents in the Gulf of Papua and western Coral Sea. P.N.G. Dept. of Primary Industry. Res. Bull., No. 27. Post Moresby, Papua New Guinea.

Macnae, W., 1968. A general account of the fauna and flora of mangrove swamps and forests in the Indo-West Pacific region. Adv. mar. Biol. 6: 73–270.

McPadden, C., 1980. Some preliminary observations on the occurrence of juvenile Penaeidae in in the Gulf of Papua. In: D.Gwyther (ed.), Purari River (Wabo) Hydroelectric Scheme Environmental Studies, Vol. 15: Possible effects of the Purari Hydroelectric Scheme on subsistence and commercial crustacean fisheries in the Gulf of Papua, Workshop 12 Dec. 1979, pp. 3–10. Office of Environment and Conservation, Waigani, Papua New Guinea.

Mohamed, K.U. and P.V. Rao, 1971. Estuarine phase in the life-history of the commercial prawns of the west coast of India. J. Mar. Biol. Assoc. India 13: 149–161.

Moriarty, D.J.W. and M.C. Barclay, 1981. Carbon and nitrogen content of food and the assimilation efficiencies of penaeid prawns in the Gulf of Carpentaria. Aust. J. Mar. Freshwater Res. 32: 245 – 251.

Opnai, L.J., 1980. The mangrove crab *Scylla serrata*. In: D. Gwyther (ed.), Purari River (Wabo) Hydroelectric Scheme Environmental Studies, Vol. 15: Possible effects of the Purari Hydroelectric Scheme on subsistence and commercial crustacean fisheries in the Gulf of Papua, Workshop 12 Dec. 1979, pp. 83 – 91. Office of Environment and Conservation, Waigani, Papua New Guinea.

Pagcatipunan, R., 1972. Observations on the culture of the alimango *Scylla serrata* at Camarines Norte (Philippines). In: T.V.R. Pillay (ed.), Coastal Aquaculture in the Indo-Pacific Region, pp. 362 – 365. FAO Fishing News Books, West Byfleet, Surrey, England.

Petr, T. and J. Lucero, 1979. Sago palm salinity tolerance in the Purari River Delta. In: T. Petr (ed.), Purari River (Wabo) Hydroelectric Scheme Environmental Studies, Vol. 10: Ecology of the Purari River catchment, pp. 101 – 106. Office of Environment and Conservation, Waigani, Papua New Guinea.

Pickup, G., 1977. Computer simulation of the impact of the Wabo Hydroelectric Scheme on the sediment balance of the lower Purari. In: T. Petr (ed.), Purari River (Wabo) Hydroelectric Scheme Environmental Studies, Vol. 2, pp. 1 – 88. Office of Environment and Conservation, Waigani, Papua New Guinea.

Pickup, G., 1983. Sedimentation processes in the Purari River upstream of the delta. This volume, Chapter I, 11.

P.N.G. Fisheries Division, 1980. Fisheries Research Annual Report 1979. The Division, Port Moresby, Papua New Guinea.

P.N.G. Fisheries Division, 1982. Fisheries Research Annual Report 1980 & 1981. The Division, Port Moresby, Papua New Guinea.

Staples, D.J., 1979. Seasonal migration patterns of postlarval and juvenile banana prawns, *Penaeus merguiensis* De Man, in the major rivers of the Gulf of Carpentaria, Australia. Aust. J. Mar. Freshwater Res. 30: 143 – 157.

Staples, D.J. and D.J. Vance, 1979. Effects of changes in catchability on sampling of juvenile and adolescent banana prawns *Penaeus merguiensis* De Man. Aust. J. Mar. Freshwater Res. 30: 511 – 519.

Stephenson, W. and B. Campbell, 1960. The Australian portunids (Crustacea, Portunidae). IV. Remaining genera. Aust. J. Mar. Freshwater Res. 11: 73 – 122.

Tabb, D.C., D.L. Dubrow and A.E. Jones, 1962. Studies on the biology of the pink shrimp, *Penaeus duorarum* Burkenroad, in Everglades National Park, Florida. Florida State Bd. Conserv., Univ. Miami Mar. Lab., Ser. 37: 1 – 30.

Tenakanai, C.D., 1980. Some aspects of the biology and fishery for endeavour prawns (*Metapenaeus* spp.) in the Gulf of Papua. P.N.G. Dept. of Primary Industry. Res. Bull., No. 28. Port Moresby, Papua New Guinea.

Thom, B.G. and L.D. Wright, 1983. Geomorphology of the Purari Delta. This volume, Chapter I, 4.

Viner, A.B., 1979. The status and transport of nutrients through the Purari River (Papua New Guinea). In: T. Petr (ed.), Purari River (Wabo) Hydroelectric Scheme Environmental Studies, Vol. 9, pp. 1 – 52. Office of Environment and Conservation, Waigani, Papua New Guinea.

Walker, R.H., 1974. State of our prawn fisheries. . . biology of major species. Aust. Fish. 33(3): 6 – 11.

Wolanski, E., M. Jones and J.S. Bunt, 1980. Hydrodynamics of a tidal creek – mangrove swamp system. Aust. J. Mar. Freshwater Res. 31: 431 – 450.

8. The importance of the Purari River Delta to the prawn trawl fishery of the Gulf of Papua.

D. Gwyther

1. Introduction

In the Gulf of Papua, a prawn trawl fishery of modest size has been in commercial operation for approximately 10 years. The banana prawn, *Penaeus merguiensis* (De Man) is the principal species caught and makes up 60% by weight of the total annual prawn catch which now averages over 1000 tonnes of tails. Other commercial species include the tiger prawns *P. monodon* (Fabricius), *P. semisulcatus* (De Haan) and the endeavour prawns *Metapenaeus ensis* (De Haan), *M. eboracensis* (Dall), *M. demani* (Roux) and *M. papuensis* (Racek and Dall).

In addition to the prawns, there is a considerable by-catch of non-commercial or 'trash' fish species. This makes up 70 – 80% of the total haul and includes sharks, rays and many teleost species with Sciaenidae, Leiognathidae and Theraponidae being common families. Many sea snakes and occasionally turtles (Green and Pacific Ridley) are also trawled. Over 300 species of fish have been recorded from the Gulf of Papua (Kailola and Wilson 1978). Some of the fish is consumed on board by the crew, some is retained for later retail in Port Moresby, and some is traded to local canoe fishermen who come alongside in exchange for fresh fruit and vegetables. However, the majority of the by-catch is thrown overboard dead after the prawns have been processed. This problem of non-utilization of such a large proportion is common to other prawn trawl fisheries.

There have been a number of plans to make greater use of this product. Some is brought back to Port Moresby and sold as crocodile food, but since freezer space on board is at a premium, only small amounts are handled. Another possibility is periodic unloading at Baimuru on the Pie River, west of the Purari Delta (see Fig. 1), but the economics and feasibility of this have so far been unfavourable.

The trawling grounds in the Gulf of Papua are situated on a narrow margin

Petr, T. (ed.) The Purari – Tropical environment of a high rainfall river basin
© *1983, Dr W. Junk Publishers, The Hague / Boston / Lancaster*
ISBN-13: 978-94-009-7265-0

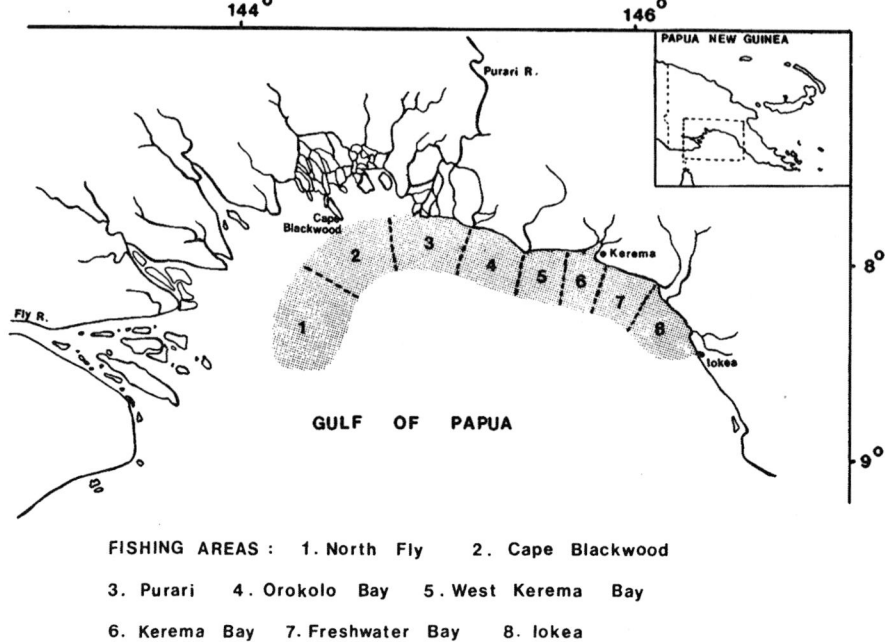

FISHING AREAS : 1. North Fly   2. Cape Blackwood
3. Purari   4. Orokolo Bay   5. West Kerema Bay
6. Kerema Bay 7. Freshwater Bay   8. Iokea

Fig. 1. The trawling grounds of the Gulf of Papua.

of continental shelf running parallel to the south coast from the Fly River Delta in the southwest to the village of Iokea in the east (Fig. 1). Part of the fishing grounds lie directly offshore from the Purari River Delta, and this is one of many large river systems draining into the Gulf from the highlands of Papua New Guinea.

Since 1974, this fishery has been shared between three Japanese fishing companies based in Port Moresby. A Japanese/Papua New Guinea joint venture company was later formed in 1977. These companies have been operating a total of 13 twin-rigged freezer trawlers of approximately 150 gross tonnes. The lack of any adjacent port facilities and inclement weather conditions for part of the year have encouraged the use of this class of vessel. Operating from Port Moresby, they usually remain at sea for up to six weeks, with trawling activities being carried out continuously through day and night. Processing the catch is done on board after each haul and this involves sorting, removing heads, grading, packing and freezing.

In comparison with the neighbouring trawling grounds of the Gulf of Carpentaria and the Arafura Sea, the Gulf of Papua prawn production figures are low. Annual catches ranging from 3000 and 12 000 t of banana

356

prawns have been landed in the Gulf of Carpentaria (Aust. Fish. 1977), although catches in recent years have been lower (Staples 1980). In the Arafura Sea annual catches of banana prawns averaged about 6000 t up to 1975 (Naamin and Sudradjat 1975).

In 1977 the Fisheries Division of the Department of Primary Industries increased its research effort into the status of the trawl fishery in the Gulf of Papua. Investigations of the prawn stocks and the collection of detailed catch statistics were aimed primarily at stock assessment and fisheries management. However, with large-scale industrial developments such as the Purari hydroelectric project always a possibility, the detailed collection of data from different regions of the fishery has enabled the relative importance of each area to be assessed. This chapter, apart from discussing the biology of the economically most important banana prawn, aims to assess the importance of the Purari Delta region to the prawn fishery from a comparison between catches taken in the grounds immediately adjacent to the Purari Delta with those from other areas of the fishery. This, together with the more direct observations on the distributions of penaeid prawns within the Purari Delta region (Frusher, this volume), forms the basis of the conclusions drawn on the possible effects of the Purari hydroelectric project upon the fishery.

2. The fishery

2.1. The biology of Penaeus merguiensis

The biology of P. merguiensis in the Gulf of Carpentaria has been described by Munro (1975). Spawning takes place in the open sea where each female can release up to 300 000 eggs. The eggs develop through a number of nauplius, mysis and zoeal stages in the plankton gradually migrating inshore as they reach the post-larval stage. The mechanisms for this immigration of juveniles inshore are not clear, but the post-larvae settle on the bottom approximately 2 – 3 weeks after hatching. As juveniles, they enter the estuarine or mangrove habitats (nursery areas), where they remain for periods ranging from a few weeks to several months (Staples 1979, 1980). Emigration of adolescent prawns back into the spawning grounds is often correlated with heavy seasonal rainfall, although it is not clear whether this is an active response to changing salinities or whether prawns are simply flushed out.

The relationship between the inshore nursery areas and the offshore trawling grounds is an important one and one on which the yield of the fishery depends. Quite how important each of the coastal areas are and which of the nursery areas along the coast provide the most significant contribution to recruitment are factors about which something must be known before an assessment of the effects of damming the Purari can be discussed. The following analysis of the commercial catches provides some indirect evidence of the importance of the Purari region to the fishery as a whole.

2.2. The collection of data

The submission of catch return forms by the fishing companies has been mandatory since commercial operations started. Records of monthly catches together with the number of hours fished are accurate and comparable from 1974 onwards. In 1979, a new catch return form was introduced on which details of each haul were recorded. These details included the time and duration of each haul, the fishing area, depth, weight (kg) of each prawn species and the weight (kg) for each size grade of banana prawns. The value of this was in providing an estimate of monthly relative abundance (as kg/h trawled) for each of the size grades and in each area and depth category. A computer programme produced a monthly print-out of catches and catch rates (kg/h) by area depth and size grade. Since all the trawlers were of a similar size and power, and used similar gear, the number of hours fished has been used as the unit of fishing effort.

Banana prawns are graded on board ship by mechanical graders according to the number of tails per pound. Normally the smallest grade was a count of 51 – 60 tails/lb, and high catches of this size grade can be considered to represent prawns newly recruited into the fishery. The monthly distribution of recruit-sized prawns during 1980 is the basis of the following analysis.

2.3. Catch statistics

Table 1 shows the annual catches and catch per unit of effort statistics in the Gulf of Papua from 1974 onwards. From this and from yield estimates determined from catch composition data, Gwyther (in press) concluded that the fishery was being more or less optimally exploited by the existing levels of fishing operations. The relationship between catch rates, fishing effort and annual catches indicates that remarkably stable conditions have occurred within the fishery, particularly over the past three years. Figure 2 gives a more detailed view of the distribution of catches within the fishery for each month of 1980. The graphs show monthly catches (tonnes) of banana prawn tails

358

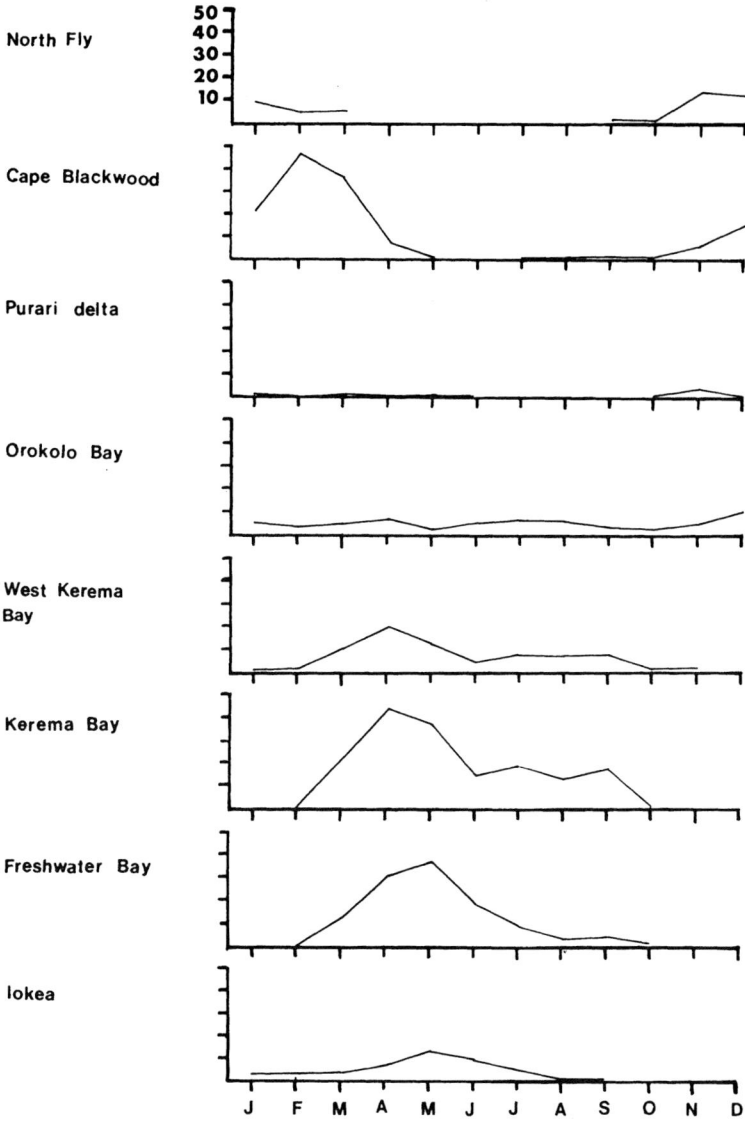

Fig. 2. Monthly catch (tonnes) of all size grades of banana prawns from each fishing area.

taken from each of the fishing areas, and this is also a reflection of the distribution of fishing effort.

For the first three months of the year, the fishing was mainly concentrated in the westerly grounds of the Gulf, west of the Purari Delta and centered

Table 1. The annual commercial prawn catches and catch per unit of effort in the Gulf of Papua (1974–1980)

Year	Number of vessels	Hours fished	Catch (t) all species	Catch (t) banana prawns	Catch rate (kg/h) all species	Catch rate (kg/h) banana prawns
1974	24	48 763	733	442	15.0	9.1
1975	6	16 902	410	258	24.3	15.3
1976	12	47 717	780	462	16.3	9.7
1977	12	56 932	842	432	14.8	7.6
1978	13	68 838	1106	548	16.1	8.0
1979	13	77 298	1174	634	15.2	8.2
1980	13	80 449	1448	668	18.0	8.3

mainly off Cape Blackwood, where catches were high. From April to October, the main emphasis of the fishery shifted eastwards into Kerema Bay and adjacent areas, from which large catches were again recorded. This period normally coincides with the period of the southeasterly trade winds, during which time, conditions are often extremely difficult for fishing in the westerly grounds. The southeasterly trades bring the major wet season to the central coastal parts of the Gulf. Broad associations between this and the high prawn catches in Kerema Bay have been made for previous years (Gwyther, in press), but a precise relationship between rainfall and offshore migration is far from clear for this fishery. By November and December the fishing fleet had again moved into the western grounds, where good catches were recorded from the Cape Blackwood grounds. This fishery therefore produces reasonable quantities of banana prawns throughout the year with peaks in April/May (Kerema area) and in December/January/February (Cape Blackwood).

This pattern of fishing during the year has become established over a number of years. It is clear from Fig. 2 that only a very small proportion of the catch actually comes from the grounds immediately offshore from the Purari Delta. This was also true for 1979 (Gwyther, in press), and the area is generally recognised by the fishing companies as being less productive (in terms of banana prawn catches) than the areas to the west and east.

2.4. Distribution of recruit prawns

If an assessment is to be made of the likely impact of damming the Purari River at Wabo, it is important to know the areas of origin and the distribution

360

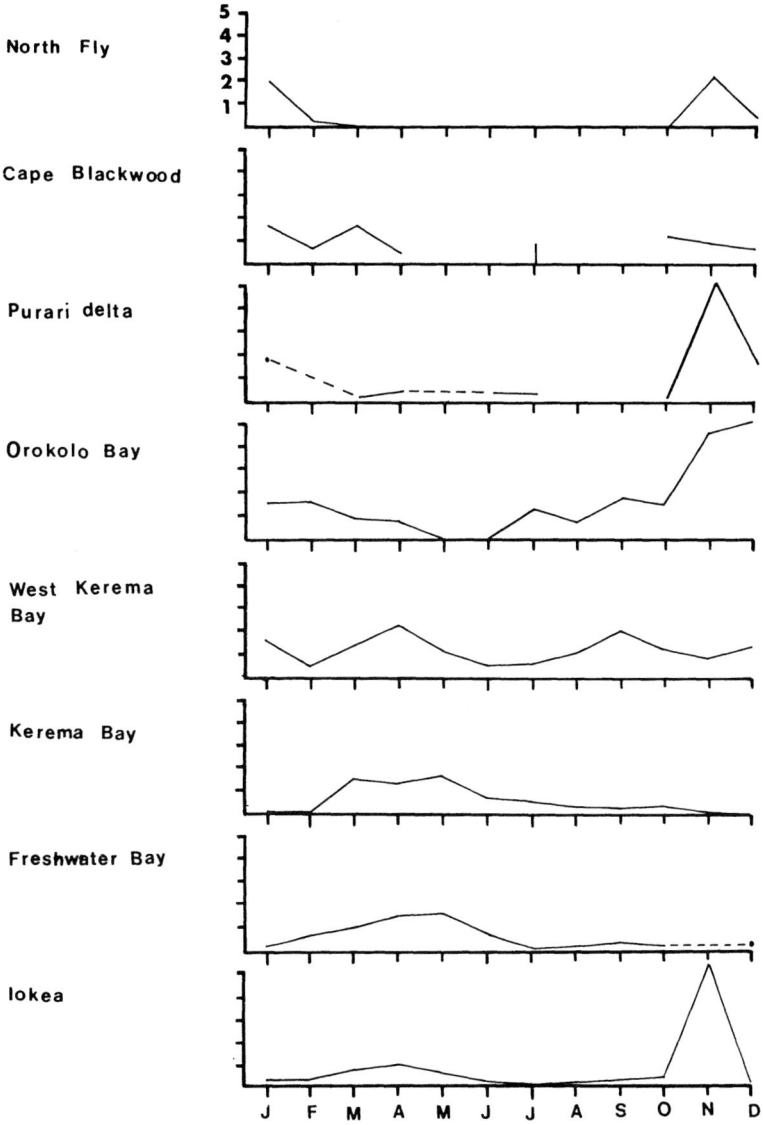

Fig. 3. Monthly catch rates (kg/h trawled) of recruit-sized banana prawns from each fishing area and taken from depths less than 20 m.

of prawns newly recruited into the fishery. The data collection system has made it possible to determine the abundance of recruit-class prawns in areas where trawling took place. The figures for monthly catch rate (kg/h) for recruit-class prawns in each of the areas and taken from depths less than 20 m are shown in Fig. 3. The results confirm that the Purari and the Orokolo Bay grounds supported large abundances of recruits at certain times of the year, particularly in November and December. Apart from the sporadic appearance of recruits in November in the Iokea area, where in fact only 22 hours of fishing took place, none of the other areas showed such high catch rates of recruits.

Similar results were found in 1979 (Gwyther 1980) with high abundances of recruits being observed from east of Cape Blackwood to the West Kerema Bay grounds. However, these were recorded during the first three months of the year so yearly differences in timing obviously occur. Nevertheless, it would seem that the main origins of recruitment into the fishery were from the areas surrounding the Purari Delta.

3. Discussion

The data from the commercial fishing records are largely circumstantial evidence for the origins and distributions of recruit prawns. Trawlers which encounter large concentrations of small prawns may well move elsewhere in search of larger and more valuable size grades of prawns and so may not locate the true distributions of recruits in the way that research fishing would. It is also not certain whether those recruits caught in the Purari grounds have come from nursery areas within the delta or from more distant areas. Nursery areas for banana prawns are known to be more extensive in the Kikori Delta further to the west (Frusher 1980). Since equivalent areas to the east of the Purari are restricted to a few isolated inlets and creeks, the inference has been that there is a southeasterly offshore migration into the grounds of Kerema Bay (Frusher 1980; Gwyther 1980). This has to some extent been supported by tag returns (Frusher, unpublished data), although some prawns were found to have moved southwestwards.

The dynamic relationship between rainfall, offshore prawn migration and lateral migration is not completely known for this fishery. Rainfall along the Gulf coast is high throughout the year. Accurate figures are scarce and rainfall may vary locally from day to day quite considerably. Local flooding of estuaries may also arise from rainfall in the highlands and therefore any correlation between rainfall and prawn migration into the fishery is difficult to detect.

In contrast, emigration of juvenile banana prawns from the rivers of the southeast corner of the Gulf of Carpentaria into the offshore fishery is more clearly related to seasonal rainfall (Staples 1979, 1980). Juveniles accumulate in estuaries and the seaward fringes of river mouths, moving offshore after rainfall. The exact nature of the stimulus for the migration is not clear but in exceptional circumstances, when rainfall is low, some prawns remain in the estuaries without migrating or maturing.

A similar relationship has been reported by Ruello (1973) for *Metapenaeus macleayi* in the Hunter River region, New South Wales, where increased river flow stimulated offshore migration. He suggested that disturbance to the bottom sediments and to the prawns' normal burying and respiration activities may act as the stimulus.

The significance of the rainfall/emigration relationship is possibly that emigration becomes coincident with additions of organic detritus to the substratum. There is also evidence to suggest that spawning and the timing of larval development in the plankton is also coincident with increased river outflow, with presumably the increase in nutrient levels a significant factor in successful development. Such a relationship has been suggested for *Penaeus esculentus* off Queensland (O'Connor 1979), for several species of penaeid prawns in the estuaries of Goa (Achuthankutty et al. 1976), for *Penaeus duorarum* off Senegal (Le Reste, undated) and for penaeid stocks in the Gulf of Mexico (Turner 1979). It may well be that in the Gulf of Papua, freshwater outflow is always high and nutrient and detritus levels are not factors which affect the timing of recruitment. It is therefore difficult to make impact assessments or predictions of effects of changes to the Purari River. One of the possible results of damming the Purari might be a change in the regime of the river at its estuary from one of deposition to one of erosion, a change which could have the most serious effects on the fishery. Dams in tropical regions have caused a number of environmental problems (Petr 1980). One direct result of the dam at Aswan was the collapse of the sardine fishery of the Nile delta (Aleem 1969). Impacts on fisheries may also result indirectly from die-back of coastal mangrove communities following the trapping of sediments by the impoundments. Such a situation is known to have occurred in the Zambesi delta (Davies 1979). The correlation between commercial penaeid yield and mangrove area has been made in a number of instances (Turner 1979; Martosubroto and Naamin 1979), and there is a possibility therefore that the prawn fishery could be impacted through loss of mangrove communities in the area surrounding the Purari Delta.

Whether the Purari Delta contains major nursery areas for juvenile banana prawns or whether adult prawns are simply migrating past, any erosion of soft mud and silt leaving a more sandy and less protective bottom for this species may impede the natural process of recruitment. Pickup (1977) predicted that approximately 53% of sediment would be retained by the proposed dam, but whether this would increase the carrying capacity of the river sufficiently to cause erosion at the delta and hence reduction of mangroves is still a matter of some speculation. However, the continuation of the present monitoring of the fishery should be able to detect any changes.

4. Summary

The main areas of recruitment of banana prawns into the fishery are the mangrove areas to the west of the Purari Delta with some isolated areas to the east. The poor fishing grounds directly offshore from these nursery areas compared with the grounds further to the west and east may indicate an adult migration away from these nursery areas. The grounds adjacent to Kerema and Cape Blackwood provide the highest commercial yield. Manipulation of the Purari watershed by trapping sediments in a reservoir may lead to a reduction of the mangrove habitat available for juvenile penaeid prawns, and such an effect would probably have considerable impact on the fishery.

Acknowledgements

I am extremely grateful to Mr Warren Klein of the Kanudi Fisheries Research Station for maintaining the efficiency of the trawl data collection system and for making available the data for 1980.

References

Achuthankutty, C.T., M.J. George and S.C. Goswami, 1977. Larval ingression of penaeid prawns in the estuaries of Goa. Proc. Symp. Warm Water Zooplankton, 17–19 Oct. 1976, Goa.

Aleem, A.A., 1969. Marine resources of the United Arab Republic. Stud. Rev. Gen. Fish. Coun. Mediterr. 43: 1–22.

Australian Fisheries, 1977. Gulf of Carpentaria fishery is complex and intriguing. Aust. Fish. 36: 4–11.

Davies, B.R., 1979. Stream regulation in Africa: a review. In: J.W. Ward and J.A. Stanford (eds.), The Ecology of Regulated Streams, pp. 113–142. Plenum Press, New York and London.

Frusher, S.D., 1980. The inshore prawn resource and its relation to the Purari Delta region. In: D. Gwyther (ed.), Purari River (Wabo) Hydroelectric Scheme Environmental Studies, Vol. 15: Possible effects of the Purari Hydroelectric Scheme on subsistence and commercial crustacean fisheries in the Gulf of Papua, Workshop 12 December 1979, pp. 11–27. Office of Environment and Conservation, Waigani, Papua New Guinea.

Gwyther, D., 1980. The Gulf of Papua offshore prawn fishery in relation to the Wabo Hydroelectric Scheme. In: D. Gwyther (ed.), Purari River (Wabo) Hydroelectric Scheme Environmental Studies, Vol. 15: Possible effects of the Purari Hydroelectric Scheme on subsistence and commercial crustacean fisheries in the Gulf of Papua, Workshop 12 December 1979, pp. 29–52. Office of Environment and Conservation, Waigani, Papua New Guinea.

Gwyther, D., in press. Yield estimates for the banana prawn (Penaeus merguiensis De Man) in the Gulf of Papua prawn fishery. J. du Conseil.

Kailola, P. and M.A. Wilson, 1978. The trawl fishes of the Gulf of Papua. P.N.G. Dept. of Primary Industry. Research Bull., No. 20. Port Moresby.

Le Reste, L., undated. The relation to the production of the penaeid shrimp (Penaeus duorarum) in the Casamance estuary (Senegal). Mimeo. Centre de Recherches Océanographiques, Dakar, Senegal.

Martosubroto, P. and N. Naamin, 1977. Relationship between tidal forests (mangroves) and commercial shrimp production in Indonesia. Marine Res. in Indonesia 18: 81–86.

Munro, I.S.R., 1975. Biology of the banana prawn (Penaeus merguiensis) in the southeast corner of the Gulf of Carpentaria. In: P.C. Young (ed.), Australian Prawn Seminar (1st), Maroochydore, 1973, pp. 60–78. Aust. Govt. Printing Service, Canberra.

Naamin, N. and A. Sudradjat, 1975. Progress report on the Arafura shrimp fishery. Laporan Penelitian Perikanan Laut 2: 45–75.

O'Connor, C., 1979. Reproductive periodicity of a Penaeus esculentus population near Low Islets, Queensland, Australia. Aquaculture 16: 153–162.

Petr, T., 1980. Possible impacts of the planned hydroelectric scheme on the Purari River deltaic and coastal sea ecosystems (Papua New Guinea). Paper presented at 2nd International Symposium on Biology and Management of Tropical Shallow Water Communities, Port Moresby, P.N.G., July 20–August 2, 1980.

Pickup, G., 1977. Computer simulation of the impact of the Wabo Hydroelectric Scheme on the sediment balance of the lower Purari. In: T. Petr (ed.), Purari River (Wabo) Hydroelectric Scheme Environmental Studies, Vol. 2, pp. 1–88. Office of Environment and Conservation, Waigani, Papua New Guinea.

Ruello, N.V., 1973. The influence of rainfall on the distribution and abundance of the school prawn Metapenaeus macleayi in the Hunter River region (Australia). Marine Biology 23: 221–228.

Staples, D.J., 1979. Seasonal migration patterns of postlarval and juvenile banana prawns Penaeus merguiensis De Man in the major rivers of the Gulf of Carpentaria. Aust. J. Mar. Freshwater Res. 30: 143–157.

Staples, D.J., 1980. Ecology of juvenile and adolescent banana prawns, Penaeus merguiensis in a mangrove estuary and adjacent offshore area of the Gulf of Carpentaria. II: Emigration, population structure and growth of juveniles. Aust. J. Mar. Freshwater Res. 31: 653–665.

Turner, R.E., 1979. Louisiana's coastal fisheries and changing environmental condition. In: J.W. Day, D.D. Culley, R.E. Turner and A.J. Humphrey (eds.), Proc. 3rd Coastal Marsh and Estuary Management Symposium, pp. 363–370. Louisiana State University Division of Continuing Education, Baton Rouge, LA.

365

9. Fish fauna and ecology

A.K. Haines

1. Introduction

In sharp contrast with the marine waters around New Guinea*, which harbor the rich Indo-Pacific fish fauna, the freshwaters of New Guinea support a depauperate fish fauna by the standards of tropical freshwaters elsewhere in the world. Zoogeographically they form part of the Australian region and are largely lacking in 'true' freshwater species. The species and families have been derived relatively recently from marine forms and contain a large component of estuarine and catadromous species.

Within Papua New Guinea there is a natural distinction between the fish faunas of the northward flowing rivers, of which the Sepik and the Ramu are the largest, and the southward flowing rivers, of which the Fly and Purari are the largest. The Sepik River contains a very limited number of indigenous species, and is much influenced by two introduced organisms which have become ecologically dominant: a fish *Sarotherodon mossambicus* and a floating plant *Salvinia molesta*. The major southern rivers in comparison support a much wider diversity of species, and no exotics have become established, except in a few smaller streams.

The waterways of the highlands represent a separate region. They are partially separated physically from the lowland rivers by rapids, and the lower temperatures and smaller size make them a distinct environment. The indigenous aquatic fauna, still not fully known, is restricted to a limited number of species and families. The highland rivers have been subjected to numerous introductions to improve the protein consumption of the people and for mosquito control (West and Glucksman 1976).

* New Guinea refers to the island, which is politically divided into Indonesian Irian Jaya in the west, and the independent state of Papua New Guinea in the east.

Petr, T. (ed.) The Purari – Tropical environment of a high rainfall river basin   367
© *1983, Dr W. Junk Publishers, The Hague / Boston / Lancaster*
ISBN-13: 978-94-009-7265-0

2. Purari River system

2.1. Highlands watershed

The Purari River drains a considerable area of the highlands, and has a catchment area of 33 670 km², an average annual runoff of 84.1 km³, and at Wabo a median flow rate of 2360 m³/sec. This is approximately equivalent to a quarter of the total runoff of the entire Australian continent (SMEC-NK 1977).

Streams in the highlands are generally shallow, rapidly flowing, and rocky with alternating pools and riffles. Water can be clear to very turbid, depending on recent rainfall. Water temperatures are low, ranging from around 10 to 15°C, depending on altitude, flow, etc. Many highland streams provide excellent conditions for trout. The larger rivers, while fast flowing, shallow and rocky, do have side pools which are quieter and warmer than average for waters in the highlands.

Hathor Gorge forms a natural boundary between the highlands and the lower Purari River. This is a chasm into which the river is compressed forming a series of gigantic rapids.

2.2. Lower Purari River

Between Hathor Gorge and the delta the Purari is a fast flowing, turbulent river with an average width of about 200 m, and an average current of 6 km per hour, which increases during times of increased flow. The water is mostly very turbid due to the high silt load. The annual sediment load transported at Wabo has been estimated at 57 million tonnes per year (SMEC-NK 1977).

Associated with the main channel are shallows, side branches, creeks and semi-isolated backwaters (billabongs), with varying degrees of flow. A noteworthy feature of the quiet, shallow backwaters is the almost complete lack of rooted aquatic vegetation.

The bottom varies from coarse sand in areas of rapid flow to fine silt in still waters. Pickup (1977; this volume) presents a detailed account of sediments in the Purari River between Wabo and the delta, and Thom and Wright (this volume) discuss the situation in the delta. Oxygen saturation of the flowing water is around 100%. Chemical analyses of Purari water are given by Petr (1976; this volume). Surface water temperature in the main channel is around 25°C, with a greater range in shallow water. In the backwaters, water temperature can rise to well above 30°C.

The river traverses land covered by dense rain forest. The deposition banks are covered with dense growths of *Phragmites* and *Saccharum*.

2.3. Purari Delta

The Purari, Pie, Era and Kikori rivers flow into a complex, interconnected deltaic system of anastomosing waterways. The deltaic system can be broadly divided into several zones based on the degree of marine influence, which is reflected in the associated vegetation.

The freshwater zone: is very similar to the river, traversing swamp forest and with *Phragmites* and *Saccharum* growing along deposition banks. The waterways consist of main channels, side branches and oxbow lakes which vary in degree of isolation from the main channels. This zone is present only in the Purari Delta. On the Kikori – Era side the rivers flow directly into the upper estuarine zones.

The Pandanus – Sonneratia zone: is typified by *Pandanus*, interspersed with clumps of *Sonneratia*, growing along the deposition banks. Swamp forest extends away from the channels. This zone is the uppermost part of the estuary, the water is almost completely fresh, but exhibits marked tidal fluctuations.

The Nypa zone: is typified by a bank vegetation which is 50 – 100% *Nypa* palm. In the Purari Delta this zone is restricted, forming usually a band between the true mangrove vegetation situated closer to the sea, and merging into the swamp forest upstream. The Kikori – Era deltas, in contrast, have a very extensive *Nypa* zone.

The lower estuarine zone: consists of a series of low-lying, tidally inundated islands covered with mangrove forest, and interspersed with tidal creeks of varying width and depth. Salinity can vary considerably from almost full seawater to almost fresh, depending on the contribution from the rivers. Where centres of larger islands are above the high tide level, the mangrove forest merges into freshwater swamp forest.

The coastal zone: includes the shoreline and its associated beaches, inlets, lagoons and tidal pools. The bottom varies from fine silt to fine sand. At certain times the salinity in this zone can be as low as 5‰.

The delta area is unstable with land being continually eroded away or built up. Often significant changes can occur within a few weeks. The water is always muddy in appearance.

2.4. Gulf of Papua

The Purari exerts a considerable influence on adjacent coastal waters of the Gulf of Papua. During periods of high outflow, well-defined sediment plumes can extend 6 – 8 kms out to sea and the salinity can be lowered markedly. Fish distribution in the Gulf is significantly influenced by the outflow of the Purari and Fly (Kailola and Wilson 1978).

A large amount of terrestrial and mangrove vegetation material is carried out to sea, and Liem and Haines (1977), Kailola and Wilson (1978) and Frusher (1980) have suggested that this forms a significant input to the marine food chain. The high turbidity of coastal waters of the Gulf of Papua has a negative effect on primary productivity, with low concentrations of phytoplankton found by Rapson (1955).

3. Purari fish fauna

There are two distinct groups of fish as determined by the physical character of the watershed. The upper part of the Purari River and its tributaries, situated above Hathor Gorge, and the Purari downstream of Hathor Gorge are believed to be almost separated by the turbulent nature of Hathor Gorge rapids, which would be insurmountable for most fish.

3.1. Highlands watershed

Indigenous species belong to the families Anguillidae, Ariidae, Plotosidae, Melanotaeniidae, Theraponidae and Gobiidae, but no complete list of species is available. The most complete coverage is in Munro (1967). The most conspicuous species are those introduced. West and Glucksman (1976) list six species which have become established in highland waterways: *Gambusia affinis*, *Salmo trutta*, *S. gairdneri*, *Salvelinus fontinalis*, *Sarotherodon mossambicus*, and *Cyprinus carpio*. Very little is known concerning the adaptations of these species to local conditions or their effects on the indigenous fauna.

While adult eels are caught throughout most of the watershed, no spectacular upstream elver migrations, as are well known in the Sepik, are known to occur in the Purari. It is possible that breeding occurs only off the north coast of New Guinea (which has deep, clear water adjacent to the coast, contrasting with the shallow, muddy Gulf of Papua). It could be suggested that for eels to enter the Purari watershed they would have to cross there from the Sepik watershed.

3.2. Lower Purari River and Delta

3.2.1. Distribution and abundance of species. The fish fauna of the Purari and related systems is richer in species along the coast and in the deltaic areas than in the rivers proper. Haines (1979) lists 49 species from 24 families from Purari freshwaters (see Table 1), while Liem and Haines (1977) list 143 species from 58 families from the estuarine area of the Purari – Kikori deltaic complex. Only a few species are confined to freshwater, most riverine species being also found in the estuarine zones. In part, it probably also reflects the marine origins of the fish fauna.

While few species are confined to any one zone, certain species are typical of a range of zones. A few species, for example *Toxotes chatareus*, *Thryssa scratchleyi* and *Parambassis gulliveri*, are almost ubiquitous.

In the Fly River, Roberts (1978) observed a number of cases where species of the same or closely related genera replace each other upstream and downstream. A similar intrageneric and ecological replacement occurs between some riverine and estuarine species in the Purari system. Some examples are listed below. Where species names are followed by a question mark the identification must be regarded as tentative:

Purari River	Purari Delta and Estuary
Carcharhinus leucas	*C. gangeticus* (?), *C. glyphis*
Thryssa scratchleyi	*T. hamiltoni*
Arius acrocephalus	*A. leptaspis*, *A. proximus*
Cochlefelis spatula (?)	*C. danielsi* or *A.* sp. (?)
Nedystoma dayi	*Nedystoma* sp.
Hemipimelodus crassilabris	*Hemipimelodus* sp.
Stenocaulus (*Strongylura*) *kreffti*	*S. strongylura*
Zenarchopterus novaeguineae	*Z. brevirostris*
Aseraggodes klunzingeri	*Aseraggodes* sp.
Liza oligolepis	*L. vaigiensis*, *L. dussumieri*, *L. tade*
Valamugil buchanani	*V. seheli*
Ambassis interruptus	*A. nalua*, *A. macracanthus*
Hephaestus fuliginosus	*H. jarbua*
Nibea soldado (?)	*N. semifasciata*
Glossogobius celebius	*G. giurus*
Oxyeleotris fimbriatus	*O. urophthalmus*, *O. lineolatus*

Table 1. Purari River fish fauna – families with number of species.

Carcharhinidae	1	Soleidae	1	Theraponidae	1
Pristidae	1	Mugilidae	3	Sparidae	1
Engraulidae	1	Melanotaeniidae	1	Sciaenidae	1
Plotosidae	1	Kurtidae	1	Toxotidae	1
Ariidae	12	Apogonidae	1	Scatophagidae	1
Anguillidae	1	Latidae	1	Gobiidae	3
Belonidae	3	Chandidae	3	Periophthalmidae	1
Hemirhamphidae	1	Lutjanidae	2	Eleotridae	6

3.2.2. Comparison of Purari with other southern rivers. The Kikori River flows into the same deltaic complex as the Purari, and while the Vailala flows directly into the sea, it is immediately adjacent to the Purari – Kikori deltaic complex. Both are broadly similar in appearance to the Purari, although there are some differences in the chemistry of the Kikori River (Petr 1978/79). Haines (1979) lists 18 species from the freshwater section of the Kikori River and 10 from the Vailala. In comparison almost 50 species occur in Purari freshwaters. Although some difference may be due to sampling intensity, the faunas of the adjacent rivers are clearly more limited. This suggests a more limited range of environments in the smaller rivers. An interesting point is that the Seribi, a clear water tributary of the Kikori, is devoid of ariid catfish. Two freshwater species in the Purari, *Melanotaenia maculata* and *Ambassis interruptus*, are intragenerically replaced in the Kikori by *Melanotaenia splendida rubrostriata* and *Ambassis agrammus*.

The Purari River fauna and the Fly River fauna are related and there is a general resemblance, although there are some significant differences. The ariid catfishes are a similarly well-represented and important group in the Fly River, but there are some differences in species composition and distribution. *A. berneyi* and the extremely large *A. augustus* are common in the middle and lower Fly respectively (Gwyther/Petr, personal communication) but absent from the Purari. It may be that in the Purari, *A. proximus* and *A. latirostris* occupy equivalent niches.

Some species are confined to estuarine zones in the Purari system but are present in freshwaters elsewhere; examples are *Kurtus gulliveri* and *Cinetodus froggatti*, which are both present in Lake Murray, a freshwater lake of the Fly River system (Moore, personal communication), while being largely or entirely restricted to estuarine waters in the Purari system. Obviously their restricted distribution in the Purari system cannot be due to limitations of salinity tolerance. The Fly River lacks any natural barrier equivalent to Hathor Gorge which might impede movement of fish. It also has extensive areas of still backwaters away from the main channel which support permanent growth of higher aquatic plants. The backwaters of the Purari, by contrast, are of limited extent with virtually no aquatic vegetation. It therefore seems that the limited range of some species in the Purari system is due to the limited range of freshwater environments available. This would also be the reason for the absence of some species from the Purari such as *Nematalosa papuensis* which is the main herbivore of the Fly River (P.N.G. Fisheries Div. 1976). Roberts (1978) listed 103 species of 33 families from the Fly, and additional species were found in 1981 during the environmental expedition for the Ok Tedi mining project (Petr, personal communication).

4. Food and feeding relationships

A notable feature of the Purari fish fauna is the paucity of herbivores and a lack of planktonic feeders. This can be directly related to the lack of plankton and aquatic vegetation to support such feeding groups.

Detritus, derived almost entirely from allochthonous sources throughout the system, forms the base of the main food chain. The Purari fish fauna follows broadly the features summarized by Lowe-McConnell (1975) for tropical freshwater fish faunas of three continents (Asia, Africa, South America), i.e. showing (1) the importance of allochthonous vegetable material as direct food for species, (2) the role of insects as fish food, (3) the importance of mud and detritus as food for fish, (4) the presence of large numbers of individuals and kinds of piscivorous fishes. However, among the Purari fish there are more prawn-eating species than piscivores, and prawns play an extremely important role in the food chain. In New Guinea freshwaters prawns appear to fill many of the roles of detritophages which are occupied by fishes in other continents.

The role of mud-eating fishes and prawns is particularly significant, as mud is among the major sources of nutrients for the food chain of the Purari River Delta.

Table 2 lists the number of fish species of each ecological zone according to feeding categories. Prawns are the major constituent of fish food, and hence the major contributors to fish biomass, particularly in the lower mangrove, *Nypa* and *Pandanus – Sonneratia* zones. In the freshwater part of the delta and the lower river, while prawns remain important, a greater proportion of food, and hence fish biomass, is derived directly from organic matter from mud, and a small but significant contribution comes directly from terrestrial sources (insects and fruit). Further upstream, in the Wabo area immediately

Table 2. Number[a] of fish species of each ecological zone according to feeding category.

Trophic niche	Zone							
	1	2	3	4	5	6	7	8
Omnivores	3(2)	–	2(1)	2(1)	2(1)	2(1)	2(1)	3(2)
Fish eaters	6(3)	1(0)	7(3)	5(2)	8(2)	10(2)	15+(2)	14+(1)
Mollusc eaters	–	–	4(1)	–	4(1)	4(1)	4(1)	6(0)
Prawn eaters	12(5)	–	14(4)	11(4)	15(3)	20(4)	20(4)	15+(1)
Crab eaters	–	–	–	–	–	8(2)	11(3)	6(2)
Insect eaters	5(1)	4(0)	3(0)	4(0)	4+(0)	4+(0)	4+(0)	–
Fruit eaters	3(2)	–	2(1)	2(1)	2(1)	2(1)	3(2)	3(2)
Other plant eaters	3(1)	–	–	–	1(0)	1(0)	2(0)	2(0)
Detritus eaters	8(4)	4(0)	4(2)	4(2)	5+(1)	9+(0)	9+(0)	13+(2)

[a] Some species included in several categories as appropriate.

Zones: 1 = River: Hathor Gorge to effluence of delta; 2 = small creeks entering river; 3 = freshwater section of delta – flowing water; 4 = freshwater section of delta – still water; 5 = *Pandanus* – *Sonneratia* zone of delta; 6 = *Nypa* zone of delta; 7 = lower mangrove zone of delta; 8 = coastal zone (species numbers probably incomplete).
( ) = number of ariid catfish species.

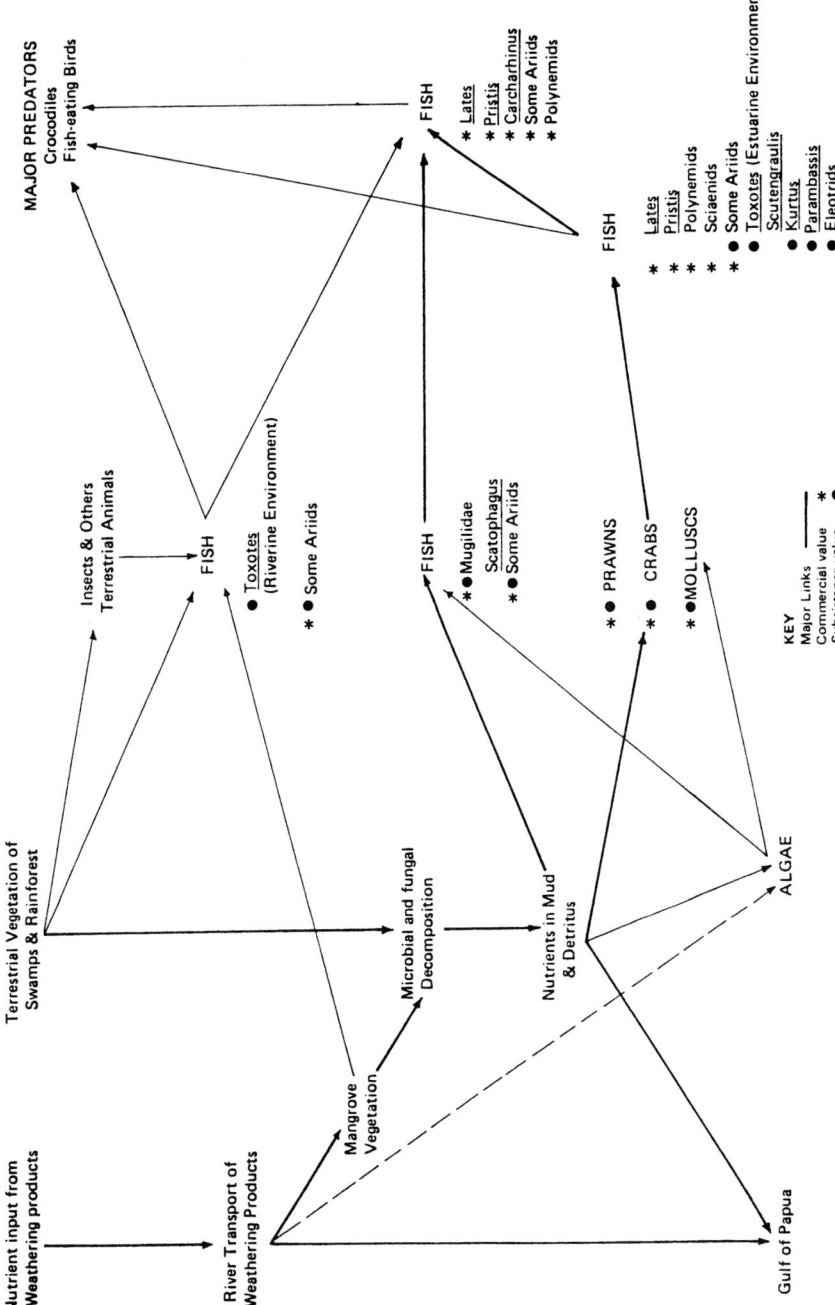

Fig. 1. Food chain and nutrient flow in the Purari system.

below Hathor Gorge, the proportion of food derived directly from terrestrial sources is greater.

Fish within each zone of the Purari system fill a similar range of trophic niches, although the relative importance of the different niches varies between zones. As examples, insectivores and frugivores are more prominent, both in species and numbers of individuals, in the river than the delta, while molluscivores are confined to the delta zones. Species may also take different food in different zones, e.g. *Toxotes chatareus* feed on insects and fruit in a riverine environment, and crabs and fruit in an estuarine environment.

Prawns are a significant food source for many fish species throughout the system − carid prawns predominate in the freshwater and upper estuarine zones and penaeids in the lower estuarine and coastal zones. Penaeids are known to be opportunistic omnivores, particularly consuming the detritus and micro-organisms present in muddy sediments (Moriarty 1977). The main carid genus present is *Macrobrachium* which is also omnivorous, feeding on detritus, plant matter and small invertebrates (Dugan, Hagood and Frakes 1975).

Figure 1 outlines the food chain and nutrient pathway of the Purari River and Delta.

A feature of the Purari system is the dominance in biomass and species of ariid catfish: various species occupy almost all trophic niches available. It would appear that the lack of a true freshwater fish fauna allowed the adaptive radiation of the family in the estuaries and rivers. Physical adaptations of the group are mainly limited to mouth structure, although there is some variation in body shape. The group generally exhibits oral incubation of eggs and hatchlings by the males, but this has not been confirmed for all species. The family is predominantly marine and estuarine with freshwater representatives. The river forms on the whole tend to show varying degrees of dietary and morhological specialization, while the marine forms appear to be mainly unspecialized.

Table 3 lists the main food and morphological adaptations of the ariids of the Purari system. Where a name is queried, or alternative names are given, the specific identification is tentative.

5. Breeding

The majority of the fish of the lower Purari system are estuarine or marine breeders. Even with the species which do breed in fresh waters, such as some ariid catfishes, there seems to be a tendency for part of the population to

Table 3. Feeding and morphological adaptations of Purari ariid catfish.

Species	Main food	Mouth	Snout	Teeth[a]	Comment
1. *Arius leptaspis*	omnivore	subterminal, wide unspecialised	unspecialised	villiform, unspecialised	generalised omnivore, abundant and widespread
2. *Cochlefelis danielsi*	prawns, crabs	subterminal, upper jaw extends beyond lower, i.e. upper teeth exposed	pointed, 'shark shaped'	villiform, unspecialised	body shape streamlined, suggesting an active predator; abundant in estuary
3. *A. latirostris* or *digulensis*	fish, prawns	very large, wide gape, terminal	broad, blunt	villiform, reduced	thick, blunt body shape, suggesting ambush-type predator; grows to large size; present river and estuary
4. *Cinetodus froggatti*	molluscs, swallowed whole	very small, subterminal, thick lips	downwardly curved	villiform, greatly reduced	specialised for exclusive diet of molluscs; abundant estuary
5. *A. carinatus*	detritus from bottom mud	small, subterminal thick lips	downwardly curved	villiform, greatly reduced	very small eye, specialised for exclusive diet of bottom mud; freshwater, very abundant in still backwaters
6. *A.* sp.	fish	wide, subterminal, upper jaw and teeth slightly protruding	slightly pointed	long, sharp, needle-like very large	strong streamlined body, grows to very large size; active predator; present, but not abundant, river and estuary
7. *A. stirlingi*	prawns	unspecialised	unspecialised	villiform, unspecialised	small marine form extending into estuary; not abundant in Purari Delta but very abundant in Gulf of Papua
8. *A. acrocephalus*	fruit, insects, prawns	unspecialised	unspecialised	villiform, unspecialised	body rather blunt, exclusively freshwater; feeds largely on allochthonous material; locally abundant
9. *A. proximus* (?)	prawns, crabs, fruit	unspecialised	unspecialised	villiform, unspecialised	small general omnivore, confined to most saline waters of estuary; not abundant
10. *Cochlefelis spatula* or *H. nudidens*	prawns, fish	very wide, terminal, large gape	elongated, flattened into 'duckbill', pointed	villiform, reduced	streamlined body suggestive of active predator; confined to river; not abundant

Table 3. Continued.

Species	Main food	Mouth	Snout	Teeth[a]	Comment
11. *A. australis* (?)	no data	unspecialised	unspecialised	villiform, unspecialised	body unspecialised suggesting a general omnivore; delta, rare
12. *Hemipimelodus macrorhynchus*	insects, fruit leaves	small, subterminal	prominant, extended, sharp	reduced	confined freshwater, more abundant in running water, allochthonous diet
13. *H. crassilabris* (?)	mud, plant matter, prawns	small, subterminal very thick, rubbery lips	downwardly curved	greatly reduced	confined freshwater, lips appear to be a suction device, more abundant in running water
14. *H.* sp.	no data	similar to 13, mouth smaller, lips not as thick	downwardly curved	greatly reduced	may be estuarine form of 13; rare
15. *Nedystoma dayi*	detritus, algae	small, subterminal prominent, overhanging upper lip	slightly downwardly curved	greatly reduced	small, specialised feeder, mainly freshwater; locally abundant
16. *N.* sp.	fruit, detritus	small, subterminal upper lip, prominent slightly overhanging	unspecialised	unspecialised	prominent longitudinal ridges along body; marine entering estuary; not abundant in estuary; seems to be mainly scavenger
17. *Arius* sp.	mud, prawns	small, terminal thin lips	very long, head heavily depressed	granular, reduced	confined to most saline waters of estuary, mainly a mud eater; not abundant
18. *Netuma sagoroides* or *mastersi*	crabs, prawns, detritus	unspecialised	unspecialised	villiform, unspecialised	appears to be a general omnivore, marine, entering estuary
19. *Arius microcephalus*	no data	unspecialised	unspecialised	granular, unspecialised	marine, rare in estuary; appears to be unspecialised

[a] The terms villiform and granular refer to palatine teeth only. Rest of description refers to both palatine and jaw teeth, where palatine teeth are absent, only general description given.

move into more saline waters to breed. The main breeding season, as determined by fish with ripe gonads or carrying young, is October – December, the end of the wet season in this area. However, as with most tropical fishes, some individuals in breeding condition can be found throughout the year. Several species, including those of commercial importance such as *Lates calcarifer* and *Polydactylus sheridani*, migrate out of the area to breed elsewhere.

The mangrove area of the delta has some significance as a nursery area judging by the numbers of juveniles present there (Table 4). It is important to note, however, that the presence of juveniles in this area does not imply that it is the only, or the most important, nursery area for the species.

6. Environmental aspects

The fauna of the lower Purari is adapted to a muddy environment, many species having sensory barbels, prominent snouts and/or thick adipose eyelids (Haines 1977). The only clear water river in the area, the Seribi, has an extremely limited fauna. A comparison with the Kikori River, with which it is confluent, suggests that turbidity, or sediment load, is extremely important in limiting fish distribution. As an example, there are no ariid catfish in the Seribi, but they are abundant in the Kikori.

The Purari fauna is deficient in small species. The small species which are present (rainbowfish, apogonids, ambassids, garfish, gobies and gudgeons) are restricted to small side streams and springs. Even large bodies of still water such as billabongs contain very few small species. The larger river species either migrate to the estuaries or sea to breed or exhibit parental care. This strongly suggests that the river areas are inimical to small fish. The strong, turbulent currents of the main streams, and the high concentrations of

Table 4. Types of fish species recorded from Purari – Kikori mangrove estuary areas.

	Estuarine species	Marine species	River species	Total[a]
Adults	63	59	15	134
Juveniles[b]	35	21	5	58
Total[a]	63	65	16	141

[a] As some species can be classified under several headings, totals given are not the sums of columns.

[b] As the specific identification of juvenile fish is difficult, the numbers of species given are approximate only.

suspended sediment, would probably discourage small fish. Possibly predator pressure would operate in quieter waters. On the other hand, some quite small juvenile fish must be capable of migrating upstream. In the Purari River, *Crenimugil labiosus* are extremely abundant below Hathor Gorge, and *Liza oligolepis* and *Valamugil buchanani* are present in the lower reaches. These mullet presumably migrate to the sea or estuaries to breed, although this has not been observed. Juvenile *C. labiosus* 10 cm in total length, however, have been captured in small side creeks between Wabo and Hathor Gorge. Either this species breeds in the area, which would be very unusual for mullet, or the small juveniles can migrate upstream against the powerful current.

All species examined had 1:1 sex ratios except for two, *Nibea soldado* (?) and *Toxotes chatareus*. These species have even sex ratios in estuaries, but a significantly higher proportion of males in the rivers. *N. soldado* (?) appear to migrate into coastal waters to spawn as none were found in breeding condition in the Purari system. Juveniles are present in coastal lagoons. *T. chatareus* breed in the estuarine areas and probably in the rivers. Although no juveniles were observed in rivers, fish with ripe gonads are sometimes present and they are known to breed in freshwater in Queensland (Queensland Fisheries Service 1978). With both species it appears that there is some sort of population imbalance in the rivers which is not present in the estuarine areas, i.e., more males move up rivers than females, or riverine populations are under some stress which reduces the proportion of females. Chadwick (1976) has discussed the ecological significance of sex ratios in fish populations and has suggested that in some cases unbalanced sex ratios may be the result of stress on a population.

A high proportion of species display some form of parental care including viviparity, particularly in the rivers. The adaptive significance of parental care has been discussed by several authors (e.g. Roberts 1972; Lowe-McConnell 1975). Predator pressure or physico-chemical factors such as low oxygen levels have been suggested as reasons for the evolution of parental care in fishes.

7. Diseases

An ulcer disease epizootic broke out among the fish populations of the Purari Delta in November 1975 and continued until mid-1976. There was an initial fish kill affecting mainly *Toxotes*, *Scatophagus* and eleotrids. Disease outbreaks occurred in all estuarine and some freshwater areas between Merauke (Irian Jaya) and Port Moresby during early 1976. All species of bony fishes were affected. In the early stages about half the fish gill-netted in the Purari Delta and near Kikori exhibited ulcerous lesions. Two months later the percentage infection of the gill-net catch had dropped to around 10, and fish were caught with what appeared to be healed lesions. Local residents had never noticed the disease before.

The most likely causative agent appears to be the bacterium *Aeromonas hydrophila* but this has not been confirmed. The disease has similarities with a recurrent ulcer disease ('Bundaberg disease') among estuarine fishes in Queensland (Australian Fisheries 1974). Outbreaks both in Papua New Guinea and Queensland appear to follow periods of unusually heavy rainfall. In the Purari – Kikori delta no outbreaks occurred in areas of consistently high salinity. It seems likely that some environmental stress weakened the fish, making them susceptible to the disease outbreak.

Since 1976 no further outbreaks have been recorded.

8. Likely effects of Wabo Dam on fish

8.1. The Wabo reservoir

If a dam is built at Wabo the fish fauna of Wabo reservoir should develop initially from those species already present which are capable of breeding in freshwater and adapting to the lacustrine environment. Up to 27 species may become established in the reservoir, although the actual number will probably be less. On the basis of the present river fauna, there will be vacant niches for herbivores feeding on aquatic vegetation, for pelagic or semi-pelagic plankton feeders and for large piscivores.

At present none of the exotic species present in the highlands watershed has become established in the lower Purari. It is unlikely that Hathor Gorge would be a barrier to the downstream movement of fish, as the much larger Victoria Falls in Africa has not prevented upstream species from becoming established in Lake Kariba (Balon 1974a). It is possible that species from the watershed, either from the limited indigenous fauna or, more likely, exotics could become established in Wabo Lake.

While at present the rapids of Hathor Gorge are an insurmountable barrier to many fish moving upstream, after the Wabo Dam is built, the new obstacle would shift the barrier about 40 km downstream from the present one. This would prevent catadromous species from entering the Wabo reservoir, but otherwise not greatly affect fish movements.

Anguilla interioris have been taken in the Purari Delta and eels are reported by village people to be present around Wabo. *A. interioris* are known from the highlands watershed (Munro 1967), but it is not known whether the Purari River is a major migration pathway. This species is probably capable of travelling overland avoiding the most turbulent parts of Hathor Gorge. It is unlikely that the dam would be a barrier to migrating eels as they are known to surmount such obstacles elsewhere, e.g., eels (*Anguilla nebulosa labiata*) were not prevented from entering Lake Kariba by the 160 m high wall of Kariba Dam (Balon 1974b). Any species which is not stopped by Hathor Gorge would be unlikely to be stopped by the dam.

8.2. Downstream effects

The fish in the lower river and freshwater section of the delta may be affected (Liem and Haines 1977). The impact on fish may be through changes in the (i) input of sediment, (ii) salinity regime, (iii) flood regime.

When an equilibrium develops after closure of the dam, Pickup (this volume) estimates that the clay input to the delta will be reduced by 22%, the silt input by 53%, and sand/gravel input by 78%. He also points out the particular importance of the Aure River as a contributor to delta sediment input.

The fish fauna of the delta is adapted to a muddy, turbid environment. In addition, many crustaceans (including prawns and crabs) and molluscs feed on and shelter in the extensive mud flats. Clear water streams in the area, such as the Seribi, have very poor aquatic fauna. Some reduction in fish and invertebrate populations could therefore be expected to result from the reduced sediment input to the delta. This would be due to the mechanical loss of sediment rendering the waters less attractive to the present species, and probable erosion of mud flats resulting from reduced sediment input.

Viner (1979), examining nutrients in the lower Purari system, suggested that after damming the nutrient input to the delta would not change to any great extent.

Below Wabo the fish fauna is an estuarine one, and hence euryhaline. At first sight, therefore, a temporary change in salinity regime during the filling of the lake would seem unlikely to affect the fish population. It is, however, worth noting that the ulcer disease epizootic discussed above did not break out in areas of consistently high salinity.

The effect of the dam would be to level out extremes of high and low flow. Saltwater would not penetrate as far inland during periods of low flow and the freshwater flushing of the delta during floods would be less pronounced. In parts of the delta, the swamp vegetation, especially mangroves, and possibly some aquatic fauna, would be stressed as a result of these changed regimes, and some adjustments of distribution and zonation would be likely. Little effect could be expected on the sago palm distribution, which is known to have a certain salinity tolerance (Petr and Lucero 1979).

After the dam is closed, the delta environment and also the inshore coastal waters will need to be closely monitored. Changes in the distributions or sex ratios of fish populations could be the first signs of major environmental stress.

Acknowledgements

My thanks are due to Dr David Gwyther for his critical comments, especially updating the nomenclature of scientific names.

References

Australian Fisheries, 1974. Bundaberg fish disease causes concern. Aust. Fish. 33(6): 36.

Balon, E.K., 1974a. Fishes from the edge of Victoria Falls Africa: demise of a physical barrier for downstream invasions. Copeia 3: 643 – 660.

Balon, E.K., 1974b. Fish production of a tropical ecosystem. In: E.K. Balon and A.G. Coche (eds.), Lake Kariba: A Man-made Tropical Ecosystem in Central Africa, Part II, pp. 253 – 676. Dr W. Junk, The Hague.

Chadwick, E.M.P., 1976. Ecological fish production in a small Precambrian shield lake. Environ. Biol. Fish 1: 13 – 60.

Dugan, C.C., R.W. Hagood and T.A. Frakes, 1975. Development of spawning and mass larval rearing techniques for brackish – freshwater shrimps of the genus Macrobrachium (Decapoda, Palaemonidae). Florida Mar. Res. Publ. No. 12. Florida Dept. Nat. Resources, St. Petersburg. 28 pp.

Frusher, S., 1980. The inshore prawn resource and its relation to the Purari Delta region. In: D. Gwyther (ed.), Purari River (Wabo) Hydroelectric Scheme Environmental Studies, Vol. 15: Possible effects of the Purari Hydroelectric Scheme on subsistence and commercial crustacean fisheries in the Gulf of Papua, Workshop, 12 Dec. 1979, pp. 11 – 27. Office of Environment and Conservation, Waigani, Papua New Guinea.

Haines, A.K., 1977. Fish and fisheries of the Purari River and Delta. In: T. Petr (ed.), Purari River (Wabo) Hydroelectric Scheme Environmental Studies, Vol. 1: Workshop 6 May 1977, pp. 32 – 36. Office of Environment and Conservation, Waigani, Papua New Guinea.

Haines, A.K., 1979. An ecological survey of fish of the lower Purari River system, Papua New Guinea. In: T. Petr (ed.), Purari River (Wabo) Hydroelectric Scheme Environmental Studies, Vol. 6, pp. 1 – 102. Office of Environment and Conservation, Waigani, Papua New Guinea.

Kailola, P.J. and M.A. Wilson, 1978. The trawl fishes of the Gulf of Papua. P.N.G. Dept. of Primary Industry. Research Bull., No. 20. Port Moresby.

Liem, D.S. and A.K. Haines, 1977. The ecological significance and economic importance of the mangrove and estuarine communities of the Gulf Province, Papua New Guinea. In: T. Petr (ed.), Purari River (Wabo) Hydroelectric Scheme Environmental Studies, Vol. 3, pp. 1 – 35. Office of Environment and Conservation, Waigani, Papua New Guinea.

Lowe-McConnell, R.H., 1975. Fish communities in tropical freshwaters: their distribution, ecology and evolution. Longman, London and New York.

Moriarty, D.J.W., 1977. Quantification of carbon, nitrogen and bacterial biomass in the food of some penaeid prawns. Aust. J. Mar. Freshwat. Res. 28: 113 – 118.

Munro, I.S.R., 1967. The fishes of New Guinea. Govt. Printer, Port Moresby.

Petr, T., 1976. Some chemical features of two Papuan fresh waters (Papua New Guinea). Aust. J. Mar. Freshwat. Res. 27: 467 – 474.

Petr, T., 1978/79. The Purari River hydroelectric development at Wabo and its environmental impact: an assessment of a scheme in the planning stage. Sci. in New Guinea 6: 105 – 116.

Petr, T., 1983. Limnology of the Purari basin. Part 1. The catchment above the delta. This volume, Chapter I, 9.

Petr, T. and J. Lucero, 1979. Sago palm (*Metroxylon sagu*) salinity tolerance in the Purari River delta. In: T. Petr (ed.), Purari River (Wabo) Hydroelectric Scheme Environmental Studies, Vol. 10: Ecology of the Purari River catchment, pp. 101 – 106. Office of Environment and Conservation, Waigani, Papua New Guinea.

Pickup, G., 1977. Computer simulation of the impact of the Wabo Hydroelectric Scheme on the sediment balance of the lower Purari. In: T. Petr (ed.), Purari River (Wabo) Hydroelectric Scheme Environmental Studies, Vol. 2, pp. 1 – 88. Office of Environment and Conservation, Waigani, Papua New Guinea.

Pickup, G., 1983. Sedimentation processes in the Purari River upstream of the delta. This volume, Chapter I, 11.

P.N.G. Fisheries Division, 1976. The Gulf of Papua marine and inland fisheries: an outline of the resources, present utilisation and a guideline for future development. The Division, Port Moresby.

Queensland Fisheries Service, 1978. Report by the Director, Queensland Fisheries Service on the activities of the Queensland Fisheries Service during the period ended June 30, 1977. Govt. Printer, Brisbane.

Rapson, A.M., 1955. Small mesh trawling in Papua. P.N.G. Agric. J. 10: 15 – 25.

Roberts, T.R., 1972. Ecology of fishes in the Amazon and Congo basins. Bull. Mus. Comp. Zool. 143: 117 – 147.

Roberts, T.R., 1978. An ichthyological survey of the Fly River in Papua New Guinea with descriptions of new species. Smithsonian Contributions to Zoology, No. 281. Smithsonian Inst. Press, Washington, D.C.

Snowy Mountains Engineering Corporation – Nippon Koei (SMEC-NK), 1977. Purari River Wabo Power Project feasibility report. 8 volumes.

Thom, B.G. and L.D. Wright, 1983. Geomorphology of the Purari Delta. This volume, Chapter I, 4.

Viner, A.B., 1979. The status and transport of nutrients through the Purari River. In: T. Petr (ed.), Purari River (Wabo) Hydroelectric Scheme Environmental Studies, Vol. 9, pp. 1 – 52. Office of Environment and Conservation, Waigani, Papua New Guinea.

West, G.J. and J. Glucksmann, 1976. Introductions and distribution of exotic fish in Papua New Guinea. P.N.G. Agric. J. 27: 19 – 48.

10. Subsistence and commercial fisheries

A.K. Haines and R.N. Stevens

1. Introduction

There are three levels of fishing in Papua New Guinea – subsistence, village level commercial and high technology commercial or 'industrial' fishing. Fishing throughout the Purari River system is largely confined to the first two categories although the processing plant at Baimuru handles product from the industrial prawn trawl fishery in the Gulf of Papua. The importance of village fishing, both subsistence and commercial, varies throughout P.N.G. although all villagers catch some fish. Fishing and fish have importance in village life in all parts of the Purari and its watershed, although the importance is greatest in the delta area.

From a fisheries point of view the Purari system can be divided into the highlands watershed, and the Purari River and delta below Hathor Gorge. The highlands as a region is distinctive socially and historically from the lower Purari River area.

2. Highlands

2.1. Capture fisheries

Highland capture fisheries are largely for subsistence and small-scale commercial purposes, in the sense that fish are often sold through markets. Traditionally, it appears that the only fish caught were eels (*Anguilla* sp.) by trapping and probably catfish (Ariidae), therapons (Theraponidae) and other species in the larger rivers, by bow and arrow, spearing and poisoning. Eels, in fact, are still highly desired as food and have ceremonial significance in the Southern Highlands Province.

Petr, T. (ed.) The Purari – Tropical environment of a high rainfall river basin
© *1983, Dr W. Junk Publishers, The Hague / Boston / Lancaster*
ISBN-13: 978-94-009-7265-0

During the 1950s and 1960s trout (*Salmo trutta*, *Salmo gairdneri* and *Salvelinus fontinalis*), tilapia (*Sarotherodon mossambicus*) and carp (*Cyprinus carpio*) were introduced into the highlands with the aim of encouraging greater protein consumption among the highland people. Trout and carp now form the basis of highland village fishing, which is mostly conducted with lines and hooks. Opinion varies to the significance of the contribution of fish in the local diet, but unquestionably there is some contribution (Kelleher and Haines, unpublished report, Department of Primary Industry, Port Moresby). Trout are known to breed in some streams, but the populations are boosted by a government fingerling seeding programme, operated from Mendi in the Southern Highlands Province.

2.2. Pond culture

At present pond culture in the highlands is limited to a single private commercial trout farm, a small number of institutional ponds in schools, missions and prisons, and a scattering of village ponds. Concerted attempts during the fifties and sixties to promote village aquaculture as a source of protein production failed for a variety of technical and socioeconomic reasons (Glucksman, Department of Primary Industry, Port Moresby, personal communication). The present institutional and village ponds are essentially remainders from that period. Most are poorly maintained, although there are exceptions. A government station at Aiyura in the Eastern Highlands Province breeds carp fingerlings which are provided to interested parties.

The whole scope of fisheries in the highlands was examined in 1979 (Kelleher and Haines, unpublished report, Department of Primary Industry, Port Moresby). It was suggested that capture fisheries provided more potential for development than pond culture, although there was scope for increased fish production by institutional pond culture. It was concluded that attempts to promote village-level pond culture would continue to be unsuccessful in present cultural circumstances; however, a practical education programme, based in primary schools, could improve the success of this approach over a generation.

3. Purari River and Delta

3.1. Subsistence

The Pawaia people of the Purari River are primarily hunters (Liem 1977) although they do catch some fish by means of fishing lines with hooks and *Derris* poison. Archer fish (*Toxotes chatareus*) are particularly important in the subsistence catch. Warrillow (1978) briefly described Pawaia fishing techniques and the role of fishing in the traditional Pawaia life-style. Lambert (1977) recorded that the Pawaia of the Wabo area did not consume much fish and that their diet generally was quite protein deficient.

Traditionally, fishing in the delta has been carried out with hand-crafted fishing gear. Fishing has been a secondary occupation to the production of sago in the subsistence diet except on Urama Island where villagers have no sago stands and trade crabs and smoked fish for sago.

A 25-day survey of fishing activities in February 1976, which investigated three villages in the Purari – Kikori deltaic system (Haines 1978/79), gives an insight into the traditional fishery in the area (Fig. 1). The villages were chosen so as to reflect different ecological zones of the delta. In the coastal village of Barea, located on a sandy beach just upstream from the river mouth with mangrove swamps extending further inland, most fishing is done either in lagoons along the seashore or in the mangroves. Morowam is within an extensive mangrove and mud-flat belt. Almost the whole village is subject to inundation during spring tides. Coconut husks are used to build up the ground around the houses and food plants are grown on burrowing crustacean mounds, which provide the only dry ground. Ravikoupara is in the predominantly freshwater section of the delta, located almost at its geographical centre. There is nevertheless a strong tidal effect and small clumps of *Sonneratia* mangroves are present. The surrounding swamp forest is rich in wildlife. Many people nominally resident in Ravikoupara live elsewhere most of the time.

The traditional methods used for catching fish are by line, hand-trap, spear or bow and arrow (Fig. 2). The hand-trap is a basket-like device, open at one end, which is used to actively catch fish after stalking. Gill nets have been introduced by Government extension staff.

Crabs are normally caught by hand or in hand-traps and finely woven triangular dip nets and used by women for prawns and small fish. Table 1 presents the relative catch of fish by the different methods. Crabs and prawns are not included.

The greatest catch is produced by gill nets. Most village gill nets are

Fig. 1. Fishing localities.

Fig. 2. An hour's fishing in the late afternoon with bow and arrow will generally produce sufficient fish for the evening meal.

5′′ − 7′′* (12 − 17 cm) mesh, 96 ply or monofilament, designed specifically for barramundi, although smaller mesh nets are used. The nets are not fully utilised, however, because people have no need for the quantity of fish they can catch, and they are frequently unskilled in net repair.

The hand-traps (Figs. 3, 4), used by women in shallow water, are more im-

Table 1. Relative importance of fishing methods: fish catch in 25-day period (excluding crustaceans) (weight in kg).

Village	Gill net		Line		Hand-trap		Spear		Bow and arrow	
	Wt.	No.	Wt.	No.	Wt.	No.	Wt.	No.	Wt.	No.
Ravikoupara	52.1	167	22.6	25	25.7	53	–	–	–	–
Morowam	95.6	149	6.5	7	5.4	59	28.1	31	–	–
Barea	164.6	36	21.7	49	28.2	46	14.2	14	3.3	11

Source: Haines 1978/79.

* In Papua New Guinea net dimensions are given in British measures.

389

Fig. 3. Hoop net used mainly for capturing freshwater prawns. The hoop is cane and the net is made from hand-spun pandanus fibre.

Fig. 4. Prawn trap. Constructed of split cane and made into a cone. The canes are bound by pandanus or nipa fibre.

portant than Table 1 suggests. They are mainly used to catch prawns, crabs and small fish. These catches went largely unrecorded during the survey.

The canoes used during fishing are always paddled. Although a few outboard motors are present in some villages, they were not used for fishing during the survey, probably because of lack of fuel.

The only village which regularly sells or trades fish is Morowam. This village is situated where the soil water is too saline for sago, the staple food, to grow. The Morowam people have only limited sago grounds elsewhere (Davey, Gulf Provincial Officer, personal communication), and therefore have to trade fish and crabs for much of their requirements. The crabs are transported live and the fish cooked and smoked so that it will last for several days. The products are either traded directly with other villages or sold in Baimuru market. During the study period 65.6 kg of crabs and 47.4 kg of fish were sold or traded.

As far as possible, all fish and crabs caught by village people were counted, weighed and measured. The total catch during the survey period is given in Table 2, and the average daily catch in Table 3.

Virtually all crabs were *Scylla serrata* (Fig. 5). Prawn identification is not complete, but *Macrobrachium* spp. predominated in most catches, with penaeids forming a larger proportion in more saline water.

Table 2. Recorded catch of fish and crabs during 25-day period (weight in kg).

Village	Fish		Crabs	
	Wt.	No.	Wt.	No.
Ravikoupara	100.3	243	2.0	21
Morowam	134.9	247	160.8	481
Barea	232.0	156	76.8	339

Source: Haines 1978/79.

Table 3. Average daily catch of fish and crabs during 25-day period (weight in kg) (Fig. 6).

Village	Fish		Crabs		Village population	Av. daily catch per person	
	Wt.	No.	Wt.	No.		Fish	Crabs
Ravikoupara	4.0	9.8	–	–	40	0.10	–
Morowam	5.4	9.9	6.4	19.2	80	0.07	0.08
Barea	9.3	6.3	3.1	13.6	240	0.04	0.01

Source: Haines 1978/79.

Fig. 5. Mud crabs are caught by hand and tied with nipa leaves to prevent escape – or injury to the captor.

As accurate identification of fish was not carried out, the data in Table 4 are therefore grouped into types of fish.

The data show that there is a wide range of difference in fishing activities between villages. During the 25-day survey period the Ravikoupara people averaged 2.7 fishing trips per person. The corresponding figure for Morowam is 0.4 and Barea 0.5, although these figures are probably depressed by poor recording of women's trips for both villages and by preparations for the start of the school year at Barea. Even at Ravikoupara the number of trips per person is low and a generally low level of overall fishing activity is apparent. This corresponds with the findings of Lambert (1977) for the nearby Ihu area

Fig. 6. The night catch from three barramundi nets.

where 75% of individuals had not consumed animal protein during the 24 hours prior to interview, despite the widespread ownership of canoes and nets.

It is interesting that Morowam, with a lower catch per head than Ravi-koupara, traded 37% of its fish and 40% of its crab catch. Most of this would have been exchanged for sago, which seems to be considered a more desirable food.

The catch figures show that the quantity caught is very low in comparison with the quantity available, as demonstrated by experimental fishing by Fisheries Division staff. Although the figures are conservative, and animal protein is also available from wildlife and domestic animals, these findings generally corroborate those of Lambert (1977) for the Ihu area, where the diets of villagers were found to be very protein deficient.

393

Table 4. Fish taken during the 25-day survey period (weight in kg).

| Type | Village | | | | | |
| | Ravikoupara | | Morowam | | Barea | |
	Wt.	No.	Wt.	No.	Wt.	No.
Barramundi (*Lates calcarifer*)	–	–	13.8	4	57.6	11
Threadfin salmon (*Polydactylus sheridani*)	–	–	27.5	22	28.6	12
Beach salmon (*Leptobrama mulleri*)	–	–	0.8	3	2.4	3
Catfish (Ariidae)	39.0	90	52.1	108	18.6	38
Sharks, rays and sawsharks (Carcharinidae, Dasyatidae, Pristidae)	17.3	2	13.0	2	84.4	13
Gobies and gudgeons (Gobiidae, Eleotridae)	17.7	32	–	–	–	–
Freshwater anchovy *Thryssa scratchleyi*	7.5	21	6.6	55	–	–
Nursery fish (*Kurtus gulliveri*)	7.2	76	0.6	1	–	–
Other	11.6	24	20.5	52	40.4	79
Total	100.3	245	134.9	247	232.0	156

Source: Haines 1978/79.

The population of the Baimuru Sub-district, excluding upper Purari, is about 10 000 (unpublished census figures from Provincial Office, Kerema).
Therefore:

the average catch per day = 800 kg of fish and 400 kg of crabs; a n d

the estimated total catch per year = 292 tonnes of fish and 146 tonnes of crabs.

The deltas of the Kikori and Era rivers are interconnected with the Purari Delta, forming a single complex system. The total population of the Kikori – Baimuru area is about 20 000. Assuming similar average catch per person throughout the area, the minimum size of the subsistence fishery in the Purari – Kikori deltas and associated waters must be in the vicinity of 600 tonnes of fish and 300 tonnes of crabs per annum.

394

Fig. 7. Barramundi (*Lates calcarifer*). On the left, buildings of the Baimuru Fish Plant (formerly Gulf Hotel).

Allowing 1000 tonnes per annum of fish and crabs for the 20 000 population of the Kikori – Baimuru area, and comparing this with an estimate from the Sepik River area of 8000 tonnes per annum for a population of 85 000 (Asian Development Bank 1976), it appears that roughly twice as much fish per person is caught in the Sepik. This ratio has probably recently changed due to the invasion of the Sepik system by *Salvinia molesta* which has made many backwater areas inaccessible to fishing (Mitchell, Petr and Viner 1980).

Even at the present level of exploitation, however, the value of the Purari Delta area as a provider of fish to subsistence fishermen is significant. Liem and Haines (1977) calculated its worth as K 1.00* per hectare per year for subsistence fishing.

* In 1977 1 kina (P.N.G. currency) = $U.S. 1.25.

395

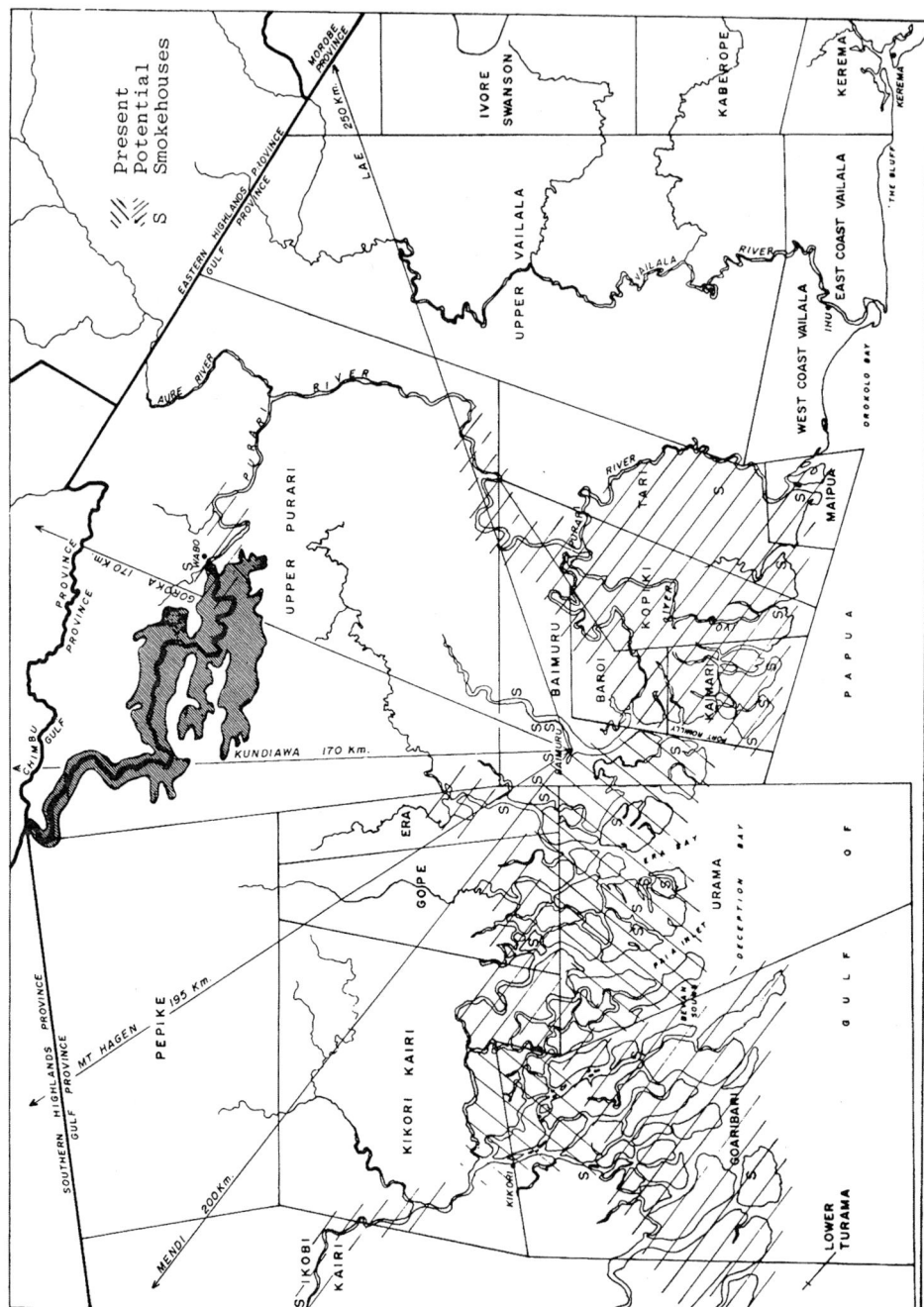

Fig. 8. Major fish-producing areas.

396

3.2. Commercial

The Gulf of Papua deltaic system forms a major nursery ground for prawns. These form the basis of a trawling industry which, however, has been of little direct benefit to the local population. Commercial fishery developments within the delta have been few, entrepreneurial in nature and short-lived. The exception to this was a venture by the Kikori Local Government Council. A 3.5-tonne freezer, which was to be provided with fish from fourteen village-based freezers, was built in Kikori in 1975. The village freezers have all failed, but the main freezer continues to receive fish with production reaching about one tonne of barramundi (*Lates calcarifer*, Fig. 7) fillets per month (Mirou, Gulf Provincial Officer, personal communication).

As a direct result of fish and crab surveys in the Purari Delta (Haines 1979) the Baimuru Local Government Council (LGC) started purchasing mud crabs for live shipment to the New Guinea highlands late in 1977. The Council continued to purchase crabs until September 1978 when recurrent financial losses forced them to cease. The Department of Primary Industry then took over the business and expanded into frozen fish production in February 1979. When the operation returned to profit in March 1979 the Gulf Investment Corporation, together with the Baimuru LGC, resumed control.

Fishing activity is limited to clan boundaries which approximately correspond to the census divisions illustrated in Fig. 8.

The major income earners have been the Urama islanders, who have a tradition of fishing commercially for crabs (Fig. 9), and villagers who can paddle to Baimuru with fresh fish, for sale to the freezer unit.

There is a possibility of major expansion in the crab fishing grounds to include the Kaimari and Koriki areas and late in 1979 the first significant catches from these areas were reported.

Smoked fish is seen as the chief means of introducing commercial fishing to the villagers. The techniques of fish smoking were first introduced to the delta in 1976 as part of a joint Department of Primary Industry – University of Technology project (Wanstall 1977). Fish may be smoked in a copra drier, and so a copra farmer can produce smoked fish with little extra effort.

A *smokpis* (smoked fish) programme was started in mid-1978 with a grant from the New Zealand High Commission. The money was used to purchase the basic requirements for a fishing group to commence making *smokpis* and was intended to cover the whole Gulf Province.

Two major drawbacks have been encountered with the *smokpis* programme. Firstly, the smoked fish to fresh fish ratio is only 1:5. However, the price of smoked fish fillet for example, in Baimuru is only three times that of

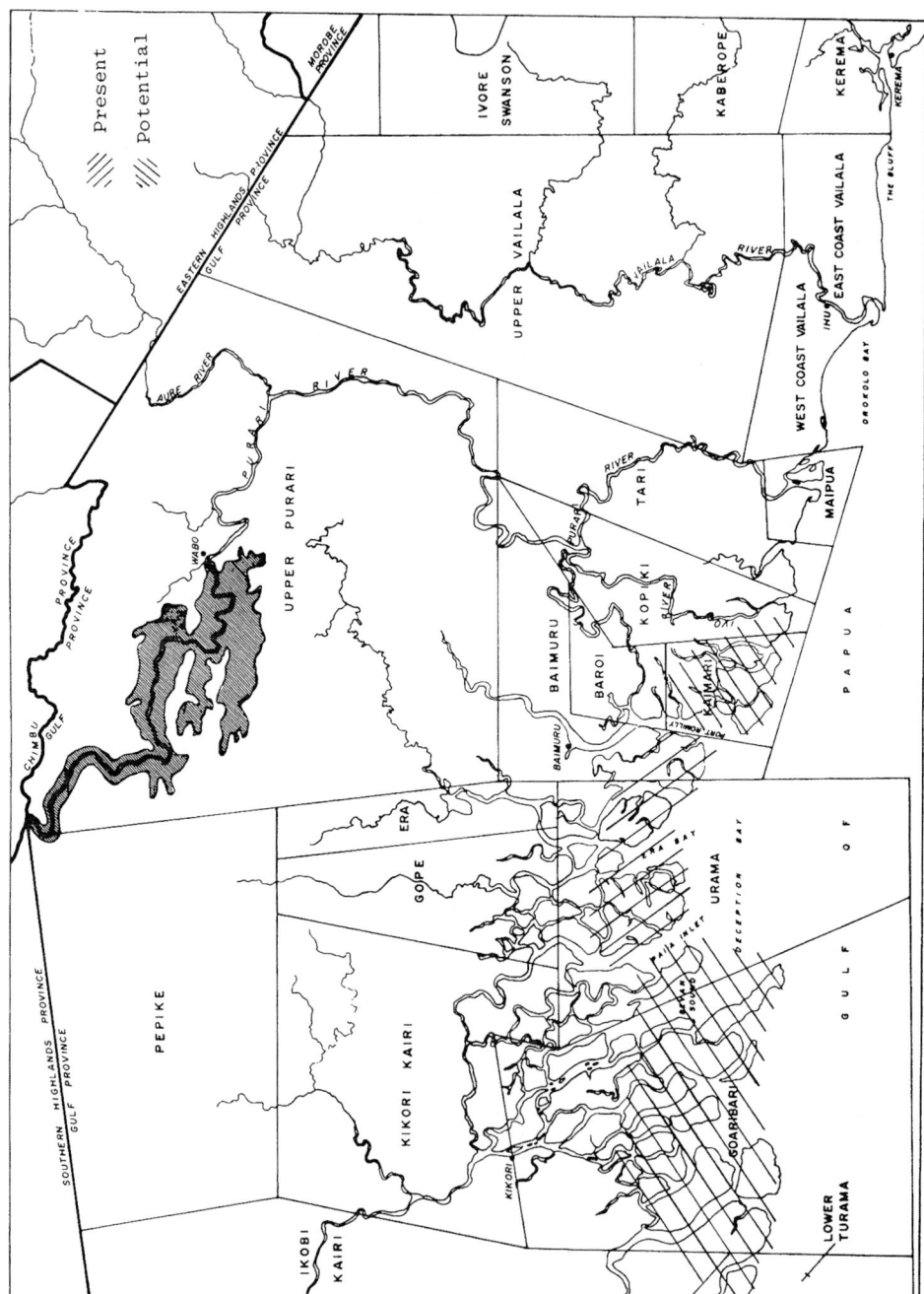

Fig. 9. Main crab-producing areas.

398

fresh fish which does not even compensate for the five-fold reduction in weight. Two hundred kilogrammes of smoked fish (one tonne fresh fish equivalent) is required to purchase a new fishing net at current prices. Secondly, internecine disputes within fishing groups have sometimes led to the gear being left unused or unrepaired for long periods. It was recognised that clan or family groups would have been more successful but it was not possible to grant gear on such a scale.

An alternative to *smokpis*, such as the installation of village freezers, is considered to be less viable in the long term than *smokpis*. This approach has proven impractical in the past because of the lack of mechanical knowledge or concept of freezing techniques among villagers and the high cost of fuel.

Smokpis also holds an important place in the solution of the malnutrition problems of mountain areas. The product is a form of dehydrated protein with an iodised salt content. This makes it an excellent proposition for transport by air to remote villages, where it has been very well received. *Smokpis* may be eaten without reconstitution and is favoured by people who may have to walk for three days back to their village after visiting the nearest government station or trade store.

For the Gulf Delta a combined *smokpis* and freezer barge fishing system is seen as the most efficient method of exploiting the stock.

It has been recognised that the chief constraint of fish production in the Purari Delta is not the lack of fish but the problem of transporting the product to Baimuru. To overcome this problem a freezer barge was chartered for four months in 1978 and stayed for four to five days in each village and during that time lent nets to villages and purchased fish from them. The villagers were thus gradually introduced to commercial fishing, without unduly disrupting their traditional activities. A top quality product was produced as the fish were frozen immediately.

Baimuru is seen as the only place in the Gulf Province where a major fishing facility can be developed. This is chiefly because the Pie River, which has a very small catchment area and a relatively large tidal range has no bar across its mouth. The station also has a 1000 m airstrip and is only one hour flying time from the provincial capitals of the highland provinces where the major market for fish products is envisaged. Baimuru is also located a maximum 12 hours steaming time from the major prawn trawling grounds. Port Moresby, from where the trawlers normally operate, is about twice this steaming time from the prawning grounds.

In 1979 the defunct Gulf Hotel in Baimuru was converted into a fish and crab processing plant (Figs. 7, 10). Tables 5 and 6 give the value of product which passed through the processing plant in its first year of operation, and

Fig. 10. 'Bushbox' freezers with self-contained diesel engine to drive compressor. Baimuru Fish Plant.

Fig. 11. Boiling of mud crabs.

Table 5. Cash flow of Baimuru Fish Plant January – November 1980 (kina)[a].

Month	Purchases	Sales	Gross income (GI)	Purchases as % of sales	Purchases as % of GI	Sales as % of GI
Jan – June	8524.40	14 469.20	22 748.89	58.91	37.47	63.60
July	3908.75	2550.65	5772.04	153.25	67.71	44.19
August	5358.48	6551.99	13 624.88	81.78	39.33	48.09
September	4162.14	5272.96	12 137.55	78.93	34.29	43.44
October	7427.43	10 601.25	19 628.70	70.06	37.84	54.01
November	5188.29	11 134.77	23 257.27	46.60	22.31	47.87
Total	34 569.49	50 580.82	97 169.33	68.35	35.58	52.06

[a] In 1980 1 kina = $U.S. 1.35.

Table 6. Fish purchases from individual villagers by Baimuru coastal fisheries station and subsidiary operations October 1978 – November 1980 (26 months) (kina).[a]

Purchasing agency	1978/9	1979/80	% Increase
Mobile freezers	3838.64	14 287.80	372.21
Fisheries station	5530.73	22 231.18	401.96
Total	9369.37	36 518.98	389.77

[a] In 1980 1 kina = $U.S. 1.35.

the percentage increase in the value of products compared with the previous year. Crabs, which were originally sold to the highlands alive, were experimentally processed to retail pack cooked meat by women workers on a piece rate basis (Figs. 11, 12).

Barramundi is frozen head off and gutted, or as fillet in 10 kg blocks or 1 kg retail packs. Threadfin salmon is frozen whole and then sawn into cutlets. Fish fillets have been sold in 10 kg blocks to 'fast food' outlets in the highlands where they compete very successfully with imported New Zealand fillets. Other species were also manufactured into fish cakes which have proved acceptable in trials by take-away food outlets.

Marketing of new products, however, has not been without difficulties. Air freight from Baimuru is expensive as it normally involves a charter, and as the aircraft flies to Baimuru empty, the whole cost of the charter has to be included when fixing the price of the fish. Regular passenger transport frequently cannot handle freight and the cost of this transport is high.

A classic case of marketing problems occurred with mangrove clam

Fig. 12. Removal of boiled crab meat by hand.

(Fig. 13); production rose so quickly that wholesalers were inundated with clams (sold alive in the shell) which they could not sell. The problem was alleviated somewhat when the clams were shelled and frozen at Baimuru, but clam purchases had to be suspended until wholesalers could exhaust their stocks.

The fish processing plant has so far proved to be successful, in that it has shown that with minimal investment, trainee management and local labour, a variety of good quality products can be produced. It has also operated profitably and the profit retained for future expansion.

A major concern of the P.N.G. National and Gulf Provincial governments is that the vast amount of the by-catch from the Gulf of Papua trawl fishery is

402

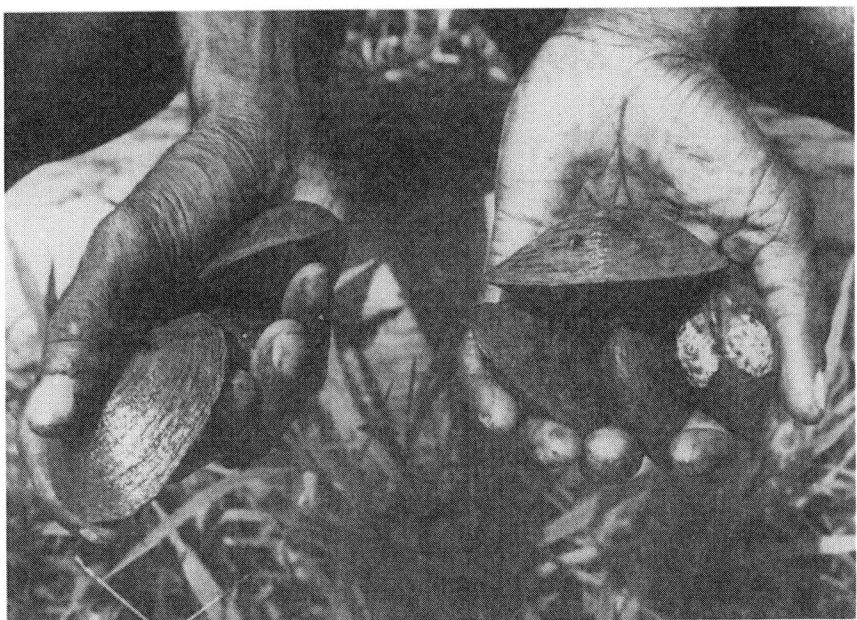

Fig. 13. Mangrove clams.

considered to be unusable and is dumped at sea. This was estimated at 9000 tonnes in 1978 (P.N.G. Fisheries Division 1979).

On 31 October 1980 the 20 m prawn trawler 'Roden Lee' landed at Baimuru its total catch from three days fishing. The trawler had been chartered by the Department of Primary of Industry to make a realistic assessment of the commercial utilisation of the catch, and the viability of making large-scale landing. The catch was processed into attractive packages which were distributed throughout the country. A breakdown of the catch and its commercial value is given in Table 7.

Samples were sent to the Papua New Guinea University of Technology in Lae, and preliminary results showed that the small, whole fish had a protein content of 15% (Ananthan, University of Technology, personal communication). This is considered adequate for livestock feed, and the product could be ground and kiln dried cheaply by burning timber off-cuts from the Baimuru sawmill. Whether such a venture would be technically and economically feasible requires further investigation.

In December 1980 a 120 tonne Taiwanese gill netter landed over five tonnes of shark carcases at the fish plant. These were filleted on a commission basis and the fillet sent to the vessel's agents in Port Moresby. The fish plant there-

403

Table 7. Breakdown of catch by-product and packaging type from F.V. Roden Lee (October 1980).

Product (see key)	Weight (kg)	% of net-landed Wt.	No. of units	Nominal Wt.	Assigned value (Kina)	Gross value (Kina)
Prawn	47.5	1.64	95	0.50	3.50/kg	166.25
Squid	11.9	0.41	18	0.66	2.00/kg	23.80
SST	483.0	16.70	1743	0.28	0.35 ea	610.05
SLT	184.0	6.36	535	0.34	0.55 ea	294.25
US (500 g)	945.5	32.68	1891	0.50	0.70/kg	661.85
US (1 kg)	335.0	11.58	335	1.00	0.70/kg	234.50
Crab	34.0	1.18	42	0.81	0.50/kg	17.00
Eel	66.0	2.28	7	9.45	1.50/kg	99.00
Sole	16.0	0.55	2	8.00	2.00/kg	32.00
Animal feed	550.0	19.01	36	15.28	0.10/kg	55.00
Discarded	220.0	7.61	–	–	–	–
Net- landed Wt.	2892.9	100.0				Ḱ 2193.70
Unaccounted	272.1					
Gross landed Wt.	3165.0					

Key:

Prawn	: Head on, green, mixed
Squid	: Includes cuttle
SST	: Selected fish, small (8'' × 5'') tray
SLT	: Selected fish, large (11'' × 5'') tray
US (500 g)	: Unselected fish, 500 g polythene bag
US (1 kg)	: Unselected fish, 1 kg polythene bag
Crab	: Three spot (*Portunus* sp.)
Eel	: Hairtail (Trichuridae)
Sole	: (Soleidae)
Animal feed	: Very small and damaged fish
Discarded	: Very spiny fish, catfish, sea snakes
Unaccounted	: Original packaging, ice, *Nypa* seeds, detritus

Table 8. Fish purchases at Baimuru Fish Plant during 1981.

Month	Barramundi		Threadfin salmon, Jewfish and mixed fish	Shark	Catfish	Crab	Total
	Fillet	Whole					
February	1032	350	1824	604	65	975	4850
March	1749.5	25	807	355.5	75	1611	4623
April	905.7	42.5	476	195	84.5	459.5	2163.2
May		Breakdown					
June		Breakdown					
July		591	505	338	46		1480
August		1172	1408	853	328		3761
September		1404	1968	705	478	592	5147
October		1962.5	1456.5	537	670	1078.5	5704.5
November		1760	818.5	694	617		3889.5
December		728.5	455.5	373	197		1754
							33 372.2

Source : J. Linton, Department of Primary Industry, Baimuru.
Note : All weights in kilogrammes; results for 1980, and January 1981 are not available; value paid to fishermen for products: K24 000.00

fore has the potential to process catch from the Gulf of Papua industrial fisheries.

The following information concerning the operation of the fish plant in 1981 has been kindly supplied by J. Linton, Coastal Fisheries Station Manager, Baimuru Fish Plant, to whom the authors are grateful (Table 8).

In January 1981 the Department of Primary Industry agreed to provide management for the Baimuru Fish Plant. During 1981 the following changes were made: (a) introduction of ice to fishermen; (b) cessation of fillet purchases; (c) increased funding for purchase of capital assets; (d) introduction of a single double entry accounting system.

As 90% of fish purchases came from an area within 15 km of the plant, it was decided to introduce ice and ice boxes in order to expand the area around Baimuru from which villagers could bring fish. This development is showing early signs of success.

In order to improve the quality control of plant output, the plant manager ceased purchase of fish fillets, which were generally of low quality.

3.3. Commercial fisheries development in the Purari Delta — reasons for successes and failures

The ownership of fishing rights or waterways by individuals, groups or villages is a deeply-held traditional concept (Haines 1982). Groups and individuals lay claim to ownership of waterways and the fish in them, sometimes to the exclusion of all outsiders and sometimes outsiders are tolerated under certain conditions.

The effects of this universally held and jealously guarded right are that each village-based fishery in the country effectively operates under a limited entry system, with further geographical restrictions within that system. This means that a fishery is in reality a series of small fisheries or subunits subject to different pressures throughout its range. One group of owners might be subjecting their subunit to heavy pressure, while another group may hardly utilize theirs at all.

Over most of the colonial period and up to the present there have been attempts to upgrade village subsistence fisheries into small-scale commercial enterprises. The attempts have generally aimed at the production of frozen fish, with the establishment of base freezers centrally located at government stations, missions or the like, sometimes with the addition of numerous small units dispersed among villages. This is what can be termed the 'freezer system' of fisheries development.

The problems with this system have been documented (see P.N.G. Fisheries Division 1978). The system largely proved unsuccessful for technical and sociological reasons; the technical ones being lack of maintenance of freezers, poor quality control over product, and expense both of fuel and its transportation. The sociological reasons are two-fold − the sporadic production from village fisheries means that periods of low production are very uneconomical, and the geographical division of fishing rights arising from the traditional concept of private ownership of these rights means that only a comparatively small number of fishermen can produce for a freezer. When their immediate monetary requirements are satisfied, production stops.

To some extent these problems can be overcome by making the freezer mobile, so that when fishermen in one area are no longer producing, the freezer plant can move to another. If the system is large enough it can afford managerial and technical expertise. This system operated with some success in the Purari Delta with the freezer barge. Formerly, expatriate operated barramundi boats successfully worked the Western Province in the same way.

Basically the problem is that large central freezers can never be supplied with sufficient input and are technically complex. Dispersed village freezers have been attempted but were uneconomic for the same reasons. Both types have been effectively rejected by villagers because there were too many new ideas cutting across too many traditions. Mobile freezers have proved more successful because they do not require infringement on local traditions for success.

The development of the mud crab fishery, and the reasons behind it, are outlined in Fisheries Research Annual Report 1978 (P.N.G. Fisheries Division 1979). The subsistence fishing survey (Haines 1978/79) showed that crabs were regularly caught in large numbers, and that these were traditionally traded in some villages. Crab catching was an activity of women who are traditionally the more active producers, and villagers were accustomed to storing and handling live crabs.

The ready acceptance of the mud crab fishery came about because of the full use of the traditional skills, to which were added several new ideas, i.e., regular production in line with plane schedules and quality control by the rejection of undersized and damaged crabs. Size limits were imposed for commercial reasons, but represent a possible conservation measure as well.

Production is decentralised, so each group can produce according to their own needs and resources. As crabs can be kept alive for long periods, distance from point of sale is no obstacle.

The crab project, once established, then became a pilot project of sufficient significance to warrant the appointment of expatriate technical expertise. The availability of technical expertise, in turn, directed village extension work along the lines described above and ensured the development of the fish processing plant. The processing plant is now enabling experimentation and the possible development of other resources, as well as accustoming more villagers to the idea of 'regular' commercial fishing.

References

Asian Development Bank, 1976. Appraisal of the East Sepik Rural Development Project in Papua New Guinea, Vol. 2: Detailed report on sub-projects. Report No. P.N.G. AP-4. Manila.

Haines, A.K., 1978/79. The subsistence fishery of the Purari delta. Science in New Guinea 6: 80–95.

Haines, A.K., 1979. An ecological survey of fish of the lower Purari river system, Papua New Guinea. In: T. Petr (ed.), Purari River (Wabo) Hydroelectric Scheme Environmental Studies, Vol. 6: pp. 1–102. Office of Environment and Conservation, Waigani, Papua New Guinea.

Haines, A.K., 1982. Traditional concepts and practices and inland fisheries management. In: L. Morauta, J. Pernetta and W. Heaney (eds.), Traditional Conservation in Papua New Guinea: Implications for Today, pp. 279–291. Institute of Applied Social and Economic Research, Boroko, Papua New Guinea.

Lambert, J.N., 1977. Purari nutrition survey. In: T. Petr (ed.), Purari River (Wabo) Hydroelectric Scheme Environmental Studies, Vol. 1: Workshop 6 May 1977, pp. 56–62. Office of Environment and Conservation, Waigani, Papua New Guinea.

Liem, D.S., 1977. Wildlife and wildlife habitat in the area to be affected by the Purari Scheme. In: T. Petr (ed.), Purari River (Wabo) Hydroelectric Scheme Environmental Studies, Vol. 1: Workshop 6 May 1977, pp. 43–45. Office of Environment and Conservation, Waigani, Papua New Guinea.

Liem, D.S. and A.K. Haines, 1977. The ecological significance and economic importance of the mangrove and estuarine communities of the Gulf Province, Papua New Guinea. In: T. Petr (ed.), Purari River (Wabo) Hydroelectric Scheme Environmental Studies, Vol. 3, pp. 1–35. Office of Environment and Conservation, Waigani, Papua New Guinea.

Mitchell, D.S., T. Petr and A.B. Viner, 1980. The water-fern *Salvinia molesta* in the Sepik River, Papua New Guinea. Environmental Conservation 7: 115–122.

P.N.G. Fisheries Division, 1978. Fisheries Research Annual Report 1976. Port Moresby, Papua New Guinea.

P.N.G. Fisheries Division, 1979. Fisheries Research Annual Report 1978. Port Moresby, P.N.G.

Stevens, R.N., 1980. The agricultural and fishing development in the Purari delta in 1978–1979. In: T. Petr (ed.), Purari River (Wabo) Hydroelectric Scheme Environmental Studies, Vol. 13, pp. 1–16. Office of Environment and Conservation, Waigani, Papua New Guinea.

Wanstall, R., 1977. Preservation of fish by salting and smoking in the delta and processing of sago. In: T. Petr (ed.), Purari River (Wabo) Hydroelectric Scheme Environmental Studies, Vol. 1: Workshop 6 May 1977, pp. 41–42. Office of Environment and Conservation, Waigani, Papua New Guinea.

Warrillow, C., 1978. The Pawaia of the Upper Purari Gulf Province, Papua New Guinea. In: T. Petr (ed.), Purari River (Wabo) Hydroelectric Scheme Environmental Studies, Vol. 4, pp. 1–88. Office of Environment and Conservation, Waigani, Papua New Guinea.

11. The status and ecology of crocodiles in the Purari

J.C. Pernetta and S. Burgin

1. Introduction

In many areas, including the Purari River basin, the extensive hunting of
crocodiles during the last three decades has drastically reduced their wild
populations. The Government of Papua New Guinea has been aware of the
pressures on wild crocodile populations and in the last five years, with the help
of the United Nations has actively encouraged and supported the establishment
of village crocodile farms. The major purpose of these farms is to produce skins
for export, and in this way to reduce the hunting pressure on the remaining
populations of crocodiles.

This chapter reviews the current status and ecology of the saltwater and
freshwater crocodiles in Papua New Guinea, and discusses possible impacts of
the proposed Purari basin development on crocodile populations of the middle
and lower Purari including the delta.

2. The biology of crocodiles in Papua New Guinea

Two species of crocodile, the estuarine or saltwater crocodile, *Crocodilus
porosus* (Schneider) and the freshwater crocodile, *Crocodilus novaeguineae*
(Schmidt) occur in Papua New Guinea. It is quite difficult to distinguish the two
species in the wild since they differ only in belly scale counts and in the arrangement of scales on the nape of the neck.

Although *C. porosus* was once thought to be restricted to coastal, tidal areas
(Neill 1971), inland populations are now known in Papua New Guinea (Lever
and Mitchell 1975) and in Australia (Webb et al. 1977; Webb and Messel 1978);
not only in association with deep, freshwater pools and rivers but also in swift
flowing rocky streams, 600 – 700 miles inland (Whitaker 1980). Typically,

Petr, T. (ed.) The Purari – Tropical environment of a high rainfall river basin
© *1983, Dr W. Junk Publishers, The Hague / Boston / Lancaster*
ISBN-13: 978-94-009-7265-0

however, this species is more abundant in brackish, coastal areas, including the tidal reaches of rivers and mangrove swamps (Bolton and Laufa 1982; Lever and Mitchell 1975). In contrast, the freshwater species is rarely encountered in estuarine areas and is most abundant in inland river systems, freshwater swamps and marshes. Thus the two species are ecologically separated, although this separation is incomplete and a wide zone of overlap in distribution occurs in the inland coastal zone.

It would appear that differently sized individuals of *C. novaeguineae* occupy different microhabitats, with the small animals being more common in grassed areas (Ross 1977) off the main rivers, while larger individuals occupy the main river channels. Lever and Mitchell (1975) suggest that juveniles of the saltwater species are found throughout the adult range, with the majority occurring nearer the coast. In this species individuals of all sizes have been observed to feed successfully in salinities ranging from $0 - 35‰$, the upper limit being around $54‰$ (Taylor 1977; Magnusson 1978). No data is available concerning the salinity tolerances of *C. novaeguineae*.

The meagre ethological evidence available suggests that *C. porosus* has complex social interactions (Magnusson 1978). Neill (1946) concluded that larger members of the freshwater species were highly territorial since he saw no interactions over an eighteen-month period of observation. Lang (1980) recently conducted behavioural studies of captive animals at the Moitaka Crocodile Farm near Port Moresby, which indicate that under farm conditions both male *C. novaeguineae* and male and female *C. porosus* maintain territories throughout the year. The intensity of intraspecific aggression in both species increases following the first rains of any one year. Position in the territorially-based dominance hierarchy is related to size.

Breeding in both species commences with mating at the start of the wet season (October to April in the Papuan region) followed by nesting and egg laying in November and December, although this may be extended and depends upon the timing of the wet season (Hill 1979; Lang 1980; Neill 1971; Webb et al. 1977). This pattern differs from the behaviour of *C. porosus* in Sri Lanka where animals breed during the hottest and driest time of the year, hatching being timed to coincide with the start of the wet season (Deraniyagala 1939; Honneger 1975). Courtship and mating have not been observed for either species in the wild, although Lang (1980) reports that under captive conditions courtship sequences were variable and consisted of 'snout lifting, circling, head contacts, bubbling and periodic submergence by the female and the male'.

The nests of *C. novaeguineae* are always found in shade between 6 and 9 metres away from adjacent lagoons or sluggish channels. Nests are roughly hemispherical about 0.6 m high and constructed of soil, litter, or available plant debris and lined with relatively soft material (Lang 1980; Neill 1971). In *C. porosus*, nests are similar in construction and are usually placed just above the high water mark (Neill 1971; Lever and Mitchell 1975; Magnusson 1978). Magnusson (1978) suggested that nesting sites in Australia may be chosen through the avoidance of areas of unsuitable plant species. Such an avoidance of certain plant associations has not been demonstrated in Papua New Guinea.

Eggs of both species are laid in a central chamber (De Vos 1979; Medem 1976; Youngprapakorn, undated) and the temperature is maintained at a constant level through the decomposition of the organic materials which produce heat, and through heat losses by evaporation and convection. Magnusson (1978) records that a drop in environmental temperature of 3 °C caused a change of 1 °C in the egg chamber of the nest. Temperatures of wild *C. porosus* nests have been recorded between 27.5° and 30°C (Braizatis 1973; Deraniyagala 1939; Lever and Mitchell 1975; Medem 1976; Webb et al. 1977). Data on clutch size, incubation time and egg dimensions of the two species are given in Table 1 (data from Hill 1979; Greer 1975).

Nests of both species are guarded by the female and excavated when hatching begins. Both male and female *C. novaeguineae* have been observed to transport eggs/hatchlings by mouth. The female and dominant male in farm pens respond aggressively in defense of young hatchlings, excluding other adult crocodiles from the water. Such a response is also stimulated by the vocalisations of hatchlings when they are handled (Lang 1980).

Sexual maturity in *C. porosus* occurs between 10 and 15 years, thus maturity may occur over a range of body sizes (Lever and Mitchell 1975; Youngprapakorn et al. 1971), although Webb et al. (1978) estimated that in Australia females reach maturity at around 1.1 m snout vent length (approximately 2.2 m total length), males at 1.6 m (approximately 3.2 m total length). In Papua New Guinea the smallest breeding female recorded has a total length of 2.2 m and the smallest male, a total length of 3.0 m (Whitaker 1980). The size of age of maturation of *C. novaeguineae* is unknown although a female of approximately 1.5 m total length and a male of approximately 2.4 m total length have been reported to have bred (Neill 1946, 1971; Whitaker 1980).

Table 1. Egg numbers, dimensions and incubation times for New Guinea crocodile species.

	C. porosus	C. novaeguineae
Weight (gm)	49.6 – 144.8	49.8 – 150.5
Length (mm)	57.7 – 94.9	60.0 – 116.0
Diameter (mm)	37.7 – 59.3	37.5 – 52.3
Incubation time (days)	77	77 – 87
Clutch size	20 – 150	10 – 48

411

Movements and dispersion of hatchling and subadult animals of both species are unknown in Papua New Guinea. In Australia Webb and Messel (1978) undertook a mark – release – recapture programme on 1, 2 and 4-year old *C. porosus*, which were recaught one year later. 93%, 73% and 57% of each age class respectively, were recaptured within 10 km of the point of release. A 'homing instinct' in animals that were experimentally relocated after capture was also demonstrated. A fraction of the population, which increases with age, were observed to be 'long-distance movers'. Messel et al. (1979a, 1979b, 1979c) suggested that such animals were responding to increasing saline penetration of riverine environments during the dry season.

Messel et al. (1980) report that for *C. porosus* in Australia, individuals of 0.9 – 1.0 m total length are excluded from breeding areas by the breeding adults which tolerate the presence of individuals of smaller size than this. These subadults move further seaward, returning to the breeding areas only after the onset of maturity. During this period between 60 and 70% of the individuals are lost, either through mortality or through dispersal to other areas.

Both species of crocodiles are reported as being opportunistic feeders (Whitaker 1980), although records of the natural diet in Papua New Guinea are scant. The stomach contents of hatchling *C. novaeguineae* examined by Ross (1977) contained mosquitos (*sic*), grasshoppers and waterbugs but no vertebrates. Neill (1946, 1971) reported that 'small' individuals of this species feed mainly on waterfowl, particularly rails and occasionally other species such as wagtails and the water lizard *Lophognathus temporalis*. One individual killed in the Laloki near Port Moresby contained a water snake (*Amphiesma mairii*), a rail and grasshoppers (Callis, personal communication). Neill (1946) also reported twice seeing a crocodile, presumed to be of this species, guarding a mound of dried grass which contained a rotting young pig. Ross (1977) found no food remnants in the stomachs of larger animals examined, although adults are reputed to eat insects, fish, frogs, turtles and ducks.

In contrast the data on the diet of *C. porosus* in Australia is more extensive. Taylor (1977, 1979) working on the stomach contents of subadults found that they fed predominantly on crabs of the subfamily Sesarminae and shrimps of the genus *Macrobrachium*. Crustaceans predominate in the diet during both the wet and dry seasons. In the upper reaches of rivers the diet of juveniles contained proportionally more insects than in the lower reaches. Other workers also report crustaceans as being common dietary items, but small fish, spiders, frogs and other vertebrates are also listed (Allen 1974; Johnson 1973; Taylor 1977, 1979; Webb et al. 1978; Webb and Messel 1978). The diet of individuals between 1.5 and 2.0 m also contains relatively few vertebrates although adults take proportionally more fish, reptiles and birds (Brazaitis

412

1973; Caras 1975; Cogger 1960; Deraniyagala 1939; Ditmars 1944; Lever and Mitchell 1975; Neill 1971; Pope 1956; UNDP/FAO 1978; Youngprapakorn, undated).

Mortality in populations of both species appears to be heaviest during the egg stage, with flooding of nests a major factor causing failure to hatch. Submergence of eggs for 8 hours still permits development, but submergence for 13 hours kills all eggs (Magnusson 1978). Webb et al. (1977) report that of thirty *C. porosus* nests visited during one season in northern Australia, 80% were flooded and the embryos killed. The effects of this mortality are felt mainly in riverside habitats and are less important in swamp areas. Varanids may predate nests; they certainly eat the eggs of *Carettachelys insculpta* in the Purari area (Pernetta and Burgin 1980), but more probably as Magnusson (1978) observed, they are unsuccessful in attacking guarded nests and are usually found eating eggs from flooded and deserted nests. Predators of hatched individuals include varanid lizards, accipiterids and sharks (Ross 1977; Webb et al. 1978). Mortality is heaviest during the young stages with adult animals suffering little mortality except for that caused by man (Ross 1977; Schultze-Westrum 1970)

3. The impact of man

The impact of man on wild populations of crocodiles in New Guinea was probably not severe in pre-contact times. Although crocodiles were hunted along the Papuan coast over the last two thousand years (Allen 1977), they have been protected as totemic animals elsewhere. The overall impact of hunting traditionally was probably small, and as Hope (1977) remarks, there is no evidence of any significant impact by man on crocodiles prior to European contact. During the 1950s a rapidly increasing demand on the world market for crocodile skins resulted in widespread, uncontrolled hunting. By 1965 this had greatly depleted the numbers of animals throughout all navigable waterways within the country (Behler 1976; Downes 1969; Lever 1974; Puffett 1972; Schultze-Westrum 1970; Tago 1977).

This decline is reflected in the export revenues derived from skins. In the year of peak production (1965–1966) skins to the value of 1 000 000 Australian dollars were exported (Cogger 1972); by 1970–1971 the value had dropped to A$ 254 179 (Puffett 1972) despite a marked increase in the skin prices during this period (Cogger 1972). In 1966 a private members bill 'The crocodile protection ordinance, 1966' was enacted although regulations under the act were not introduced until 1969 (Abal 1969). Amendments have occurred during the interim as refinements of policy, including the 'Crocodile trade (Protection) bill' of 1974. Broadly, national policy objectives seek to develop the crocodile skin industry as a locally-controlled and village-based industry. The farming policy seeks to protect wild populations from further exploitation, 'with rearing farms as a tool to this end' (Downes 1974). Downes considered that a network of farms with a total capacity of between 50 000 and 100 000 individuals was essential to achieve this objective. Downes later predicted (Downes 1978) that in 3 – 4 years this network would have an annual production of 30 000 skins from farm-raised animals.

To assist in the implementation of this programme the government sought technical aid from the United Nations, and the resultant UNDP/FAO project 'Assistance to the crocodile industry' came into existence in 1977 (Bolton and Laufa 1979, 1982; Downes 1978). At the time the project became operational there were already approximately 124 village-based farms in the Sepik, Gulf and Western Provinces as well as three government farms. By December 1979 the total number of farms had subsided from a peak earlier in the year of 200 to around 150. Unfortunately the prognosis for the farming scheme is not good; the average number of crocodiles held in village-based farms from July 1977 to December 1979 shows a highly significant decline from around 50 to 32 per farm. This factor combined with the decline in total numbers of crocodiles held in farms of all types; from over 16 800 in June 1979 to around 15 000 in December 1980 suggests that the resource continues to be over-exploited in this country (Burgin 1980). Latest available figures show no improvement and the farming project may well have increased the pressure on the resource rather than acting as a brake on cropping wild animals.

4. Utilisation of the crocodile resource in the Purari

The effects of this resource decline are marked in the Purari area. The failure of many village-based farms in this area is blamed upon the paucity of crocodiles, at least in part. Table 2 details the stock held on village-based farms in the Purari area; the total numbers of animals have remained fairly constant following the initial impetus provided by the UNDP/FAO crocodile project. 468 individuals of the 489 held in farms in the area were examined during the course of the present study and the species determined; of these 327 were *C. porosus* and 141 were *C. novaeguineae*. This reflects the greater skin value of *C. porosus* rather than the relative abundance of the two species in the area. Indeed around Wabo and the proposed dam site, as far downstream as the Varoi River junction all individuals are probably *C. novaeguineae* since

414

no skins of the saltwater species are traded out of this area. The mean number of crocodiles held in village-based farms in the delta area is 17 compared with the national average of over 30. Three business groups in the area hold stock of 30, 64, 215 individuals.

In the case of five of the individually owned farms, stock added during the 6 months up to December 1979 were reliably ascertained. 38 animals were added representing 32% of the total stock; comparative figures for the whole area may be as low as 6% (Bolson, personal communication). The present authors prefer to accept the higher figure since stock losses are high, resulting from farm deaths, usually from inadequate feeding, and stock escapes in the delta resulting from pen flooding. All the farm stock is replenished directly from the wild by capture of young individuals.

In addition to the replenishment of farm stock, wild crocodiles are hunted for skins throughout the area. During the present study, hunter interviews were conducted in 8 villages. The results of this survey are presented in Table 3. Whilst it is probably true that all people in the area will catch or kill a crocodile if the opportunity presents itself during the course of their other subsistence activities, few individuals would describe themselves as crocodile 'hunters' in the real sense. The mean number of hunters per village visited (8)

Table 2. Record of stock held at different villages within the Purari area from 1977 to 1979. Figures in parentheses give gross stock changes between the 6 monthly stock takes.

	June '77	Dec. '77	Jan. '78	Dec. '78	June '79	Dec. '79
Ara'ava	0	0	0 (+58)	58 (−30)	28 (+2)	30
Auma	25 (+20)	45 (+1)	46 (−41)	5 (+5)	10 (+8)	18
Apiope	0 (+80)	80 (−20)	60 (−4)	56 (−1)	55 (+22)	77
Bekoro	0	0 (+31)	31 (−12)	19 (+9)	28 (−2)	26
Kararua	0	0	0	0 (+5)	5 (+22)	27
Kinipo	20 (+20)	40 (−4)	36 (−10)	26 (−11)	15 (+20)	35
Korovaki	0	0	0	0 (+16)	16 (−16)	0[a]
Maipenaru	4 (+6)	10 (+2)	12 (+16)	28 (+19)	47 (+3)	50
Mirimarua	0	0 (+12)	12 (+11)	23 (−8)	15 (+19−34)	0[a]
Old Iari	70 (+145)	215 (+5)	220 (+28)	248 (+12)	260 (−34)	226
Wabo	0	0	0	0 (+16)	16 (+12−28)	0[a]
Total	119 (+271)	390 (+27)	417 (+46)	463 (+32)	495 (−6)	489

[a] Farms which closed between June and December 1979. Korovaki stock was sold to the business group at Old Iari; the stock at Mirimarua escaped following high tides which flooded the pens and caused their collapse; the stock at Wabo died.

Table 3. Hunter exploitation of wild crocodiles in the Purari area.

| Village | Number of hunters | Hunting method[a] | Crocodiles caught | | | Skins | Live | %Salt | %Fresh | Habitats[b] |
			night	wet season	dry season					
Apiope	7 (6)	S/H/G	up to 4	14–20	14–20	-	all	90	10	M/T
Old Iari	6 (5)	S/H	1–2	6	0	4'+	< 4'	50	50	M/T/G
Kinipo	4 (1)	S/H	2–3	16	13	-	all	0	100	M
Mirimarua	1 (1)	S/H	6	7	8	-	all	100	0	M/T
Maipenaru	2 (1)	S/H/H'	2–3	5–6	9–20	-	all	100	0	M/T
Wabo	7 (2)	S/H/N	3–4	few	90	all	-	10[c]	90	M/T
Kararua	? (1)	G/H	10–15	few	180	10''+	< 10''	0	100	M/T
Korovaki	3 (1)	S/H/H'	up to 20	-	400	4'+	< 4'	50	50	M/T

a S = spearing; H = by hand; H' = hand and forked stick; G = gun; N = net.

b M = main channels; T = side channels and tributaries; G = grass swamp.
   Numbers in parentheses below the column indicating numbers of hunters in each village are the numbers of hunters present during the interviews.

c On the basis of skin returns C. porosus probably does not occur in the area. In addition this was the only village where the people distinguished the species purely on the basis of colour and took no account of the neck scale differences which are used by hunters in coastal areas.

was 4, of whom an average of 2.5 were present during the interview. Ignoring the anomalous data from Korovaki and Kararua, the mean number of crocodiles caught per hunter per night is approximately 2.96 and per hunter per annum, 34. In reality these estimates may be high since hunters frequently work in pairs, although non-hunters may accompany a hunter to handle the canoe. Assuming that these figures have some validity approximately 1300 crocodiles are caught annually in this area.

When interviewed, the main skin buyer in the area, Evara Trading in Baimuru, claimed to purchase between 20 and 30 skins per month, the bulk coming from Wabo and Poroi on the Purari River. All sizes between 5 and 20 inches belly width (the legally permitted limits) are traded, but approximately 80% were between 5 and 12 inches. Of the skins purchased, 80% are from *C. porosus* and only 20% from *C. novaeguineae*. The only other licensed skin buyer in the area is based at Old Iari.

During 1978 and 1979 the Wildlife Department purchased 95 skins from these two sources, of which 77 came from Old Iari and 18 from Baimuru. Since Wildlife only purchases skins between 8 and 20 inches belly width (in an attempt to encourage the rearing of smaller animals), the bulk of the skins are traded through private buyers and are between 5 and 8 inches belly width. Data was obtained from the two licensed buyers and exporters in Port Moresby who handle crocodile skins. One company purchased 140 skins from Old Iari and 2 from Baimuru in 1978; 104 and 10 respectively during 1979. The second company's figures were 0 and 7 in 1978; 0 and 177 in 1979 for the two sources respectively.

5. Field survey results

The field survey was completed during three visits to the area in September, November and December 1979. Survey techniques for monitoring crocodile populations in Papua New Guinea have not yet been standardised. Techniques employed during this study were modified from those of Messel et al. (1978) and those described by Woodward and Marion (undated) and are fully described in Pernetta and Burgin (1980).

During daytime surveys the riverside vegetation was recorded, crocodile sign, both tracks and slides, were examined and an estimate of size made. Following each sighting, the animal's size, the time, air and water temperatures and the water velocity were recorded. Since all measurements were taken close to the bank of the river they cannot be taken as representative of the temperature and current values for the whole river at that point:

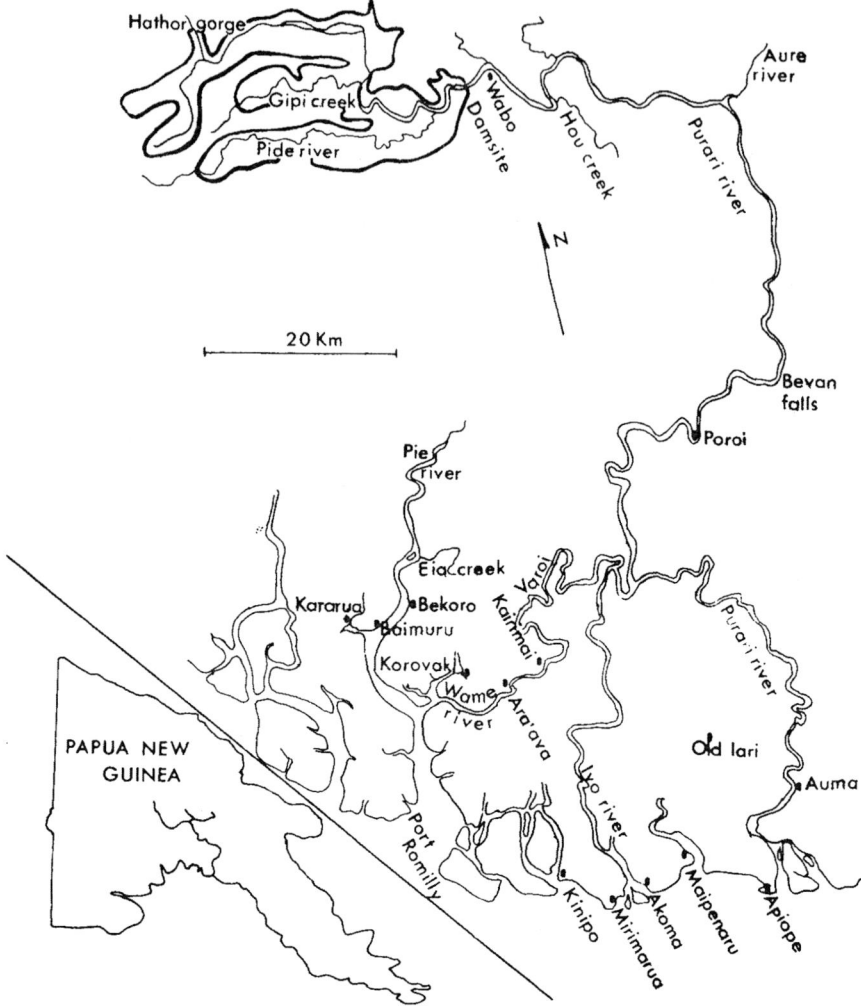

Fig. 1. Location of the study area, proposed dam and lake, villages and waterways named in the text.

velocities are likely to be below and water temperatures above the mean value at each point along the river.

Crocodiles were found throughout the area studied (Fig. 1) in all the different riverside habitats present. Of the 101 sightings of animals, 68 were encountered in obvious depositional areas of low slope (less than 25°), 11 in association with erosional banks of 75° plus and 22 close to banks with intermediate slope. With one exception all animals were observed singly.

418

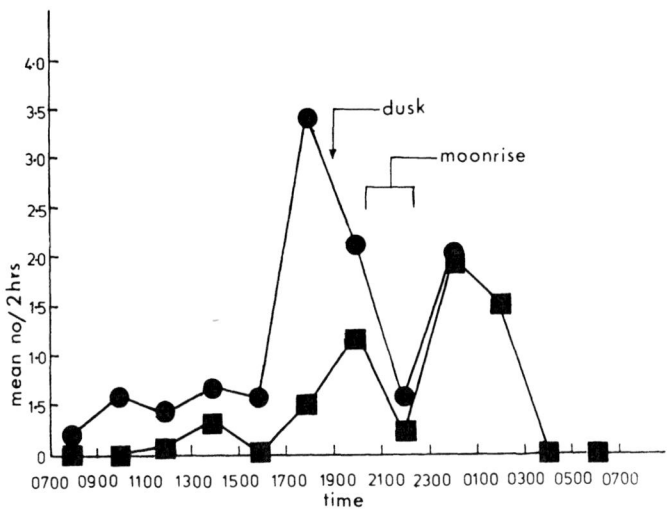

Fig. 2. Mean number of crocodiles seen per 2 hours.
■ delta area; ● inland area.

The mean number of animals seen per hour rises after sunset in both the delta and the inland areas (Fig. 2). The peak prior to dusk in the case of the inland survey is probably an anomalous result due to the low number of hours on which this point is based. The low value in both areas around 2200 hours corresponds to the period of moonrise during the two survey periods. It is interesting to note the close agreement between the proportion of small and medium-sized animals found in both survey areas (Table 4).

The relationship between crocodiles location in and out of water and the air

Table 4. Size categories of crocodiles observed during the survey (figures in parentheses are percentages).

Area	Small	Medium	Large	Hatchling	Total
Coast & Delta	15 (51)	12 (41)	1 (3)	1 (3)	29
Inland river & creek systems	38 (53)	24 (33)	10 (14)	0 (0)	72
Total	55	36	11	1	101

Small   = hatchling to 3 feet total length.
Medium = 3 – 6 feet total length.
Large   = greater than 6 feet total length.

and water temperatures is displayed in Table 5. When water temperature exceeds air temperature, significantly more animals were seen in the water than out ($Chi^2_{2(p < 0.1)} = 10.2558$).

In a number of instances sections of main rivers or side creeks were surveyed on more than one occasion. In most instances multiple daytime surveys yielded comparable results on each occasion. The only major disparity is seen in the case of the September and December surveys of the Pie River. In September 15 animals were seen, whilst in December only a single individual was observed. In most cases nighttime counts were higher than those made during the day, although the difference shown by the data is not as great as one might expect from other surveys of crocodilians (Whitaker, personal communication).

In a sense the number of crocodiles sighted may be taken as a minimum estimate of density although they are presented here in order that variations over the study area might be examined. With the exception of the nighttime figures for the Purari above Wabo (low value) and the Varoi (high value), the data are surprisingly consistent throughout the area (Table 6).

In no cases were crocodile slides or tracks noted in areas where live animals were not observed. These data do not therefore extend the known area of crocodile distribution; they do indicate however that at least in Hou Creek and the lower Purari the minimum density estimates given in Table 5 are probably lower than the actual density.

Table 5. Location of animals in relation to air/water temperature differentials at different times of the diurnal cycle.

Relative temperature Animal sighted in:	Air = Water		Air > Water		Air < Water	
	Water	Air	Water	Air	Water	Air
Morning 0700 – 1300	3	1	4	7	0	1
Afternoon 1300 – 1900	0	1	4	6	1	0
Night 1900 – 0200	6	1	4	3	31	10
Total	9	3	12	16	32	11

6. Status of crocodile populations in the area

Unfortunately it was not possible to distinguish between the two species during the field survey. Whilst the survey shows that crocodiles are found throughout the area, discussions with skin buyers and hunters indicate that contrary to the suggestion of farm stock figures, the saltwater species is both less numerous and more restricted in its distribution than the freshwater species. *C. novaeguineae* is present throughout the area from above Wabo in the north down to the swamp forest of the delta. The saltwater crocodile is apparently restricted to the area south of Poroi on the Purari, to the coast and main channels of the delta area. Hunters in the Wame and Pie areas agreed that *C. novaeguineae* was the only species found in Eia creek and the creeks to the north of the Wame/Varoi river systems.

In discussing the status of crocodiles in this area from the viewpoint of their current abundance, one is at a disadvantage in that the absence of standardised survey techniques makes comparison between this and other surveys difficult. The only other surveys conducted in this area are those of Whiteside (1979), who conducted a 3-hour survey at Wabo and surveys at Akoma and Old Iari. At Old Iari he observed 12 crocodiles in 5 hours (2.4/hr), at Akoma, 10 in 6 hours (1.7/hr). Both of these values are higher than the mean figure for the delta area (0.8/hr) obtained during the present survey. During the present survey, however, 2.7 animals were shined per hour around Old Iari. Thus if both surveys were conducted at similar speeds, these data would suggest that crocodile distribution in the delta is 'patchy', that the status of the population

Table 6. Numbers of crocodiles sighted in relation to distance surveyed.

Location	Day			Night		
	No.	km	No./km	No.	km	No./km
Purari above Wabo	3	74	0.041	1	13	0.077
Purari above Bevan rapids	8	70	0.114	5	25	0.200
Lower Purari	8	82	0.098	9	30	0.300
Varoi	0	85	0.000	9	13	0.692
Wame	1	86	0.012	5	34	0.147
Pie	16	113	0.142	15	77	0,208
Ivo	4	98	0.041	7	57	0.123
Delta & swamp forest	0	123	0.000	10	34	0.294
Total	40	731	0.053	61	283	0.212

around Old Iari is unchanged and at higher density than elsewhere in the delta.

A number of hunters complained that the abundance of animals had declined over the last ten years. An examination of skin exports from this area would be necessary to support this claim. No substantive information was available to the authors which might throw light on this matter.

In considering the results of this survey, it must be strongly borne in mind that only navigable channels were surveyed. Hunters using canoes frequently penetrate areas of grass swamp not visited by the survey teams, and the area was extensively hunted by European hunters in the 1960s (Tetley, personal communication). However, despite such penetration of swamps, large areas of impenetrable swamp remain unexploited, and the relationship between crocodiles in channels, main rivers and swamps is unknown. A number of hunters suggested that crocodiles move into and out of the swamp areas in response to seasonal changes in water level, breeding activities and movements related to age. However, the delta area, having a larger number of navigable channels, has correspondingly smaller areas of swamp which might act as refuges or reservoirs from which crocodiles might seed surrounding areas. In that the saltwater species *C. porosus* is more commonly encountered in the delta, one might suggest that it is currently under the greatest exploitation pressure.

No evidence actually exists to suggest that substantial numbers of animals move from the inland swamps to coastal deltaic environments, and the role of the inland swamps in the maintenance of crocodile populations remains unresolved. The work of Magnusson (1978) suggests that hatchling movements are slight; of the sample studied 26% moved less than 100 m in inland areas with a mean distance moved of 1.26 km. Downstream only 2 individuals moved less than 1 km, and the mean distance travelled was 3.53 km. Any reseeding of over-exploited areas (as was suggested to occur by Downes 1976) would have to take place through the movements of older animals.

7. Current levels of utilisation

Data collected during the survey, together with information from skin exporters, shows that 387 skins were purchased from this area during 1979. The data presented in Table 2 suggests that approximately 132 animals were added to village farms during this period. Allowing for a proportion of farm-derived skins in the total sold from this area, the overall rate of capture of wild crocodiles is around 480/year for the whole area. Of these between 10 and

20% are *C. porosus* in contrast with the proportions on village farms, where 70% are *C. porosus* and only 30% *C. novaeguineae*.

The stability displayed in total stock held in the area over the two-year period suggests that the limits in terms of supply/maintenance of the farming system have been reached.

Levels of utilisation as indicated by hunter interviews are higher than the above figures suggest, with a corrected minimum of 866 and a corrected maximum of 1200 (Pernetta and Burgin 1980). Even accepting the lowest possible estimate based on hunter interviews, a considerable disparity exists between these figures and the data on skins sold from the area. This disparity may reflect any or all of the following: wastage through incorrect skinning or preservation; animals outside the legally permitted limits of 5'' and 20'' belly width; animals used to replace pen deaths which were not noted at the six month censuses; over-estimate of hunting success by the hunters; disposal of skins through some illegal outlet. It seems probable that the figures claimed by hunters are realistic, and since the existence of an illegal outlet in this area is highly improbable, the answer must lie in the other three possibilities.

8. Implications of the dam for crocodile populations

If a dam is constructed at Wabo, this will divide the populations above and below the dam site, and the possibility presents itself for management of the crocodile population in the future lake in an open farming system. Most of the area above the dam site is currently of marginal suitability for crocodiles thus a lake will greatly increase the carrying capacity of the area by increasing suitable habitat particularly the land/water interface and large bodies of slowly moving water. Such an increase in population size will be moderated by changes in the prey species composition and alteration of the fish and crustacean standing crops.

Haines (1979) is optimistic about the ultimate fish-carrying capacity of the lake, however during the process of lake filling and decomposition of the drowned terrestrial vegetation the lake may go through a severe phase of deoxygenation with consequent reduction in fish stocks, and effects upon the crocodile population. The magnitude of this effect will depend upon the length of time taken by the fish community to stabilise following natural or artificial colonisation of the lake. Both Petr (1978/1979) and Haines (1979) suggest this anaerobic phase of lake development may be short due to the high projected turnover time for the proposed lake of 2.5 months.

Haines (1979) makes various suggestions concerning the structure of the fish community of the lake which he suggests will have a substantial standing crop but a reduced species diversity. It is interesting to note that a large fish piscivore will be absent from the lake, and Haines suggests that suitable introductions would be saratoga (*Scleropages jardini*) or Murray River catfish (*Tandanus tandanus*). An alternative would be to encourage crocodiles to fill this role in the community and to crop the consequent production. Whilst Guggisberg (1972) suggests that Nile crocodiles do not compete with mans' fisheries in Africa this is in contradiction to the study of Cott (1961) which shows that at least in some areas one of the predominant food items of these animals is *Tilapia*, a genus of fishes widely used for food by man in Africa and elsewhere. It is regrettable that no data exist concerning the natural diets of crocodiles in Papua New Guinea, however it is likely that different-sized crocodiles will have substantially different diets and that smaller individuals feeding on insects, crustaceans, frogs and smaller fish might not compete with any commercial food fishery established on the reservoir.

Below the dam major changes projected are in the rates of flow and sediment loading of the river. Provided that water levels in the swamps and delta are not significantly lowered on a permanent basis it is unlikely that such changes will have a major effect on crocodiles. Changes in flow rate and erosional characteristics might alter the suitability of particular areas whilst a reduction in the probability of flooding during the breeding season might enhance egg survival through reduction of nest loss by flooding. Alternatively the modulating of flooding by the dam may adversely affect the accessibility of swamp habitats and thus reduce the total carrying capacity of the area. Most of the adverse changes are likely to be more directly felt by the *C. porosus* population which is also the population most threatened at the present time.

Haines (1979) predicts some reduction in fish productivity below the dam with associated changes in species composition. The retention of organic debris in the impoundment may have an effect on the detrital-based food chains below Wabo which contain both fish and prawns. As the top carnivore in the system crocodiles are thus likely to be affected. Similarly it is possible that changes in water turbidity in the lower Purari may enhance water clarity and hence predation on species such as prawns (Petr, in press). Any such increase in predation pressures on the hatchling crocodiles' food resources is likely to be detrimental to the population. However, since the current crocodile population levels in the delta are well below the carrying capacity of the system, this effect may well be masked.

Should the hydroelectric scheme proceed it is likely that exploitation of crocodiles in the area will increase as a result of increased human population density, an increase in liquid capital leading to more outboard motors and money for fuel and increased access through road and river transport.

At the present time however crocodiles represent a significant source of cash income to the people of the area (Liem and Haines (1977) estimate 20t/ha/annum*) and as such should be conserved and managed alongside future developments. The creation of the Wabo reservoir might well provide the opportunity for expansion of the freshwater crocodile resource in the impoundment area.

References

Abal, T., 1969. Crocodile trade (protection) ordinance – official statement. In: Territory of Papua New Guinea House of Assembly Debates, Second House, 5th Meeting of the First Session, 16 – 27 June 1969, 3(5): 1375.

Allen, G.R., 1974. The marine crocodile, *Crocodylus porosus*, from Ponape, East Caroline Islands, with notes on food habits of crocodiles from the Palau Archipelago. Copeia 2: 553.

Allen, J., 1977. Management of resources in prehistoric coastal Papua. In: J.H. Winslow (ed.), The Melanesian Environment, pp. 35 – 44. Australian National University Press, Canberra.

Behler, J.L., 1976. The crocodile industry in Papua New Guinea. A report to the United Nations. Wildlife in Papua New Guinea, mimeo 77/25. Wildlife Division, Konedobu, Papua New Guinea.

Bolton, M. and M. Laufa, 1979. The crocodile project in Papua New Guinea. FAO/UNDP Assistance to the Crocodile Skin Industry. Mimeo. Wildlife Division, Konedobu, Papua New Guinea.

Bolton, M. and M. Laufa, 1982. The crocodile project in Papua New Guinea. Biol. Conservation 22: 169 – 179.

Brazaitis, P., 1973. The identification of living crocodiles. Zoologica, Fall: 59 – 101.

Burgin, S., 1980. A review of crocodile farming in Papua New Guinea. Science in New Guinea 7(2): 73 – 88.

Caras, R.A., 1975. Dangerous to man, rev. ed. Barrie and Jenkins, London.

Cogger, H.G., 1960. Crocodiles and their kin. Australian Museum Mag., June: 200 – 204.

Cogger, H.G., 1972. Crocodiles. In: Encyclopaedia of Papua New Guinea, Vol. 1, pp. 221 – 222. Melbourne University Press.

Cott, H.B., 1961. Scientific results of an inquiry into the ecology and economic status of the Nile crocodile (*Crocodilus niloticus*) in Uganda and Northern Rhodesia. Trans. Zool. Soc. Lond. 29: 1 – 356.

Deraniyagala, P.E.P., 1939. The tetrapod reptiles of Ceylon, Vol. 1: Testudinates and Crocodilians. Colombo Museum, Ceylon.

De Vos, A., 1979. A manual on crocodile management. Food and Agriculture Organization of United Nations, Rome. FO: Misc/79/30, W/N 4886, Draft.

Ditmars, R.L., 1944. Reptiles of the world. MacMillan, New York.

Downes, M.C., 1969. Protection for the crocodile skin trade. Extract from a report to the House

---

* P.N.G. currency unit Kina = 100 toia(t). In 1977 one Kina = U.S.$ 1.25.

of Assembly, Territory of Papua New Guinea. Mimeo. Wildlife Division, Konedobu, Papua New Guinea.

Downes, M.C., 1974. An explanation of the national policy on crocodile farming. Wildlife in Papua New Guinea, mimeo 2/74. Wildlife Division, Konedobu, Papua New Guinea.

Downes, M.C., 1978. Notes on the national crocodile policy and the preliminary objective of the UNDP/FAO project. Prepared for discussion with Mr M. Bolton and Mr E. Bolson (Wildlife Managers UNDP/FAO). Wildlife in Papua New Guinea, mimeo 78/17. Wildlife Division, Konedobu, Papua New Guinea.

Greer, A.E., 1975. Clutch size in crocodilians. J. Herpetology 9 (3): 319 – 322.

Guggisberg, C.A., 1972. Crocodiles: their natural history, folklore and conservation. David and Charles, Newton Abbot, Devon.

Haines, A.K., 1979. An ecological survey of the fish of the lower Purari river system, Papua New Guinea. In: T. Petr (ed.), Purari River (Wabo) Hydroelectric Scheme Environmental Studies, Vol. 6, pp. 1 – 102. Office of Environment and Conservation, Waigani, Papua New Guinea.

Hill, L. 1979. *Crocodylus porosus* and *C. novaeguineae*, preliminary breeding data for 1978 – 79 breeding season at Moitaka. Typescript. Biology Department, University of Papua New Guinea.

Honnegger, R., 1975. Red data book, Vol. 3: Amphibia and Reptilia. IUCN Survival Service Comm., Morges, Switzerland.

Hope, J., 1977. The effect of prehistoric man on the fauna of New Guinea. In: J.H. Winslow (ed.), The Melanesian Environment, pp. 21 – 27. Australian National University Press, Canberra.

Johnson, C.R., 1973. Behaviour of Australian crocodiles, *Crocodylus johnstonii* and *Crocodylus porosus*. J. Linn. Soc. Zool. 52: 315 – 336.

Lang, J.W., 1980. Reproductive behaviours of New Guinea and saltwater crocodiles. Paper presented at SSAR symposium 'The reproductive biology and conservation of crocodilians', 6 – 8 August, 1980, Milwaukee.

Lever, J., 1974. Wastage in the crocodile skin industry. Wildlife in Papua New Guinea, mimeo 74/3. Wildlife Division, Konedobu, Papua New Guinea.

Lever, J. and G. Mitchell, 1975. Crocodile industry training manual. Wildlife Manual, 75/1. Wildlife Division, Konedobu, Papua New Guinea.

Liem, D.S. and A.K. Haines, 1977. The ecological significance and economic importance of the mangrove and estuarine communities of the Gulf Province, Papua New Guinea. In: T. Petr (ed.), Purari River (Wabo) Hydroelectric Scheme Environmental Studies, Vol. 3, pp. 1 – 35. Office of Environment and Conservation, Waigani, Papua New Guinea.

Magnusson, W.E., 1978. Nesting ecology of *Crocodylus porosus* Schneider in Arnhem Land, Australia. Ph.D. Thesis. University of Sydney.

Medem, F., 1976. Report on the survey carried out in Papua New Guinea in 1976. Wildlife in Papua New Guinea, mimeo 77/26. Wildlife Division, Konedobu, Papua New Guinea.

Messel, H., A.G. Wells and W.J. Green, 1978. *Crocodylus porosus* population studies – survey techniques in tidal river systems of northern Australia. Mimeo. 4th working meeting of IUCN Survival Services Comm., Crocodile Specialist Group, Madras, India.

Messel, H., A.G. Wells and W.J. Green, 1979a. Surveys of tidal river systems in the Northern Territory of Australia and their crocodile populations, monograph 5: The Goomadeer and King River systems and Majarie, Wurugoij and All Night Creeks. Pergamon Press (Aust.), Rushcutters Bay, N.S.W.

Messel, H., A.G. Wells and W.J. Green, 1979b. Surveys of tidal river systems in the Northern Territory of Australia and their crocodile populations, monograph 6: Some river and creek systems on Melville and Grant Islands. Johnston River, Andranangoo, Bath, Dongau and Tinganoo Creeks and Pulloloo and Brenton Bay Lagoons on Melville Island; North and South Creeks on Grant Island. Pergamon Press (Aust.), Rushcutters Bay, N.S.W.

Messel, H., A.G. Wells and W.J. Green, 1979c. Surveys of tidal river systems in the Northern Territory of Australia and their crocodile populations, monograph 7: The Liverpool and Tomkinson Rivers system and Nungbulgarri Creek. Pergamon Press (Aust.), Rushcutters Bay, N.S.W.

Messel, H., G.C. Vorlicek, A.G. Wells and W.J. Green, 1980. Surveys of tidal river systems in the Northern Territory of Australia and their crocodile populations, monograph 9: Tidal waterways of Castlereach Bay and Hutchinson and Cadell Straits. Bennett, Darbitla, Djigagila, Djabura, Ngandadauda Creeks and the Glyde and Woolen Rivers. Pergamon Press (Aust.), Rushcutters Bay, N.S.W.

Neill, W.T., 1946. Notes on *Crocodylus novae-guineae*. *Copeia* 1: 17 – 20.

Neill, W.T., 1971. Last of the ruling reptiles. Columbia University Press, New York.

Pernetta, J. and S. Burgin, 1980. Census of crocodile populations and their exploitation in the Purari area (with an annotated checklist of the herpetofauna). In: T. Petr (ed.), Purari River (Wabo) Hydroelectric Scheme Environmental Studies, Vol. 14, pp. 1 – 44. Office of Environment and Conservation, Waigani, Papua New Guinea.

Petr, T., 1978/9. The Purari River hydroelectric development at Wabo and its environmental impact: an assessment of a scheme in the planning stage. Science in New Guinea 6(2): 105 – 116.

Petr, T., in press. Possible impact of the planned hydroelectric scheme on the Purari River deltaic and coastal sea ecosystems (Papua New Guinea). Presented at 2nd International Symposia on Biology and Management of Mangroves and Tropical Shallow Water Communities, Papua New Guinea (Port Moresby), July 20 – August 2, 1980.

Ross, C.A., 1977. *Crocodylus novaeguineae*, natural history and morphology. In: Crocodile Extention Report, UNDP Crocodile Project, Papua New Guinea, pp. 17 – 27. Unpublished. FAO/UNDP Crocodile Project File, PNG/74/029.

Schultze-Westrum, T.G., 1970. Conservation in Papua New Guinea. WWF Mission Report. Mimeo.

Tago, S., 1977. The crocodile industry and crocodile laws in Papua New Guinea. FAO/UNDP File PNG/74/028 – 14, Waigani, Papua New Guinea.

Taylor, J.A., 1977. The foods and feeding habits of sub-adult *Crocodylus porosus* Schneider in northern Australia (Crocodilia: Reptilia). M.Sc.Thesis. University of Sydney.

Taylor, J.A., 1979. The foods and feeding habits of sub-adult *Crocodylus porosus* (Schneider) in northern Australia. Aust. Wildl. Res. 6: 347 – 359.

UNDP/FAO, 1978. Recent crocodilians of the world and their conservation. Wildlife in Papua New Guinea, mimeo 78/13. Wildlife Division, Waigani, Papua New Guinea.

Webb, G.J.W. and H. Messel, 1978. Morphometric analysis of *Crocodylus porosus* from the north coast of Arnhem Land, northern Australia. Aust. J. Zool. 26: 1 – 27.

Webb, G.J.W., H. Messel and W. Magnusson, 1977. The nesting of *Crocodylus porosus* in Arnhem Land, northern Australia. Copeia 2: 238 – 249.

Webb, G.J.W., H. Messel, J. Crawford and M.J. Yerbury, 1978. Growth rates of *Crocodylus porosus* (Reptilia: Crocodilia) from Arnhem Land, northern Australia. Aust. Wildl. Res. 5: 385 – 399.

Whitaker, R., 1980. The status and distribution of crocodiles in Papua New Guinea. Food and Agriculture Organization of the United Nations. FO:DP/PNG/74/029 Field Document 1. Wildlife Division, Port Moresby, Papua New Guinea.

427

Whiteside, D., 1979. General survey of crocodiles in the Gulf Province. Unpublished typescript, 3 pp. Crocodile Project Survey File, Konedobu, Papua New Guinea.

Woodward, A.R. and W.R. Marion, undated. An evaluation of factors affecting night counts of alligators. Mimeo. Journal Series, No. 1400. Florida Agric. Exp. Stn., Gainesville, Florida.

Youngprapakorn, U., undated. The breeding of crocodiles in capacity at Samut Prakan, Thailand. Borpit, Thailand.

Youngprapakorn, U., J.A.McNeeley and E.W. Cronin, 1971. Captive breeding of crocodiles in Thailand. IUCN Publ. (N.S) Suppl. 32: 98 – 101.

III. Social environment and developmental aspects

1. A short history of the upper Purari and the Pawaia people

C. Warrillow

1. Introduction

One day an area of up to 290 km² may become inundated by the Wabo hydroelectric reservoir on the Purari River. This area of extremely high rainfall covered by a lowland humid tropical forest is owned exclusively by clans of the Pawaia-speaking people. The Pawaia, currently about 2000 strong and occupying parts of three provinces, claim rights over perhaps some 10 000 km² of land. In the Upper Purari Census Division, where the Wabo project is planned, about 700 Pawaia speakers live permanently in an area of approximately 4150 km². Apart from them, this area is also inhabited by a few Polopa and Daribi speakers who have been recruited through marriage or who live from time to time with their relatives. As a matter of convenience, the author considers the upper Purari to consist of Pawaia speakers only. Their area is inhabited by the people of Gurimatu, Kairuku, Koni, Pawaia No. 1, Pawaia No. 2, Tatu, Uraru, Uri, Weijana, Weme and to some extent by Lake Tebera (Tebora) Polopa speakers (see Fig. 1 and 2). In recent years this village has been physically amalgamated with Gurimatu, but still retains its identity as a separate census unit.

Other Pawaia speakers are located in the Karimui (Sena River area), Pio and Tura Census Divisions of the Simbu Province (Fig. 2). The Weme people are partly in the South Fore Census Division of the Eastern Highlands Province although some of these are still living in the Upper Purari Census Division. Weijana are still censused in the Upper Purari although most of them have now moved to the Haia area of Tura Census Division. A small group of Pawaia live at Keka in the Upper Vailala Census Division of the Ihu Sub-District.

The National Census of 1980 recorded only 644 persons as being residents of the Upper Purari Census Division. The census was taken at Wabo, Poroi No. 1 and No. 2 (Pawaia No. 1 and No. 2), Weijana (Eijana), and Lake Deba (Lake Tebera). The resulting number of Pawaia is lower than the 784 listed for the 1973

Petr, T. (ed.) The Purari – Tropical environment of a high rainfall river basin
© *1983, Dr W. Junk Publishers, The Hague / Boston / Lancaster*
ISBN-13: 978-94-009-7265-0

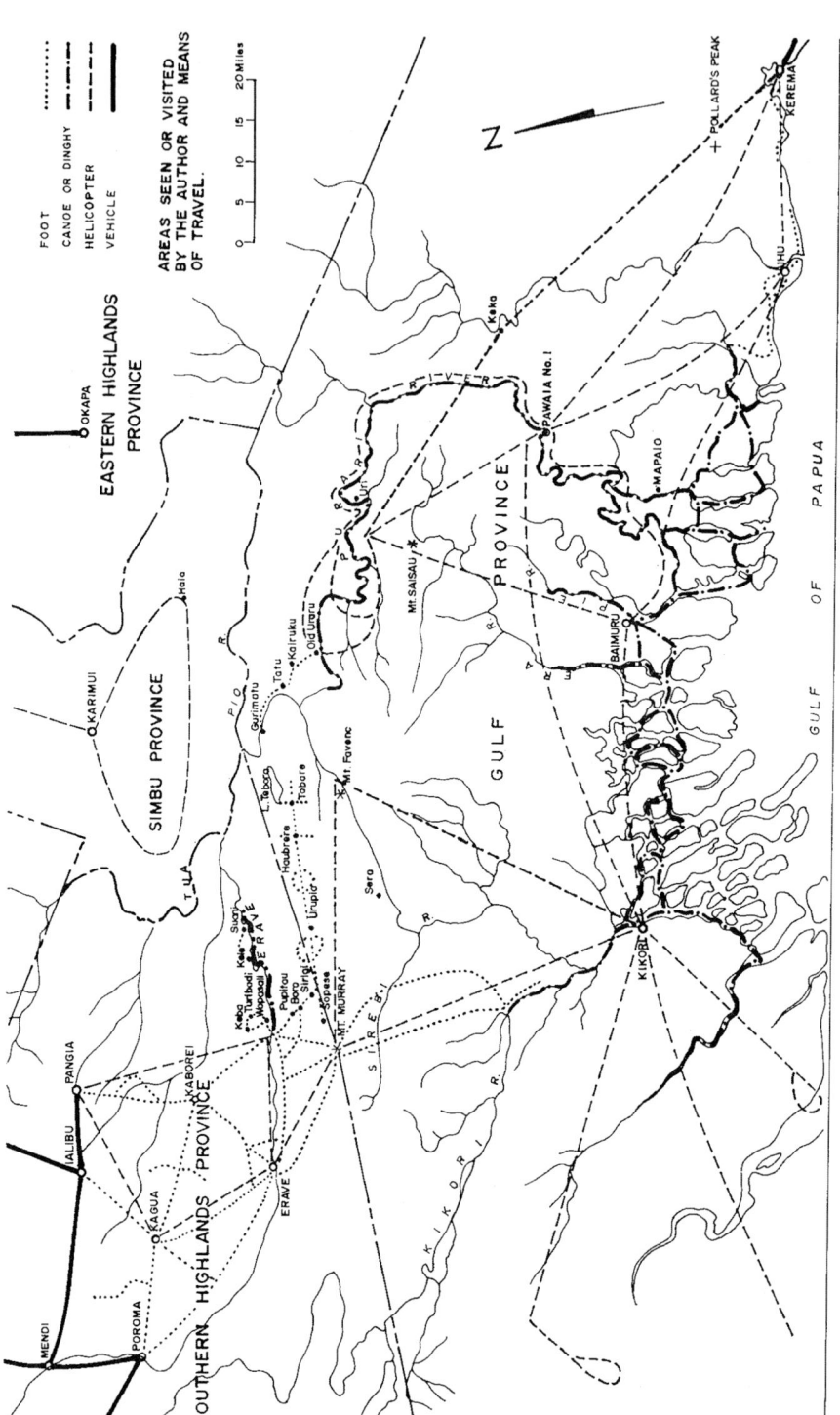

Fig. 1. The study area with locations visited by the author, and means of travel.

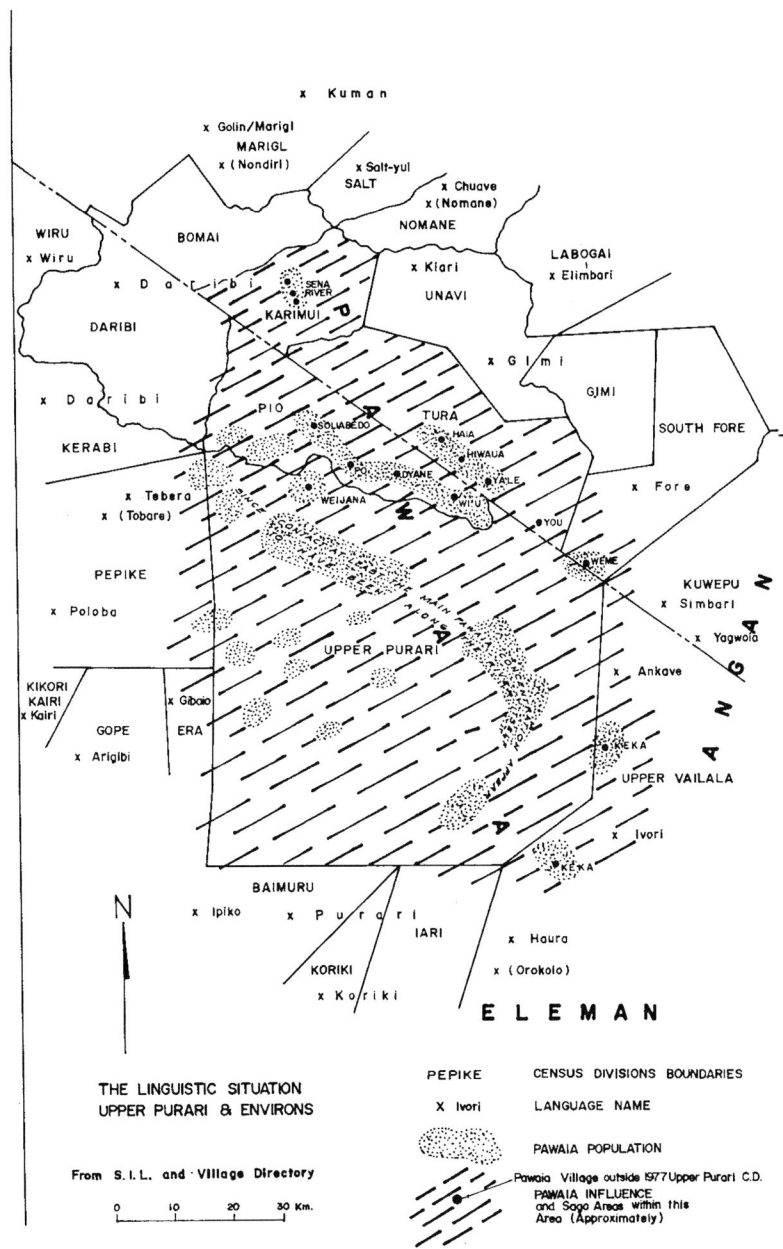

Fig. 2. The linguistic situation in the upper Purari and adjacent areas (based on information provided by the Summer Institute of Linguistics and extracted from the Village Directory of Papua New Guinea).

431

Fig. 3. The Purari River, Wabo airstrip, and part of Wabo camp. (Photo T. Petr.)

census (Warrillow 1978), and one reason for this could be that the 1980 census did not cover all census units of 1973.

For the Pawaia, being semi-nomadic people, changing settlements, or abandoning them completely is nothing new. In the past, frequently neither the Kairuku, nor the Uraru people lived on their traditional land at Uraru village, which itself has since been abandoned; all were squatting on Koni land. Koni village no longer exists. Gurimatu people have in the past experienced the loss of a village and perhaps dozens of people through flooding when, perhaps less than fifty years ago, a major landslide blocked the Purari temporarily and a secondary slide, very soon after, caused a huge wave to race back up the valley (Warrillow 1978). The relatively recent establishment of the semi-permanent Wabo village in the dam site area vicinity should not be considered as an introduction of permanency to settlement in their life, as today's Wabo village would have to be vacated if the dam project were to go ahead, as the site, although downstream, would most likely be the location of power house and tail races (Figs. 3 – 8).

The present study summarizes the current knowledge of the Pawaia history and is based on a review of literature, government reports, personal communications, and on author's extensive travel through the area, as well as his long-term contacts with the Pawaia speakers.

Fig. 4. Wabo camp, established for the engineering feasibility studies for the Wabo hydroelectric scheme. (Photo T. Petr.)

Fig. 5. Aerial view of Peri settlement near Wabo airstrip. (Photo T. Petr.)

433

Figs. 6 and 7. Two types of houses in the Wabo settlement. (Photos T. Petr and A. Hall.)

434

Fig. 8. Dug-out canoes on Wabo Creek at Wabo. (Photo A. Hall.)

2. Early contact

The Pawaia, as with most pre-literate societies in Papua New Guinea, can seldom trace back more than four generations. Therefore their oral history and traditions, unless corroborated by other evidence, leave much to doubt as to what is myth and what is authentic history. Western 'discovery' of the upper Purari came about as a result of both missionary zeal and of adventurism, the latter explorers sometimes having a sense of duty to drive them on, but more frequently the incentive of perhaps a discovery of riches to reward them for their efforts and privations.

That the population of the upper Purari was sparse would have been obvious to the early explorers. Thus, perhaps, there was little incentive for missionaries, in search of souls to convert to Christianity, to concentrate on the area. Similarly the doubtful agricultural potential of the upper Purari, its isolation and again low population did not warrant early extension of Government influence into the area. The upper Purari therefore could well have escaped the further attention of westerners for many years to come had it not been for the latters' search for energy sources.

First to ascend the Purari River (the mouths of which he had discovered in 1879) was missionary Rev. James Chalmers in 1883 when he sailed up the Wame – Varoi branch naming it the Wickham. However it was April, 1887 before the first white man managed to sail to anywhere near the upper Purari. Theodore Bevan in his book (1890) gives a brief description of the journey which he claims took him 100 miles upstream from the coast. However the map, in Bevan's book, which indicates he perhaps travelled as far as 15 miles upstream of today's Pawaia No. 1 (Poroi), does not show McDowell Island nor the Aure River which are some 150 km and 165 km respectively, upstream from where the Wame flows into Port Romilly and a similar distance up the main Purari from where it flows into the Gulf of Papua via Alele and Aievi passages.

It was left to none less than the then Lieutenant-Governor himself, Sir William McGregor, to pioneer the way upstream to the upper Purari in 1893 when, during the month of February, he steamed five miles further up the river than the Aure River confluence (B.N.G.A.R. 1892 – 93). Returning a year later Sir William reported (B.N.G.A.R. 1893 – 94) reaching as far upstream as today's Uraru which he shows on his map as Biroe. The map accompanying the report clearly indicates that he also camped at the location of modern Wabo village. Upon returning downstream the party stopped at Abukiru (Abukiri) Island 'some half score of miles above the Purari bifurcation' (in fact 7 kilometres). Here fragments of coal were discovered and specimens collected for later examination.

The first recorded land expedition in the upper Purari did not take place until some fourteen years later when in 1908 a party led by Donald Mackay and the Hon. W.J. Little of the Papuan Legislative Council (Carne 1913) went in search of gold and the deposits from which McGregor's coal had originated. It seems this party travelled up the Purari as far as Wai'i Creek, near Hathor Gorge (Fig. 9), and thence struck inland following Wai'i Creek where coal seams were indeed discovered, thence into the Irou Valley (shown on the expedition's map and modern maps as Samia River) which they followed westwards to the eastern slopes of Mt. Murray. Several coal occurrences were

Fig. 9. The Pawaia country near Hathor Gorge. (Photo T. Petr.)

also discovered along this river. This expedition abandoned its aim of travelling west to the Strickland River and returned by the same route it had gone in by. Constant rain, malaria and harassment by the local people, aggravated by lack of supplies and the lack of food availability en route, combined with the terrain and distance to be covered to defeat the original goal.

The journey left one carrier killed by arrow fire near Hathor Gorge, (for which six houses were burnt in retaliation), and at least one local man dead from gunshot wounds in a separate skirmish (Clune 1942).

It would appear that the administrators of this time (Resident Magistrates and Assistant Resident Magistrates) were too fully occupied establishing Government posts and extending government influence along the coast to follow-up McGregor's patrols of the previous decade. In fact use of the Lieutenant-Governor's vessel the 'Merrie England' was afforded a later expedition to the 'Purari coalfield' whilst it was engaged in transporting men and supplies for the establishment of a government station on the Kikori River late in 1911.

Missionaries, also tied down consolidating their earlier work along the coast, were still interested in the hinterland. J.H. Holmes of the London Missionary Society, managed to travel as far upstream as the Aure junction in 1908.

437

Here, however, the current beat his vessel and he headed back downstream covering in eight and a half hours the same distance of his six-day ascent of the river (Holmes 1924).

Unlike gold discoveries elsewhere in Papua coal occurrences caused no rush. Thus there was no real pressure on administrators or missionaries to move in to either 'protect' the miners or save the 'heathen'. Even so, by the end of 1909 'several applications were received by the Warden at Kerema to prospect for coal in lands adjoining the headwaters of the Purari River, but none were executed' (Carne 1913).

In 1910, following more coal discoveries (on the lower Kikori River) the Lieutenant-Governor, J.H.P. Murray, sought an easier route to the Purari coal finds and climbed Aird Hills, in the Kikori Delta, with the object of viewing the intervening country (P.A.R. 1909 – 10). Carne reports that this route was successfully tested by Little who, on 17th March, 1911, started on a further examination of the 'Purari coalfield'. Large samples of coal were secured and 'with the exception of one carrier, killed by the Purari natives, the party returned safely and expeditiously to the coast' (Carne 1913).

March, 1912 saw yet another expedition set off from Kikori for the Purari coal fields. Led once again by Little, and accompanied by Carne (N.S.W. Dept. of Mines) and others, the party followed the Kikori – Seribi (Sirebi) route, thence up the Curnick River (on today's maps a tributary by which the Wao, Kuru and Sire rivers join the Sirebi) from where they travelled overland to what had now become known as Coal Creek (Wai'i Creek) which was followed to its confluence with the Purari before the party retraced its steps.

The conclusion reached, following all this work, was that 'so far as at present known, the inaccessible nature of the intervening country, to say nothing of the character, thinness, and disturbed condition of the coal, renders it of no present or prospective value to the Territory' (Carne 1913).

3. Consolidation of government influence

It seems that the failure of the coal fields to warrant further investigation, the tasks at hand consolidating government and missionary influence along the coast and the drain on (and loss in action of some) manpower due to World War I left the upper Purari and its people untouched by western influence for most of the second decade of the twentieth century. Yet still little was known of the people who inhabited this vast area of country.

Holmes (1924) made no mention of the Pawaia although he did, in 1908, contact 'some men of the Kaura tribe' (Eleman language stock) just above Bevan Rapids. On 2nd October, 1908 further upstream he recorded in his diary 'the only signs we have seen today of natives in the vicinity were on McDowell's Island; a few huts and a banana garden'.

Carne (1913) was camped, on April 25th, 1912, at the spot where Mackay and Little had camped in 1908 where Wai'i Creek joins the Purari and records in his journal:

'Whilst lunching at the river the party was evidently watched by natives, whose tracks were frequent and plain; on starting up stream again to reach camp, they came out on points of jungle fronting the stream, calling loudly. As the previous expedition had suffered from being friendly, no notice was taken of them, which made them bolder in their approach and louder in their shouting. Being armed with bows and arrows it was deemed advisable to give them a check. On their next disappearance for the purpose of making a detour through the jungle to get nearer, the police were made to present arms, and on their reappearance to fire a volley over their heads, which had the desired effect; they vanished like magic, were not further seen.'

It was government patrols which later started to learn and record something concrete about the people with whom these early expeditions had clashed.

With the war ended, the administration of Papua slowly returned to normal. The Papuan Annual Reports of 1917 – 18 and 1918 – 19 make mention of patrols up the Vailala River to Keki (Keka?) and overland towards Wai'i Creek (where Ipigo and Polopa* people were contacted). But still no mention was made of Pawaia people.

In 1922 R.A. Woodward (A.R.M.) ascended the Purari to above the Aure junction and contacted people whom he believed were then trading with their former enemies, the Kairi, from the Kikori area. About this time too Patrol Officer Lambden patrolled up the Vailala River, past the Ivori junction and met people of the Nahikai'a tribe, 'who came from the Purari but were now living north of Kerema'. Soon after, the name Hawaiu and Hawoiu started to appear in reports of patrols which ascended the Vailala and also moved north of Kerema to contact the Kukukuku tribes. There were suggestions that these people (Hawaiu) originated from the Purari and it is interesting to note that two small rivers which eventually flow into Kerema Bay from the northwest are known (even today) as the Hawaiu and Purari.

H.M. Saunders, a government officer, spent time in the upper Purari area in 1923. He was escorting a party of Anglo-Persian Oil Company geologists who were studying the country between the Purari and Kikori and who remained in the area for three months. Leo Austen relieved Saunders, who was required elsewhere, and writing of a group of people he contacted south of Wai'i Creek stated 'I feel inclined to believe Ope means a bushman and is not

* Also spelled Poloba.

the real name of the tribe' (Hope 1979). Actually Ope (or Opei) is a clan of the Uraru people whose land boundaries do extend south of Wai'i Creek (Warrillow 1978).

A year later two government officers (F.R. Crawley and S.H. Chance) examined the country between the Purari and the Era following application by Maira Estates Ltd. to prospect for oil (Hope 1979).

Mention is made in the 1927/28 Papuan Annual Report of a patrol led by R. Speedie who went in search of the Namaina tribe 'supposed to be living between the Era and Purari'. We now know these people to be the Koni of Wabo (Warrillow 1978). Speedie failed to find these people.

Exploration and contact from the opposite direction to those patrols, which ascended the Purari itself or the Kikori – Sirebi thence inland, came later in 1929 and 1930. The first exploration to travel downstream into the upper Purari was conducted by government officers, B.W. Faithorn and Claude Champion who started from Kikori in 1929. They struck north to the Erave River, thence heading east they came upon a village now known as Suani (Daribi* speakers) on the lower Erave where they camped. It is of interest to note Faithorn records the village name as Udunebi whilst Wagner (1967) refers to these river-dwelling Daribi as the Urubidi. Following the river downstream they came to the Purari and were guided around Hathor Gorge by the Pawaia (Hughes 1977), probably following the same route used by present day patrols via Tatu, Kairuku and down Mua (Moa) Creek to Uraru (Fig. 10).

A year later Pawaia territory was visited from the north when gold prospectors Leahy and Dwyer left the Ramu Valley and followed the Dunantina downstream to the Tua, then turned eastward down the Purari as their predecessors had done a year before. In 1936 government officers Champion and Anderson on their Bamu – Purari patrol entered the Pawaia country from the northwest having patrolled from Lake Kutubu to Mt. Ialibu thence along the then Territory border. They headed southeast through Daribi country and possibly met the Sena River Pawaia before meeting the Purari well downstream of Hathor Gorge.

The name Pawaia and the people themselves, appear to have first warranted attention and earned the displeasure of Government in 1931 when the 'Pawaia nomads' who inhabited the area downstream from the Aure (today's Pawaia No. 1 and No. 2) attacked and killed a number of Uri people who were living further downstream. The raid was carried out with the assistance of the 'Turoha' who are described as neighbours of the Pawaia (and in fact are today's Koni), who lived between the Aure junction and Uraru (Hides 1938).

---

* Also spelled Dadibi.

Fig. 10. A group of Pawaia at the confluence of Moa Creek with the Purari. (Photo T. Petr.)

Ron Speedie led the first patrol in from Kikori (by river) and arrested the ringleaders, then Jack Hides followed some four or five months later to round up the rest, including the fight leader named Omaka who had earlier escaped from Speedie's patrol (Hides 1938; Sinclair 1969).

Foldi led the last formal government patrol through the area before World War II disrupted administration of Papua. A report on this patrol, Kikori No. 11 of 1937/38 is held in the Government Archives at Waigani. There is mention in records at Kerema of one patrol to the upper Purari during the war. This was in 1944, led by an Angau Officer Lieutenant Ross. However, no report is available.

Prior the war, from 1937, the upper Purari was visited by geologists employed by companies engaged in oil exploration. The Australian Petroleum Company held permits for the area (it still does) and a young geologist S.W. Carey was most enthused by his findings which he felt indicated the possibility of the presence of hydrocarbons beneath the surface. Further exploratory work continued after the war and finally resulted in drilling in and around the Census Division which lasted into the early 1960s. One well (Puri No. 1) in the Era River area did in fact produce an oil flow before being plugged and

441

abandoned after saltwater was encountered. Bwata, in the Pide valley, is a proven gas field.

4. Establishment of western influence

Post-war administration of the upper Purari resumed in 1948 with Beara (this station was later relocated to Baimuru) patrol number 4 of 1948 – 49 led by Patrol Officer G.D. Colins. During the period 19th October to 12th November, 1948 this patrol visited villages between Wai'i Creek just below Hathor Gorge and the present day Pawaia No. 1. The patrol met some Gurimatu men at Uraru.

Most of the early post-war patrols to the upper Purari entered via the head-waters of the Pide River, from the headwaters of the Era River. It was not until the 1960s that more reliable outboard motors became available which made it possible to reach the upper Purari by river rather than by foot, although some patrols, particularly those travelling as far only as Pawaia No. 1 and Pawaia No. 2. did travel up the river. During the 1950s, 1960s and 1970s, government patrols visited the Upper Purari Census Division more frequently, often once a year, for routine administration and census revision purposes.

During the 1950s and 1960s extensive patrolling was conducted throughout the highlands and many blank spaces and unplotted river courses were filled in on maps. Some of these exploration, routine administration, and even medical patrols passed through parts of the Upper Purari Census Division from several directions (Warrillow 1978).

As more became known of the area and its people so it became evident that Pawaia-speaking people inhabited areas as far south and east as the lower Vailala, west to the Era River and north towards Okapa (Eastern Highlands) and Mt. Karimui (Simbu). Up to 2000 Pawaia people had been contacted by the early 1960s, living in an area which covered as much as 10 000 square kilometres. Of these some 800 lived in the Upper Purari Census Division of the Gulf Province – an area of approximately 4150 square kilometres.

Meanwhile, throughout the 1950s, yet another search for energy was underway in the upper Purari following recognition of the hydroelectrical potential of the river, especially in the Wabo and Hathor Gorge areas. With the discovery of huge deposits of bauxite at Gove and Weipa, in northern Queensland, Australia, British Aluminium were investigating sources of plentiful and cheap electricity. New Guinea Resources (N.G.R.) Prospecting Company Limited was formed to undertake the necessary feasibility studies.

In 1952, a main camp was established near Wabo and this was occupied, on

442

and off, for periods of up to two years until 1960 whilst teams camped, walked and helicoptered throughout the area doing mapping and hydrological work. British Aluminium and other companies formed a consortium with Comalco in 1958 and two years later the proposed project was abandoned in favour of Bell Bay in Tasmania and the Bluff in New Zealand.

A member of the team who spent many years working with N.G.R. recalls using Pawaia labour and guides. He found them good workers and fairly law-abiding people. The only serious troubles encountered during the entire operation were when work parties were forced to withdraw by Daribi men from the Karimui area and on another occasion Polopa men from the direction of Lake Tebora who launched raids against the Pawaia in the Gurimatu area and threatened work teams. A government officer, Gus Bottril, was attached to N.G.R. for 3 or 4 weeks in order to restore order in the area, after one incident involving Daribi warriors, to enable field work to resume. My informant told me that the Pawaia were terrified of the invaders and 'the first to run' (David 'Jack' Sarjeant, personal communication).

Missionaries, particularly the London Missionary Society (L.M.S.) (now the United Church), visited the upper Purari fairly regularly in the 1950s and finally established a post at Uraru in 1961. This post was manned more or less continuously with a pastor and often an aid post orderly until March 1972 when the New Tribes Mission attempted to move in from the Karimui area. This mission did not last long in the area and left after only a few months. It established a station at Haia in the Tura Census Division of the Simbu Province and some upper Purari people, especially the Weijana, have been attracted to it and have 'settled' nearby – if one can ever describe the Pawaia as being 'settled'. The Seventh Day Adventist Mission had a school at Koni between 1963 – 65 and European Bahai Missionaries made annual visits to Pawaia No. 1, Pawaia No. 2, and Uri between 1969 and 1972. The United Church (formerly L.M.S.) is the only mission currently active in the Upper Purari Census Division, but has no personnel living there.

An Authorised Landing Area (i.e., an airstrip limited to restricted operations) was completed at Uraru Village in late 1966. It was 420 m long and 42 m wide. Lack of maintenance led to its early abandonment but United Church Missionaries continued to visit the area, including medical teams from Kapuna mission hospital, by flying in aboard the Mission Aviation Fellowship float-Cessna. With the completion of an all weather 800 metre long airstrip at Wabo in May, 1976, speedy and reliable access to the upper Purari was at last guaranteed.

5. The Pawaia: their eastward migrations and western origins

Prior to the current environmental studies there had been very little study done of the Pawaia. Government officers recorded their impressions and some oral history. Studies of the language were conducted by linguists attached to the Summer Institute of Linguistics (S.I.L.), but mainly in the Karimui area amongst the Sena River Pawaia (Trefry 1969; MacDonald 1973, 1977). No anthropologist has yet lived and worked amongst the Pawaia for any length of time. However, more studies were carried out within the framework of the Purari River basin environmental studies, coordinated by the Office of Environment and Conservation and Department of Minerals and Energy (Warrillow 1978; Eggloff and Kaiku 1978, this volume; Nurse 1980; Toft 1980, this volume).

Oral history of a generally eastwards migration, from coastal rather than highland areas to the west, is corroborated to some extend by both linguistic and medical evidence. It seems probable that the Pawaia once gained occupation of land as far east as Murua* River, north of Kerema, before being forced back to the Vailala by the Angan people, whilst at a similar time their southward push was thwarted by the coastal Elema (Fig. 11). However, I do not believe these most easterly gains were as a result of victories in tribal warfare. Rather, it appears, the Pawaia just moved into virtually unoccupied territory until meeting resistance.

Pawaia northeasterly migration too appears to have been into virtually uninhabited country until they met highland types such as the Fore, Gimi and Kiari and pressed no further. It was only to the north itself that the Pawaia appear to have settled almost shoulder to shoulder with another linguistic group − the Daribi (or Dadibi). However, it is noted elsewhere (Nurse 1980) that there appear to be fairly close links between the two groups − not that this prevented continual fighting between them (Warrillow 1978).

Turning to the linguistic situation, MacDonald (1973) summarises his studies of the Pawaia and adjacent language groups as follows:

'Wurm has already shown that Daribi and Pawaia are not closely related to the East New Guinea Highlands Stock to their north, nor to the Wiru to the west. Lloyd in Chapter 2 of this volume, demonstrates that Pawaia is not related to the Angan family to the east.'

'Pawaia appears to be closer phonologically to the Teboran family (Daribi and Polopa) than any other neighbouring language. It is the eastern − most language featuring nasalised words in a belt stretching far to the west. Languages to the south of Pawaia and Polopa lack this feature. Pawaia and Daribi share a number of structural features.'

* Also spelled Muroa or Muro.

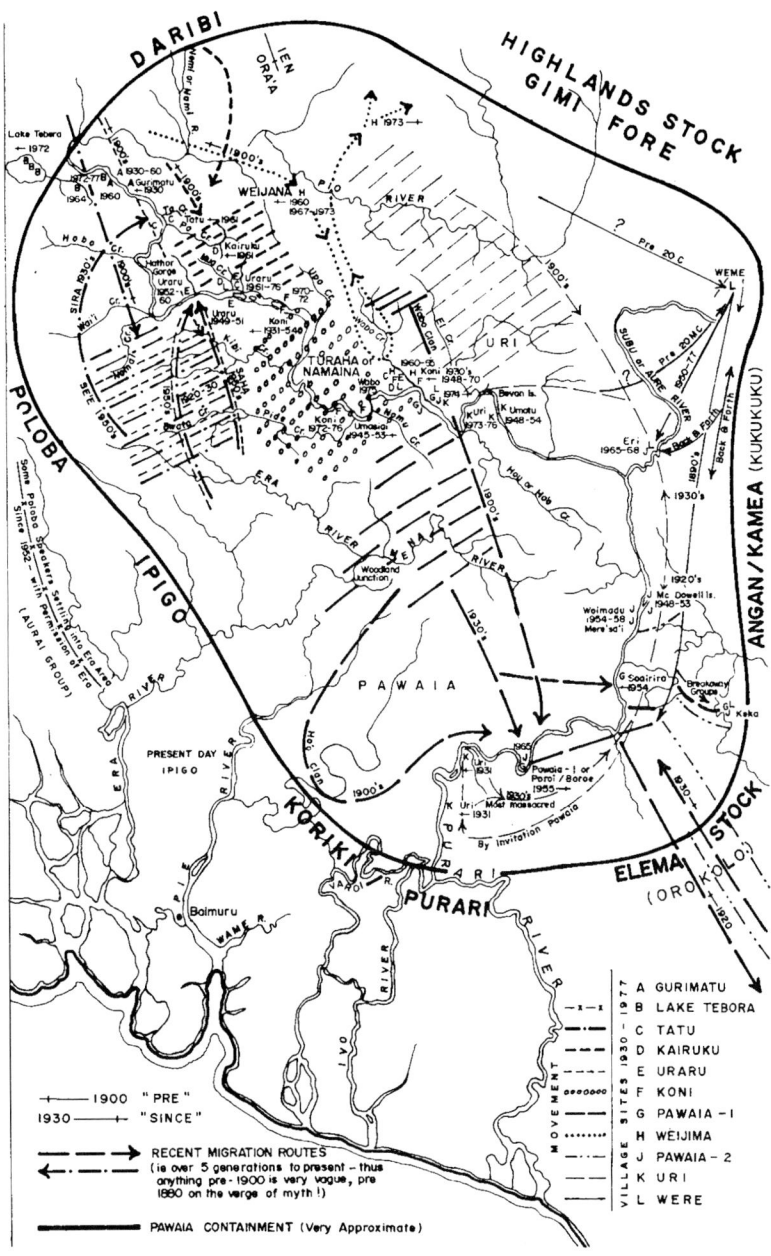

Fig. 11. Upper Purari Census Division. Marked are approximate locations of previous 'village' sites, areas of occupation by individual groups, and probable migration patterns.

445

'Pawaia, hitherto classified as an isolate, shows more relationship to the Teboran family than to anything else with which it has been compared. Since it is lexically closer to Polopa than to Daribi, further studies in Polopa in particular may well uncover additional links in both structure and lexicon between it and Pawaia.'

In 1982, Mr and Mrs N. Anderson of S.I.L. have completed ten years study of Polopa, having lived in Pupitau, the largest village of that group. Perhaps future publications by the Andersons may throw further light on the relationship between Pawaia and Polopa.

Of interest too is the fact that MacDonald has compared notes with another S.I.L. linguist (Shaw) who has worked amongst the Kubo Samo near Nomad in the Western Province and each have found a number of similarities in the languages (MacDonald, personal communication).

The earliest recorded medical research carried out in the upper Purari was undertaken by C. Gajdusek in 1957 when he patrolled from Okapa (Eastern Highlands Province) into the Gulf Province by way of Weme and the Aure River to the middle Purari. The main purpose of this patrol was to check on the spread of kuru disease.

Later the Chief Government Leprologist summarised his work in the Karimui area, including Sena River Pawaia, as follows:

'Results of a blood group survey of the Karimui people, together with tests for salivary ABH secretion over a wider area, are reported. Certain aspects of these findings point to coastal affinities of the subjects. These are the higher B than A gene frequency, the occurrence of Le (a +) individuals of ABH non-secretors, and the identification of thalassaemia haemoglobin H disease in the population. Thus to some extent the oral traditions of these people, which indicate a general eastward and later northerly movement, are supported.' (Russell 1971).

Writing on his short research in the upper Purari, Nurse (1980) concludes that:

'In the course of their evolution as a people, and despite their apparent isolation, they (the Pawaia) appear to have received genetic contributions both from the Dadibi of the highland fringe and from the peoples of the Gulf Coast.'

Pawaia origins thus remain obscure. It is not, at this time, possible to determine exactly where they came from and via which route. We can by no means be certain, either, as to what period in time they entered the upper Purari.

Karl Franklin wrote to me in 1975, when he was Director of the Summer Institute of Linguistics:

'It is obvious that . . . the Pawaia groups have occupied these areas for a very ancient period of time, and that (formerly) the Pawaia have roamed down much further south than the area (which) they now occupy. If there can be any migration theory at all, it would be that the northernmost penetrations are more recent, but beyond that there is not a great deal that we can say at present.'

446

However, as Egloff and Kaiku (this volume) point out, the Pawaia have oral traditions of claiming a recent arrival. But then one should always treat oral history of time periods with caution. Many times have I heard, or read, of something which happened with or to a P.N.G. group 'within the last hundred years or so'. As pointed out earlier, P.N.G. pre-literate societies can seldom trace back more than four generations.

6. The Pawaia and their neighbours – boundaries and relationships

We have already discussed the probability of the Pawaia having moved generally eastwards into relatively unpopulated country. Populations to their west, south and east are relatively sparse and there is much 'no man's land' (which was, and is, exploited only on a hunting and gathering basis) between the Pawaia and most of their neighbours (Warrillow 1978). Exact boundaries are not known in most cases and generally remain fluid.

Even so, there were of course some marriage and trade links between the Pawaia and most, if not all, of their neighbours. There was also traditional ('tribal') warfare between the Pawaia and their neighbours (Warrillow 1978; Egloff and Kaiku 1978; Nurse 1980; Toft 1980). But so too was there fighting (and still much today in other areas) between the different clans of the same linguistic groups.

Nurse (1980) believes that:

'The most satisfactory reconciliation of the evidence from both language and oral history would make the Pawaia products of a very ancient fission of an original stock with the Dadibi (Daribi); who then later divided into northern and southern wings, both moving eastward, the northern to become the modern Dadibi and the southern the Polopa.'

My own research into Pawaia history, based on the evidence then available but relying quite heavily on what I had learnt from George MacDonald (personal communication) led me to the following conclusions, which were detailed in Warrillow (1978).

The Pawaia moved generally eastwards, perhaps along or south of the lower Erave valley, and exploited uninhabited country above and immediately west and south of Hathor Gorge in the upper Purari. They then moved in several directions. Uraru clans moved into the Era and Mena areas. With canoe technology various clans of Pawaia No. 1 and No. 2 moved directly down the Purari itself, some of which headed east to the Vailala and, for a time, beyond. Moves to the north of Hathor Gorge appear to have been later thwarted by the Daribi and the Tatu and Kairuku clans were pushed back south of the Pio. The Gurimatu clans are the only ones which appear to have remained permanently on the river above the gorge.

There was, however, little resistance to the east up the Pio. This migration split with some clans again moving north whilst others continued further east. The latter (including the Weijana) claimed land in the upper Pio and its northern tributaries with a few even crossing the divide into the upper Aure tributaries (the Weme). The former claimed land along the Sena River and northeast slopes of Karimui from where they expanded westward to the Boisa River where again the Daribi were met.

From the Boisa these Pawaia retreated, back eastwards, to where they have remained settled in the Sena River area. These Pawaia (the 'Tundawe') are rapidly changing, both culturally and linguistically, as they increasingly come under Daribi influence. MacDonald (1977) noted that 'the marriage pattern seems to involve more Daribi women marrying into Pawaia villages than vice versa, with the result that their children learned Daribi first . . . '.

Warrillow (1978) touches on each of the Pawaia neighbours, and discusses in some detail the similarities between the Daribi and Pawaia. There is a similarity in their food and their housing. Both the Pawaia and Daribi have sweet potato as their staple, and the majority of both do not live on a major river, navigable by canoe. However, for many Pawaia sago is the staple food and canoe technology has been mastered, the same being valid for some Daribi (the 'Uribidi' of Kele and Suani), although the latter did not use paddles until at least 1965 – an anomaly noted by Faithorn's patrol in 1929. Toft (1980) suggested, however, that the fact that the Pawaia are sago eaters and river dwellers whilst the Daribi are sweet potato eaters and non-riparian, indicates major cultural differences between these two peoples.

Another common life-style suggesting further evidence of past cultural similarities of the Pawaia and Daribi is their housing. Glasse (1965) suggested that Daribi double-storied quonset hut type houses (*sigibe*) might be unique in Papua New Guinea. The author in his earlier paper (Warrillow 1978) cited a 1938 patrol report by Foldi which describes a similar type of house in the upper Purari village of Sira (today's Tatu). As early as 1894 Sir William McGregor described houses on what we now know to be Koni land, as being 10.5 m by 6 m with a ground floor and two upper storeys, the third floor being some 6 m above the ground (B.N.G.A.R. 1893 – 94). This type of house, called a *bouhabo* by the Pawaia, is certainly different to anything built by neighbouring groups, including the Polopa who build very large 'long-houses' (men's houses) under which pigs are sometimes housed. Polopa women occupy small houses around the main one. Both Pawaia and Daribi have not traditionally, it seems, had the same strict segregation rules for men and

Table 1. Land and population patterns in the area of the future Wabo impoundment.

Village	Approximate area of land owned (km^2)	Population May, 1977	Land to be inundated (km^2)	Total % of lake	Total % of land owned	Area available after inundation (km^2)	Per person now(km^2)	Per person after inundation (km^2)
Gurimatu	not known	89	less than 1	–	–	land loss minimal	–	–
Kairuku	not known	130	nil	–	–	–	–	–
Koni	600	80	260	83	40	360	7.5	4.5
Pawaia No. 1	1300	89	6	2	0.5	1294	14.6	14.5
Tatu	not known	80	less than 1	–	–	land loss minimal	–	–
Uraru	300	53	44	15	5	256	5.7	4.8

449

women. Both could live under one roof – either on separate floors or in a separate half of a suitable partitioned building. I have never seen, or heard of, a Polopa woman entering a long-house.

7. The future

The formation of a new reservoir at Wabo would lead to a minimal displacement, and to a limited loss of land, in spite of flooding some 260 to 290 km^2 of land. Only Koni village of perhaps 80 people would lose much land. 83% of the land to be inundated would be Koni owned (Table 1), and their present holdings of 7.5 km^2 per person would be reduced to 4.5 km^2 per person.

The Pawaia of the upper Purari would not lose their means of subsistence. But will they want to continue on, as throughout history, a semi-nomadic people locked into a subsistence economy? History to date has been fairly kind to the Pawaia with slow, gentle changes over several generations. The construction of the Wabo Dam would bring traumatic change in less than a generation.

References

Bevan, T.F., 1890. Toil, travel and discovery in British New Guinea. Kegan Paul, Trench, Truber and Co., London.

B.N.G. A.R.: British New Guinea Annual Reports, 1892/93, 1893/94.

Carne, J.E., 1913. Bulletin No. 1. Notes on the occurrence of coal petroleum, and copper in Papua. N.S.W. Govt. Printer, Sydney.

Clune, F., 1942. Last of the Australian explorers: the story of Donald Mackay. Angus and Robertson, Sydney.

Egloff, B.J. and O. Kaiku, 1983. Prehistory and paths in the upper Purari River basin. This volume, Chapter III, 3.

Glasse, R.M., 1965. Leprosy at Karimui. Papua and New Guinea Med. J. 8: 95 – 98.

Hides, J.G., 1938. Savages in serge. Angus and Robertson, Sydney.

Holmes, J.H., 1924. In primitive New Guinea. Seely Service and Co., Ltd., London.

Hope, P., 1979. Long ago is far away. Australian National University Press, Canberra.

Hughes, I.M., 1977. New Guinea stone age trade. Terra Australis 3. Australian National University Press, Canberra.

MacDonald, G.E., 1973. The Teberan language family. In: K. Franklin (ed.)., The Linguistic Situation in the Gulf District and Adjacent Areas, Papua New Guinea, pp. 111 – 148. Pacific Linguistics Series C, No. 26. Australian National University, Canberra.

MacDonald, G.E., 1977. The language situation in the Karimui areas. Mimeo. Summer Institute of Linguistics, Papua New Guinea.

Nurse, G.T., 1980. The Pawaia as agents and patients. In: M. Alpers (ed.), Purari River (Wabo) Hydroelectric Scheme Environmental Studies, Vol. 8: Viral and parasitic infections of the people of the Purari River and mosquito vectors in the area, pp. 67–84. Office of Environment and Conservation, Waigani, Papua New Guinea.

P.A.R.: Papuan Annual Reports, 1909/1910, 1917/18, 1918/19, 1922/23, 1927/28.

Russell, D.A., S.C. Wigley, D.C. Vincin, G.C. Scott, P.B. Booth and R.T. Simmons, 1971. Blood groups and salivary ABH secretion of inhabitants of the Karimui Plateau and adjoining areas of the New Guinea highlands. Human Biology in Oceania 1: 79–89.

Sinclair, J., 1969. The outside man: Jack Hides of Papua. Lansdowne Press, Melbourne.

Toft, S., 1980. Social study of the Pawaia, Papua New Guinea. In: T. Petr (ed.), Purari River (Wabo) Hydroelectric Scheme Environmental Studies, Vol. 12, pp. 1–94. Office of Environment and Conservation, Waigani, Papua New Guinea.

Toft, S., 1983. The Pawaia of the Purari River. Social aspects. This volume, Chapter III, 2.

Trefry, D., 1969. A comparative study of Kuman and Pawaian. Pacific Linguistics Series B, No. 13. Australian National University Press, Canberra.

Village Directory, 1973. Department of the Chief Minister and Development Administration, Port Moresby, Papua New Guinea.

Wagner, R., 1967. The curse of the Souw: principles of Daribi clan. Definition and alliances in New Guinea. University of Chicago Press.

Warrillow, C., 1978. The Pawaia of the upper Purari, Gulf Province, Papua New Guinea. In: T. Petr (ed.), Purari River (Wabo) Hydroelectric Scheme Environmental Studies, Vol. 4, pp. 1–88. Office of Environment and Conservation, Waigani, Papua New Guinea.

2. The Pawaia of the Purari River: social aspects

S. Toft

1. Introduction

A select group of 613 Pawaia (Table 1), inhabiting the area along the Purari between Gurimatu and Pawaia No. 2 were studied for two months by a research group of 6 people from the Papua New Guinea Administrative College. After a period of interview and participant observation a questionnaire was compiled which included direct questions seeking Pawaia opinions on the proposed hydroelectric scheme. Questionnaires were answered by 42% of the adult population (over 18 years of age). For details see Toft (1980).

A number of these Pawaia have land and houses in the proposed Wabo Dam inundation area. During engineering studies for the hydroelectric scheme most of the Pawaia moved to the dam site where they established three distinct settlements linked by canoe. The largest settlement, referred to as Wabo, is occupied mainly by people who originate from the proposed dam inundation area upstream. The smaller settlements known as Abe and Peri lie on opposite sides of the main river downstream of Wabo just below Gleeson Island. People originally from below the inundation area live there.

In 1977, following the completion of engineering feasibility studies, the research on which this chapter is based took place. Its purpose was to identify characteristics of Pawaia daily life and social organisation. This would form a basis for decisions regarding the socio-economic future of the Pawaia should the dam be built, and would provide an ethnographic record of a social group not previously studied and whose traditional way of life is swiftly changing.

Petr, T. (ed.) The Purari – Tropical environment of a high rainfall river basin
© *1983, Dr W. Junk Publishers, The Hague / Boston / Lancaster*
ISBN-13: 978-94-009-7265-0

Table 1. 1977 (May) census, villages of the Wabo Dam inundation area and other Pawaia villages of the upper Purari River included in the survey (Fig. 1).

Village	Totals (excluding absentees)				Absentees (resident outside electorate)				1977 Grand total
	Under 18 years		18 years and over		Under 18 years		18 years and over		
	M	F	M	F	M	F	M	F	
Gurimatu	17	11	28	27			3	3	89
Kairuku	18	23	40	43		3	2	1	130
Koni	19	14	13	21			11	2	80
Pawaia No. 1	19	10	26	33			1		89
Pawaia No. 2	8	8	20	12			6		54
Tatu	16	14	20	21		5	3	1	80
Uraru	13	10	13	16			1		53
Uri	7	3	14	13		1			38
Total	117	93	174	186		9	27	7	613

Survey sample was 42% of adult population of 360 (174 males + 186 females = 360).

454

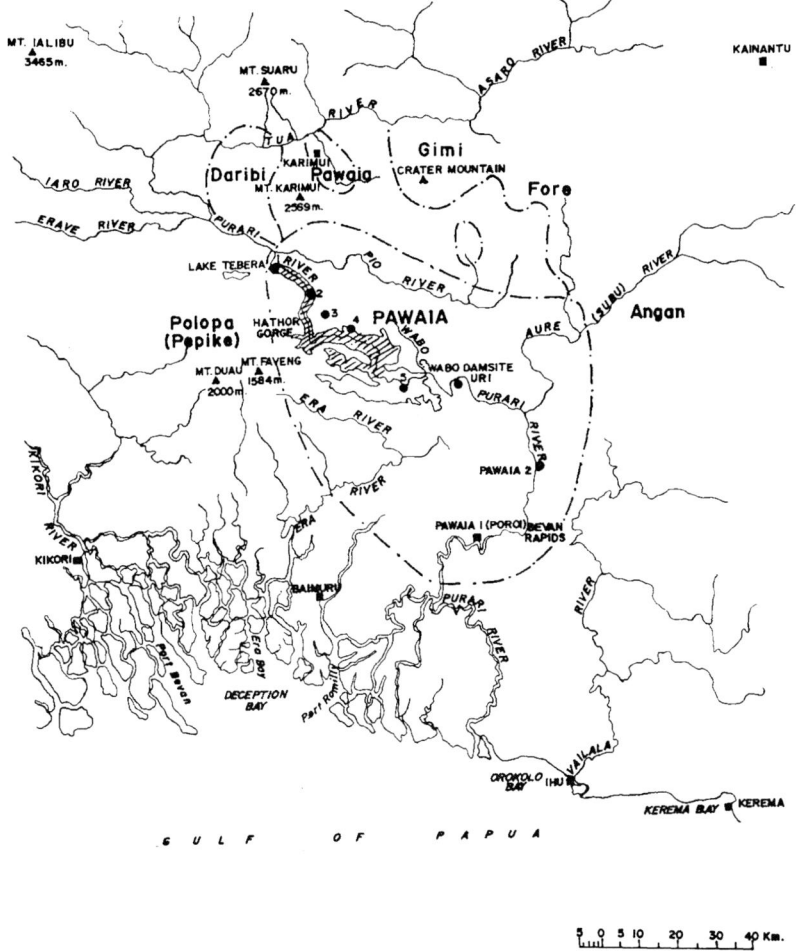

Fig. 1. Approximate group boundaries (after Egloff and Kaiku 1978). Key: 1 – Gurimatu; 2 – Tatu; 3 – Kairuku; 4 – Uraru; 5 – Kone.

2. Daily life 1977

The Pawaia day usually begins in mist or drizzle. Each family has slept in its own house (Fig. 2), men occupying one part and women another. Small children still at the breast sleep with their mother.

The separation of sexes is upheld in many different aspects of daily life. Women are generally expected to avoid contact with men to whom they are

455

Fig. 2. A typical Pawaia house, near Uraru. (Photo T. Petr.)

not closely related, even to avoid being seen by them. Male visitors are not usually invited into a house with women present, but if men do enter, women are expected to sit to one side hidden by a bark-cloth cloak. Although this behaviour demonstrates respect for the visitor it is also believed that if a visitor admired a woman he might become envious and use magic to seduce her or to harm her husband. Another reason for the custom is women's fear that they will be assaulted. It is believed that women can pollute men and this is particularly so during menstruation. For instance they may not help in the preparation of food during that time. Childbirth is another time of danger through pollution. It is believed that if contaminated the child's father might become weak and grow old prematurely.

Boys are taught at a young age to avoid women, and a strict separation of men and women occurs between the time a boy reaches puberty and his marriage. The boy moves into a communal house for unmarried men and mixes exclusively with males, leading a celibate life until marriage, which may not take place until he is as old as thirty. Food is delivered to the communal house from family houses.

It is customary for women to reside with their husband's family, often from the age of eight or nine years. After a man marries he considers establishing

456

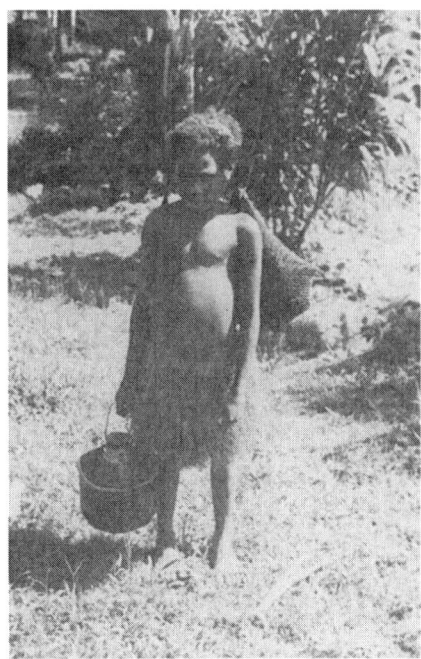

Figs. 3 and 4. Modern and traditional dress at Wabo. Note the quill nose ornament. (Photo A. Hall.)

his own household, though he may wait until his wife bears a child. In cases where polygamy is practised the co-wives share the same house.

Pawaia women have responsibility for preparing food and looking after children. Traditionally food is baked either by being placed directly on the fire or by being placed first into a bamboo segment. Sago powder is dropped into a bamboo segment, occasionally layered with another ingredient such as fish. The bamboo is then placed on smouldering embers until it has charred and the sago has coagulated into a dry though slightly glutinous mass. Traditionally, bamboo was also used for carrying and storing water. Today, although bamboo is used widely, most households also own metal pans in which food can be boiled.

Two other cooking styles are smoking and mu-mu, when food is wrapped in leaves and baked among hot stones which are often covered with earth. Mu-mu is a preferred way to cook meat but smoking is commonly used to cook or preserve fish or meat. The most commonly eaten foods are sago, cooking banana, cassava (tapioca), taro, birds and lizards. The Pawaia have no set meal

457

Fig. 5. An older man smoking a bamboo pipe. (Photo T. Petr.)

times. They eat when they are hungry, but the early morning and evening are times when families may chance to eat together.

Female responsibilities around the home include weaving bags for processing sago, making skirts (Fig. 3) (fibres for both bags and skirts are strands from the sago palm leaf), bark-cloth capes, string bags (from tree bark) and personal decorations: earrings from cassowary quills, shell necklaces and bone ornaments. Some western style dresses are worn (Fig. 4), usually over the traditional skirts. During the survey men were only seen in western style clothes.

Men who are around the home area during the day build canoes, decorate bamboo pipes (Fig. 5), make bows and arrows, haft axe heads and make their own dress ornaments. Young men sometimes make Jews' harps out of bamboo and bush string.

Household possessions are few, and the total Pawaia material culture is limited in a way typical of people who are not sedentary and who therefore keep a minimum number of objects to carry around. However, non-traditional possessions have entered the area and Table 2 gives an idea of these

458

Table 2. Traditional and non-traditional household possessions of the Pawaia.

Traditional	Imported and now commonly owned	Imported and owned by only a few people
stone blade axe	metal blade axe	outboard motor (3)
stone blade adze	metal blade adze	shotgun
spear	bush knife	coleman lamp
bow & arrows	hand knife	metal file
sago beater	coconut scraper	patrol box (1)
sago bag	clothes	mosquito net (1)
sago filter	pans	watch
string-bag	bowls	saw
articles of apparel	blanket	cups
smoking pipe	pillow	plates
	pamphlets & posters	floor-mat
		radio (1)
		fishing net (1)

items plus the usual traditional household possessions.

Although the Pawaia have perhaps experienced less colonial or western contact than most other Papua New Guinea tribal groups, as was the case with those other groups, one of the first aspects of social organisation to be affected by outside contact was the traditional political system. In the old days, leaders were usually good warriors and had killed, preferably an opposing leader, in battle. This showed not only strength and bravery but also the possession of supernatural powers. Another essential for a traditional leader was an ability to accumulate wealth in wives, shell-money and pigs. Strong leaders also had sufficient influence in society to organise or prevent pay-back killings, decide whether or not to go to war and give safe passage to outsiders in their area. They were admired for abilities in oratory, organisation and daily skills. Today status is given to those men who have managed to assimilate aspects of western culture, yet who also operate successfully within the traditional framework.

3. Subsistence

The basic Pawaia household group is also the basic unit of economic production. Nearly all routine daily activities are directed towards the production and consumption of food. The Pawaia subsist to a large extent on food from the forest, and the population fragments into mobile family groups which move into the bush for weeks at a time. Sago is the dietary staple, and its processing, done at the palm site in the forest by family members, necessitates the group living in bush camps and leading in many respects a hunter – gatherer style of life. It is common to see a man, his wife and small children, paddling upstream from Wabo, their canoe loaded for their forest camp and one person holding a smouldering stick to start their fire. Often there will be two or three canoes of an extended family travelling together. Most of the sago camps are located near the banks of waterways – creeks or the main river. At the camp site a temporary shelter is built.

Most of the Pawaia sago palms are planted, though wild ones are also used. Apart from the infrequent clearing of undergrowth whilst the tree is a sapling, there is no cultivation. It is preferable not to fell a sago tree until it is twenty years old*, by which time it has reached a good size, usually over ten metres long, and can be expected to yield five bundles of processed sago. Each bundle weighs several kilos and will feed a man, two wives and four children for five to six days. An average nuclear family will consume a tree in five to six weeks. Sago production is arduous and lengthy. It is usual to process two trees on each expedition and two women, working intermittently and helped by children, take six weeks to complete this task**. So ten bundles, taking six weeks to produce, last only ten weeks – four weeks in excess of the period of labour. Thus the Pawaia are required to spend at least 60% of their year working in sago camps. Whilst in the forest the women also take time to gather wild vegetables and fruits which supplement the diet.

It is sometimes possible when a sago palm grows near a waterway to fell it and then float the trunk downstream to a convenient site for working, but a majority of trees are not so well placed. Men fell the palms and strip bark from the upper side of the fallen trunk. Women, sometimes helped by men, then begin to hack out the sago pith from the trunk interior with special beaters. Leaching the sago from the pith is women's work and requires a specially constructed apparatus which includes the woven sago bags mentioned before (Figs. 6 – 11).

Women were not seen unchaperoned at sago camps, but whilst based at the camps men make the opportunity to move out into the forest to hunt. The hunting catch is considered second to sago in importance in the diet and time spent at the forest camp provides an opportunity to stock up on smoked meat

---

* It is possible that at Wabo conditions do not allow for optimal growth of sago palms, as all reliable documentation indicates that *Metroxylon* palms flower and die long before reaching this age (editorial note).

** For the Baroi of the delta it takes 10 days to complete the same task – see Ulijaszek, this volume (editorial note).

Fig. 6. Planted sago palms (lower) and coconut palms (taller) in the Purari Delta near Kapuna. (Photo T. Petr.)

Fig. 7. Sago stems partially worked out for sago. (Photo T. Petr.)

461

Fig. 8. The sago pith is pounded with wooden tools. (Photo B. Egloff.)

Fig. 9. The apparatus used for excavating sago. (Photo B. Egloff.)

Fig. 10. Washing and filtering the sago pith. (Photo B. Egloff.)

Fig. 11. The sago starch settles in tank. Note the filter bag to the right to tap the pith. (Photo B. Egloff.)

to take back to Wabo. Traditionally men could only hunt within strict territorial limits. Now it is said that permission is given to most Pawaia who do not have hunting rights in the Wabo area to hunt nearby, though this is only likely to be through kinship ties. Hunting trips are on a two to three day basis and are directed towards small objects, such as bandicoots, lizards, birds and grubs. Large hunting trips in preparation for a feast may last for one or two weeks in which case the game is butchered and partly cooked for preservation. Wild pig and cassowary are highly sought on such occasions. Hunting methods are based on trapping, tracking (often with dogs) and shooting, both with bow and arrow and with guns.

The Pawaia make little use of the rich supply of fish in their area. As spearing with bow and arrow is the usual method of catching fish and prawns, it is predominantly in the shallow river tributaries where any fishing is done.

Although the Pawaia say that gardens are as important a source of food as the sago palm, observations did not support this claim. Compared with people who subsist almost entirely on garden produce, the Pawaia spend only a small part of the working year in their gardens. Gardens are mainly dotted sparsely along the banks of the waterways. The cultivation system is based on the slash-and-burn technique, but the cleared bush is often allowed to rot or is dumped in the waterways when it is too wet to burn. It is men's responsibility to help women clear the bush, but women are the cultivators and should initiate any work. Weeding is done periodically by the women but gardens are not well maintained. The most important crops are cooking bananas, taro and cassava, but yam and, in well-drained areas, sweet-potato are also grown. Various other plants are included in gardens, but the Pawaia do not depend on gardens to supplement the sago in their diet as they also forage for wild plants in the forest.

4. Descent and land rights

The Pawaia practise patrilineal descent. Thus, nuclear families are almost always tied to the man's line of descent: a woman leaves her own family through marriage to legally and physically join her husband's kin group and her children will belong to her husband's family. Inheritance passes through men and the most important legacies relate to subsistence: land-use rights for gardens, hunting rights, fishing rights, sago stands and other food-yielding trees and plants. When a woman marries she shares all her husband's rights.

The Pawaia identify social grouping most readily in two ways, with regard to marriage and with regard to land: people who they may or may not marry

464

and people with whom they do or do not share land-use rights. There was no evidence that the Pawaia acknowledge the concept of clan. There is no equivalent for the word clan in the Pawaia language (MacDonald 1973, 1977), and the terms kin group or lineage are used in this report.

Kin groups share villages with other kin groups. Each group uses land in the surrounding area which belongs exclusively to its own lineage. Confusion concerning land occurs when those with legal rights choose to assert them over those who claim rights through usage over time (people who have 'borrowed' land over an extended period). Land disputes are a major cause of conflict, second only in importance to disputes over women. Although the inheritance of land rights is technically handed down through men only, a married woman can be given permission by her brothers to use their land for temporary gardens. A widow can have clear land rights through her own family if her husband dies without achieving a satisfactory level of bride-service, in which case the children belong to her and not her husband's family.

5. Marriage

A man who wishes to marry approaches the bride of his choice by presenting her parents with a small, stylised, ornamental arrow, on the shaft of which is tied a small cluster of sago fibres symbolising a grass skirt. The arrangement is sealed by a gift of yams from the groom's family to the bride's. That presentation symbolises the bethrothal or 'marking' of the bride, but what binds the relationship is continued service and payment. From then onwards the man renders services to the bride's family until he or she dies. These services include the periodic supply of food or objects, such as weapons or grass skirts, and work for the benefit of the bride's household depending on its needs.

The marking of a bride often takes place at her birth. The choice of partners is limited by kin relationships. When a man reaches maturity he may mark a girl for himself but it cannot be done beforehand on his behalf by his family. Consequently, men of eighteen years and older are marking babies, and the normal age gap between spouses is at least a generation. The reason given for this is that the custom of polygamy causes a shortage of women. There are hardly ever any girls who are still available for marriage by the time they reach puberty. When the girl reaches seven or eight years of age she goes to live with her future husband's family where she is trained by her future mother-in-law until she is nubile.

In addition to the usual form of marriage just described there are two other types. One is marriage by exchange, when two men exchange their sisters as wives. The other is where women are given as compensation, either in exchange for land-use rights, and in some cases land ownership, or in peace negotiations as compensation for a death. In these cases bride-service is usually waived.

Marriage between lineage groups is finely prescribed. The term intermarriage refers to marriage between members of different residential groups. Intermarriage between Pawaia village groups is increasing, to some extent because of a desire to claim land-use rights at Wabo through kinship ties. Marriages take place occasionally between Pawaia and their neighbours in the Karimui, Okapa and Vailala areas. Marriage, it is maintained, causes conflict and fighting and the majority of all Pawaia disputes are over women.

Separation and divorce are uncommon. If a husband leaves his wife he will cease to perform bride-service and is not entitled to the return of payments made, though he may accept a substitute wife from the same kin group. If a woman leaves her husband repayments of gifts must be made. Traditionally a man could be entitled to kill his wife for committing or being suspected of committing adultery. This could in turn provoke her family into a pay-back killing. However, the girl's family may accept that rights in her have been transferred to the husband's group and may not attempt to revenge her death at his hand. It would seem to depend on the standing of the husband with the girl's family. When a woman is aware of her husband's adultery, she often remains silent for fear of arousing his aggression.

Because of the disparity in ages usually found between a woman and her first husband it is common for a woman to experience widowhood during her child-bearing years. There then follows a mourning period of up to two years during which time the widow is free to mix in society but does not take a man or remarry. A widow usually has no choice regarding her second marriage as it is in the hands of her two kin groups, her own and her dead husband's. It is usual for her to pass to a brother or other close kin of the husband. The status of children whose mother remarries in this way does not alter. However if a widow marries outside her husband's family it is usual for the children to remain with their father's family, in which case there is no refund of bride-service payments.

If a child is orphaned, it is absorbed into the adopting family, taking a formal place as a member of that kin group with all the attached rights and obligations. The adopting family is often closely related to the orphan.

6. Beliefs

The Pawaia consistently claim that there are no Pawaia sorcerers. Sorcerers are male and they come from the Irohi area of Polopa speakers to the west, but they can be employed by the Pawaia. If a sorcerer wishes to kill someone he first paralyses his victim through use of a certain type of stone. Whilst paralysed the victim is hit across the throat with a club to suffocate and thus kill him. The sorcerer then resurrects the victim, giving him a specific lease on life which can range from two days to a month. The limit always expires as forecast and the victim duly dies. After a death attributed to sorcery the dead person's relatives attempt to identify the 'murderer' through a form of divination. Sorcery in general is blamed for deaths. It is also used to attract or seduce women. Jealousy is seen as a reason for someone to resort to sorcery, hence leaders, or 'big-men', are vulnerable. Fear of sorcery is thus a social leveller, a means of maintaining egalitarianism, because to avoid being victims of jealousy the successful share their benefits.

The traditional type of Pawaia magic most openly referred to is for hunting. Gardening magic usually relates to spells spoken at the time of planting. There is little doubt that Christian ideas have permeated and influenced beliefs, or, at least, what people say about their beliefs. There is still however observance of beliefs in ancestor and other spirits, especially those of the forest. These spirits can have both good and evil influences within the environment and on the lives of the living.

Traditional rites are also performed, particularly to mark events concerning the passage from one phase of life to another such as births, or marriage. These ceremonies comprise feasting and dancing over variable periods of time but initiation rights have shown a sharp decrease in frequency. Initiation ceremonies used to take place at intervals of a few years. The initiation group of boys and girls who had reached the age of puberty would be celebrated by feasting and dancing over a period of twelve months during which time tribal wars would be planned. Little attention was paid to gardening at this time and the people ate mainly wild leaves and the bark of certain trees.

7. Discussion

The Pawaia have accepted and adapted to change in the past, most of which has been imposed by neighbours who have driven them by degrees to their present environment (Warrillow 1978, this volume). More recently changes have been brought about by contact with the west, first through government administration and then by missionaries. Government contact was limited to infrequent visits by patrol officers and mission contact, when it came, was spasmodic and confusing. The London Missionary Society, now known as the United Church, first visited the area in the 1950s, and established a post and small airstrip at Uraru in 1961. Between 1963 and 1975 the Seventh Day Adventist Mission ran a school at Koni, whilst between 1967 and 1972 expatriate Bahai missionaries made annual visits to the villages downstream of Wabo (Warrillow 1978). At the time of study, in 1977, the United Church was flying medical aid into the area by float-plane. This was to bring nurses for a few hours to hold a maternal and child health care clinic. There was also a New Tribes Mission station at Weijana, a Pawaia village to the north of the main inundation area at Haia in the Simbu Province.

Major changes which government and mission agents introduced were: a village style of settlement and the coastal design for houses which physically reshaped community life, a reduction in warfare and therefore easier contact with neighbours, a different set of values and beliefs some of which seem to have been accepted, a different style of dress (especially for men), new technology (including medicine), new languages (Motu and Pidgin), and a cash economy. Some of these changes, for example the settlement pattern, were imposed in such a way that they were soon adopted. But it is impossible to judge the extent of changes in social organisation generated by this because of the lack of detailed information about the pre-contact system. Some of the new concepts, such as Christian beliefs, have been to some extent adopted and incorporated within the traditional system of beliefs. Other changes have not had time to show their effect and the full impact of many changes is still to be realised.

Engineering studies for the hydroelectric scheme which took place between 1971 and 1977 brought the Pawaia further western contacts. A number of Pawaia men worked spasmodically as casual labourers for the engineering companies, capital from compensation paid by the government for land was used to establish a trade store which stocked items not previously available in the area and, in 1976, the government opened a primary school and medical aid-post. There were also influences from infra-structural changes caused by the engineering activities, including frequent flights in and out of the area, the arrival of vast quantities of strange cargo, and an influx of strange people with strange ways.

An important area of change which is currently influencing the Pawaia is related to cash. The desire for cash and a felt need for it is an important break from traditional subsistence economic values. Papua New Guineans in most areas have traditionally used money (shell) and traded goods inter-tribally.

Many of them, including the Pawaia, have also been achievement oriented: a man could achieve status through his own endeavours and personal qualities. Magic, or the fear of it, and competition were the levelling agents used to prevent individuals from becoming too powerful. The obligation to share wealth once it had been accumulated also helped to equalise society, whilst gaining status for the giver. Consequently it is not altogether surprising that Pawaia men should feel it necessary to acquire cash in order to compete and achieve within the modern world. The desire for cash can thus be interpreted as reflecting a desire to join the new system, or at least as an acceptance of the inevitable rise of the cash economy. During the past two decades there have been odd opportunities for Pawaia men to earn cash within their area through government and other development activities, and the possibility of employment at the dam was an incentive to move to Wabo. Some Pawaia men have also moved temporarily out of their own area for periods of employment. The men are more cash oriented than the women who prefer barter. Women have no opportunity to make use of cash except for the purchase of odd items at the recently established trade store. A man however can today include cash in bride-service payments and use money to travel outside his area.

Probably the greatest change which the Pawaia have experienced is their move to Wabo. The Wabo conglomerate of three settlements (Wabo, Abe and Peri) brings the people together into what is probably their largest ever residential group.

The traditional semi-sedentary life style also means that the Pawaia are accustomed to travel away from their base for food. Now they are travelling further. The fact that there is no change in the basic procedure of food production could well be an important reason why the Pawaia have made the move to Wabo without undue complaint. The present situation regarding the need to move away from the village for food collection does not therefore seem to pose a problem, but this could change as travelling will increase with the need to procure a higher percentage of the food supply out of the Wabo area. By the end of 1977 the land around Wabo was being more heavily used than ever before. People were not having to return to their old subsistence areas as often as they will have to once the carrying capacity of the land at Wabo is exhausted under the traditional system of exploitation.

Government records claim that there is no land shortage in the upper Purari River Pawaia area, but although the total area of land may seem adequate for a population density of approximately 1 person per 5 square kilometres, its carrying capacity is low. There is no doubt that the Pawaia are striving in a harsh environment to subsist from their land. It is true that they could be better nourished if they caught more of the available fish. But frequent fish eating is not a tradition. Habits die hard, and it appears that traditional foodstuffs are in short supply both because of the inadequacy of traditional environmental conditions and because of the present crowding at Wabo.

It is difficult to identify or assess social change occurring at Wabo today as a result of the resettlement. This is partly because the resident groups had been there only two years before the study took place, so that the effects of change were not fully manifested, and partly because of the short duration of the study. It would seem however that as pressure on land becomes greater, disputes over entitlements to its use will increase. It also seems likely that disputes over women will become even more frequent because the social distance women have traditionally kept from men is not as easy to maintain or regulate in a larger community. It would then follow that accusations of sorcery will increase in instances where it is believed to be motivated by jealousy over women. In fact an overall increase in sorcery accusations could possibly occur as an expression of the increased friction in inter-personal relations arising within the new settlement. It must be remembered that misfortunes have traditionally been attributed to and explained by beliefs in magic and religion, but an increase of accusations on the more personal level of sorcery would not be surprising in the new social situation.

In contrast, a positive effect of the coming together of the Pawaia at Wabo is that in the face of change they may well benefit from the opportunity to increase group identity and solidarity. Increased social interaction between relatives and friends could be an attraction of the new settlement.

In conjunction with the move to Wabo at the end of 1975, a new school was established there which opened with one grade in March 1976. In November 1977 there were thirty-eight students in two grades, only three of whom were girls. This small number of girls is not surprising in view of the traditional customs regarding the seclusion of females and marriage, including a young girl's early residence with her prospective parents-in-law. Examination of the school attendance sheets did not reveal absences of the sort that implied children were being kept out of school to move with their family on trips into the forest. An aid-post was also established in 1976 but, probably for reasons relating to the belief that illness is usually due to magic, up to late 1977 had not been seen by the Pawaia as an important advent (for medical aspects, see Hall, this volume).

A significant development in the area is the all-weather airstrip which was completed in 1976. Although planes were not new to the Pawaia the all-weather strip created more frequent contact with the outside world. Cargo was moved into Wabo by barge from Baimuru on a large scale since 1974 for the purpose of dam feasibility studies including the airstrip construction.

Planes and barges also brought in people. These included expatriates of Nippon-Koei and the Snowy Mountains Engineering Corporation (the companies involved in the hydroelectric project) but also Papua New Guinean strangers, who appeared to the Pawaia as sophisticated and superior. It is perhaps the influx of strange people with their bewildering activities, equipment and habits which has had the strongest impact on the Pawaia. Amongst these strangers are people of the Purari Delta region, mainly from Mapaio, who have formed a Purari Action Group, expressing an aim to safeguard the interests of the people and the area, but also to claim land they maintain was originally owned by their ancestors. If the claims are upheld financial compensation for inundated land would be payable.

Those Pawaia men who have travelled out of their local area, who have visited towns in the Gulf Province or in the highlands, and the handful who have been to Port Moresby, compare what they have seen there with the upper Purari and are depressed. They feel they are a neglected and an underprivileged group. They stated clearly in interview that they want their share of national development.

If the proposed hydroelectric scheme goes ahead and inundation takes place, much of the present means of subsistence in the upper Purari will disappear. The hydroelectric scheme would flood 260 to 290 km^2 of some 10 000 km^2 claimed by the Pawaia and not contested by other ethnic groups (Warrillow 1978). However the land that would be effected today lies along the main stream of the river and has been at the nucleus of Pawaia activities. There seem to be two alternatives open to the Pawaia, one is resettlement away from the upper Purari, the other is planned development within the developing upper Purari area, either along the reservoir shore or at Wabo.

Although the Pawaia believe in forest and ancestor spirits, there is no detectable attachment to land for mystical of supernatural reasons. The Pawaia are attached to their land because it is theirs and because it is at present their only source of subsistence. Most people return to their old village area for sago processing and hunting, however there is no reason to believe that if adequate and suitable land in another area was fairly distributed, it would not be a viable substitute. But what have the Pawaia to gain from this? They would be transferring to another area which would be strange and yet show no significant improvement on their present one. Although fatalistic in approach, partly as a result of increasing outside contact which provides a basis for comparison, the people feel that life is already hard and that it could be improved.

The Pawaia are shown to be poorly nourished (Lambert, this volume). They display this in their physical appearance, lethargic manner, limited energy and high mortality rate. In addition, sago stands take a minimum of seven years to mature, and should only be planted by a person on his own land. This could be a considerable problem in the organisation of resettlement. The move would have to be phased-in over a minimum of seven years. Overall, resettlement 'to preserve traditional culture' is not to be recommended in the Pawaia context.

An alternative to resettlement would be to remain in the area of the upper Purari. The Pawaia could be left alone to subsist along the lake shores of the impoundment area, but they would doubtless feel very strongly the impact of general physical and social changes introduced at Wabo by dam construction. It would seem impossible that they could remain in isolation within the dam area.

The hydroelectric project would also result in an influx of outsiders: consultants, managers, skilled staff, migrant labourers and personnel. Where a population is concentrated the government can provide better services more cheaply. The situation would need to be carefully monitored and accommodated by an effectively co-ordinated system of social services in which Pawaia interests are safeguarded. Health, agriculture and education extension services need to be interrelated, and operated in consultation with welfare and community development experts. A group of specialists should be formed when a firm decision to build the dam is made, with the responsibility of integrating the Pawaia into the development scheme.

96% of the Pawaia questioned said they would be happy to stay at Wabo if a dam is built. Only one respondent gave a negative answer and there were three people who did not respond. 46% of the people questioned said that if the dam is not built they would prefer to return to their own area over which they exercise customary land rights — this is in fact a high figure when correlated with the information that of the 52% opting to stay at Wabo, many of them have their origins near there, at Koni or Uri for example. If the dam is built and if the Pawaia then stay in the upper Purari area, individuals will be able to choose whether to pursue a subsistence economy on carefully reallocated land above the waters of the inundation area, or whether to participate in the construction of and services for the hydroelectric scheme.

A township at Wabo, near the hydroelectric scheme, could provide employment for men and a cash income to which many of them are already attracted. Problems could arise in a town settlement because of the social distance customary between men and women. The number of disputes over women would be expected to become more frequent and women would have difficulty in adjusting to a more exposed environment. However if women could be channelled into useful production their new role could be made easier. The introduction of a high protein crop, such as the winged bean, would supplement the diet whilst adding valuable nutrients. In this way the women could make an important improvement in the living standards of their people. They might also begin to appreciate any cash earned by surplus crops. Pawaia women have not had as much good gardening experience as many other Papua New Guinea women, and change to full-time gardening might be difficult. It can take a long time for agricultural innovations to be accepted, but if skilled and experienced extension workers were to promote innovation on an intensive level there would be some chance of success.

Fishing would perhaps be an alternative and preferable form of agricultural production for women. It would be a source of protein for the home diet and cash if developed into an industry. The extension agent would again play a vital role. The new reservoir at Wabo would provide a fine opportunity for such activity (Haines 1979).

The present Pawaia staple is sago. The arduous and lengthy work involved in sago production was frequently referred to with resignation. A liking for rice was evidenced by attempts to barter for it with members of the survey team, and in a cash economy there is reason to believe it could substitute for sago in the diet, even though it involves boiling, which is not a traditional method of cooking. There would probably always be a limited demand for sago. People everywhere tend to retain a genuine or sentimental liking for traditional foods, but it would become a special food eaten only on occasions. Sago production in the Wabo area would have to be safeguarded to this extent, but it would seem unlikely and undesirable, both from nutritional and practical points of view, that sago could be retained as the staple food in a developed Wabo.

The Papua New Guinea government is promoting change in the name of economic development, and if the economic climate is favourable the Wabo Dam scheme could be one result of that policy. Whatever the decision for the future of the scheme, the Pawaia have already started to depart from their traditional way of life. Outside contacts are already changing their socio-economic and physical environment. The financial compensation which the Pawaia would be paid and the government obligation to provide services, if the dam is built, would give an opportunity for the Pawaia to embark on a new life which might offer a future brighter than was their past. It is not possible to predict exactly how the psyche would respond in the face of major change. Even to stay continuously in one place could cause stress for people who have been used to constant movement. But it is surely rational in present day Papua New Guinea to direct the Pawaia towards the type of development the government and people are promoting in other parts of the country, rather than away from it.

Acknowledgements

I thank my research assistants Michael Bite, Kila Kini, James Koibo, Albert Rovi and John Yauwi who took part in the field work whilst students at the Administrative College of Papua New Guinea. The results are theirs, gained through observation, interview and the labourious completion of question- naires. I am also very grateful to Mark Turner and David Doulman who kind- ly spared time to comment on the draft of this chapter.

References

Egloff, B.J. and R. Kaiku, 1978. An archaeological and ethnographic survey of the Purari River (Wabo) dam site and reservoir. In: T. Petr (ed.), Purari River (Wabo) Hydroelectric Scheme Environmental Studies, Vol. 5, pp. 1 – 53. Office of Environment and Conservation, Wai- gani, Papua New Guinea.

Hall, A., 1983. Health and diseases of the people of the upper and lower Purari. This volume, Chapter III, 4.

Lambert, J., 1983. Nutritional study of the people of the Wabo and Ihu areas, Gulf Province. This volume, Chapter III, 9.

MacDonald, G.E., 1973. The Teberan language family. In: K. Franklin (ed.), The Linguistic Situation in the Gulf District and Adjacent Areas, Papua New Guinea, pp. 111 – 148. Pacific Linguistics Series C, No. 26. Australian National University, Canberra.

MacDonald, G.E., 1977. The language situation in the Karimui areas. Mimeo. Summer Institute of Linguistics, Papua New Guinea.

Toft, S., 1980. A social survey of the Pawaia of the upper Purari River. In: T. Petr (ed.), Purari River (Wabo) Hydroelectric Scheme Environmental Studies, Vol. 12, pp. 1 – 94. Office of Environment and Conservation, Waigani, Papua New Guinea.

Warrillow, C., 1978. The Pawaia of the Upper Purari, Gulf Province, Papua New Guinea. In: T. Petr (ed.), Purari River (Wabo) Hydroelectric Scheme Environmental Studies, Vol. 4, pp. 1 – 88. Office of Environment and Conservation, Waigani, Papua New Guinea.

Warrillow, C., 1983. A short history of the upper Purari and the Pawaia people. This volume, Chapter III, 1.

3. Prehistory and paths in the upper Purari River basin

B.J. Egloff and O. Kaiku

1. Introduction

One hundred kilometres inland from the Gulf of Papua, on the very fringes of the highlands of Papua New Guinea reside the Pawaia. The dense rain forest bordering the upper Purari River basin is the homeland of approximately 800 Pawaia-speaking people. From Gurimatu to the north of the Hathor Gorge to Uri just south of the proposed Wabo Dam site, the population is very sparse indeed (Figs. 1, 2). The everyday affairs of the Pawaia revolve around the *kombati* or temporary camp which is established as the need arises to exploit a variety of resources. Sago stands are of paramount importance to the Pawaia; however, these are interspersed with gardens and stands of tended nut trees. *Kombati* structures are fragile and would leave little evidence for the prehistoric record. No prehistoric habitation site had been recorded in the basin. The rough terrain and lush vegetation when coupled with the imperma- nent material culture of the Pawaia people would not be conducive to the preservation of prehistoric sites. The activities carried out on the site of the *kombati* would also leave very little physical evidence behind insofar as the materials used in these daily activities are for the most part equally perishable. One might expect at the best to find some scattered charcoal, perhaps a few bits of stone tools, discarded animal bones and the occasional shell ornament, lost or broken. Nor large midden heaps of shell fish would be found, nor would one expect to find deposits of discarded pottery nor sizeable quantities of waste flakes remaining from flaked tool production.

The traditional life-style of the Pawaia is marked by features which could be said to be associated with people on a threshold. They are not truly seden- tary nor are the Pawaia nomadic. Agriculture is present but the planting and tending of the sago palm is apparently of greater importance. Ethnographical- ly little is known in detail of this life-style nor is it known, although it is some-

Petr, T. (ed.) The Purari – Tropical environment of a high rainfall river basin
© *1983, Dr W. Junk Publishers, The Hague / Boston / Lancaster*
ISBN-13: 978-94-009-7265-0

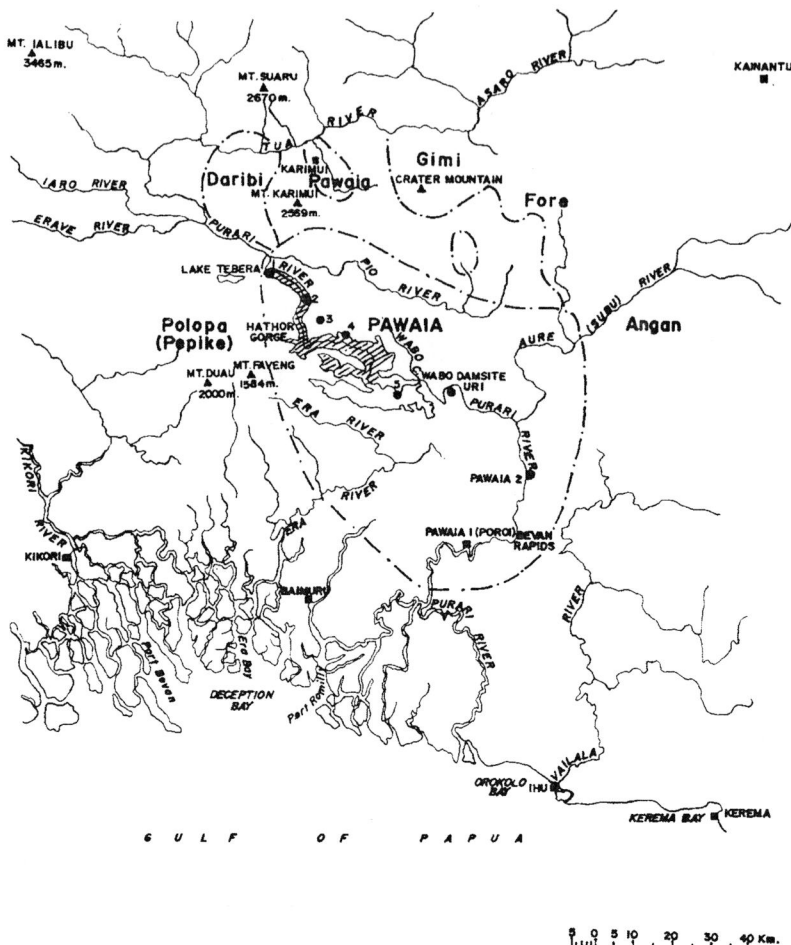

Fig. 1. Approximate group boundaries. 1. Gurimatu 2. Tatu 3. Kairuku 4. Uraru 5. Kone.

times assumed, that this pattern would have been common during prehistoric times, prior to the introduction of tropical root crops.

The survey party was interested in obtaining information which would lead them to prehistoric sites as well as information which could be used to interpret sites if they were located. A wide range of material culture was collected as well as lists of useful plants and animals whose remains might be found in an archaeological site. While collecting data outlining environmental exploitation an attempt was made to formulate some sort of a plan of the daily pattern of life. This led to eliciting information about social patterns which might be

476

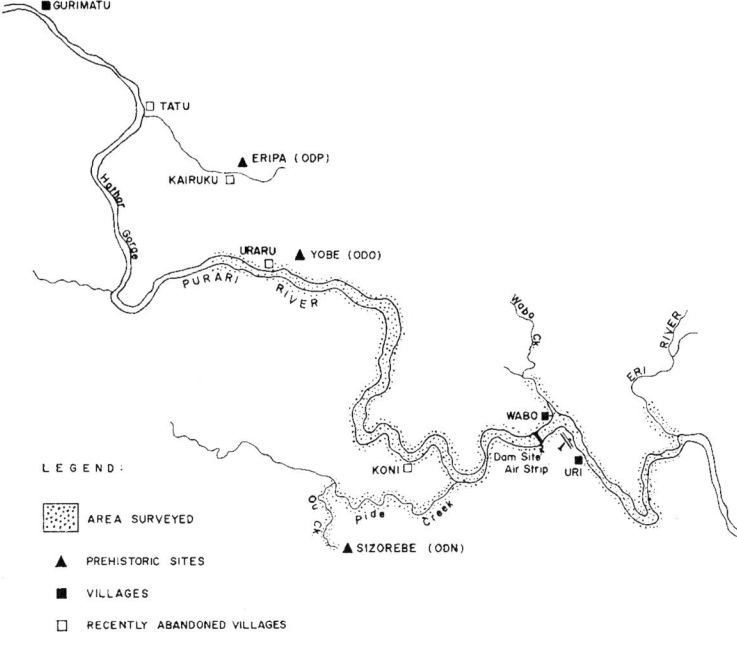

Fig. 2. The Purari between Hathor Gorge and the Eri River. Survey area and prehistoric sites (ODP, ODO, and ODN NIUGINI are survey codes for archeological sites).

related to environmental exploitation. Throughout this process, moving from the collection of material culture to detailed lists of exploited flora and fauna followed by a review of social patterns, it was apparent that external contacts were of importance. Artifacts were collected from Pawaia which often had an external source. Plant and animal products were often traded as well as individuals being recruited into Pawaia society through marriage. Many men had travelled on traditional routes to trade with adjacent groups. To some extent Pawaia culture can only be understood as that part of a melange of cultures which lies between the coast and the highlands, receiving goods from both as well as interacting with adjacent piedmont peoples to the east and west.

Swadling (1975) compiled a listing of all known prehistoric sites within the greater Purari River basin. Only one site is known within the area to be affected by the Wabo hydroelectric dam. That site is a salt seep opposite the village of Gurimatu, at the end of the planned Wabo reservoir.

2. The archaeological survey

The survey commenced on the 21st of October and finished on the 8th of November 1977. The archaeological aspect of the survey lasted until the end of November. At this time emphasis was shifted from following bush tracks into fairly inaccessible locations, distant from the proposed impoundment area, to a careful inquiry with informants in their basin settlements. This shift became imperative as sites were not being located in the impoundment area and as what information remained pointed towards locations within extremely inaccessible areas three to four days walk from the basin. Home-base was established at Wabo camp and contacts made with villagers across the Purari in Wabo. A canoe was procured from one individual and a motor from another. A limited amount of petrol and oil was acquired from Wabo camp supplies. Thus mobility of some sort was assured upon the first day. Thereafter it was a matter of forging upstream against constant flood waters, dodging mud banks and huge logs, consuming quantities of petrol in the process. In this fashion the river was travelled from below Wabo Creek up to and beyond Uraru (Fig. 2). Tributaries of the Purari were investigated on a number of occasions. Travel was limited to following positive leads which on one occasion took the survey up Pide Creek and on another venture to Uraru and thence overland to the abandoned village of Kairuku. Other trips searched for informants in an attempt to interview elders from all Pawaia groups holding land rights within the basin.

Three shelters were located during the course of the survey.

2.1. Sizorebe (Fig. 3, 4)

Upstream from the dam site, approximately 7 km, is the confluence of Pide Creek and the Purari. The lower reaches of the Pide meander in a sluggish fashion, the banks dotted with occasional sago stands and gardens. Further upstream to the west, the current quickens as the river bed becomes quite shallow. About midway in the Pide, Ou Creek joins it from the south. Following heavy rains the Pide is deep enough to easily canoe as far as the Ou and to continue for a short distance up the Ou. The Ou meanders between incised banks two or three metres in height. Approximately one kilometre up the Ou, on the left hand bank, is the *kombati* belonging to the survey guide, Kenome (of Koni).

This camp consists of a small shelter about 3 metres wide by 6 metres long, floor raised about 40 cm above the ground with a thickly thatched roof peak-

478

Figs. 3 and 4. Sizorebe shelter on the Ou, photographs taken from the top of the talus towards the cliff face.

ing at about 2.6 metres. Rain began to fall at about 1500 hrs and continued until 0900 hours the following day. It is significant to note that this camp was the only habitation sighted on Ou Creek aside from the *kombati* at its junction with the Pide.

After a short canoe ride up the Ou from the *kombati* of Kenome, the canoe was tied to the bank and four hours of walking commenced; up the river, over small ridges which the Ou meandered, gorges were encountered and the bed of the Ou changed from gravel to a jumble of large rocks. As Kenome had not been to the Sizorebe shelter since he was a small child, he had some difficulty in locating it, entailing an exhausting climb of the wrong mountain.

Sizorebe shelter lies high above Sizorebe Creek, a tributary of the Ou. Scrambling up a boulder strewn slope one is faced by a large lunate shaped scarp with a marked overhang. To stand and gaze at the shelter is dangerous as a spray of water cascades over the edge and large pieces of shale are continually being dislodged. Small seams of coal are visible in the cliff face.

The scarp is approximately 100 metres long, having a height of 12 to 14 metres from the base to the lip. A continuous flow of water cascades over the edge weakening and dislodging the soft sandy shale thus producing a ridge of talus with a height of 2 to 3 metres. The upper layer of the talus, appears to be recently deposited not being covered by moss or small plants.

Kenome claims that as a child when he sheltered with his parents, the floor was flat, the ground was soft underneath, and the great quantity of fallen shale was not present.

The overhang is sizeable and could have sheltered a number of people. The only way to determine if the shelter was regularly used would be to remove the boulders with block and tackle while working under a protective timber shield. This shelter probably lies above the reservoir.

2.2. Yobe (Fig. 5)

While camping at a *kombati* east of Uraru on the trail to Kairuku, Mr Maruwai Puri led the survey party up the steep slopes of nearby Pubel Mountain to an area of karst limestone near the crest. Within this jumble of fissures and standing limestone blocks is a very small overhang named Yobe. Apparently at some time in the past this spot had been used as a refuge by the father of Maruwai.

The floor area is approximately 3 metres wide by 2.5 metres deep. A small test pit was dug yielding cranial bones at a depth of circa 25 cm. A hard ash layer is present on the surface of the shelter floor. Underneath this thin layer is

Fig. 5. Excavating at Yobe shelter, site ODO.

a sterile reddish to buff-coloured sandy soil. The site would appear to have little potential for excavation.

2.3. Eripa (Figs. 6 – 8)

Approximately twenty minutes hike following a small river in a northwesterly direction from the abandoned village of Kairuku, is a pronounced limestone formation. This formation is clearly exposed at the crest of the ridge which forms the western wall of the valley. Rock overhangs at a lower level close to the river are filled with flood sediments.

Eripa is the name given to a complex of caves which have formed within this formation (Erebe of Warrillow 1978). Three large entrances are present, each opening is a considerable distance from the other. Only the first entrance encountered appeared to be suitable for excavation. The second entrance was located after descending a steep slope into the cave, reaching a series of well-developed passage ways. Approximately midway along this passage system is a large cavern with a ceiling height on the order of 20 metres. The cavern displays elaborate drip-stone 'cathedral' formations. Circa 20 minutes of walk-

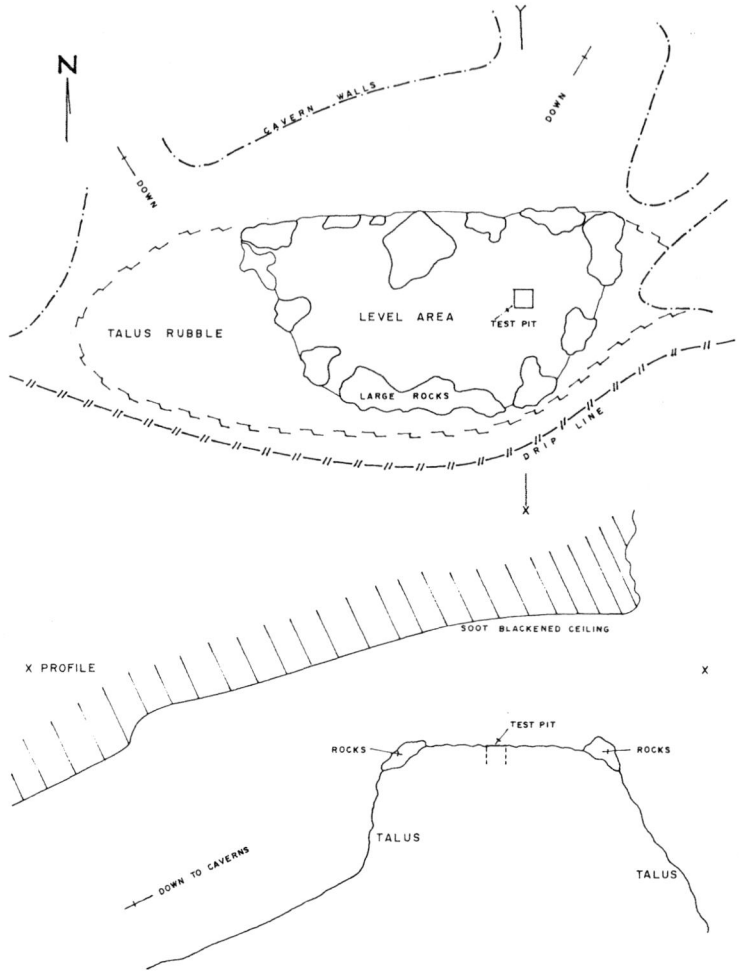

Fig. 6. Excavation location at Eripa (ODP), first entrance.

ing and a steep climb of a talus slope brings one to the second mouth. Directly outside of the mouth is a vertical drop to the valley below. The third entrance is reached by descending into the cave and taking a second passage from the main cavern, and proceeding along passages again climbing a steep rubble slope.

The three entrances to Eripa are similar in form, each having a large open mouth at least 15 to 20 m wide and 3 to 15 m high. There is a large ridge of talus partially filling each mouth with the passage ways of the cave system situated a considerable distance below the crest of the talus slope. In all cases the talus is quite steep, only at the first entrance is there enough level ground for a

482

Figs. 7 and 8. Excavations at Eripa, site ODP.

person to lie down. At the first and the third entrance human skulls are tucked in niches. The second entrance is partially blocked by a small stone wall which Maruwai said was built during times of tribal war.

An excavation was dug at the crest of the first entrance. A 50 cm by 50 cm excavation was dug by trowel to a depth of 40 cm, all soil was screened. At that depth the soil fell through crevices in the underlying talus and only a loose jumble of rocks was encountered.

To a depth of 11 to 15 cm small chips of a dark chert were found as well as small bird or bat bones. Small bones were found to a depth of approximately 23 cm. Below 28 cm nothing of significance was recovered. Although charcoal was present the confused and shallow nature of the deposit would render any date determination tenuous.

The presence of small chert chips indicates that some form of activity took place at the cave mouth. Today, the caves passages are well known and frequently used when hunting bats, the permanent inhabitants of the cave. That the cave mouths were used as burial places is evidenced by the presence of human bones.

The three shelters described above share the following characteristics or features: (i) they were used ethnographically as shelters; (ii) traces of casual use are present; (iii) they are formed in unstable rock formations and evidence continual roof collapse; (iv) because of factors three and four, they present a relatively difficult situation with respect to excavation; (v) extensive prehistoric remains (i.e. depth of deposit) were not located.

2.4. Further considerations

Two shelters were mentioned by informants but were not investigated by the survey party.

When at Uraru, where only a few dilapidated buildings now stand, an elderly man, Iero Touai, said that he knew of a shelter over the mountain and down by the river. This was interpreted as being in the Hathor Gorge. Iero Touai seemed to be reluctant to discuss the matter further stating only that the place was 'nogut tru'. This information could be similar that R. Wagner received concerning shelters in the gorge of the Tua River (Egloff 1977). As the Purari was in full flood it was not possible to enter the gorge. The gorge is an unstable geological formation, witness a recent landslide presumably causing a blockage in the gorge (Warrillow 1978).

A man at Uri named Pawai Eri now living on the Aure (Subu) River but formerly of Wea'mi (Weme) discussed a shelter commonly used by trading parties heading northwards towards the Karimui Plateau. The shelter is named Neaweimijoe. It is said to have a level ground floor which will accommodate 'many' sleeping men and 'rocks do not fall from the roof'. The shelter is said to be about two hours walk from Wea'mi (Weme) or about two days distance from Wabo via the Aure River by motor canoe and then walking, stopping one night on the trail. The shelter could then be within the reservoir area of the Aure and should be investigated if action is taken towards the construction of the dam.

3. Discussion

The picture which emerges is not one which gladdens the heart of an archaeologist. The Purari has hidden its prehistory and only through chance or a very time-consuming survey will it be revealed.

Continual questioning of Pawaia failed to produce information locating any significant sites within the impoundment area. As we were led to shelters there can be no doubt that informants knew what the members of the survey were looking for. It is always possible to argue that locations may not have been discussed as they were regarded as sacred. Although as informants readily revealed the location of a burial cave it is doubtful if information was withheld on this basis. Quite a number of factors no doubt combined to produce a minimal input: (i) the Purari basin is relatively uninhabited and is densely vegetated. Visibility is low and perhaps sites are not known to the Pawaia as they themselves have oral traditions claiming a recent arrival and there has been continual movement of groups within the basin (Warrillow 1978); (ii) the guides and informants utilised could have been only selectively knowledgeable on the terrain of the basin, or may not have revealed the location of sensitive sites; (iii) that due to the instability of rock formations all shelter sites have gradually been destroyed by natural forces or perhaps were not used as it is quite easy within this environment to construct a shelter of bush material.

To the southwest, prehistoric research 25 km upstream on the Kikori River has suggested a more or less stable land use pattern over the last one and a half thousand years (Rhoads 1980).

A detailed pattern of traditional Pawaia land uses seems to have disappeared along with their traditional communal houses just at the time of European contact in the 1930s (Warrillow 1978). It is then difficult to compare the study by Rhoads of the Kikori peoples with the Pawaia research. It does seem as if the Kikori peoples are definitely more coastal than piedmont and in that respect evidence a more stable prehistory than the Pawaia who would appear to have been in a state of constant flux.

Trade contacts throughout this period for the people of the Kikori River were both with the inland as well as the coastal people. There appears to be a marked intensification involving highland axes and coastal pots relatively recently perhaps just before European contact (Rhoads 1980).

Some attempt was made by the survey party to record the intensity and nature of external contacts. It is reasonable to assume that contemporary traditional contact would in some way resemble or maintain vestiges of patterns which operated during prehistoric times. 'Pax Britannica' has enabled contact to be intensified between previously hostile groups, still the Pawaia of the upper Purari definitely maintain their closest traditional links with the people of the highlands. Contact with the comparatively 'rich' cultures of the Gulf is fairly limited.

The epic legend of the Daribi (Wagner 1967) which narrates the adventures of the culture hero Souw is mirrored in the Pawaia myth of Soi (see Egloff and Kaiku 1978: Appendix 1). Soi and Souw provide the Pawaia and Daribi with a perspective of life and death. While the Daribi epic journey of Souw lies to the north and east, the trail of Soi goes south. Here he meets with coastal groups and encourages them to attack the Pawaia, thus institutionalising the hostility between these groups.

Although there was continual hostility between neighbouring peoples there was also considerable traffic in goods.

Trade to the peoples of the upper Purari basin is an important facet of life. Pawaia in the north at Gurimatu, and formerly at Tatu and Kairuku, traded northwards on to the Karimui Plateau. Uraru (now almost abandoned) in a central position traded eastward into the Weijana (Weijano) area on the Pio River and westward with scattered isolated groups of people referred to as Pepike (Polopa speakers). Pawaia further south on the Purari traded westward with the Pepike at the headwaters of the Pide as well as southwards to Aure. At the eastern extreme the groups at or near Weme on the upper Aure did not trade with their hostile neighbour, the Angan speakers to the east, rather with Pawaia to the south at Soairira (Pawaia No. 2), and also with Pawaia groups to the west in the area of Hau. Apparently some axe blades and salt came into the Purari basin from the groups to the northeast, the Fore.

An outline of sources of goods and exchange commodities follows. Note that no distinction has been made between trade and exchange.

Karimui: those Pawaia speakers living on the Karimui Plateau in the area near the Patrol Post (Warrillow refers to these Pawaia as Sena River). On occasion the term is used to include the Weijana (Pawaia groups) on the Pio River.

given	received
cuscus fur	tobacco
crocodile teeth	women (betrothed)
bird-of-paradise feathers	stone axes
red and black parrot feathers	arrows
eagle feathers	kina shells
pig meat	net bags
cassowary (dead & alive)	nassa shell ornaments

Polopa (*Pepike*): area west of the Purari River, headwaters of Pide Creek. For more information see Warrillow (1978).

given	received
black bird-of paradise feathers	stone axes
tapa cloth capes	women (betrothed)

Aurai: a Polopa (Pepike)-speaking village to the southwest of the Purari, on a tributary of the Era.

given	received
arrows	shell nose ornaments
tobacco	cone shell discs
net bags	women (betrothed)
stone axes	
women (betrothed)	

Hau: Pawaia groups at the headwaters of the Pio traded with people of Wei'me (Weme).

given	received
women (betrothed)	body ornaments
bird feathers	women (betrothed)
	kina shells
	stone axes
	white cockatoo feathers

Soairira: (referred to as So'o, which Warrillow notes as a clan of Pawaia No. 2), Pawaia group on the lower Purari south of Wei'me (Weme). Objects traded to Soairira would then go further south to Mapaio. Pre-1940 objects from the coastal *hiri* trade were filtering into the Pawaia.

given	received
women (betrothed)	shell nose ornaments
stone axes	shell discs
shell ornaments	shell ear ornaments

tobacco
arrows
fire makers
string bags

Much of the exchange of goods is linked with marriage payments. A man marks a young girl as a wife to be and then proceeds to make a number of payments to her kin. A good example comes from a young man living in Wabo. Apparently, about 20 years ago the father of Mathew's mother was killed while travelling from Tatu to Wen'a'ajoe (near Karimui) to collect a girl he had marked. Following this treachery, the people of Wen'a'ajoe owed a woman to the Tatu people. Mathew's betrothed will balance the affair. The young girl sought after by Mathew was born in 1966. In 1973, 1974 and 1975 Mathew made trips from Wabo north to Wen'a'ajoe. On these three trips he took a bed sleeve, 80 kina in currency, 3 bush knives, 3 axes, 6 bird-of-paradise, 7 pigs (probably refers in part to wild pigs and meat, Warrillow, personal communication), 1 dead cassowary and a live cassowary chick. In 1976 he posted 20 kina in currency to the girl's father. The girl's father gave them three bundles of tobacco and three net bags. On the first two trips about 12 men went with Mathew while on the third only his brother accompanied him. He still has more to pay and will apparently do so throughout his life (see Warrillow 1978).

Trips to the south have been regularly taken by Horiaiwai whose family has strong ties in Aurai, at the head of the Era River. The informant comes from the group which holds land on Pide Creek and regards itself as Koni people. They traditionally had links to the south. The father of Horiaiwai has a sister who was married and living in Aurai. Her husband died and she has since returned to Wabo. The first trip Horiaiwai took was as a young lad with his father and four other men. They took tobacco, sago bags, tapa, and net bags. They brought back European cloth, shell discs and shell nose ornaments. The second time, Horiaiwai went with four men and one woman. They took the same goods as before and received five shell discs and three shell nose ornaments. The third trip was in 1972 when he went with three men. They took three bundles of tobacco, three net bags, three sago bags and some arrows. In return they received three nose ornaments, two shell discs and three bags. In 1975 Horiaiwai and his wife and three men went to tell the people at Aurai (actually only went as far as Era Maipua) that his father had died. No goods were taken or received during this trip.

An interesting facet of the trip between the Pide and the Era is that canoes can only be taken as far as the mid-waters of the Pide; at that point they are left behind and the party walks overland to the headwaters of the Era. There they build canoes and make sago; if no women are in the party, the sago is made by men. They then proceed down the Era in the new canoes.

Perhaps one of the reasons for taking walks overland to headwaters of the Era is to exploit sago stands cultivated on the land owned by the party members. En route to these sago stands, the party is often involved in hunting and in gathering edible plants.

Another interesting point relates to arrows which are widely traded. From a strictly functional point of view there seems to be no purpose in this exchange. However it is not unusual for functionally similar items to be exchanged, perhaps in some cases they are regarded as being better because of their exotic status or as having additional magical powers. Besides being better because of their exotic status or as having additional magical powers, the exchange of arrows between the Pawaia and other tribal groups serves to strengthen their social relationship. This is evident from a number of arrows given in return to some of the Pawaian informants who had claims over women from the groups concerned.

Axes are said to have entered the upper Purari basin from a number of sources. Salt is said to have come only within recent times through the Pawaia at Weme who obtained it from the previously hostile Angan speakers to the east or the Fore to the northeast.

Travel time varied considerably. The trip from the Pide to the Era took an unspecified time as sago and canoes were made. Three days were spent on the trail from Wabo to Haia while it took two nights on the trail from Wei'me (Weme) to Hau. The people from Hau used to make this trip to attend *singsings* in Wei'me.

Ian Hughes in his epic study of traditional trade in the highlands of New Guinea regards trade as flowing up hill (Hughes 1977). It is interesting to note that women figured highly in terms of 'contact stimulus' as recorded by Toft (1980). That goods flowed into the upper Purari from the middle Wahgi Valley in the central highlands is confirmed by an axe collected in 1915 below the Aure, on the Purari. That axe has been identified as coming from the Abiamp quarry (Hughes 1977).

The Pawaia are a low consumption society when compared with the cultures of the Gulf and highlands, and as such appear to have few external needs and indeed have little to offer other than forest produce. They definitely have not played the role of middlemen in their strategic position on a major waterway between the highlands and the Gulf. Traffic between these two major culture areas was limited and certainly did not flow against or for that matter with the current of the Purari River.

Newton (1961) speculates that settlement of the Gulf may have been over the central range from the Sepik River basin on the north coast. If that was the case then little remains within recent traditional practices that would support such a movement. Warrillow does observe a pattern of settlement which is very indecisive, moving towards the Gulf and then retracting (Warrillow 1978).

4. Summary

Criteria for evaluating the success of an archaeological survey vary. Although no significant sites were located within the impoundment of the proposed Wabo hydroelectric dam, three sites were located outside the impoundment. However to excavate these sites would present considerable mechanical as well as logistic problems. The current survey lists sites of potential archaeological significance, as they appeared from a single limited field trip. More intensive research could possibly discover more sites. It is also possible that no sites of significance have been missed during the present survey, although that is very doubtful.

The Pawaia follow a pattern of life which is neither truly that of sedentary agriculturists nor is it typical of hunters and gatherers. It is an adaptation of available technology to the exploitation of an environment which lies somewhere between these two classic patterns. The Pawaia way of life is of interest to prehistorians as archaeological sites are excavated elsewhere in the country which appear to reflect just such a pattern. If archaeological sites were found within the impoundment and firm associations could be drawn between the pattern of life being expressed by the prehistoric remains and the traditional Pawaia pattern of exploitation, the results would be invaluable. There are few places in this world where a comparison between people following a semi-sedentary life can be made directly with archaeological data, both expressing the same environmental content. Information from the sites located in the area when integrated with the Pawaia ethnographic data provides for a deepening insight into a previously unfathomed part of Papua New Guinea.

Acknowledgements

The Office of Environment and Conservation sponsored the fieldwork as part of the Purari River (Wabo) Hydroelectric Scheme Environmental Studies. The National Museum provided financial assistance and later, direction by Pamela Swadling, Curator of Prehistory. Sincere thanks go to Mr Heni Ipai for narrating 'The curse of Soi' myth, and also to those who have been mentioned in the text for their contribution.

Three weeks is an impossibly short time to conduct a survey within such a rugged and extensive river basin. Fortunately C. Warrillow introduced us to Maruwai Puri who recruited Davey Wo'wai and the Sori brothers, Kenome and Horiaiwai. As informants and guides these four individuals made the survey possible, providing a maximum amount of mobility by acting as the motor canoe crew.

References

Egloff, B.J., 1977. Ancestral and prehistoric sites at the Wabo dam site and the future impoundment area. In: T. Petr (ed.), Purari River (Wabo) Hydroelectric Scheme Environmental Studies, Vol. 1: Workshop 6 May 1977, pp. 65–66. Office of Environment and Conservation, Waigani, Papua New Guinea.

Egloff, B.J. and R. Kaiku, 1978. An archaeological and ethnographic survey of the Purari River (Wabo) dam site and reservoir. In: T. Petr (ed.), Purari River (Wabo) Hydroelectric Scheme Environmental Studies, Vol. 5, pp. 1–53. Office of Environment and Conservation, Waigani, Papua New Guinea.

Hughes, I., 1977. New Guinea stone age trade. Terra Australis 3. Australian National University, Canberra.

Newton, D., 1961. Art styles of the Papuan Gulf. The Museum of Primitive Art, New York.

Rhoads, J.W., 1980. Through a glass darkley. Ph.D. Thesis. Australian National University, Canberra.

Swadling, P., 1975. Ancestral and prehistoric sites in the Purari River Basin. University of Papua New Guinea, Port Moresby.

Toft, S., 1980. A social survey of the Pawaia of the Upper Purari River. In: T. Petr (ed.), Purari River (Wabo) Hydroelectric Scheme Environmental Studies, Vol. 12, pp. 1–94. Office of Environment and Conservation, Waigani, Papua New Guinea.

Wagner, R., 1967. The Curse of Souw: principles of Daribi clan definition and alliance. The University of Chicago Press, Chicago.

Warrillow, C., 1978. The Pawaia of the Upper Purai (Gulf Province), Papua New Guinea. In: T. Petr (ed.), Purari River (Wabo) Hydroelectric Scheme Environmental Studies, Vol. 4, pp. 1–88. Office of Environment and Conservation, Waigani, Papua New Guinea.

4. Health and diseases of the people of the upper and lower Purari

A.J. Hall

1. Introduction

This chapter deals with the health status of the riparian people living along the Purari River and inhabiting the area bordered in the north by Hathor Gorge, and by the Gulf of Papua in the south. The ethnic groups investigated comprise the Pawaia living around Wabo, and the Namau who live in the Purari Delta. If in the future a hydroelectric dam is to be constructed at Wabo (Fig. 1), these people will be directly affected by changes caused by this project to the environment.

2. Sources of medical information and population dynamics

The census, records of the health services and surveys carried out as part of the routine work of the Gulf Province Health Department have provided the baseline information for medical considerations.

While at the beginning of the century the population of the area was estimated to be about 20 000 (Annual Report for Papua 1907 – 1908), the first complete census data (1955 – 56, Table 1) for the area covering five census divisions (with the exception of the upper Purari, inhabited by the Pawaia) give much lower numbers. However, since 1956 there has been a steady slow rise in population.

The decline of population of the Namau in the first half of the century has been followed by a decline in the numbers of Pawaia, as described by Warrillow (1978), to a total of some 650 in 1978. The reasons for these declines in population are not clear. Sterility from venereal disease has been suggested subsequent to the oil exploration in the adjacent Kikori Delta. There is no evidence to support this. It seems most likely that the decline is related to the

Petr, T. (ed.) The Purari – Tropical environment of a high rainfall river basin
© 1983, Dr W. Junk Publishers, The Hague / Boston / Lancaster
ISBN-13: 978-94-009-7265-0

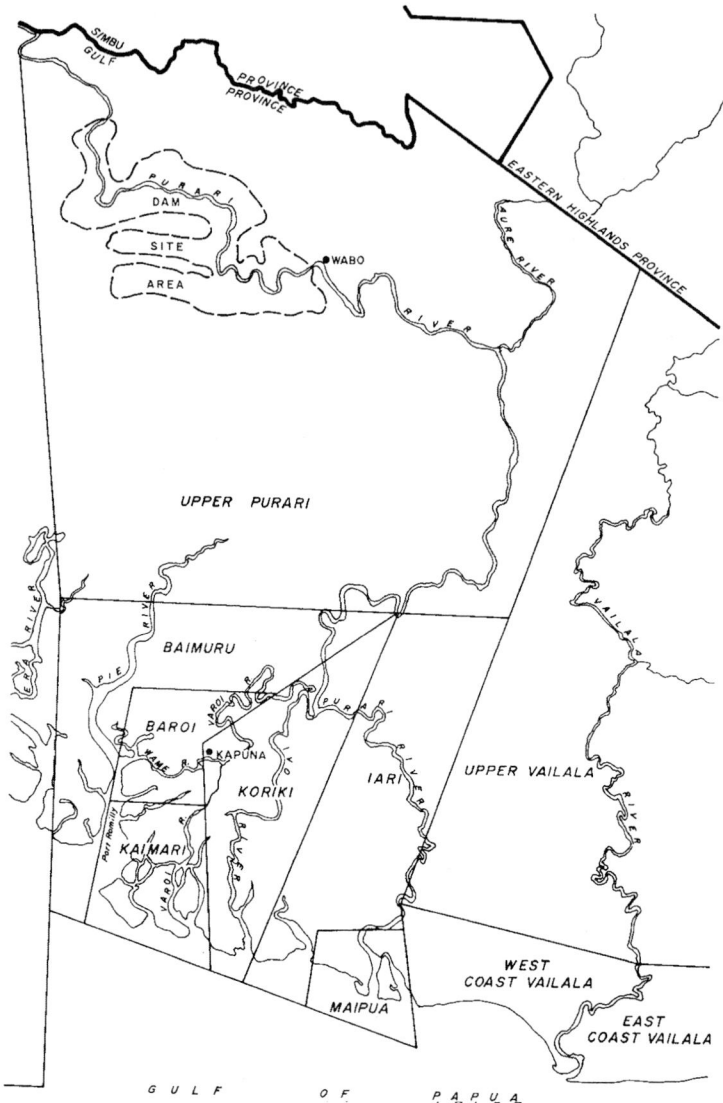

Fig. 1. Upper Purari and delta census divisions.

introduction of epidemic infectious diseases to a susceptible population. For example the Annual Report for Papua (1931 – 1932) records a mortality of 1% in coastal populations due to an influenza epidemic. This would explain why the Pawaia decline has been more recent, coinciding with their more recent contact with westernisation.

494

Table 1. Delta census data.

Census division	1948 – 49	1951 – 52	1955 – 56	1961 – 62	1965 – 66	1972 – 73
Baroi	481	500	488	563	624	655
Koriki	1666	1641	1760	1950	1941	2116
Iari	n.d.	n.d.	1751	1814	1910	1807
Kaimari	n.d.	n.d.	807	834	832	795
Maipua	n.d.	n.d.	240	281	316	403
Total	–	–	5046	5442	5623	5776

The rise in population of the Namau since the 1950s can be credited largely to the health services in the area. These consisted of infrequent medical patrols until 1949 when the London Missionary Society established a hospital at Kapuna on the Wame River (Figs. 2, 3). Doctors Peter and Lyn Calvert have worked there since 1953. The network of health services which they have developed is a reflection of the national policy. The basis of the service is the aid-post orderly, who has two years training in simple health and disease prevention. He provides primary health care in the village (Figs. 4, 5). He is also responsible for treating simple diseases; when patients require assistance, he refers them to the local health centre. In addition to the above care, mother and child care is carried out by community nurses. They are based at the health centre or health sub-centre and visit villages to carry out deliveries, immunisation programmes and advise mothers.

Kapuna acts as the health centre for the delta. The Calverts also train aid-post orderlies and community nurses. In 1980 there were seven aid-posts in the delta and one at Wabo serving the Pawaia and Namau. This last has only been open since 1975. A previous post serving the Pawaia existed at Uraru, upstream of Wabo, from 1965 until 1970 but received little community support. Even now Wabo aid-post closes during school holidays because the Pawaia go collecting sago, then leaving the orderly alone in the village. Community nurses travel by canoe from Kapuna to nearby villages monthly; more distant ones are visited by floatplane every two months.

Records kept at Kapuna form the basis of most of the statistics available and considered in this chapter. Although these records relate only to patients, births and deaths at the Kapuna hospital, they at least give a minimum estimate of vital rates. There are particularly marked underestimates for the Pawaia since it takes them at least a day to reach Kapuna. The vital statistics outlined in Table 2 are also affected by the large number of young males who

Fig. 2. Kapuna Hospital, Purari Delta.

Fig. 3. Inside a Kapuna ward.

temporarily migrate out of the delta to seek work.

This is clearly shown in the population pyramids in Fig. 6 where dotted lines represent absentees from the delta. It is interesting to see the change in population structure that has occurred over the thirty years that health services have been available, in particular the expansion of the proportion of elderly and very young.

496

Fig. 4. Mapaio aid post and ordely.

Fig. 5. Wabo aid post and orderly.

3. Dominant diseases

All causes of deaths at Kapuna were recorded for 1977 and 1978. These are compared in rank order to national causes in Table 3.

Respiratory disease, and particularly tuberculosis, are clearly an important cause of death in the delta. The importance of infection in these populations is emphasized by the fact that half of the deaths occurred in children under 5

Table 2. Vital statistics.

	Delta	Pawaia	National average
Crude birth rate per 1000 population	37.1	27.7	43.9
Crude death rate per 1000 population	17.8	38.7	17.3
Growth rate per 1000 population	+ 19.3	− 11	+ 26.6
Infant mortality per 1000 live births	66	114[a]	66
Toddler mortality per 1000 children per year (1 − 4 years of age)	11	35[a]	14
Neonatal mortality per 1000 births	24	not available	33

[a] The figures used to compute these statistics are so small as to make these highly inaccurate. National figures are taken from Bell (1973).

Table 3. Rank order of causes of death.

Rank	Kapuna	National
1	Pneumonia	Pneumonia
2	Tuberculosis	Gastroenteritis
3	Cor pulmonale	Birth conditions
4	Birth conditions	Neoplasms
5	Gastroenteritis	Meningitis
6	Whooping cough	Tuberculosis
7	Neoplasms	Accidents, poisoning, violence
8	Malaria	Heart, circulatory disease
9	Encephalitis	Malnutrition
10	Meningitis	Whooping cough

years of age and 80% of these were due to infections.

The dominance of infections in disease was also shown by morbidity data. Half of the patients seen at the aid posts at Wabo and Mapaio suffered from febrile illness, presumably mostly due to malaria. The conditions treated in hospital at Kapuna can be compared to the national figures (Table 4). Although these reflect to some extent diagnostic preference and referral bias, there are some large differences which it is difficult to explain except on the basis of differences in disease incidence.

Pneumonia and skin disease are particularly common in the delta, whereas the prominence of anaemia and otitis media may be due to diagnostic and recording differences.

Special surveys have also contributed to the morbidity picture. All villagers

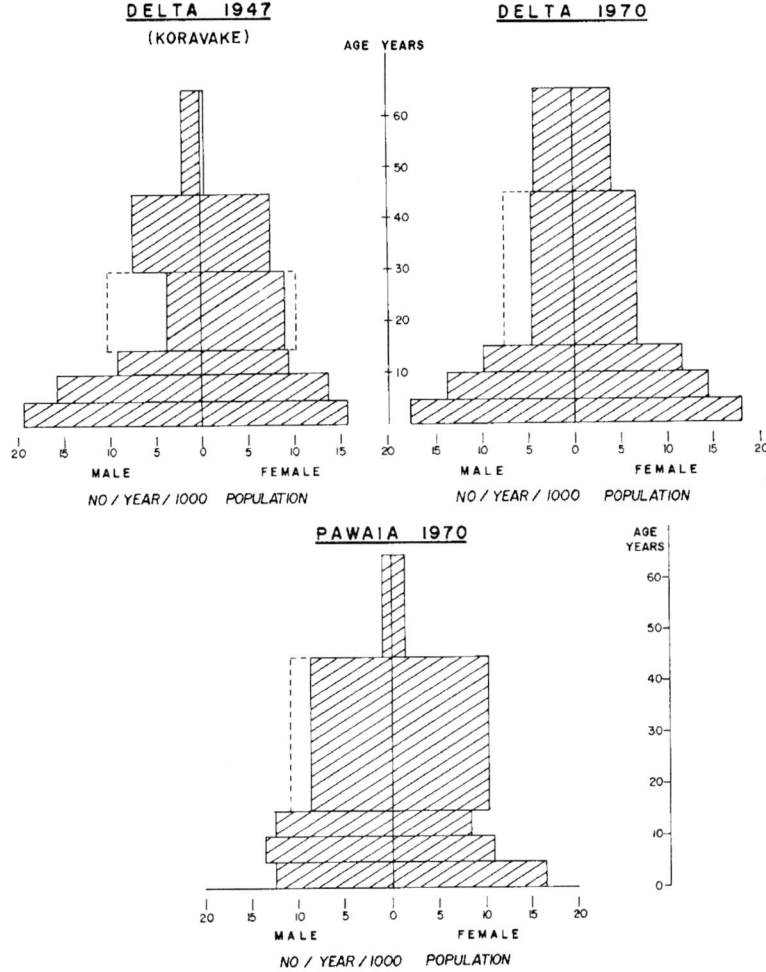

Fig. 6. Population pyramids.

at Wabo were fully examined, and it was found that all males over the age of 30 years had clinical evidence of chronic obstructive airways disease. This gives some idea of the respiratory disease problem in the area. The area has very high rainfall and all people are exposed to dense wood smoke in the houses, two factors which may contribute to the problem. Smoking is widespread, mainly of high tar tobacco, including children aged five and six. An epidemic of carcinoma of the bronchus thus seems certain.

The high prevalence of tuberculosis is of interest in view of Wigley's (1977) suggestion that Samoan pastors of the London Missionary Society may have introduced it. Hipsley (1950) in the Nutrition Survey of 1947 found that by the

Table 4. Conditions treated in Kapuna hospital as compared with national statistics.

Kapuna (%)	National (%)
Pneumonia (30)	Pneumonia (14.2)
Skin (13.5)	Gastroenteritis (9.1)
Gastroenteritis (11.5)	Malaria (7.4)
Anaemia (7.4)	Trauma (5.7)
Malaria (5.9)	Bronchitis (4.7)
Otitis media (4.2)	Skin (4.1)
Trauma (4.1)	Skin ulcer (2.0)
Tuberculosis (2.2)	Tuberculosis (1.7)

age of 29 years 79% of the population of Karavagi village in the delta had positive tuberculin tests.

The Pawaia demonstrate another observation of Wigley, that it is young migrant workers who are particularly susceptible to tuberculosis. Nine Pawaia were diagnosed as having tuberculosis between 1969 and 1979; four of these were males who had worked outside the area and four more were females of the same age in the same families. This may have important implications for the phase of dam construction.

A survey of community schools showed the marked difference in the prevalence of skin diseases between Mapaio in the delta and Wabo (Table 5).

The high frequency of tinea imbricata at Wabo (Fig. 7) was borne out in the examination of adults: eight out of forty-seven had the condition. As well as the conditions found in the schools, molluscum contagiosum (Fig. 8) is universal in the two-to-four-year olds but is ignored by parent and child. Tropical ulcer is the most sinister of the infections because of the

Table 5. Skin disease survey of Wabo and Mapaio community schools, May 1979.

Tinea versicolor	21 (16%)	6 (17%)
Tinea imbricata	1 (0.7%)	12 (33%)
Tinea corporis	9 (7%)	5 (14%)
Tropical ulcers	9 (7%)	3 (8%)
Scabies	4 (4%)	0
Albinism	1	2
Digit amputation	1	2
Number examined	135	36

Note: Percentages calculated from total children examined.

500

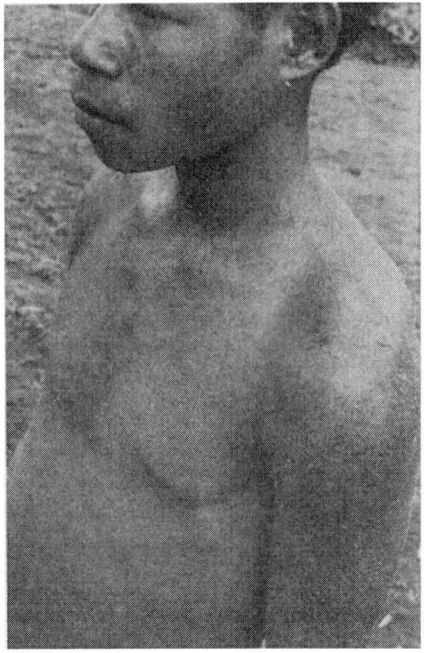 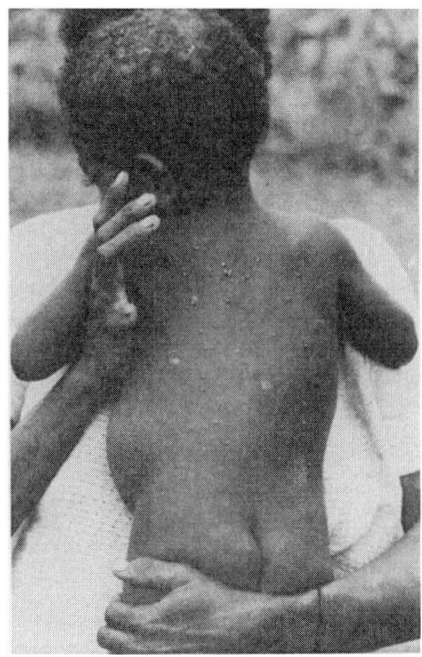

Fig. 7. Tinea imbricata.

Fig. 8. Molluscum contagiosum in a Pawaia. Note: mother had finger amputated as an infant.

occurrence of squamous cell cancer in the scars later in life. This form of cancer is the most common in Papua New Guinea.

Leprosy has a very low prevalence in the delta, as judged by treated cases; it is about 0.5/1000 population. The Pawaia have a higher prevalence but of course absolute numbers are small (Fig. 9).

Gastrointestinal disease is also common. There have been no studies of the causes of gastroenteritis in the area, although this is a major cause of morbidity and mortality. The water supply is from the river or rain catchment, so this is hardly surprising. Hepatitis A is ubiquitous judging from studies in adjacent parts of Gulf Province, as is Hepatitis B. The latter is most probably spread perinatally and is the major factor in the high incidence of hepatocellular carcinoma in the Gulf, as in the rest of Papua New Guinea. Two surveys of intestinal parasites in the delta have shown how common they are (Tables 6 and 7) but have not related this to morbidity.

Neither the delta villages nor the semi-nomadic Pawaia afford large enough

501

Fig. 9. Leprosy at Wabo.

populations to support the childhood infectious diseases in endemic form. These pass through the communities in epidemic waves as has been shown by the serological studies of Hazlett, Blake and Nurse (1980). Nevertheless, whooping cough and measles are important causes of mortality and probably play a part in the respiratory disease in later life. Poliomyelitis still occurs in this population, and lameness surveys in Gulf have shown a prevalence of 4 per 1000 school children due to this cause (Hall 1982).

Table 6. Intestinal parasites at Koravake 1947 (Bearup and Lawrence 1950).

Parasite	Number positive	%
Hookworm	120	96
Trichuris	102	81.6
Ascaris	24	19.2
Enterobius	3	2.4
Total examined	125	100

Table 7. Purari intestinal parasites 1978 (Ashford and Babona 1980).

Parasite	Wabo		Kapuna		Mapaio	
	No. Exmd.	% Pos.	No. Exmd.	% Pos.	No. Exmd.	% Pos.
Hookworm	69	78	37	25	122	45
Ascaris	69	25	37	50	122	35
Strongyloides	69	42	37	5	122	0
Trichuris	69	3	37	35	122	20

Anaemia is a commonly diagnosed problem in both the Pawaia and the Namau. Population surveys of haemoglobin levels by the author showed mean levels for the Pawaia of 12.75 gm/dl for males and 9.40 gm/dl for females, although these are based on only small sample sizes (18 and 28 respectively). In the delta the levels were 11.10 for males and 9.54 for females (sample sizes 566 and 27 respectively). The effect of this degree of anaemia on health is not clear, nor is the aetiology of the anaemia apparent. Many factors probably play a part: haemolysis from malaria, dietary iron and folate deficiencies (particularly in pregnancy), hookworm infestation, chronic infection and thalassaemia. The contribution of each of these is not clear although both β-thalassaemia and α-thalassaemia appear to be common in this population. Thus 36% of cord bloods from Gulf neonates were found positive for haemoglobin Barts, indicating α-thalassaemia.

The management of childbirth in the delta is unremarkable. Prematurity has fallen markedly since the introduction of malaria prophylaxis in pregnancy by the Calverts in 1969. Multiparity and its complications are of course common. Most interestingly, however, pre-eclamptic toxaemia is virtually unknown.

Burkitt's lymphoma occurs in this population as in other parts of lowland New Guinea. Trachoma is a mild disease in this population; follicles are found on the tarsal conjunctiva but no pannus, and it rarely if ever progresses to chronic scarring. Similarly the late stages of vitamin A deficiency have never been diagnosed at Kapuna.

Snakebite was the commonest cause of emergency evacuation from Wabo during the Nippon Koei-SMEC feasibility study, although less than one in ten of those complaining of a bite developed envenomation.

The major group of diseases which are likely to be affected by the building of the dam are those borne by arthropods. In this environment these are malaria, arboviruses and filariasis.

Malaria is holoendemic in this population. A number of surveys have shown spleen rates of 80 – 100% in children aged 2 – 9 years. The pattern is similar to the rest of New Guinea in that adults retain high spleen rates of 60 – 70% of the population (Table 8). The most frequent species is *Plasmodium falciparum*, so it seems likely that there is a high mortality in young children which is not reflected by hospital statistics. No longitudinal surveys, either parasitological or entomological, have been carried out. The rainfall patterns do not make prediction of peaks of transmission easy, but on available data, Wabo probably has a peak from April to June, and lower levels of transmission in August, September and October, whereas in the delta the peak is probably November to January. All three of the major vectors, *Anopheles farauti*, *A. koliensis* and *A. punctulatus*, in New Guinea have been found in the delta (Marks, this volume). However, the relative importance of each has not been assessed.

There are no control measures except for the availability of treatment. Chloroquine resistance is an increasing problem, and Fansidar resistance was recently reported from New Guinea.

Clearly if a dam is built at Wabo, a great deal of attention should be given

Table 8. Malaria splenometry.

Survey	Wabo village			
	Population	Number	Spleen rate %	Average enlarged spleen
'Tropic Hat'*	2 – 9 years	8	88	2.2
Hall	2 – 9 years	50	90	2.2
Hall	Adult	47	66	2.1
	Delta area villages			
	Village (2 – 9 years)	Number	Spleen rate %	Average enlarged spleen
'Tropic Hat'	Koravake	36	84	2.0
'Tropic Hat'	Evara	34	97	2.5
'Tropic Hat'	Mapaio	92	72	2.1
'Tropic Hat'	Maipenairu	91	68	2.0
'Tropic Hat'	Mariki	62	74	2.2
'Tropic Hat'	Mean	–	79	2.16
Hall	Mapaio	135	96	2.34

* Surveys conducted by the Australian Army and P.N.G. Defence Force in September – October 1975.

504

to malaria control at the dam site. Abate spraying or a similar form of larviciding would probably be most appropriate.

Arbovirus disease in the Gulf appears to be due to three viruses: Murray Valley encephalitis, Ross River virus and dengue. Serology has shown them to be widespread in the populations. Clinical evidence of infection is of polyarthritis, encephalitis or dengue fever respectively. Polyarthritis is a relatively common finding in the delta, usually of an acute self limiting form which could well be viral in origin. Encephalitis is a more difficult clinical diagnosis, and the exact viral aetiology almost always remains unknown. Nevertheless, 8% of all under five-year old deaths are due to it. This could be a serious threat to a susceptible population. Dengue is rarely diagnosed and has never been confirmed serologically as the cause of disease.

Nocturnal periodic *Wuchereria bancrofti* causes filariasis throughout the delta and at Wabo. Ashford and Babona (1980) found 8 out of 51 children and 5 out of 9 adults positive for microfilaraemia at Wabo. Sixteen percent of delta high school children were positive. Yet morbidity rates judged by admissions to Kapuna were low. This low morbidity presumably reflects a low intensity of transmission, although no entomological studies directly related to filariasis have been performed.

4. Discussion

What would be the possible health effects of the dam? These can be considered in two parts: effects of the reservoir and change in the river, and effects of a major construction programme.

In connection with the establishment of a reservoir, the major effect is likely to be a change in the man – animal – mosquito balance. The information available on the present situation is too sparse to make predictions, and will require intensive study and monitoring to ensure that the already bad situation does not worsen. There is the possibility that resistant malaria could become common, making treatment difficult. Even more care needs to be taken to ensure that schistomiasis is not introduced from South East Asia, as this would flourish in some slow moving delta waters, but especially in the new reservoir. There is no data on the presence of possible snail hosts for this disease at present.

Socioeconomic considerations play a part here too, and it is likely that the introduction of western style foods and the diversion of men to dam construction from agriculture could worsen the nutritional status of the population. This is critical in a situation where infection is the main cause of ill health

505

because of the interaction between nutrition and immunological defences. Rapid changes in diet secondary to construction activities may also lead to the introduction of new diseases such as hypertension and ischaemic heart disease, at present virtually unknown in the community.

The major conclusion from studying the situation in the delta is that too little is known about the complex interaction of man and this environment. Highly detailed baseline studies need to be performed prior to construction, followed by monitoring of the environment and disease surveillance after construction.

Acknowledgements

I would like to thank Lyn Calvert and her husband, the late Peter Calvert, for making their records and memories freely available to me, and for all that they taught me.

References

Annual Report for Papua, 1907 – 1908. Government Press of Victoria, Melbourne.

Annual Report for Papua, 1931 – 1932. Commonwealth Government Press, Canberra.

Ashford, R.W. and D. Babona, 1980. Parasites of Purari people. In: M. Alpers (ed.), Purari River (Wabo) Hydroelectric Scheme Environmental Studies, Vol. 8: Viral and parasitic infections of the people of the Purari River and mosquito vectors in the area, pp. 47 – 53. Office of Environment and Conservation, Waigani, Papua New Guinea.

Australian Army and P.N.G. Defence Force, 1975. Operation 'Tropic Hat' 1975: a malaria survey of Gulf Province by the Australian Army and P.N.G. Defence Force. Department of PublicHealth, Port Moresby, Papua New Guinea.

Bearup, A.J. and J.J. Lawrence, 1950. Parasitological report. In: Report of the New Guinea Nutrition Survey Expedition 1947, pp. 177 – 200. Government Printer, Sydney.

Bell, C.O., 1973. The disease and health services of Papua New Guinea. Department of Public Health, Port Moresby, Papua New Guinea.

Hall, A.J., 1982. A school lameness survey in Gulf Province. Papua New Guinea Med. J. (in press).

Hazlett, D.T.G., N.M. Blake and G.T. Nurse, 1980. Red cell enzymes, serum proteins and viral antibodies among the Pawaia. In: M. Alpers (ed.), Purari River (Wabo) Hydroelectric Scheme Environmental Studies, Vol. 8: Viral and parasitic infections of the people of the Purari River and mosquito vectors in the area, pp. 55 – 65. Office of Environment and Conservation, Waigani, Papua New Guinea.

Hipsley, E.H., 1950. Report of the New Guinea Nutrition Survey Expedition 1947. Government Printer, Sydney.

Marks, E.N., 1983. Mosquitoes of the Purari River lowlands. This volume, Chapter III, 7.

Warrillow, C., 1978. The Pawaia of the upper Purari, Gulf Province. In: T. Petr (ed.), Purari River (Wabo) Hydroelectric Scheme Environmental Studies, Vol. 4, pp. 1–88. Office of Environment and Conservation, Waigani, Papua New Guinea.

Wigley, S.C., 1977. The first hundred years of tuberculosis in New Guinea. In: J.H. Winslow (ed.), The Melanesian Environment, pp. 471–484. Australian National University Press, Canberra.

5. Serological studies of influenza, measles and mumps in the Purari Pawaia

D.T.G. Hazlett and M.P. Alpers

1. Introduction

If and when the dam is constructed on the Purari River above Wabo Creek, it will be the lands of the Pawaia which will be inundated. Until recently, the Pawaia have been among the least known of the peoples of Papua New Guinea. It is only with the study by Warrillow (1978) and the ethnographic survey conducted by Egloff and Kaiku (1978) that they have been placed in adequate perspective.

Their linguistic affinities (MacDonald 1973) have shown to be with the better-known Dadibi* and the even less studied Polopa, and their social and cultural characteristics, though dictated to a large extent by their situation along the upper reaches of the Purari and their dependence on the river and its creeks for their transport and subsistence, also indicate likely connections with the Dadibi (Warrillow 1978; Egloff and Kaiku 1978). The small sample investigated by us in 1978 suggested that marriage alliances are more numerous to the north than to the south.

The present study was aimed at obtaining evidence of the viral diseases to which the Pawaia have been exposed. Nurse (1980) has elsewhere discussed at length the question of the extent to which the Pawaia could function in the transmission of genes and the spread of infectious disease between the highland and the coast.

---

* Also spelled Daribi.

Petr, T. (ed.) The Purari – Tropical environment of a high rainfall river basin
© *1983, Dr W. Junk Publishers, The Hague / Boston / Lancaster*
ISBN-13: 978-94-009-7265-0

Fig. 1. Medical expedition on the upper Purari. (Photo T. Petr.)

2. Materials and methods

Specimens of venous blood were collected into evacuated plain glass and lithium heparin tubes from 38 adult individuals who stated that they were from Wabo village, and 31 who gave their home village as Gurimatu. The main collection took place at Wabo aid post, across the river from the camp, but most of the Gurimatu sample was collected at a number of *kombati*, or riverside hamlets, upstream from the dam site (Figs. 1 and 2). At a *kombati* below Wabo Creek a further 8 specimens were taken from 3 adults and 5 children from Poroi, or Pawaia No. 1, village. In addition, 42 specimens of plain blood only were taken from pupils at Wabo school; 13 pupils were from Poroi, 12 from Kone, 11 from Uraru, 4 from Weijana and 2 from Uri. These specimens were examined only virologically. Thin smears were made of the blood of the 77 subjects in the serogenetic sample, and fixed immediately. A further 99 smears were obligingly collected and fixed at a later date by Dr A. Hall. Failure of the refrigeration system at the camp made it inadvisable to collect further serogenetic specimens, and all cooling resources feasible were used in keeping those already collected at a reasonable temperature. When an

510

Fig. 2. Collecting venous blood near Uraru. (Photo T. Petr.)

aircraft became available, they were flown to Port Moresby, recooled, and thence sent to Goroka for repacking and transmission to Canberra. Sera separated from the plain blood specimens were used in the search for viral antibodies.

Variations in the red cell enzymes 6-phosphoglucomutase and dehydrogenase, acid phosphatase, first and second locus phosphoglucomutase and indophenol 'oxidase', lactate dehydrogenase, soluble malate dehydrogenase, esterase D, glutamic-pyruvic transaminase, phosphoglycerate kinase and adenosine deaminase were investigated in the laboratories of the Department of Human Biology of the John Curtin School of Medical Research, Australian National University, Canberra, by standard electrophoretic methods. The thin blood smears were screened unstained at the Institute of Medical Research, Goroka, for ovalocytosis.

Gene frequencies were estimated in the first instance by gene counting; but the small numbers of certain genotypes, due to small population size, necessitated maximum likelihood calculations in several cases. These were carried out by iteration and minimization of X^2. All sera, diluted 1:8 in Veronal buffer, pH 7.2, were screened for the presence of complement-fixing antibodies to influenza A, influenza B, measles, and mumps viruses by means of the microtitre technique described by Sever (1962). Sera were scored as negative, weakly positive, or strongly positive for the presence of complement-fixing antibodies to each of the four virus types.

The nomadic life of the Pawaia, from one sago stand to another, building scattered short-lived *kombati* and only slightly less ephemeral villages, makes them a peculiarly difficult people to sample representatively. We were able to take samples from only three villages, one high on the Purari (Gurimatu), one in the upper-middle reaches (Wabo), and one relatively far south (Poroi; see map in Warrillow, this volume).

3. Results and discussion

Clinical evidence of infectious diseases was collected simultaneously with blood specimens. One subject had elephantiasis of the legs and one Gurimatu man, originally from Lake Tebera, had tuberculoid leprosy. Among the children examined without venepuncture, there was one case of molluscum contagiosum and three of tinea imbricata, a common skin disease at Wabo (Hall, this volume). Tinea imbricata was seen in seven of the 77 people from whom blood was taken, which would suggest a frequency of at least 0.302 for the autosomal recessive gene conferring susceptibility to this condition (Serjeantson and Lawrence 1977), if indeed, this was the source of susceptibility in this population. Seven of the 77 blood slides examined showed ovalocytosis; the probability that this might have been confused with oval poikilocytosis due to iron deficiency anaemia is slight as, on Giemsa staining, remarkably few slides (six), one of them from an ovalocytic, showed any morphological evidence of iron deficiency. This suggests a frequency of 0.302 ± 0.052 for the presumed recessively inherited gene for hereditary ovalocytosis, a trait which may be protective against malaria (Serjeantson et al. 1977).

Local informants, and the information we secured about the birth-places of the mothers of the subjects, convinced us that movement and mate-giving were sufficiently general for us to be able to pool the samples from the three villages to give some idea of their profile. An unexpected consequence was that all seven genetic systems, in which it was possible to test for it, appeared to be in Hardy-Weinberg equilibrium (that is, at a theoretically ideal genotype frequency, which would be expected if mating were random within the pool). Whether we are justified in assuming that this fact, coupled with the geographical spread of the origins of the subjects, indicates that the sample is representative after all, may be doubted.

One reason for the doubt is the extent to which a direction of spread may be ascribed to certain alleles. Of the seven ovalocytics (see below) five had at least one, and two had both, parents from Gurimatu (see map in Warrillow, this

512

volume). Several mothers were said to come from Hola; if we are right in identifying this as Haia, on the Pio River, there is a suggestion that the PGM_2^{10} allele may have come quite recently from the north as, of the four subjects possessing it, two had mothers who were Dadibi, while one other mother came from Haia and the fourth from Karovae, which we are unable to identify. There appeared to be no connection of any of the 10 subjects with PGM_2^9 with anywhere to the north; in fact, no less than six of them were among the eight children from Poroi, and the mother of two of these was from Kerema.

Of the 119 sera examined for viral antibodies, 10 possessed high levels of anticomplementary activity, even after adsorption with guinea pig serum (Taran 1946), and were, therefore, excluded from the study. The distribution by age of antibodies to influenza A and B, measles, and mumps viruses is indicated in Table 1.

All but one (aged 48 years) of 59 individuals 16 years of age or over possessed antibodies against mumps virus, whereas 13 of 46 (28%) in the 7 to 15 year age group were seronegative. Thus, the age distribution of antibodies to mumps was found to be similar to that found in other parts of the world where the disease is endemic (Feldman 1976). However, although all children under 10 years of age and adults 17 years of age and over possessed antibodies against measles virus, 11 of the 42 children (26%) in the 10 to 16 year age bracket lacked such antibodies. Therefore, serological evidence suggests that, until recently, a relatively large number of individuals reached adolescence before acquiring measles infection, a phenomenon also observed in other dispersed populations, presumably on account of the reduced probability of virus transmission (Black 1976). As distinct from other parts of the tropical world, in Papua New Guinea measles is a relatively mild disease, with the case fatality rate being high only in malnourished children.

Table 1. Age distribution of complement-fixing viral antibodies in Pawaian sera.

Age groups (years)	No. of sera	% positive for antibodies against			
		Influenza A	Influenza B	Measles	Mumps
7– 9	8	62.5	25.0	100	66.7
10–14	35	42.9	42.9	74.3	71.4
15–19	10	30.0	0	80.0	90.0
20–29	25	54.2	25.0	100	100
30–39	13	50.0	46.2	100	100
≥40	18	43.8	23.5	100	93.8
7–54	109	44.8	30.8	89.5	86.7

The highest prevalence of antibodies to influenza type A occurred in the 7 to 9 year old children, with a smaller peak occurring in those aged 20 to 29. Peak levels of antibodies to influenza B, however, were found in older individuals, occurring in those in the 10 to 14 and 30 to 39 year age groups. In Gurimatu village, only individuals older than 18 and 10 years of age possessed serological evidence of previous infection with influenza viruses types A and B, respectively; in Wabo, sera were not positive for antibodies to either agent in individuals less than 22 years of age.

Therefore, although the number of sera examined was small, there is some evidence that the inhabitants of Wabo and Gurimatu have not had contact with influenza virus types A or B for one or two decades. However, 22 of the 42 (52%) children attending the school at Wabo were seropositive for one or both agents, indicating that these infections were introduced into and spread through the school community, but not into the more scattered population.

The association of each variant allele with antibodies for influenza types A and B was tested by the relative risk test of Woolf (1954) and by X^2. The nearest approaches to significance appeared in the lowness of the figures for the P^b allele for the red cell enzyme, acid phosphatase, with both types of antibody: 0.12 (95% fiduciary limits $0.04 - 0.38$, $X_1^2 = 3.49$) with that to influenza A and 0.19 (95% fiduciary limits $0.08 - 0.45$, $X_1^2 = 3.83$) with that to influenza B. The suggestion that a relative inability of bearers of this allele to produce antibodies to influenza (or, conversely, their relative ability not to contract the disease) may be concerned, probably combined with other factors, in the maintenance of this polymorphism is worth following up.

Both malaria and filariasis have long been known to occur here (Ashford and Babona 1980). In a study of red cell enzymes, the rather high frequency of the Hp 0 phenotype of acid phosphatase, found in eight of 72 adults and two of five children, may be ascribed to the holoendemicity of malaria; and, of 16 children examined clinically, only one did not have splenomegaly (Hazlett et al. 1980). Finally, suitable vectors for the transmission of malaria abound along the Purari (Marks, this volume; Work and Jozan, this volume). Seven people had ovalocytosis, a malaria-associated polymorphism.

Ashford and Babona (1980) have also reported on the prevalence and distribution of other parasitic diseases in the area. In general, hookworm infections were more abundant in the Wabo area and *Ascaris* and *Trichuris* commoner in the delta, while *Strongyloides* spp. eggs were found at Wabo only. All infections were reported to be light. Microfilariae and, to a lesser degree, malaria parasites were found in the blood of Wabo residents. They observed that the Purari people seem to live in some sort of immunological harmony with the common parasites and that the greatest danger will probably come with population movements and immigration through the introduction of different strains of the parasites already present. Immigrants too will be at risk from locally prevalent organisms. Thus preventive measures such as spraying and wire gauzing of quarters and the administration of antimalarial prophylactics to non-immune migrants into the area seem to be indicated. Marks (this volume) has studied the vector mosquito population of the area and Work and Jozan (this volume) have reported on the prevalence of antibodies to arboviruses amongst the Pawaia.

Though the Pawaia are close to being an isolate geographically, it is obvious from their marriage patterns and the distribution of genetic markers in them (Hazlett et al. 1980; Nurse 1980) that socially and biologically they are related to the peoples who adjoin them. They plainly constitute, at present, a link in the transmission of genes and of infective agents up and down the Purari River. The extent of their importance, however, cannot be gauged until similar studies have been carried out on their neighbours and kin, the Polopa and Dadibi, and their traditional enemies, the Koriki. The age stratification evident in the possession of viral antibodies may be of particular significance in this respect. Construction of the dam would destroy their function as genetic and epidemiological agents of transmission by limiting their wandering, but would also probably expose them both to increased genetic inflow and to more frequent and varied assaults by introduced infections.

Acknowledgements

The authors thank Dr George Nurse for his part in the planning and execution of these studies; and Dr Tomi Petr for logistic support in the field and constant encouragement; Mr C. Warrillow and Dr A. Hall of Gulf Province for field support; Mr Tom Tiki, Aid-Post Orderly, Wabo, for valuable help in arranging the collection of blood samples and in gathering social particulars; and the Pawaia people for their cooperation.

References

Ashford, R.W. and D. Babona, 1980. Parasites of Purari people. In: M. Alpers (ed.), Purari River (Wabo) Hydroelectric Scheme Environmental Studies, Vol. 8: Viral and parasitic infections of the people of the Purari River and mosquito vectors in the area, pp. 47–53. Office of Environment and Conservation, Waigani, Papua New Guinea.
Black, F.L., 1976. Measles. In: A.S. Evans (ed.), Viral Infections of Humans: Epidemiology and Control, pp. 297–316. Plenum Publ. Corp., New York.

Egloff, B.J. and R. Kaiku, 1978. An archaeological and ethnographic survey of the Purari River (Wabo) dam site and reservoir. In: T. Petr (ed.), Purari River (Wabo) Hydroelectric Scheme Environmental Studies, Vol. 5, pp. 1 – 53. Office of Environment and Conservation, Waigani, Papua New Guinea.

Feldman, H.A., 1976. Mumps. In: A.S. Evans (ed.), Viral Infections of Humans: Epidemiology and Control, pp. 317 – 336. Plenum Publ. Corp., New York.

Hall, A., 1983. Health and diseases of the people of the upper and lower Purari. This volume, Chapter III, 4.

Hazlett, D.T.G., G.T. Nurse and N.M. Blake, 1980. Red cell enzymes, serum proteins and viral antibodies among the Pawaia. In: M. Alpers (ed.), Purari River (Wabo) Hydroelectric Scheme Environmental Studies, Vol. 8: Viral and parasitic infections of the people of the Purari River and mosquito vectors in the area, pp. 55 – 65. Office of Environment and Conservation, Waigani, Papua New Guinea.

MacDonald, G.E., 1973. The Teberan language family. In: K.J. Franklin (ed.), The Linguistic Situation in the Gulf District and Adjacent Areas, Papua New Guinea, pp. 111 – 148. Research School of Pacific Studies. Dept. of Linguistics. Pacific Linguistics Series C, No. 26. Australian National University, Canberra.

Marks, E.N., 1983. Mosquitoes of the Purari River lowlands. This volume, Chapter III, 7.

Nurse, G.T., 1980. The Pawaia as agents and patients. In: M. Alpers (ed.), Purari River (Wabo) Hydroelectric Scheme Environmental Studies, Vol. 8: Viral and parasitic infections of the people of the Purari River and mosquito vectors in the area, pp. 67 – 84. Office of Environment Conservation, Waigani, Papua New Guinea.

Serjeantson, S., K. Bryson, D. Amato and D. Babona, 1977. Malaria and hereditary ovalocytosis. Human Genetics 37: 161 – 167.

Serjeantson, S. and G. Lawrence, 1977. Autosomal recessive inheritance of susceptibility of tinea imbricata. Lancet 1: 13 – 15.

Sever, J.L., 1962. Application of microtechnique to viral serological investigations. J. Immunol. 88: 320 – 329.

Taran, A., 1946. A simple method for performing Wassermann test on anticomplementary serum. J. Lab. Clin. Med. 31: 1037 – 1039.

Warrillow, C., 1978. The Pawaia of the upper Purari, Gulf Province, Papua New Guinea. In: T. Petr (ed.), Purari River (Wabo) Hydroelectric Scheme Environmental Studies, Vol. 4, pp. 1 – 88. Office of Environment and Conservation, Waigani, Papua New Guinea.

Warrillow, C., 1983. A short history of the upper Purari and the Pawaia people. This volume, Chapter III, 1.

Woolf, B., 1954. On estimating the relation between blood groups and diseases. Ann. Human Genetics 19: 251 – 253.

Work, T.H. and M. Jozan, 1983. Human arbovirus infections of the Purari River lowlands. This volume, Chapter III, 6.

6. Human arbovirus infections of the Purari River lowlands

T.H. Work and M. Jozan

1. Introduction

Of the many major water impoundment schemes being planned or already implemented in the wet tropics, the proposed Purari River hydroelectric project in the Gulf Province of Papua New Guinea is among the larger ones. It is planned to occupy a markedly underpopulated area, and, as a result of this to lead to an increase in population density. This will result in a variety of sociological and health problems which must be anticipated to avoid untoward or even disastrous consequences such as those experienced elsewhere (Freeman 1974).

Engineering feasibility studies show that the Wabo Dam would have an output of some 1800 MW. The expected increase in population associated with the development makes difficult a health impact assessment based on data from the present sparse, mobile population in a largely virgin forested habitat that certainly would be considerably altered. Among health problems which may be magnified by population increases, are those of human arbovirus infections. Because of the nature of the habitat, supporting widespread dissemination of vector-borne diseases such as malaria and filariasis (see Ashford and Babona 1980; Hall, this volume), it was essential also to investigate the presence and prevalence of arthropod-borne virus (arbovirus) infections and their arthropod vectors.

Part of the Purari River drainage covers the more densely populated highlands which are not likely to be ecologically affected by impoundments at much lower elevations far downstream. Therefore, settlements were selected in the area to be immediately affected such as Wabo dam site and in the delta at Mapaio, Kapuna and Ara'ava (Figs. 1 and 2).

Wabo is characterized by a heavy rainfall throughout the year. Records show that the highest precipitation, exceeding 800 mm per month extends

Petr, T. (ed.) The Purari – Tropical environment of a high rainfall river basin
© *1983, Dr W. Junk Publishers, The Hague / Boston / Lancaster*
ISBN-13: 978-94-009-7265-0

Fig. 1. Wabo camp on the Purari River. (Photo T. Petr.)

Fig. 2. Kapuna hospital on the Wame arm of the Purari Delta. (Photo T. Petr.)

518

from June through September. Monthly rainfall for the other months of the year ranges from 500 to 700 mm. This not only supports rain forest but also leaches the soil and causes substantial rises and falls in the Purari and its tributaries, producing steep, eroded banks and limiting the extent of swampy mosquito breeding habitats. The flushing action of such deluges also limits build up of mosquito populations. Slowing of the river flow by impoundment would have a drastic effect on mosquito breeding potential, which could be partly anticipated by studying areas with slower moving water such as those in the Purari Delta.

This delta is situated about 120 km downstream from Wabo, and includes the large village of Mapaio where the level fluctuates considerably, reflecting upstream periodic deluges. About 10 km to the northwest is Kapuna, with the only hospital in the area, to which patients are referred and brought from the Purari Delta and some adjacent areas. Neighbouring Kapuna downstream is Ara'ava village. Brackish water from the substantial tidal differential of the Gulf of Papua reaches this part of the delta. This habitat supports a range of mosquito species different from the Wabo area (Marks, this volume).

2. Mosquito collections

Two methods for capture of adult mosquitoes were used in both the Wabo and Purari Delta areas. California State Health Department, Bureau of Vector Control (BVC) modifications of the CDC mosquito light-trap were used for capturing live insects (Fig. 3) (for more details see Work and Jozan 1980). These traps continuously release carbon dioxide as an attractant from desiccating solid CO_2 (dry ice) contained within. Insects were removed from the trap and quick-frozen in a container of dry ice. Subsequently, on a dry-ice-refrigerated plate, the insects were sorted, the mosquitoes being combined into pools by species in plastic vials for storage in liquid nitrogen, in which they were transported to laboratories in Brisbane and Perth for long-term storage at $-70°C$ until processing for virus isolation.

A variety of sites were selected for sampling, ranging from rain forest edge to peridomestic premises within and at the edge of villages. Additional mosquitoes were obtained by distributing plastic vials to residents to bring live specimens. The species taken in adult collections and by sampling breeding places are discussed by Marks (this volume).

Anopheles farauti was the dominant mosquito at Wabo village (Table 1), which explains the occurrence of hyperendemic malaria and filariasis there. Marks (this volume) found additional species. Shortage of dry ice prevented

Fig. 3. Modified CDC trap, using carbon dioxide to attract mosquitoes. (Photo T. Petr.)

extension of collecting sites further into the forest. A greater variety of habitats were sampled in the Purari Delta at Mapaio because of the level landscape and more open access along the river bank into sago palm swamps and into banana and coconut cultivations. Here the predominant mosquito caught was *Mansonia papuensis*. Lesser catches of anopheline mosquitoes than at Wabo were reflected in less evidence of severe malaria in the village population. The evidence of arboviral infections found in the serological survey suggests that the very abundant *Mansonia papuensis* might be a vector.

520

Table 1. Mosquitoes[a] collected by CDC light-traps[b] on the Purari River in August 1978.

Species	Number of mosquitoes[a]			
	Wabo	Kapuna	Mapaio	Total
Anopheles farauti	106	0	0	106
Anopheles longirostris	0	0	40	40
Mansonia papuensis	0	27	1756	1783
Mansonia septempunctata	0	0	22	22
Coquillettidia spp.	0	0	3	3
Culex pullus	0	0	2	2
Not identified	0	20	1	21
Total	106	47	1824	1977

[a] Species pooled for viral isolation; additional species from these traps are discussed by Marks (this volume).
[b] Traps baited with CO_2 at Wabo and on one night at Mapaio; without CO_2 on one night at Mapaio and at Kapuna.

3. Arbovirus antibodies in Wabo area residents

To measure arbovirus transmission, a determination of antibodies in the blood of indigenous residents was carried out by serological survey. Collected serum specimens were later analysed in the Microbiology Laboratory of the University of Western Australia in Perth.

There, the sera were screened by haemagglutination inhibition (HI) against alphavirus Ross River and flaviviruses Murray Valley Encephalitis (MVE) and Kunjin (KUN), according to standard procedure (Work 1964). One hundred and sixteen sera were collected in the region of the Wabo dam site. While the populations of this area are traditionally hunters and gatherers, the government concentrated them recently into Wabo, where medical services and schooling can more effectively be delivered. The most comprehensive serum collections were made at Wabo Medical Aid Post and at the school.

Results of the initial HI screening are shown in Table 2 according to reported locality of residence. They are displayed by age group to determine whether infection was cumulative and the trend continuous. The residents of small settlements at a distance from Wabo have a higher antibody prevalence to both alphaviruses and flaviviruses. This may reflect closer proximity to bush and rain forest or increased occupational exposure. While the prevalence of antibodies to flaviviruses is higher than to alphaviruses, the increasing trend with age is similar. This indicates persistent or cyclic virus transmission. With high rainfall throughout the year, this could mean continuous breeding

521

Table 2. Arbovirus haemagglutination inhibiting antibodies in village people at the Wabo dam site on the upper Purari River of Gulf Province, August 1978.

Locality or residence	Age group							Total	M	F	%+
	0–9	10–14	15–24	25–34	35–44	45–54	55+				
A. HI antibodies to alphavirus Ross River											
Johnmatu[a]	0/1	0/4	3/8	2/13	0/4	0/5	NS	5/35	2/14	3/21	14
Uraru	NS	4/10	NS	NS	NS	NS	NS	4/10	4/9	0/1	40
Wegena[a]	0/2	1/2	NS	NS	NS	NS	NS	1/4	1/3	0/1	25
Wabo/Kone	0/1	4/12	2/9	1/17	0/5	1/8	NS	8/52	7/43	1/9	15
Poroi	0/2	4/8	3/3	NS	NS	NS	NS	7/13	7/13	NS	54
Uri	NS	0/2	NS	NS	NS	NS	NS	0/2	0/2	NS	0
Total	0/6	13/38	8/20	3/30	0/9	1/13	–	25/116	21/84	4/32	
%+	0	27	40	10	0	7	–	22	25	12	22
B. HI antibodies to flaviviruses Murray Valley encephalitis and/or Kunjin											
Johnmatu	0/1	1/4	2/8	6/13	2/4	1/5	NS	12/35	7/14	5/21	34
Uraru	NS	3/10	NS	NS	NS	NS	NS	3/10	3/9	0/1	30
Wegena	1/2	0/2	NS	NS	NS	NS	NS	1/4	0/3	1/1	25
Wabo/Kone	0/1	2/12	5/9	10/17	1/5	1/8	NS	19/52	17/43	2/9	37
Poroi	1/2	5/8	3/3	NS	NS	NS	NS	9/13	9/13	NS	69
Uri	NS	2/2	NS	NS	NS	NS	NS	2/2	2/2	NS	100
Total	2/6	13/38	10/20	16/30	3/9	2/13	–	46/116	38/84	8/32	
%+	33	34	50	53	33	15	–	39	45	25	39

Note: Number with HI antibodies (numerator) in sample (denominator); NS = no sample.
a Name given by the persons examined.

and activity of virus vectors. Some other vertebrate reservoirs – perhaps domestic dogs, pigs, fowl and cassowaries – should be sought because it does not appear that this human population density would support a man – mosquito – man cycle unless there is consistent periodic reintroduction of the aetiological viruses.

High spleen rates and positive malaria slide rates indicate chronic febrile disease, which would probably mask the occurrence of illness due specifically to arbovirus infection. This would obscure an epidemic excursion of such viruses into these people. Antibodies in young children indicate infection at an early age, which supports the hypothesis that there are sources of arboviruses maintained within the Wabo region. In such an apparent enzootic situation, collections of sera from non-human vertebrates should be considered.

4. Prevalence of arbovirus infection in villagers of the Purari River Delta

Mapaio, Kapuna and Ara'ava are situated in a delta habitat with differences in mosquito fauna (cf. Table 1 and Marks, this volume). While results for Ara'ava in Table 3 show a significant prevalence of alphavirus antibodies, it is markedly lower than in Mapaio. The substantially higher percentage of positive reactions to flaviviruses reflects a similar difference between the two populations only a dozen kilometres apart. The differences in vector populations between those in the lowland, freshwater forest at Mapaio and those of the predominantly mangrove habitat at Ara'ava deserve detailed investigation over a longer period of time. Numerous flavivirus positives from Mapaio indicate that this would be a good location for further field collections to elucidate which flaviviruses and mosquito vectors are in fact of greatest importance in arbovirus transmission to man in the Purari Delta.

It is fundamental in arbovirus epidemiology that the HI test is the most sensitive detector in alphavirus and flavivirus infection, but HI antibodies are least specific concerning the particular type of infecting virus. It is also almost axiomatic that MVE virus has the widest cross reactivity of any of more than forty flaviviruses in an HI test of human sera. Antibody for the specific infecting virus will usually neutralize that virus in tissue culture neutralization or mouse protection tests.

To determine specificity of the flavivirus HI antibodies, all HI positive sera were subjected to a tissue culture neutralization test in a series of test runs at a time when the number of MVE and Kunjin specific sera from an epidemiological study of the 1978 epidemic of MVE and Kunjin virus infection in tropical Western Australia were also being tested. These tests were run

Table 3. Arbovirus haemagglutination inhibiting antibodies in village people in the Delta, August 1978.

Locality or residence	Age group							Total	M	F	%+
	0 – 9	10 – 14	15 – 24	25 – 34	35 – 44	45 – 54	55 +				
A. HI antibodies to alphavirus Ross River											
Ara'ava	0/13	1/13	3/15	0/2	NS	NS	0/1	4/44	1/23	3/21	9
Mapaio	2/19	4/7	3/9	2/3	NS	0/2	3/5	14/45	11/26	3/19	31
Total	2/32	5/20	6/24	2/5	NS	0/2	3/6	18/89	12/49	6/40	
%+	6	25	24	40	–	0	50	20	24	15	20
B. HI antibodies to flaviviruses Murray Valley encephalitis and/or Kunjin											
Ara'ava	3/13	4/13	4/15	0/2	NS	NS	1/1	12/44	7/23	5/21	27
Mapaio	8/19	2/7	7/9	2/3	NS	1/2	4/5	24/45	16/26	8/19	53
Total	11/32	6/20	11/24	2/5	NS	1/2	5/6	36/89	23/49	13/40	
%+	34	30	46	40	–	50	83	40	47	33	40

Note: Number with HI antibodies (numerator) in sample (denominator); NS = no sample.

524

with 35 TCD50 doses of MVE and Kunjin viruses in continuous porcine kidney cells. Results are presented in Table 4.

Only a few of the flavivirus HI positive sera neutralized MVE or Kunjin or both viruses. It can therefore be concluded that there are one or more other flaviviruses being transmitted in both the Wabo and Purari Delta habitats. This requires intensive search because of its implications in regard to future disease impact on arrival of new non-immune workers and settlers in the area.

Much additional information could doubtless be obtained by extensive neutralization testing of these sera positive for alphavirus and flavivirus against a variety of viruses known to be active in adjacent Australian tropical habitats (Doherty 1972). But such testing would not mitigate the field investigations necessary to isolate arboviruses, which are obviously active and extensive in this region.

5. Discussion

The substantial prevalence of alphavirus and flavivirus antibodies in populations of the upper and lower Purari River establishes that arbovirus diseases

Table 4. MVE and Kunjin virus neutralization by flavivirus-HI-positive human sera collected along the Purari River, August 1978.

Location	Flavivirus HI-positive	MVE NT-positive	KUN NT-positive	MVE/KUN NT-positive	MVE/KUN NT-negative	%HI+ NT-negative
Wabo area						
<15 years	15/43	3/15	0/15	0/15	12/15	80
>15 years	31/73	4/30[a]	1/30	1/30	24/30	80
Total	46/116	7/45	1/45	1/45	36/45	80
Lower Purari						
<15 years	17/52	0/17	0/17	0/17	17/17	100
>15 years	19/37	3/19	0/19	1/19	15/19	79
Total	36/89	3/36	0/36	1/36	32/36	89

HI = haemagglutination inhibition; NT = neutralization.
[a] One serum had inadequate volume for NT.

are an important constituent of the infectious disease problems which must be anticipated and dealt with as a result of any development in the Gulf Province of Papua New Guinea.

An earlier serological survey by Wisseman et al. (1964) showed HI antibodies to alphaviruses and flaviviruses in lowland populations of Papua New Guinea. In contrast, when tested against MVE virus, prevalence of flavivirus antibodies was low in the highlands of Papua New Guinea, but the alphavirus HI-positives remained at significant levels. Later systematic studies along the Sepik River by Woodroofe and Marshall (1971) isolated a strain of Sindbis, which was distinguishable from the Australian type from Queensland (Doherty 1972). A geographically extensive serological survey for alphavirus infections in populations of Southeast Asia and Australasia by Tesh et al. (1975) demonstrated that Chikungunya specificity terminated at Wallace's Line. East of Borneo the predominant alphavirus reactions were against Ross River virus (Doherty, Carley and Best 1972), with few reactions to Sindbis.

While Sindbis, in its four-continent distribution, is recognized to be an occasional human pathogen (Berge 1975), Ross River virus has been incriminated repeatedly as a cause of severe illness, frequently associated with polyarthritis and rash, resulting in lengthy and painful convalescence (Anderson and French 1957; Anderson, Doherty and Carley 1961; Doherty et al. 1961, 1971; Clarke, Marshall and Gard 1973).

Experimental pathogenesis of Ross River virus in extraneural tissues, skeletal muscle, smooth muscle and brown fat (Murphy et al. 1973) helps explain the severity of the human disease, which can masquerade as malaria and which has been reported from most states of Australia. The discovery by Woodroofe, Marshall and Taylor (1977) that there are geographically separable strains of Ross River virus similar to the alphavirus complex of Venezuelan equine encephalitis in the New World (Work 1972), requires location and definition of epizootic foci from which epidemic excursions periodically ensue.

Preliminary detection of alphavirus HI antibodies reported here was accomplished with only one Ross River virus strain, T48, the original Queensland isolate from *Aedes vigilax* mosquitoes collected at Townsville in 1959 (Doherty et al. 1963; Doherty, Carley and Best 1972). It is therefore imperative that long-term efforts be directed toward isolation of additional local strains of Ross River and other alphaviruses from mosquitoes and human case sera collected in the Purari River drainage and delta to determine more accurately the ecology and transmission patterns of these pathogens.

Clinical disease in Irian Jaya (Essed and Van Tongeren 1965) and fatal encephalitis from an exposure west of Port Moresby (French et al. 1957) document disease already recognized as resulting from Murray Valley Encephalitis (MVE) virus infection in New Guinea. The prevalence of flavivirus neutralizing antibodies beyond those specific for MVE and Kunjin detected in this study emphasize the importance of the search for and elucidation of additional flaviviruses by implementation of field studies of febrile disease and mosquito vectors.

While continuing such assessment of arbovirus disease to be anticipated in human intrusion into the Purari Delta, the present paucity of population, and an ecology that would be drastically changed by construction of a dam at Wabo, require consideration of alternative study sites. Fortunately, a semipermanent natural impoundment of water, Lake Tebera, lies just west of the Purari River. Because it is separated by only a ridge from the Purari River, it can be considered a microcosmic sample of what would occur in the lake that would be created by the dam.

Since there are no permanent human settlements at Lake Tebera, other indigenous vertebrates – wild birds, mammals and reptiles – should be captured for serum samples and tissue specimens for virus isolation. Careful consideration should be given to selection of sentinel animals for studies such as those that have been so productive in the rain forest of Amazonia (Causey et al. 1961).

6. Summary

A preliminary survey of mosquitoes and human arbovirus antibodies in two areas of the Purari River drainage was undertaken to elucidate present and potential vectors and the arbovirus infections that might result from human population increase and development following construction of a dam at Wabo.

Anopheles farauti was the prevalent mosquito light-trapped at Wabo, while *Mansonia papuensis* was most prevalent in the delta.

There is a high prevalence of antibodies to both alphaviruses and flaviviruses in populations at Wabo and Mapaio, which indicates frequent or constant transmission of a variety of pathogenic arboviruses in both areas.

Anticipated ecological changes and increase in human population resulting from construction of the dam are likely significantly to increase infection and diseases caused by a variety of mosquito-borne arboviruses.

Limited information derived from these studies to date suggests the importance of increased and continued field and laboratory studies. These should extend testing of specimens already collected and establish long-term field data collections at already identified and suggested additional sites.

Acknowledgements

The authors wish to thank Dr E.N. Marks and Dr M.P. Alpers for their critical comments on the manuscript. The assistance given during the field survey by Tom Tiki, the Wabo Aid-Post Orderly, the late Dr P. Calvert, and his wife Dr Lyn Calvert of the Kapuna Hospital, is gratefully aknowledged.

References

Anderson, S.G. and E.L. French, 1957. An epidemic exanthem associated with polyarthritis in the Murrey Valley, 1956. Med. J. Aust. 1: 478 – 481.

Anderson, S.G., R.L. Doherty and J.G. Carley, 1961. Epidemic polyarthritis: antibody to a group A arthropod-borne virus in Australia and the island of New Guinea. Med. J. Aust. 1: 273 – 276.

Ashford, R.W. and D. Babona, 1980. Parasites of Purari people. In: M. Alpers (ed.), Purari River (Wabo) Hydroelectric Scheme Environmental Studies, Vol. 8: Viral and parasitic infections of the people of the Purari River, and mosquito vectors in the area, pp. 47 – 53. Office of Environment and Conservation, Waigani, Papua New Guinea.

Berge, T.O. (ed.), 1975. International catalogue of arboviruses: including certain other viruses of vertebrates, 2nd ed. U.S. Dept. of Health, Education and Welfare. Public Health Service. Publication, No. (CDC) 75 – 8301. Washington, D.C.

Causey, O.R., C.E. Causey, O.M. Maroja and D.G. Macedo, 1961. The isolation of arthropod-borne viruses, including members of two hitherto undescribed serological groups in the Amazon region of Brazil. Am. J. Trop. Med. Hyg. 10: 227 – 249.

Clarke, J.A., I.D. Marshall and G. Gard, 1973. Annually recurring epidemic polyarthritis and Ross River virus activity in a coastal area of the New South Wales. I. Occurrence of the disease. Am. J. Trop. Med. Hyg. 22: 543 – 550.

Doherty, R.L., 1972. Arboviruses of Australia. Aust. Vet. J. 48: 172 – 180.

Doherty, R.L., S.G. Anderson, K. Aaron, J.K. Farnworth, A.F. Knyvett and D. Nimmo, 1961. Clinical manifestations of infection with group A arthropod-borne viruses in Queensland. Med. J. Aust. 1: 276 – 279.

Doherty, R.L., E.J. Barrett, B.M. Gorman and P.H. Whitehead, 1971. Epidemic polyarthritis in eastern Australia, 1959 – 1970. Med. J. Aust. 1: 5 – 8.

Doherty, R.L., J.G. Carley and J.C. Best, 1972. Isolation of Ross River virus from man. Med. J. Aust. 1: 1083 – 1084.

Doherty, R.L., J.G. Carley, M.U. Mackerras and E.N. Marks, 1963. Studies of arthropod-borne virus infections in Queensland. III. Isolation and characterization of virus strains from wild-caught mosquitoes in North Queensland. Aust. J. Exp. Biol. Med. Sci. 42: 149 – 164.

Essed, W.C.A.H. and J.A.E. van Tongeren, 1965. Anthropod-borne virus infections in western New Guinea. I. Report of a case of Murray Valley encephalitis in a Papuan woman. Trop. Geogr. Med. 17: 52 – 55.

Freeman, P.H., 1974. The environmental impact of tropical dams. Guidelines for impact assessment based upon a case study of Volta Lake. Smithsonian Institution, Washington, D.C.

French, E.L., S.G. Anderson, A.V.G. Price and E.A. Rhodes, 1957. Murray Valley encephalitis in New Guinea. I. Isolation of Murrey Valley encephalitis virus from the brain of a fatal case of encephalitis occurring in a Papuan native. Am. J. Trop. Med. Hyg. 6: 827 – 834.

Hall, A.J., 1983. Health and diseases of the people of the upper and lower Purari. This volume, Chapter III, 4.

Marks, E.N., 1983. Mosquitoes of the Purari River lowlands. This volume, Chapter III, 7.

Murphy, F.A., W.P. Taylor, C.A. Mims and I.D. Marshall, 1973. Pathogenesis of Ross River virus infection in mice. II. Muscle, heart, and brown fat lesions. J. Inf. Dis. 127: 129 – 138.

Tesh, R.B., D.C. Gajdusek, R.M. Garruto, J.H. Cross and L. Rosen, 1975. The distribution and prevalence of group A arbovirus neutralizing antibodies among human population in Southeast Asia and the Pacific Islands. Am. J. Trop. Med. Hyg. 24: 664 – 675.

Wisseman, C.L., D.C. Gajdusek, F.D. Schofield and E.C. Rosenzweig, 1964. Anthropod-borne virus infections of aborigines indigenous to Australia. Bull. WHO 30: 211 – 219.

Woodroofe, G.M. and I.D. Marshall, 1971. Arboviruses from the Sepik district of New Guinea, pp. 90 – 91. Report. John Curtin School of Medical Research, Australian National University, Canberra.

Work, T.H., 1964. Isolation and identification of arthropod-borne viruses. In: E.H. Lennette and N.J. Schmidt (eds.), Diagnostic Procedures for Viral and Rickettsial Diseases, 3rd ed., pp. 312 – 355. Am. Publ. Hlth. Assoc., New York.

Work, T.H., 1972. On the natural history of Venezuelan equine encephalitis: conclusions and correlations. In: Venezuelan Encephalitis, pp. 333 – 349. Pan American Organization Sc. Publ. No. 243. Washington, D.C.

Work, T.H. and M. Jozan, 1980. Vector mosquitoes and human arbovirus infections of the Purari River drainage. In: M. Alpers (ed.), Purari River (Wabo) Hydroelectric Scheme Environmental Studies, Vol. 8: Viral and parasitic infections of the people of the Purari River and mosquito vectors in the area, pp. 29 – 45. Office of Environment and Conservation, Waigani, Papua New Guinea.

7. Mosquitoes of the Purari River lowlands

E.N. Marks

1. Introduction

The proposed Purari River (Wabo) hydroelectric scheme involves the building of a dam to produce hydroelectric power at Wabo (7° 00′ S, 145° 04′ E), 100 km inland on the Purari River, Papua New Guinea, the transmission of this power to a port site on the coast of the Gulf of Papua and the development of an industrial complex at this port site.

Alpers (1977) outlined plans to establish baseline studies of the health problems and disease patterns of people living in the region affected by the Purari scheme, and then to monitor the changes that take place during and following construction of the dam. The health of workers coming into the area would also be considered. Of particular concern in this project are the arthropod-borne virus (arbovirus) diseases. Isolation of viruses would be attempted from mosquitoes caught in the area, after they had been identified by the entomologist in the research team. A search for avian and mammalian reservoirs for the viruses, and serological tests for the presence of antibodies to arboviruses in the human population would be undertaken.

The initial arbovius – mosquito investigation fieldwork for this project was undertaken in August 1978 by arbovirologists Drs Telford H. Work and Martine Jozan (see Work and Jozan, this volume) and the author. This paper reports on the mosquitoes collected, their potential as disease vectors, and some possible effects on the establishment of the dam.

Steffan (1966) listed 241 species of mosquitoes from the New Guinea mainland, and others have been described since. In addition, many as yet undescribed species have been recognised in collections made during the past 25 years, most notably among those now housed in the Bernice P. Bishop Museum, Honolulu, and the University of Queensland Insect Collection. It is thus probable that few, if any, of the undescribed species reported here are entirely new to science.

Petr, T. (ed.) The Purari – Tropical environment of a high rainfall river basin
© *1983, Dr W. Junk Publishers, The Hague / Boston / Lancaster*
ISBN-13: 978-94-009-7265-0

There appear to be no published records of mosquitoes from the Purari area and indeed there are few exact locality records from the Gulf and Western Provinces as compared with the rest of Papua New Guinea. This contrasts with extensive collecting carried out by Dutch workers before and after the 1939–45 war in the south of what is now Irian Jaya. Although no comparisons are drawn from the material now available, it seems likely that the mosquito fauna of the Purari will be found to be similar to that of the Digoel River.

Collections were made at five localities during a 12-day period. The aim of the fieldwork was to collect and identify specimens for arbovirus studies and in addition to obtain as representative a sample of the mosquito fauna as possible.

2. Mosquito collections

2.1. Localities (Fig. 1)

Mosquitoes were collected at Wabo (altitude ca. 70 m) 17–23 August, at both Wabo village and Wabo camp, and on a short visit to Uraru about 12 km upstream from Wabo on 20 August. In the Purari Delta, collections were made at Kapuna, 23–28 August (with a small day-biting collection from the nearby village of Ara'ava on 24 August) and at Mapaio, 25–27 August.

Collecting sites at Wabo and Uraru were close to primary rain forest, and near the margin of or within the zone of secondary regrowth. Both areas appeared generally well drained, with small flowing streams in the gullies. Ponds of standing water, probably the result of blocked drainage or excavations, were found at Wabo. Natural containers were prolific and at Wabo artificial containers also provided breeding places.

At Kapuna and Mapaio the ground was flat and poorly drained with many shallow water-holding depressions. These became more numerous in the vegetable gardens behind the villages, which merged into swamp forest. Natural containers were plentiful.

2.2. Collecting methods

Man-biting collections were made at all locations with the help of local people who were issued with vials. Two small light-traps designed by Mr H.A. Standfast were operated at Wabo and Kapuna. Several modified CDC traps (see

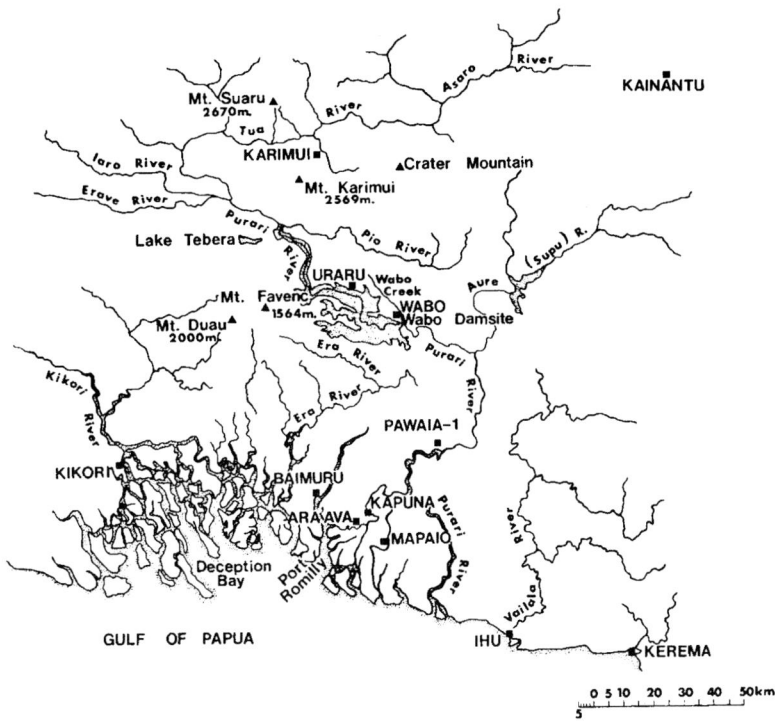

Fig. 1. The Purari River and some of its tributaries.

Work and Jozan 1980, this volume) were used at Wabo, Mapaio and Kapuna. For five nights at Wabo and one at Mapaio these were baited with dry ice (referred to hereafter as CO_2 traps). This attracted much greater numbers of species and of specimens at Mapaio than when the traps operated only as light-traps on one night at Mapaio and one at Kapuna (see Work and Jozan, this volume, Table 1, which lists specimens frozen for virus isolation).

Approximately 40 species of *Culicoides* (biting midges of the family Ceratopogonidae) collected in light and CO_2 traps are being studied by Mr A.L. Dyce, CSIRO Division of Tropical Animal Science.

Sampling of larvae and pupae from a wide range of breeding places yielded 8 positive collections from Wabo, 10 from Uraru, 12 from Kapuna and 8 from Mapaio. Adult mosquitoes were reared to confirm larval identifications and approximately 80 link-bred specimens (associating at least two of the three stages, larva, pupa, and adult) were obtained. Local school children assisted in collecting.

2.3. Identifications

These are based on the published literature, on the author's notes on types and other specimens in Australian and overseas collections, on manuscript keys drafted by the author and by other mosquito taxonomists, and on direct comparison with specimens in the University of Queensland Insect Collection.

Specimens to be frozen for arbovirus study were identified in the field as far as possible. However, speed was essential and precluded the extensive examination of single specimens necessary for specific identification of *Aedes* (*Verrallina*) and *Tripteroides* species.

The species collected are listed in several ways, the significance of which is explained by the following examples.

(a) *Anopheles* (*Cellia*) *longirostris*. Positive identification of a named species.
(b) *Aedes* (*Finlaya*) 'Marks sp. No. 104'. An undescribed species which has been studied in some detail by the author.
(c) *Tripteroides* (*Rachisoura*) sp. nr *bisquamatus*. An undescribed species which will key out (either as larva or adult) to the species it is 'near'.
(d) *Uranotaenia novaguinensis*? Identification of females or larvae tentative; specific distinctions best shown by males and/or larvae.

The collections have been studied further since reported by Marks (1980a), resulting in some alterations to the previous list.

3. Biology of Purari mosquitoes

3.1. Anophelines

Anopheles farauti was the only mosquito trapped in numbers at Wabo, the largest single CO_2 trap catch being 40. Larvae were found in a partly shaded pond near the boat landing, on the landward side of the formed track from Wabo camp (Figs. 2, 3). This depression, about 5 m in diameter, appeared to be either a borrow pit from which soil had been taken to build the track, or an old drainage channel dammed by the raised track. Aquatic plants were plentiful and larvae were found mainly among their horizontal sub-surface roots and among algae. All specimens of *An farauti* retained for taxonomic study were identified as *An farauti* No. 1, following Bryan (1974).

Anopheles punctulatus was taken in a light-trap at Kapuna and *An koliensis* and *Bironella travestita* in CO_2 traps at Mapaio.

Anopheles longirostris was also taken in CO_2 traps at Mapaio, the largest

534

Fig. 2. Wabo. The two drums in the foreground are at the road edge; the two behind are in a depression where *Anopheles farauti, Culex squamosus* and *Culex* 'Marks sp. No. 68' were breeding.

single trap catch being 28. Larvae were found in shallow partly shaded pools amongst fallen *Pandanus* fronds at the margin of a sago swamp (Figs. 4,5).

3.2. Aedes species

Species of subgenera *Huaedes* and *Leptosomatomyia* and most species of subgenus *Finlaya* breed in natural water-holding containers of plant origin;

Fig. 3. Wabo. Pond in depression beside road. Breeding place of *Anopheles farauti, Culex squamosus* and *Culex* 'Marks sp. No. 68'.

Fig. 4. Mapaio. At the sago swamp margin shallow pools contained larvae of *Aedes carmenti, Aedes neomacrodixoa, Anopheles longirostris* and *Culex pullus. Aedes* 'Marks sp. No. 108' was breeding in axils of the long-leafed plant on left.

Fig. 5. Mapaio. Shallow pool among fallen *Pandanus* leaves in sago swamp. Breeding place of *Aedes carmenti, Aedes neomacrodixoa, Anopheles longirostris* and *Culex pullus*.

Fig. 6. Uraru. Groove in log at river edge. Breeding place of *Aedes novalbitarsis*.

537

some can also colonize man-made containers such as tins.

At Uraru *Aedes* (*Huaedes*) 'Marks sp. No. 146' was breeding in bamboo stumps and *Aedes* (*Finlaya*) *novalbitarsis* in a groove in a fallen tree trunk (Fig. 6); neither species is known to bite man. *Aedes* (*Leptosomatomyia*) *aurimargo*? (a tree-hole and bamboo breeder) bit man at dusk in the bush at Mapaio.

The following species belong to the *kochi* group of *Finlaya*, members of which breed in the leaf axils of plants. *Aedes kochi* was taken in light-traps at Kapuna and in CO_2 traps at Mapaio; at both places it was breeding in banana axils (Fig. 7) (it frequently breeds in *Pandanus* axils and Assem (1959) recorded it in sago axils). *Aedes wallacei* larvae were collected from pineapple axils at Kapuna; two adults trapped at Mapaio may represent a second *wallacei*-like species. At Mapaio *Aedes* 'Marks sp. No. 104' was taken in CO_2 traps; it is known to breed in axils of *Crinum*-like plants and is the undescribed species recorded as a biting pest on Frederik Hendrik Island by Assem (1959). Larvae of *Aedes* 'Marks sp. No. 108' were found in axils of an unidentified long-leafed plant at Mapaio (Fig. 4) and adults were taken biting at Kapuna, where *Aedes* 'Marks sp. No. 106' was taken in a light-trap.

Species of subgenus *Geoskusea* breed in crab or crayfish burrows; some species bite man. *Aedes* sp. nr *longiforceps* was taken in CO_2 traps at Mapaio.

Species of subgenus *Verrallina* breed in temporary ground pools in rain-filled depressions; most species bite man. Larvae of *Ae carmenti* and *Ae neomacrodixoa* were found in shallow pools amongst banana, taro and sweet potatoes in garden areas at Kapuna and Mapaio (Figs. 4, 5, 7, 8). Both species entered CO_2 traps at Mapaio, and *Ae carmenti* bit at dusk at Kapuna. A day-biting catch at Ara'ava included *Ae carmenti*, *Ae parasimilis* and a distinctive new species *Ae* 'Marks sp. No. 171' (listed as *Ae sentanius*? by Marks (1980a), whose records of *Ae vanapus* were misidentifications of *Ae carmenti*).

3.3. Armigeres species

Armigeres papuensis bit man outdoors in the shade in late afternoon at Wabo and Kapuna and was taken in a CO_2 trap at Mapaio; *Ae milnensis* was biting at Kapuna. Larvae were not found, but these species breed in putrid water in such sites as sago palms and drains.

Fig. 7. Mapaio. *Aedes kochi* was breeding here in leaf axils of banana, *Malaya leei* in axils of banana and taro, *Uranotaenia atra* and *Uranotaenia obscura* in fallen banana leaves, and *Aedes carmenti* and *Aedes neomacrodixoa* in shallow ground pools.

3.4. Coquillettidia and Mansonia species

Adults only were collected. In these two genera, larvae and pupae attach their breathing tubes to the roots or stems of aquatic plants from which they obtain oxygen. Hence they do not come to the water surface to breathe and are very difficult to find. Six species entered CO_2 traps at Mapaio (life-histories of the first four are unknown): *Coquillettidia* sp. nr *memorans*, two species of the *giblini* complex (*Cq* sp. nr *nigrochracea* and *Cq* 'Marks sp. No. 170'), *Mansonia papuensis, Ma uniformis* and *Ma septempunctata. Mansonia papuensis* was by far the most numerous, over 500 being taken in a single CO_2 trap. It also entered light-traps and bit man at dusk at both Mapaio and Kapuna. *Mansonia uniformis* was trapped at Kapuna.

Fig. 8. Mapaio. A breeding place of *Aedes carmenti and Aedes neomacrodixoa* in shallow pools amongst yams in a banana patch where *Aedes kochi* bred in banana axils, *Malaya leei* in axils of banana and taro, and *Uranotaenia atra* and *Uranotaenia obscura* in fallen banana leaves.

3.5. *Culex species*

Three species of subgenus *Culex* were taken at Wabo. There must be a slight reservation about the single female of *Cx annulirostris* from a CO_2 trap, as these traps had been recently used in Western Australia. However, the record is discussed here as a valid one. Larvae of this species were not found; it breeds in ground pools, usually with floating or emergent vegetation, and bites a wide range of hosts including man. *Culex* 'Marks sp. No. 68', a member of the *vishnui* group, was taken in CO_2 traps; its biting habits are unknown. It was breeding, together with *Cx squamosus*, in the same pond as *An farauti* (Figs. 2, 3). Larvae of *Cx squamosus* are usually associated with filamentous algae; females feed on birds and mammals and occasionally attack man (Kay, Boreham and Williams 1979).

The three species of subgenus *Culiciomyia* collected breed in ground pools and natural and artificial containers, often in polluted water. They are not known to bite man. At Wabo *Cx pullus* and *Cx fragilis* were breeding in a

540

Fig. 9. Wabo. Cassowary pen. Woman squatting is feeding bird on greenstuff. Rainwater from sheet of corrugated iron collect in 44-gallon drum in right hand corner of pen, breeding place of *Culex fragilis, Culex pullus, Tripteroides bimaculipes* and *Uranotaenia atra.*

44-gallon drum in a cassowary pen (Fig. 9) and *Cx pullus* in another drum containing a drowned rat. At Mapaio *Cx pullus* larvae were found in shallow pools (Figs. 4, 5), among *Pandanus* leaves and adults entered CO_2 traps. Both *Cx pullus* and *Cx papuensis* were taken in light-traps at Kapuna.

Females of subgenus *Lophoceraomyia* are in most cases not readily identifiable to species. Many species breed in ground pools, some in rock pools, tree holes or plant axils. Some species bite man. Adults only were collected. Males of *Cx lakei* entered a CO_2 trap at Mapaio; *Cx ornatus* was identified from a light-trap at Kapuna and probably about six other species are represented in trap collections from Kapuna and Mapaio.

3.6. Hodgesia and Uranotaenia species

Hodgesia spoliata, a tiny swamp-breeding mosquito, was taken biting and in CO_2 traps at Mapaio and biting at Kapuna.

Mosquitoes of genus *Uranotaenia* do not bite man. Some species feed on

541

frogs (Australian records summarised in Marks (1980b) include *Ur argyrotarsis* and *Ur novaguinensis*).

Species of subgenus *Uranotaenia* breed in swamps and ground pools. No early stages were found but light-traps collected *Ur novaguinensis*? at Wabo and *Ur argyrotarsis* and *Ur* sp. nr *tibialis* at Kapuna, while *Ur moresbyensis, Ur paralateralis, Ur* sp. nr *tibialis* and *Ur* 'Marks sp. No. 139' were taken in CO_2 traps at Mapaio.

Species of subgenus *Pseudoficalbia* breed in plant containers or leafy ground pools. *Uranotaenia atra* larvae were collected from a cut bamboo at Uraru, from a 44-gallon drum in a cassowary pen (Fig. 9) at Wabo, and, together with *Ur obscura* (= *Ur papuensis*), from fallen banana leaves at Kapuna and Mapaio; *Ur* sp. nr *diagonalis* was breeding with *Ur atra* at Kapuna. *Uranotaenia atra* was also taken in a light-trap at Wabo. *Uranotaenia obscura* and *Ur hirsutifemora* were taken in CO_2 traps at Mapaio and the latter species in a light-trap at Kapuna.

3.7. Malaya and Topomyia species

Larvae of *Malaya leei* and *Topomyia* 'Marks sp. No. 144' were collected from leaf axils. Neither of these small mosquitoes bites man. *Malaya leei* was breeding in taro at Wabo, Kapuna and Mapaio, banana at Wabo and Mapaio (Fig. 7), and pineapple at Kapuna, and the *Topomyia* species in taro at Uraru and Wabo, taro and pineapple at Kapuna.

3.8. Toxorhynchites species

These large mosquitoes cannot suck blood; their larvae prey on other mosquito larvae. The genus is currently being revised by Dr W.A. Steffan, and a specific identification is not attempted for the two larvae collected, one from a bamboo stump at Uraru and one from a sago stump at Mapaio (Fig. 10).

3.9. Tripteroides species

Species of *Tripteroides* breed in plant containers and some also colonize artificial containers. Many species bite man. Larvae of *Tripteroides (Rachisoura)* sp. nr *bisquamatus*, *Tp (Rac)* 'Marks sp. No. 60', *Tp (Rac) longipalpatus*, *Tp (Tripteroides) bimaculipes* and *Tp (Polylepidomyia)* sp. nr

Fig. 10. Mapaio. *Toxorhynchites* sp. and *Tripteroides bimaculipes* were breeding in the sago stump in the axils of the fronds.

standfasti were collected from bamboo stumps at Uraru. At Wabo *Tp bimaculipes* was breeding in a 44-gallon drum in a cassowary pen (Fig. 9) and *Tp* 'Marks sp. No. 60' in bamboo stumps (the latter species, previously identified as *Tp* sp. nr *fuscipleura* (Marks 1980a) runs to *Tp felicitatis* in some keys but differs from the holotype). *Tripteroides* sp. nr *bisquamatus* bit in the bush by day at Uraru and Wabo and indoors in artificial light at Wabo. *Tripteroides (Pol) argenteiventris*? was taken biting indoors and outdoors by day and in bedrooms overnight at Wabo and at dusk in the bush (this was possibly a second species) at Mapaio and Kapuna. *Tripteroides (Trp) littlechildi* bit indoors by day at Kapuna and with *Tp bimaculipes* in the bush at dusk at Kapuna and Mapaio; both species entered CO_2 traps at Mapaio where *Tp bimaculipes* larvae were found in a sago stump (Fig. 10).

4. Purari mosquitoes in relation to mosquito-borne diseases

4.1. The diseases

The mosquito-borne diseases of concern or potential concern in the Purari area have in common that the causative organisms undergo an obligatory period of development or replication in the mosquito. They are: (a) filariasis, caused by the nematode worm *Wuchereria bancrofti*; (b) malaria, caused by protozoan parasites *Plasmodium* spp.; (c) arboviral diseases, exemplified by (i) dengue fever; (ii) Murray Valley encephalitis (MVE) = Australian encephalitis; (iii) epidemic polyarthritis caused by Ross River virus (RR).

Filariasis and malaria have in common that man is the only vertebrate host and the cycle is man – mosquito – man. As far as is known this is also the cycle for dengue in New Guinea.

Murray Valley encephalitis and epidemic polyarthritis, on the other hand, are zoonoses, diseases whose normal cycle does not include man; in the case of MVE it probably involves amplifying cycles in birds and mammals with occasional transmission to man when infective mosquitoes become numerous.

4.2. Some attributes of a vector

Where the association between mosquito and man is very close, transmission may be maintained by a comparatively small mosquito population. In the case of a zoonosis, the vector to man may often be a species that will feed on a range of hosts and its effectiveness may be due to the large size of its populations increasing the chance of man receiving an infected bite. Other zoonoses may have separate maintenance and epidemic vectors.

The time required for the disease organism to complete development in the mosquito (extrinsic incubation period) depends on the temperature. It is 8 – 19 days for dengue (Mackerras 1946), 7 or more days for MVE (Kay, Carley et al. 1979) and, in coastal areas of New Guinea, 10 – 12 days for *Plasmodium* spp. (Assem and Bonne-Wepster 1964) and 12 days or more for *Wuchereria bancrofti* (de Rook 1959).

Proven vectors obviously survive for the required period. Amongst other common biting species, Mackerras (1946) reported *Aedes aurimargo* and species of *Armigeres* and *Aedes* (*Verrallina*) surviving 15 days or more, but there appears to be no information on longevity of *Tripteroides*.

4.3. Vector potential of the species collected

4.3.1. Malaria. Anopheles farauti, *An koliensis* and *An punctulatus* are the major malaria vectors in New Guinea; *An punctulatus* prefers ephemeral freshwater collections, *An koliensis,* less temporary breeding sites, and *An farauti,* more permanent water collections, including those that are brackish (van Dijk and Parkinson 1974).

Anopheles longirostris has no practical importance as a malaria vector in the better known parts of New Guinea, but it has survived to the infected stage in the laboratory and its role needs investigating in places where anopheline catches are almost exclusively *An longirostris* (Assem and Bonne-Wepster 1964).

4.3.2. Filariasis. In New Guinea *Wuchereria bancrofti* shows a nocturnal periodicity (i.e. the infective stage is in the peripheral blood stream at night). No filarial vectors have been found among species biting in daylight or early evening hours. The relative importance of different species as vectors varies in different localities. Of the species collected in the Purari area, *Anopheles farauti, An koliensis, An punctulatus, Aedes kochi, Culex annulirostris* and *Mansonia uniformis* have been found infective in nature in New Guinea, as have two indigenous species likely to occur in the area, *Anopheles bancrofti* and *Culex bitaeniorhynchus* and one species likely to be introduced by man, *Culex quinquefasciatus* (*=fatigans*) (Assem and Bonne-Wepster 1964).

Mansonia papuensis can readily be infected experimentally, but de Rook and van Dijk (1959) consider that more evidence is needed on its hospitability, since the mortality under experimental conditions was very high and only one specimen survived 12½ days and developed mature *W. bancrofti* larvae. These authors regard *An farauti* as the most important vector among the anophelines, with which *Ma uniformis* and *Cx annulirostris* compare in efficiency, while *Cx quinquefasciatus* is an increasing danger.

4.3.3. Dengue. Mackerras (1946) reported transmission of dengue to volunteers in Sydney by *Aedes scutellaris* that had fed on patients in New Guinea. Results of similar experiments with *Armigeres milnensis, Ae similis* and *Ae aurimargo* were inconclusive. *Armigeres milnensis,* as recognised at that time, included also *Ar papuensis.*

Aedes scutellaris, a widespread indigenous species, probably already occurs in the Purari area. In any case both it and two introduced vector species, *Ae aegypti* and *Ae albopictus*, are likely to be transported there from other localities. These three species breed in artificial containers, including tyres, in which their eggs may be carried from place to place. *Aedes aegypti* occurs in many urban areas of Papua New Guinea. *Aedes albopictus*, reported from the Madang area by Schoenig (1972), has since spread to the Solomon and Santa Cruz Islands (Elliott 1980).

4.3.4. Murray Valley encephalitis. Culex annulirostris has been taken naturally infected with MVE in Australia on numerous occasions and is considered the major vector. Other species, including *Cx quinquefasciatus* and several species of *Aedes*, can transmit the virus experimentally (McLean 1957; Kay, Carley et al. 1979).

Culex squamosus was found in Australia naturally infected with the related Kunjin virus (Doherty et al. 1968). This virus also is known to cause encephalitis (Stanley 1982).

4.3.5. Epidemic polyarthritis. Culex annulirostris is a proven vector, as is *Aedes vigilax* (which breeds in intertidal areas and is likely to occur near the mouth of the Purari and along the adjacent coast), but the number of potential vector species is likely to be much greater (e.g. *Aedes polynesiensis*, a member of the *scutellaris* group, was shown to be an efficient vector in Fiji (Mataika 1982)).

4.3.6. General considerations. The competence of many New Guinea mosquitoes to transmit arboviruses is unknown. In situations where man-biting species breed abundantly close to human habitation, these merit investigation. In the Purari area this would apply to *Ma papuensis*, to species of *Aedes* (*Verrallina*), and possibly to *Armigeres* and *Tripteroides* species.

5. Some ways in which establishment of Wabo Dam may affect mosquito populations

5.1. Construction stages

The species most advantaged by large-scale engineering activities are those that breed in ground pools and those that colonize man-made containers. The sorts of breeding places provided are comparable to those that were produced by operations during the 1939–45 war.

Species of the *Anopheles punctulatus* group will quickly colonize open pools in shallow excavations and tracks of heavy vehicles as well as the larger and more permanent ponds resulting from soil borrow pits or blocked drainage. *Anopheles punctulatus* is especially favoured by open sunlit pools, whereas *An farauti* will colonize the more permanent vegetated ones (as exemplified at Wabo); however, there is no line of distinction. Thus an upsurge in anopheline breeding can be expected, which, unless controlled, may lead to intensified transmission of malaria and filariasis. Vegetated pools are also likely to provide for increased breeding by the MVE vector, *Culex annulirostris.*

Associated with construction works are many water-holding containers such as tyres, drums, tins, cavities in unused machinery, sagging tarpaulins, plastic containers and other rubbish. Mosquitoes which may breed in pest numbers in these sites are *Armigeres* species and four important disease vectors not yet recorded from the area but very likely to be transported to it. These are *Aedes scutellaris, Ae aegypti* and *Ae albopictus* (potential carriers of dengue), and *Culex quinquefasciatus*, which also thrives in imperfectly sealed septic tanks and in polluted effluents from sewage treatment works. *Culex quinquefasciatus* is a filariasis vector.

5.2. The established reservoir

Shallow edges with emergent and surface vegetation may provide for large-scale breeding by *Culex annulirostris* and *Culex bitaeniorhynchus,* although this might be controlled to a large extent by larvivorous fish.

Mansonia larvae, which do not swim to and from the water surface but remain relatively motionless attached to underwater plants, are much less subject to predation (Assem 1958b). Assem and Metselaar (1958) listed five host plants for *Mansonia uniformis* from the vicinity of Merauke, *Hydrocharis* (?) *parvula, Monochoria vaginalis, Utricularia* sp., *Ipomoea aquatica* and a small water fern, probably *Azolla* sp. *Pistia stratiotes* was a common host plant near Hollandia (Jayapura), but in experiments larvae showed no marked preference for it over roots of Gramineae. It was concluded that in New Guinea *Ma uniformis* larvae probably try to pierce the submerged plant tissues in their breeding places without any preference and stay attached to the kinds that are soft enough to be penetrated.

Assem (1958a) recorded *Mansonia uniformis* and *Ma bonnewepsterae* from roots of *Pistia stratiotes* and *Hydrocharis asiatica* on the shore of Sentani Lake. He noted that local distribution of *Mansonia* larvae was not identical with these plants but was restricted to an area behind a village where there was heavy organic pollution. Conn (1977) discussed aquatic plants from the Purari River watershed and aquatic weeds which might be introduced. It seems likely that favourable breeding places for *Ma uniformis*, and probably *Ma septem-punctata*, could become established at the margins of Wabo reservoir in the vicinity of villages, with implications for filariasis transmission.

Mansonia papuensis appears to be mainly associated with brackish swamps (Peters and Christian 1963) and swamp forest and is unlikely to colonise the reservoir.

6. Conclusions

Over 50 species of mosquitoes were collected from five localities on the Purari River. Much more extensive collecting during several seasons would be needed to obtain a fair assessment of the total mosquito fauna of the area which might comprise two or three times that number of species. However, it is reasonable to assume that the collection included the principal potential disease vectors active at the time of the visit. *Anopheles farauti, An koliensis* and *An punctulatus* are the three most important vectors of malaria in New Guinea. They are also efficient vectors of filariasis, as are *Culex annulirostris, Mansonia uniformis* and *Aedes kochi. Culex annulirostris* is the vector of Murray Valley encephalitis and epidemic polyarthritis. Malaria and filariasis are already established diseases in the Purari area.

It is possible to predict with confidence some of the effects on mosquito populations of environmental changes associated with large-scale construction work, for similar changes occurred elsewhere on the New Guinea mainland during the 1939 – 45 war. The disturbance of ground and drainage provides greatly expanded areas of ground pools favourable to anopheline breeding, leading to increased opportunities for malaria and filariasis transmission. The large number of artificial water-holding containers associated with construction and camp sites promotes breeding by biting pests *Armigeres* and *Aedes* spp., including potential dengue vectors. Containers and polluted effluents may provide for large populations of *Culex quinque-fasciatus,* a filariasis vector. On the other hand, construction works are likely to destroy many plants that are breeding places of another filariasis vector, *Aedes kochi.*

The opportunities that the established dam would provide for extensive mosquito breeding can only be tentatively assessed by comparison with large freshwater New Guinea lakes. There is little information about these. Possibly *Culex annulirostris* and another filariasis vector, *Culex bitaeniorhynchus,* would find favourable breeding sites amongst aquatic vegetation at the margins. Findings from Lake Sentani suggest that *Mansonia uniformis* and

Ma septempunctata could build up large populations where there are plentiful aquatic weeds associated with polluted water close to villages. Investigation of mosquito-breeding in Lake Tebera, the nearest freshwater lake to Wabo, should provide an indication of species likely to colonise the reservoir.

Acknowledgements

I wish to thank Dr M. Alpers for the invitation to take part in this investigation, Dr T. Petr for organizing the Purari patrol, Drs T.H.. Work and M. Jozan for their collaboration in the fieldwork, Dr G. Nurse for collecting specimens, Dr L. Calvert and her husband, the late Dr P. Calvert, for hospitality and help at Kapuna, the Deputy Premier, Gulf Province, and the Deacon, Mapaio, for facilitating our visit to Mapaio, and the many people of Wabo, Kapuna, Mapaio and Ara'ava who helped in collecting and in other ways. I am grateful also to Mr T.L. Fenner, formerly of Port Moresby, and the late Dr J.L. Gressitt, Wau Ecology Institute, for providing facilities for rearing the Purari larval collections, and to Mrs A. Pound for assistance in this task. This study was financed and supported by the Queensland and Papua New Guinea governments.

References

Alpers, M., 1977. Health aspects of the Purari Scheme with particular reference to arthopod-borne diseases. In: T. Petr (ed.), Purari River (Wabo) Hydroelectric Scheme Environmental Studies, Vol. 1: Workshop 6 May 1977, pp. 63 – 64. Office of Environment and Conservation, Waigani, Papua New Guinea.

Assem, J. van den, 1958a. *Mansonia (Mansonioides) bonnewepsterae*, spec. nov. (Culicidae), with notes on habits and breeding place. Trop. Geogr. Med. 10: 205 – 212.

Assem, J. van den, 1958b. Some experimental evidence for the survival value of the rootpiercing habits of *Mansonia* larvae (Culicidae) to predators. Ent. Exp. Appl. 1: 125 – 129.

Assem, J. van den, 1959. Some notes on mosquitoes collected on Frederik Hendrik Island (Netherlands New Guinea). Trop. Geogr. Med. 11: 140 – 146.

Assem, J. van den and J. Bonne-Wepster, 1964. New Guinea Culicidae, a synopsis of vectors, pests and common species. Zool. Bijd., No. 6.

Assem, J. van den and D. Metselaar, 1958. Host plants and breeding places of *Mansonia (Mansonioides) uniformis* in Netherlands New Guinea. Trop. Geogr. Med. 10: 51 – 55.

Bryan, J.H., 1974. Morphological studies on the *Anopheles punctulatus* Donitz complex. Trans. Roy. Ent. Soc. Lond. 125: 413 – 435.

Conn, B.J., 1977. The proposed Purari River hydroelectric scheme – notes on aquatic and semi-aquatic flora. In: T. Petr (ed.), Purari River (Wabo) Hydroelectric Scheme Environmental Studies, Vol. 1: Workshop 6 May 1977, pp. 21 – 27. Office of Environment and Conservation, Waigani, Papua New Guinea.

Dijk, W.J.O.M. van and A.D. Parkinson, 1974. Epidemiology of malaria in New Guinea. Papua New Guinea Med. J. 17: 17 – 21.

Doherty, R.L., R.H. Whitehead, E.J. Wetters and B.M. Gorman, 1968. Studies of the epidemiology of arthropod-borne virus infections at Mitchell River Mission, Cape York Peninsula, north Queensland. II. Arbovirus infections of mosquitoes, man and domestic fowls, 1963 – 1966. Trans. Roy. Soc. Trop. Med. Hyg. 62: 430 – 438.

Elliott, S.A., 1980. *Aedes albopictus* in the Solomon and Santa Cruz Islands, South Pacific. Trans. Roy. Soc. Trop. Med. Hyg. 74: 747 – 748.

Kay, B.H., P.F.L. Boreham and G.M. Williams, 1979. Host preferences and feeding patterns of mosquitoes (Diptera: Culicidae) at Kowanyama, Cape York Peninsula, northern Queensland. Bull. ent. Res. 69: 441 – 457.

Kay, B.H., J.G. Carley, I.D. Fanning and C. Filippich, 1979. Quantitative studies of the vector competence of *Aedes aegypti, Culex annulirostris* and other mosquitoes (Diptera: Culicidae) with Murray Valley encephalitis and other Queensland arboviruses. J. Med. Ent. 16: 59 – 66.

Mackerras, I.M., 1946. Transmission of dengue fever by *Aedes* (*Stegomyia*) *scutellaris* Walk. in New Guinea. Trans. Roy, Soc. Trop. Med. Hyg. 40: 295 – 312.

Marks, E.N., 1980a. Mosquitoes of the Purari River area. In: M. Alpers (ed.), Purari River (Wabo) Hydroelectric Scheme Environmental Studies, Vol. 8: Viral and parasitic infections of the people of the Purari River, and mosquito vectors in the area, pp. 3 – 27. Office of Environment and Conservation, Waigani, Papua New Guinea.

Marks, E.N., 1980b. Mosquitoes (Diptera: Culicidae) of Cape York Peninsula, Australia. In: N.C. Stevens and A. Bailey (eds.), Contemporary Cape York Peninsula, pp. 59 – 76. Royal Society of Queensland, Brisbane, Australia.

Mataika, J.U., 1982. Ross River virus in Fiji. In: T.D. St. George and B.H. Kay (eds.), Arbovirus Research in Australia. Proc. 3rd Symposium, 15 – 17 Feb. 1982, pp. 179 – 183. CSIRO and QIMR, Brisbane, Queensland, Australia.

McLean, D.M., 1957. Vectors of Murray Valley encephalitis. J. Inf. Dis. 100: 223 – 227.

Peters, W. and S.H. Christian, 1963. The bionomics, ecology and distribution of some mosquitoes (Diptera: Culicidae) in the Territory of Papua and New Guinea. Acta Trop. 20: 35 – 79.

Rook, H. de, 1959. Filariasis in the village of Inanwatan (south coast of The Vogelkop, Netherlands New Guinea). Trop. Geogr. Med. 11: 313 – 331.

Rook, H. de and W.J.O.M. van Dijk, 1959. Changing concept of *Wuchereria bancrofti* transmission in Netherlands New Guinea. Trop. Geogr. Med. 11: 57 – 60.

Schoenig, E., 1972. Distribution of three species of *Aedes* (*Stegomyia*) carriers of virus diseases on the main island of Papua and New Guinea. Philippine Scientist 9: 61 – 82.

Stanley, N.F., 1982. Human arbovirus infections in Australia. In: T.D. St. George and B.H. Kay (eds.), Arbovirus Research in Australia. Proc. 3rd Symposium, 15 – 17 Feb. 1982, pp. 216 – 226. CSIRO and QIMR, Brisbane, Queensland, Australia.

Steffan, W.A., 1966. A checklist and review of the mosquitoes of the Papuan subregion (Diptera: Culicidae). J. Med. Ent. 3: 179 – 237.

Work, T.H. and M. Jozan, 1980. Vector mosquitoes and human arbovirus infections of the Purari River drainage. In: M. Alpers (ed.), Purari River (Wabo) Hydroelectric Scheme Environmental Studies, Vol. 8: Viral and parasitic infections of the people of the Purari River, and mosquito vectors in the area, pp. 29 – 45. Office of Environment and Conservation, Waigani, Papua New Guinea.

Work, T.H. and M. Jozan, 1983. Human arbovirus infections of the Purari River lowlands. This volume, Chapter III, 6.

8. Nutritional status of the people of the Purari Delta

S.J. Ulijaszek and S.P. Poraituk

1. Introduction

The subsistence economy of the people of the Purari Delta is based on sago cultivation and fishing, with the growing of root and tree crops being a subsidiary activity. In 1947, the New Guinea Nutrition Survey Expedition (Hipsley and Clements 1950) chose Koravake as their village dependent on sago as the staple food, and collected much valuable information on disease patterns, agriculture and food beliefs of the people of the area. However, child growth and dietary considerations were not covered; the present authors returned to Koravake and Mariki villages for a one-month period during September of 1980 to study sago production and dietary intake. Responsibilities for the various aspects of the project were shared by the authors to allow efficient use of time and to minimise disruption of village activities. Three additional team members were called in to assist in the collection of dietary data.

Weights and heights of 166 children, representing 23% of the population, aged zero to five years, were measured in nine Purari Delta villages (Koravake, Ara'ava, Kaimari, Mariki, Varea, Kinipo, Mapaio, Maipenairu and Urika (Fig. 1) in April 1980 by the senior author in order to determine the levels of undernutrition as indicated by retardation in growth.

The data presented here should be considered in conjunction with the studies by Hall (this volume) and Lambert (this volume).

2. Survey area and methods

The anthropometric survey of children under five years of age was carried out as part of a routine maternal and child health clinic run on a regular basis to

Petr, T. (ed.) The Purari – Tropical environment of a high rainfall river basin   551
© *1983, Dr W. Junk Publishers, The Hague / Boston / Lancaster*
ISBN-13: 978-94-009-7265-0

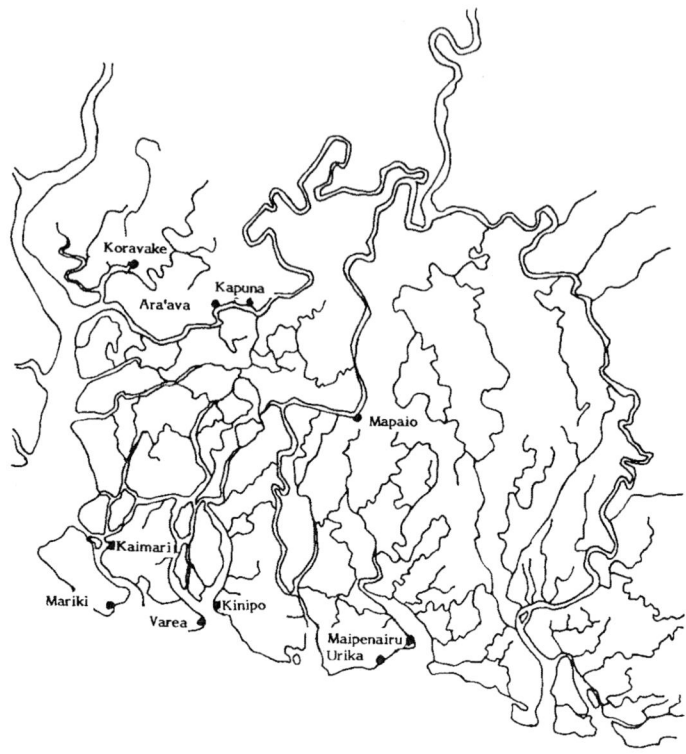

Fig. 1. Villages visited during the 1980 anthropometric survey, Purari Delta.

all delta villages by Kapuna Christian Mission. The methodology used was as recommended by Jelliffe (1966). All clinic attenders had maternal and child health register cards from which accurate dates of birth were obtained.

A weighed food intake study was carried out in an attempt to obtain a profile of nutrient intake for members of Koravake village of all ages, with specific reference to energy and protein, and subclinical nutrient deficiencies. In all, food intakes of 39 individuals from 5 nuclear families (mother, father, and their offspring only) were measured for a three-day period; this was 15% of the total Koravake population.

The amount of each food type eaten was recorded by weighing both the food offered and the unused plate waste. In all cases raw weights of food were used. Food composition tables (FAO/United States Department of Health, Education and Welfare 1972; Norgan, Durnin and Ferro-Luzzi 1979; Peters 1957; Hipsley and Clements 1950) were used to calculate the three-day average of nutrient intake for each individual.

552

Adequacy of the diet was measured by comparing the actual intake to the Recommended Daily Intake (RDI) per individual for each nutrient. World Health Organisation Western Pacific RDI's were used as reference values (FAO/WHO 1974), acknowledging that the true requirements of the Koravake people may be less for some nutrients. Few physiological studies exist which attempt to determine the nature of metabolism of Melanesian people, and it is unlikely that such studies could be extrapolated to coastal Papuan people. Thus, for the purpose of the present study, the following definitions were adopted:

Nutrient requirements are the minimum amounts of nutrients needed to maintain normal health and growth. Requirements may be affected by age, sex, body weight, energy expenditure and certain physiological states such as pregnancy and lactation.

Recommended intakes are equal to the nutrient requirements plus a safety margin to allow for individual variation. This safety margin is approximately 'plus 2 standard deviations of the mean requirement'. The requirements of the majority of the population are therefore covered by the recommended intakes.

Energy requirements depend on body size, age, sex and activity. Requirements in Koravake were based on light activity in males, and moderate activity in females.

Protein requirements depend on the protein quality of the total diet. Protein quality is measured in terms of Net Protein Utilisation (NPU) and is the proportion of protein intake which is used in maintenance and building of body proteins. (NPU is 100% when all of the protein intake is used in maintenance and anabolism; in practice, no diet has an NPU of 100%.) Animal proteins have higher concentrations of essential amino acids than most vegetable proteins, and diets containing even small amounts of animal protein have a higher NPU than those whose sources of protein are almost totally of vegetable origin. The Koravake diet is a mixed one in which animal source foods provide between 15% and 20% of total energy intake. FAO/WHO (1974) recommend that the protein requirements of people on such a diet be based on an NPU of 80%.

Protein and energy requirements are interrelated, inasmuch as at low levels of energy intake, protein deficiency is a result of preferential metabolism of protein for energy (Platt and Miller 1959). Models proposed by Beaton and Swiss (1974) and Payne (1975) consider variability in protein and energy requirements, expressed as Protein Energy Percent (PE%). The minimum percent of total energy provided by protein varies with body size, age and energy intake, and can be expressed by the following relationship:

$$PE\% = \frac{\text{Protein requirement / kg body weight}}{\text{Energy intake / kg body weight}} \times 400$$

(from Beaton and Swiss 1974).

As energy intake is increased, the PE% is reduced as protein is spared for body muscle maintenance and/or anabolism. PE% is used as an indicator for estimating the potential of the diet to satisfy protein needs.

The ratio of fat energy to total energy intake (FE%) is used for estimating the potential of the diet to meet energy requirements. A value of between 20% and 30% is recommended (FAO/WHO 1977).

The ratio of fat energy to protein energy (FE/PE) is a supplementary indicator used here to avoid false interpretations of the PE% and FE% information. A value of 2.0 has been proposed as the minimum standard (Araya and Arroyave 1979).

Calcium requirements are uncertain because absorption is regulated to some extent by the bodies' needs and there appears to be adaptation to diets supplying low quantities of this mineral. Vitamin D is needed for calcium absorption and utilisation, whilst the presence of phytates and oxalates inhibits absorption.

Iron absorption is related to the proportion of the diet coming from animal sources. Where such items provide a high proportion of the dietary energy, a lower intake of iron is needed than when they provide a low proportion. High intakes of vitamin C facilitate iron absorption, whereas the presence of phytates in the food inhibits it.

Vitamin A is fat soluble and stored in the liver, so daily fluctuations in intake are less important than with the water soluble B group vitamins and vitamin C. Extra vitamin A is needed by lactating women to build up liver stores of the vitamin in their breast feeding infants.

Vitamin C requirements, as conventionally calculated, provide for a safety margin of about 10 mg. Ascorbic acid in food is quite unstable and very soluble, and the substantial losses that occur in cooking should be taken into account when estimating intakes.

3. Results and discussion

The period of greatest nutritional stress is in the age group 1 to 1.9 years, when 15% of children are undernourished at the time of survey (low W/H), and 17% have undergone stunting of growth (low H/A). In the age group 2 to

5 years, only 4.5% are undernourished, and 21% have had their growth stunted (Table 1).

Other sets of data add to this picture. Records at Kapuna hospital in the Purari Delta show that there has been no significant increase in birth weight over the period 1974 to 1980 (Table 2) and that an average of 19% of babies are born weighing less than 2.5 kg. Although it is not known what proportion of these children were prematurely born, these figures are too high to be attributed to prematurity alone, and are indicative of some level of foetal malnutrition.

Secondly, a survey of 672 community school students aged 7 to 15 years in the Baimuru District (Eng and Leonard 1978) showed 33% of males and 20% of females to have H/A below 90% and only 1% with W/H below 80%.

The three sets of data taken together indicate that there is very little nutritional deprivation in this later age group, and that nearly all the stunting of

Table 1. Purari Delta Nutrition Survey. Classification by weight for height (W/H) and height for age (H/A) – percent (sexes pooled).

Age (years)	H/A W/H	$\geq 80\%$	$< 80\%$
0 – 0.9	$\geq 90\%$	86.2%	10.3%
		$n = 29$	
	$< 90\%$	3.5%	–
1 – 1.9	$\geq 90\%$	75.0%	8.3%
		$n = 48$	
	$< 90\%$	10.4%	6.3%
2 – 5.0	$\geq 90\%$	74.2%	3.4%
		$n = 89$	
	$< 90\%$	21.3%	1.1%

Note: Retardation of growth in height and weight is a widely used index of undernutrition throughout the world. Weight for height (W/H) is a measure of current nutritional status; children with W/H below 80% of the Harvard mean are considered undernourished at the time of survey. Height for age (H/A) is a measure of past nutritional status; children with H/A below 90% of the Harvard mean are considered to have undergone a period of undernutrition during their lifetime, though not necessarily at the time of survey. Children with low W/H and H/A are considered to have suffered nutritional deprivation both in their past and at the time of survey (Waterlow 1976).

Table 2. Mean birth weights of Purari Delta children born between 1974 and 1980, sexes pooled.

Year	Size of sample	Birth weight (kg)	Standard error	%Below 2.5 kg
1974	10	3.19	0.31	10%
1975	10	3.00	0.32	10%
1976	19	2.85	0.38	16%
1977	22	2.84	0.54	27%
1978	39	2.81	0.37	23%
1979	50	2.90	0.39	30%
1980	35	2.89	0.29	17%

growth, a result of interacting undernutrition and infection, occurs before children reach the age of five years.

Table 3 gives the mean daily intake of nutrients according to age and sex, and Table 4 gives these values as a percentage of the FAO/WHO (1974) recommended daily intakes for the Western Pacific.

Amongst the Purari Delta people, protein and energy requirements are met for the first three to four years of life, with subsequent fall-off from requirement, stabilising in adolescence. Adult male intakes are higher in both protein and energy than in adolescence, whereas adult female intakes are both lower.

When comparing average protein and energy adequacy for all age groups and both sexes with figures derived from a sago economy in the Sepik Province (Salfield 1973) and a sweet potato economy in the Simbu Province (Bailey and Whiteman 1963), we find that the main deficiency in the Simbu sweet potato eaters is protein, and in the Sepik sago eaters it is energy. In the Purari Delta the deficiencies are of both energy and protein (Table 5).

In all cases except adult males, fat energy percent (FE%) is below that recommended (Table 6). FE% increases with age, males having marginally higher values than females. In all age groups, protein energy percent (PE%) is too low at existing levels of energy intake, although much higher than might be expected in a community whose main staple consists of almost pure starch, and no other nutrients whatever. If males of all age groups could acquire 80% of the recommended daily intake of energy, and females 90%, then the existing PE% would be adequate to meet all physiological needs. In the age group 2 to 10 years both males and females have a lower FE/PE ratio than the minimum recommended value. The older age groups, both male and female, have FE/PE ratios above 2.0.

Taken as a whole, the figures indicate that the present diet of the Purari Delta people is deficient in energy for children 2 to 10 years of age, and defi-

Table 3. Mean daily intake of nutrients per capita.

Age (years)	Sex	Size of sample	Energy (Kcals)	Energy (MJ)	Protein (grams)	Fat (grams)	Fibre (grams)	Calcium (milligrams)	Iron (milligrams)	Vitamin A (micrograms)	Vitamin C (milligrams)
2 – 3.9	Male &	2	1196	5.00	16.3	16.3	2.4	120	5.4	803	38
4 – 6.9	female	7	1004	4.20	17.4	14.9	2.1	132	6.6	748	49
7 – 9.9	combined	4	1020	4.27	14.1	12.0	2.3	161	5.9	741	74
10 – 12.9		4	1162	4.86	18.9	20.7	3.4	166	8.3	1004	59
13 – 15.9	Male	1	1181	4.94	32.2	30.6	3.3	181	10.7	901	75
16 – 19		1	915	3.83	14.8	18.7	3.0	129	6.2	947	53
10 – 12.9		1	1665	6.97	38.2	24.9	3.3	219	11.8	1065	93
13 – 15.9	Female	5	1132	4.74	17.6	23.5	3.1	153	7.7	794	60
16 – 19		3	998	4.18	18.7	19.1	3.1	167	7.1	1199	75
Adult	Male	5	1344	5.62	29.1	31.1	4.5	246	12.8	1515	87
Adult	Female	6	1132	4.74	22.9	24.7	3.7	193	8.7	1343	68

Table 4. Percent adequacy of intake per capita.

Age (years)	Sex	Size of sample	Mean weight	Energy	Protein	Calcium	Iron	Vitamin A	Vitamin C
2 – 3.9	Male &	2	11.0	108%	100%	30%	77%	321%	190%
4 – 6.9	female	7	14.9	74%	93%	33%	94%	249%	245%
7 – 9.9	combined	4	20.1	65%	64%	40%	84%	185%	370%
10 – 12.9		4	30.2	54%	62%	28%	119%	175%	295%
13 – 15.9	Male	1	36.0	58%	99%	30%	119%	124%	250%
16 – 19		1	53.0	35%	37%	29%	103%	126%	177%
10 – 12.9		1	28.5	94%	141%	37%	236%	185%	465%
13 – 15.9	Female	5	39.5	57%	56%	26%	43%	109%	200%
16 – 19		3	41.5	56%	65%	33%	37%	160%	250%
Adult	Male	5	53.1	60%	77%	62%	256%	202%	290%
Adult	Female	6	49.2	58%	72%	48%	46%	179%	227%

cient in energy and, to a lesser extent, protein for adolescents and adults of both sexes. Increased consumption of energy-dense foods such as coconut cream and more frequent meals or snacks during the day by children under 10 years of age would result in an increased FE%, FE/PE ratio and hence total energy intake. Increased consumption of the present diet by older children and adults, by more frequent meals or snacks would raise their intake of protein and energy to recommended levels.

However, there appears to be no impairment of physiological function and there has been a vast improvement in the general health status since 1947 (Hall 1979). As food supplies are plentiful and much more time could be devoted to food-getting activities than at present, it is likely that nutrient requirements in most age groups are being met and this study has duplicated findings of earlier workers, of intakes of energy, protein and other nutrients amongst physically well-developed Papua New Guineans (Oomen 1958; Hipsley and Kirk 1965; Norgan, Ferro-Luzzi and Durnin 1974). Thus, more knowledge is needed concerning human adaptation in Papua New Guinea before accepting the WHO recommended daily intakes for the Western Pacific region as being valid for Melanesian peoples.

Table 5. Dietary adquacy of protein and energy in three different subsistence cultures.

Economy	Dietary adequacy		
	Protein	Energy	Source
Sweet potato (Simbu)	59%	95%	Bailey and Whiteman 1963
Sago (Sepik)	87%	60%	Salfield 1973
Sago (Purari Delta)	75%	64%	Present study

Table 6. Proportion of energy intake provided by protein and fat, and fat energy/protein energy ratio, by age and sex.

Age (years)	Size of sample	Protein energy %	Fat energy %	Fat energy/ protein energy ratio
Males				
2 – 9.9	4	7.3%	12.4%	1.71
10 – 19	6	7.8%	17.7	2.28
Adult	5	9.0%	20.8%	2.31
Females				
2 – 9.9	9	6.3%	12.3%	1.95
10 – 19	9	6.9%	17.6%	2.56
Adult	6	7.8%	19.6%	2.53

559

Low energy intakes deny children the opportunity of catch-up growth, lower their resistance to infection, and may prohibit the attainment of full genetic potential for stature and build. However, it is doubtful whether such attainment of full genetic potential would confer any adaptational advantages for life in the swamps.

Intakes of calcium are low, but there is no clinical evidence of deficiency in the people of the Purari Delta. The age group 2 to 10 years has an intake which is one of the lowest recorded in the world, and despite adaptation to low intakes, gives some cause for concern, because the main source of this mineral is the green leafy vegetable *Gnetum gnemon* which has sufficient oxalic acid content to immobilise up to 50% of available calcium (Langley 1950). It would be worthwhile conducting a calcium balance study on growing children to see whether there is indeed an adaptation to these very low intakes. Children under 2 years of age are likely to receive adequate calcium from breast milk, even if the mother is deficient. Many children over 10 years of age, adolescents and adults meet their calcium needs from the lime (calcium hydroxide) obtained from the slaking of mollusc shells used as part of the process of betel nut chewing, during which much of the betel nut, pepper-vine and lime mixture is swallowed. Vitamin D intakes are probably adequate, because most of the people are continually exposed to sunlight.

Intake of fibre is generally low, largely because the main staple, sago, is low in fibre. This would indicate that the delta people are potentially susceptible to diverticular disease and cancer of the large intestine, but there is no evidence to support this view.

Iron requirements are met only in adolescent and adult males but not in children of 2 to 10 years of age, and in adolescent and adult females, indicating that iron deficiency may be a major cause of anaemia in the Purari Delta. These figures support Hall's (1979) observation that the marked difference between adult male and female haemoglobins are due in part to marginal iron intake, in addition to increased susceptibility of pregnant women to malaria.

Intake of vitamin A during the period of study was exceptionally high in all age groups, due mainly to the high consumption of green leafy vegetables. This study has not revealed any vitamin A deficiency, which supports the results of previous studies: the 1947 New Guinea Nutrition Survey Expedition found no xerophthalmia or follicular hyperkeratosis, and Dr Peter Calvert (personal communication) has seen no keratomalacia during his twenty-six years of working in the Purari Delta.

B vitamin status was not examined, although it is acknowledged that deficiency diseases such as beri-beri occasionally present themselves in older people (Calvert, personal communication). This is due to subsistence on an almost total sago diet during the wettest time of year when other food supplies are limited, and there is no means of buying store foods such as rice and tinned fish, whose B vitamin contents are good.

Intake of vitamin C was exceptionally high, and there is no doubt that requirements are met, even taking into account cooking losses. This high intake is again mainly due to large consumption of green leafy vegetables.

There is little variation in the proportion of energy derived from animal sources between the age groups 2 to 9 years and 10 to 19 years, although this increases in adulthood (Table 7). In all cases a bigger proportion of dietary energy is obtained from foods of animal origin by males than by females. These intakes can be judged to be high by almost any standards.

The proportion of total dietary energy derived from sago falls with age, from 47.5% to 38.5% in males, and from 46.0% to 37.4% in females. On average, sago contributes 43% of the energy intake in Koravake, compared with 72% in Lumi (Corden 1970), 93% in Waropen, Irian Jaya (Oomen and Malcolm 1958), and 85% in the Sanio Hiowe of the Sepik (Townsend 1974). It can be inferred that there is greater diversity in the diet of the people of the Purari Delta than in other sago-eating areas studied in Papua New Guinea.

4. Conclusions

The weight and height data indicate, in agreement with Hall's (1979) conclusion, that up to one third of the population has suffered from malnutrition by the age of seven, resulting in stunting of growth. Indeed, it is believed that the main periods of nutritional deprivation occur before birth, and between one

Table 7. Proportion of energy intake provided by sago and animal sources, by age and sex.

Age (years)	Size of sample	Percentage of energy intake from:		
		Sago	Animal sources	Other sources
Males				
2 – 9.9	4	47.5%	19.7%	32.8%
10 – 19	6	45.8%	19.1%	35.1%
Adult	5	38.5%	25.2%	36.3%
Females				
2 – 9.9	9	46.0%	14.8%	39.2%
10 – 19	9	43.8%	14.8%	41.4%
Adult	6	37.4%	18.0%	44.6%

and two years of age. Thus, in order to achieve child growth comparable to Western (Harvard) standards it would be necessary to improve maternal nutritional status and vigorously promote high energy dense supplementary foods such as coconut cream and more frequent meals or snacks for children from about one year of age.

The dietary evidence indicates that the present diet of the people of the Purari Delta is deficient in energy for children aged two to ten years, and deficient in energy and, to a lesser extent, protein for adolescents and adults of both sexes. This is entirely in line with most similar studies carried out in Papua New Guinea (Oomen 1958; Hipsley and Kirk 1965; Norgan et al. 1974).

As the diet in the delta is more varied and plentiful than amongst the Pawaia at Wabo and the infant and toddler mortality rates in the delta are no greater than the national average (Hall 1979), it may be that we are witnessing an environmental adaptation in which the rate of growth and intake of food is a part-function of the total life system and is but one part of the total health situation (Hipsley 1973). During the authors' stay in the Purari Delta, they observed an efficient subsistence system in which neither time nor land were limiting factors in food production, and plate waste was recorded more often than not during the weighed dietary intake. Study of energy balance (intake of energy versus expenditure) on Purari Delta individuals could shed further light on the validity of international standards for energy requirements as applied to Papua New Guinea, and could have wider significance for other parts of the world, such as Ethiopia where similar phenomena have been recorded (Miller et al. 1976).

Malaria is hyperendemic in the Purari Delta, with morbidity and mortality highest amongst the one to five year olds (Hall 1979) and it is probable that this is a major contributing factor to low nutritional status, and hence of slow growth, amongst children of this age group. However, improved nutrition would undoubtedly increase resistance to infection, and any programme designed to increase and improve the diet would be an important part of the overall health programme for the Purari Delta.

The proposed Purari Hydroelectric Scheme is unlikely to have any nutritional consequences on the people of the Purari Delta. However, wage-earning opportunities could lure men away from active cultivation of gardens and sago (Ulijaszek and Poraituk, this volume), potentially resulting in shortfalls in quantity and range of available food. Nutritional problems created would thus be of sociological origin.

Acknowledgements

The authors would like to express their thanks to the people of Koravake village, without whose help and understanding this study would not have been possible. In addition, we would like to thank the following:

Drs Peter and Lyn Calvert, for sharing some of their vast experience of work in the Purari Delta.

Dr Edmundo Alvarez, WHO Inter-Country Programme Nutritionist for the South Pacific Region, Ms Jean Eng, Health Department, Konedobu, and Ms Cherry Farrow, for assistance in the weighed dietary intake.

Dr Peter Heywood, Institute of Medical Research, Madang, Dr Patricia Townsend, Institute of Applied Social and Economic Research, Port Moresby, Professor George Beaton, University of Toronto, and Mr Derek Miller, University of London, for their helpful advice and criticism at various stages of the project.

Dr Damien Jolley, Institute of Medical Research, Madang, for help in the analysis of the dietary data.

References

Araya, H. and K. Arroyave, 1979. Relación del contenido energético proveniente de grassas y como indicador de la potencialidad energético-proteínica de las dietas de poblaciónes. Archiv. Latinoamer. Nutr. 29: 103 – 112.

Bailey, K.V. and J. Whiteman, 1963. Dietary studies in the Chimbu (New Guinea Highlands). Trop. Geogr. Med. 15: 377 – 388.

Beaton, G.H. and L.D. Swiss, 1974. Evaluation of the nutritional quality of food supplies: prediction of 'desirable' or 'safe' protein – calorie ratios. Am. J. Clin. Nutr. 27: 485 – 504.

Cordon, M.W., 1970. Some observations on village life in New Guinea. Aust. Inst. Anatomy. Food and Nutrition Notes and Reviews 27: 77 – 82.

Eng, J. and D. Leonard, 1978. Nutrition survey of community school students in Baimuru District. Mimeo. Department of Public Health, Konedobu, Papua New Guinea.

FAO/United States Department of Health, Education and Welfare, 1972. Food composition table for use in East Asia. DHEW Publication No. (NIH) 75 – 465. Washington.

FAO/WHO, 1974. Handbook on human nutritional requirements. FAO Nutritional Studies, No. 28. Rome.

FAO/WHO, 1977. Dietary fats and oils in human nutrition. Food and Nutrition Paper, No. 3. Rome.

Hall, A.J., 1979. Health and disease patterns of the Purari people. In: T. Petr (ed.), Purari River (Wabo) Hydroelectric Scheme Environmental Studies, Vol. 7, pp.1 – 61. Office of Environment and Conservation, Waigani, Papua New Guinea.

Hall, A.J., 1983. Health and diseases of the people of the upper and lower Purari. This volume, Chapter III, 4.

Hipsley, E.H., 1973. The nutritional state of the population of Papua New Guinea. In: C.O. Bell (ed.), The Disease and Health Services of Papua New Guinea, pp. 87–95. Department of Public Health, Port Moresby, Papua New Guinea.

Hipsley, E.H. and F.W. Clements, 1950. Report on health and nutritional status in New Guinea. In: Report of the New Guinea Nutrition Survey Expedition, 1947, Part 5, pp. 143–176. Government Printer, Sydney.

Hipsley, E.H. and N.E. Kirk, 1965. Studies of dietary intake and the expenditure of energy by New Guineans. South Pacific Commission. Technical Paper, No. 147. Noumea.

Jelliffe, D.B., 1966. The assessment of the nutritional status of the community. WHO Monograph, No. 53. Geneva.

Lambert, J., 1983. Nutritional study of the people of the Wabo and Ihu areas, Gulf Province. This volume, Chapter III, 9.

Langley, D.M., 1950. Food consumption and dietary levels. In: Report of the New Guinea Nutrition Survey Expedition, 1947, Part 5, pp. 92–142. Government Printer, Sydney.

Miller, D.S., J. Baker, M. Bowden, E. Evans, J. Holt, R.J. McKeag, I. Meinertzhagen, P.M. Mumford, D.J. Oddy, J.P.W.R. Rivers, G. Sevenhuysen, M.J. Stock, M. Watts, A. Kebede, Y. Wolde-Gabriel and Z. Wolde-Gabriel, 1976. The Ethiopia applied nutrition project. Proc. R. Soc. Lond. 194: 23–48.

Norgan, N.G., J.V.G.A. Durnin and A. Ferro-Luzzi, 1974. The composition of some New Guinea foods. P.N.G. Agric. J. 30: 25–39.

Norgan, N.G., A. Ferro-Luzzi and J.V.G.A. Durnin, 1979. The energy and nutrient intake and the energy expenditure of 204 New Guinean adults. Phil. Trans. R. Soc. Lond. 268: 309–348.

Oomen, H.A.P.C. and S.H. Malcolm, 1958. Nutrition and the Papuan child. South Pacific Commission. Technical Paper, No. 188. Noumea.

Payne, P.R., 1975. Safe protein–calorie ratios in diets. The relative importance of protein and energy intake as causal factors in malnutrition. Am. J. Clin. Nutr. 28: 281–286.

Peters, F.E., 1957. Chemical composition of South Pacific foods. South Pacific Commission. Technical Paper, No. 115. Noumea.

Platt, B.S. and D.S. Miller, 1959. The net dietary protein value (NDpV) of mixtures of foods – its definition, determination and application. Proc. Nutr. Soc. 18: 7–8.

Salfield, J.R., 1973. Nutrition survey, Sepik District. Mimeo. Provincial Health Office, Wewak, Papua New Guinea.

Townsend, P.K., 1974. Sago production in a New Guinea economy. Human Ecol. 2: 217–236.

Ulijaszek, S.J. and S.P. Poraituk, 1983. Subsistence patterns and sago cultivation in the Purari Delta. This volume, Chapter III, 10.

Waterlow, J.C., 1976. Classification and definition of protein energy malnutrition. In: G.H. Beaton and J.M. Bengoa (eds.), Nutrition in Preventive Medicine, the Major Deficiency Syndromes, Epidemiology and Approaches to Control, pp. 530–553. WHO Monograph Series, No. 62. Geneva.

9. Nutritional study of the people of the Wabo and Ihu areas, Gulf Province

J.N. Lambert

1. Introduction

Until very recently, little was known about the nutritional status of the people inhabiting the Purari River banks and its delta. The Purari River, which forms at the confluence of the Erave and Tua rivers, is in its upper part virtually uninhabited, except very few Pawaia settlements, such as Gurimatu (cf. Warrillow, this volume). Downstream of Gurimatu the river enters a series of gorges, to emerge from Hathor Gorge upstream of Uraru, a village which nowadays is virtually abandoned. Thus, it is only at Wabo where there is the first large settlement on the Purari River. Wabo has been projected to become a site of a hydroelectric dam, to be erected there sometime in the future. In 1976, during engineering feasibility studies for this scheme, a nutrition survey was carried out by the author and other members of the Nutrition Section of the Health Department of the Papua New Guinea Government. Apart from the Wabo and Uraru group of the predominantly Pawaia people, studies extended also to the mouth of the Vailala River at Ihu, east of the Purari Delta.

Around Wabo, 213 persons were surveyed. In the coastal area 962 persons were surveyed of the total population estimated at 10 000. The basic purpose of the survey was to assess the nutritional status of the population with reference to other population groups in Papua New Guinea.

2. Methodology

The weight and height of each individual was recorded, the latter using a Microtoise anthropometer. Mid upper arm circumference was measured using a fibre glass tape, and triceps skinfold thickness was measured by taking the mean of three readings using Harpenden skinfold calipers. All the anthro-

Petr, T. (ed.) The Purari – Tropical environment of a high rainfall river basin 565
© *1983, Dr W. Junk Publishers, The Hague / Boston / Lancaster*
ISBN-13: 978-94-009-7265-0

pometric measurements were performed by the same individual to eliminate variations and reduce observer error, in the manner as prescribed by Jelliffe (1966).

The age of infants and children, when not known accurately by parents, was obtained from one of the following sources:
- Maternal and child health register cards, where the exact date of birth was recorded.
- Mission or school records, which usually recorded the age of the child to the nearest year, or occasionally the nearest month.
- In the absence of the above, the dental index formula was used, as described by Bailey (1963). Here the age of an infant in months is approximately equal to the number of teeth plus six. For older children other developmental indicators were used, as described by Jelliffe (1966).

Blood haemoglobin levels were determined using an A.O. haemoglobinometer. Each individual was subjected to a clinical examination by a public health nutritionist of the World Health Organisation with long-standing experience in the clinical assessment of nutritional status, in both the Middle and Far East, where vitamin deficiencies are common, and was questioned about food intake over the previous 24 hours. There was insufficient time to carry out a more detailed investigation of food intake. In view of the extremely time-consuming nature of individual weight/food intake surveys, it was decided to carry out a large number of food recall interviews. Information was also obtained concerning infant feeding practices, food taboos, and various socioeconomic indicators.

3. Results

The two survey areas were treated as separate groups, in order to ascertain whether any significant differences existed between the two populations.

Males from Wabo were consistently heavier and shorter than males from Ihu, indicating a tendency towards a highland physique as opposed to the typically slimmer coastal dwellers (Table 1). However, the females from the Wabo area were both lighter and shorter than those from the coastal area. Among adults a marked decline in weight with increasing age was observed, this decline being greater amongst the females than the males.

Mean weights for the two survey populations appeared to compare closely with the Harvard standards for the first few years of life. After four years of age an increasingly marked deficit in weight for age was observed in both survey areas.

Table 1. Mean weights and heights.

Age	Sex	Wabo population			Ihu population		
		Number	Mean weight (kg)	Height (cm)	Number	Mean weight (kg)	Height (cm)
0–5 months	M + F	1	6.5	65.5	6	6.0	63.1
6–11 months	M + F	5	7.4	70.7	16	8.5	75.2
12–23 months	M + F	8	8.9	77.2	21	8.8	79.1
24–35 months	M + F	6	10.9	83.5	19	11.0	86.9
36–47 months	M + F	2	13.9	90.0	26	12.6	91.7
48–59 months	M + F	3	15.6	102.2	30	14.5	98.0
5–9 years	M	23	22.5	117.5	141	21.2	120.6
	F	15	17.5	106.2	183	21.5	120.3
10–14 years	M	10	33.1	137.6	142	31.3	138.6
	F	10	29.0	130.0	106	34.1	140.4
15–19 years	M	7	49.8	158.8	9	46.4	159.5
	F	2	39.5	146.0	32	49.3	156.0
20–29 years	M	15	59.5	162.1	19	56.2	166.7
	F	30	46.3	148.6	49	51.6	156.4
30–39 years	M	28	56.1	157.4	24	55.9	164.6
	F	22	45.6	149.7	47	48.5	154.3
40–49 years	M	7	50.6	155.9	27	56.8	162.7
	F	8	48.1	148.8	34	48.4	153.4
50+ years	M	4	53.9	163.7	35	51.7	160.9
	F	7	41.4	149.4	44	42.9	151.7
All adults	M	54	56.2	159.0	107	54.8	163.3
	F	67	45.2	149.1	174	47.9	154.1

The mean weight for age for both the Wabo and Ihu infants from 0 – 5 months (Table 2) was close to the American Harvard standards. This indicates that, as with other populations studied in Papua New Guinea, for infants and young children, the potential exists to equal European standards of growth. Indeed the mean weight for the male Ihu infants was considerably above the Harvard standards for the first months of life. Above this age, however, a marked decline in the mean weight for age was observed with a minimum value observed in the 12 to 23-month age group. This was followed by a slight improvement up to the age of 5, followed by a further decline in school-aged children. Over 50% of the children aged from 10 to 14 years were below 80% of the standard weight for age. This was probably due to the chronic effects of undernutrition. In a nationwide nutrition survey carried out in 1978 (Lambert 1979), 61% of under fives attending maternal and child health clinic in Gulf Province were found to be below 80% of weight for age. Comparisons between the two populations reveal that generally up to age 14 a higher percentage of the Ihu population were underweight (below 80% standard weight for age) than the Wabo population.

Mid upper arm circumference (M.U.A.C.) remains approximately constant from the ages of one to five years (Table 3), and therefore provides an age-independent index of nutritional status. The numbers below 14 cm (approximately equivalent to 80% of the standard weight for age) were greater than those below 80% weight for age.

Triceps skinfold measurements were obtained in order to ascertain the degree of adiposity of the two populations. Mean values for Wabo adults were 4.2 mm for males and 5.6 mm for females. For the Ihu population mean values were higher at 5.2 mm for males and 7.5 mm for females, despite the fact that Ihu males were on average 1.4 kg heavier than the Wabo males. These results indicate very low levels of fat reserves amongst the surveyed population. Not a single case of obesity was observed amongst the 1175 individuals seen during the survey.

For the first 10 years of life the Wabo children appear to be extremely anaemic, with haemoglobin values much lower than the Ihu children (Table 4). However, among adolescents and adults the picture is reversed and the Wabo population appears to be the less anaemic. Records from the Department of Health indicated that hookworm and malaria could be expected in up to 90% of the population (Papua New Guinea. Dept. of Health. Nutrition Section, undated).

Nearly 20% of the Wabo pre-school children got only 1 meal a day, while less than 20% got 3 meals a day, the minimum requirement for a child of this age. For both populations the majority of families only consumed 2 meals each day (Table 5).

568

Table 2. Mean weight as a percentage of the weight for age (Harvard standard).[a]

Age	Mean weight for age				Percentage below 80% of standard weight for age (sexes combined)	
	Wabo		Ihu		Wabo	Ihu
	Male	Female	Male	Female		
0 – 5 months	97	–	119	96	0	17
6 – 11 months	81	81	119	80	40	31
12 – 23 months	81	79	81	80	50	57
24 – 35 months	81	82	88	80	33	32
36 – 47 months	92	–	84	79	0	42
48 – 59 months	93	–	85	80	33	43
5 – 9 years	86	80	77	82	37	57
10 – 14 years	86	72	78	82	55	61
15 – 19 years	80	75	89	92	44	17
Adults	87	81	88	80	30	22

[a] Due to the absence of any local standard, Harvard weight standards (from Jelliffe 1966) were used.

Table 3. Mean mid upper arm circumference.

Age	Sex	Wabo		Ihu		% Below 14 cm	
		Number	M.U.A.C.	Number	M.U.A.C.	Wabo	Ihu
0 – 5 months	M + F	1	12.6	6	13.3	–	–
6 – 11 months	M + F	5	12.1	16	12.8	–	–
12 – 23 months	M + F	8	13.0	21	12.8	63	76
24 – 35 months	M + F	6	14.6	19	14.4	50	37
36 – 47 months	M + F	2	14.2	26	14.5	50	23
48 – 59 months	M + F	3	14.6	28	14.8	–	13
5 – 9 years	M	23	16.6	141	16.4	–	–
	F	15	14.9	133	16.5	–	–
10 – 14 years	M	10	19.6	142	18.5	–	–
	F	10	18.6	106	19.4	–	–

569

Table 4. Mean blood haemoglobin levels.

Age	Sex	Wabo		Ihu	
		Number	Hb (g/100 ml)	Number	Hb (g/100 ml)
0 – 11 months	M + F	6	5.7	13	8.3
12 – 23 months	M + F	8	6.4	14	8.3
24 – 35 months	M + F	6	8.6	15	9.1
36 – 47 months	M + F	2	7.4	21	9.3
48 – 59 months	M + F	3	9.1	21	9.5
5 – 9 years	M	21	9.3	66	9.7
	F	15	8.7	71	10.0
10 – 14 years	M	10	10.8	82	10.2
	F	10	11.1	60	10.5
15 – 19 years	M	7	12.0	7	10.7
	F	2	11.1	22	10.7
Adults	M	54	12.5	78	11.3
	F	67	10.5	123	10.3

Table 5. Percentage frequency of daily (24 hours) food intakes.

% of Sample eating Age	1 meal per day		2 meals per day		3 or more meals per day	
	Wabo	Ihu	Wabo	Ihu	Wabo	Ihu
0 – 5 years	18%	4%	64%	41%	18%	55%
5 – 14 years	10%	4%	45%	58%	45%	38%
Adults	8%	8%	52%	55%	40%	37%

Study of frequency of consumption of various foods for the 24-hour period preceding the time of the survey has shown that in both areas by far the most important food item was sago; however, sweet potato or some other similar root crop was consumed at least once a day by 30% of the individuals in the Wabo area (Table 6). Virtually no bread or rice was consumed by the Wabo population, and less than 15% of the Ihu population had consumed any in the previous 24-hour period. No beans or nuts were consumed by either of the populations and surprisingly little fish was consumed, despite the fact that all the villages at Wabo were built on the banks of the Purari, and in the Ihu area the whole population lived in close proximity to the sea. Most of the animal protein consumed in the Wabo area was pork, for during the survey period a pig was killed and distributed (a rare event), to celebrate the opening of a new primary school. Even at Ihu, where most of the villagers had fishing boats and

Table 6. The percentage frequency of consumption of various foods.

Food items	Not consumed		Consumed once		Consumed twice		Consumed 3 times	
	Wabo	Ihu	Wabo	Ihu	Wabo	Ihu	Wabo	Ihu
Sago, cassava	4	9	15	14	56	46	25	31
Sweet potato, yam	70	85	9	11	14	3	7	1
Rice, bread	97	87	3	10	–	3	–	–
Fat, sugar	94	25	3	24	3	35	–	16
Plant protein (beans, nuts)	100	100	–	–	–	–	–	–
Animal protein	82	73	15	17	3	8	–	2
Vegetables	84	80	9	16	7	3	–	1
Fruit	100	79	–	18	–	3	–	–

nets, almost 75% of individuals had consumed no animal protein during the 24-hour period recalled.

Sugar (used mainly in tea and coffee) appeared to be very popular in the Ihu area, being consumed by 75% of individuals. This was the only purchased food consumed widely. In both areas very few fruits and vegetables were consumed; in fact in the Wabo area no fruit at all was eaten. Coconut cream was widely used in the Ihu area to cook sago, fish and vegetables with. This practice is highly desirable from a nutritional point of view, as it adds both energy and protein to the diet, particularly important for young children whose small stomachs and rapid rate of growth mean that it is almost impossible for them to meet their basic nutritional requirements on a diet based on high bulk sago or sweet potato.

Interviews with individuals during the survey indicated that the diet was less varied than it had been twenty years previously. One possible explanation for this is that the introduction of shotguns has had a dramatic effect on the wildlife population, an important part of the traditional diet.

As indicated by clinical examinations, there exists a disturbingly high incidence of clinical signs of vitamin deficiencies (Table 7). Conjunctival xerosis is an indication of vitamin A deficiency and if unchecked can lead to impairment of vision and even blindness. Angular stomatitis and magenta tongue both indicate a dietary deficiency of riboflavin, a condition that can lead to unsightly skin changes and itchy eczema, a very irritating condition. In view of the low consumption of fresh fruit and vegetables, shown in Table 6, these findings are hardly surprising. Details of these findings are filed in the Nutrition Section of the Department of Health, Konedobu.

Each mother was questioned about the outcome of each pregnancy and the

Table 7. The percentage of the population displaying the following deficiency symptoms.

Age	Sex	Conjunctival xerosis		Angular stomatitis		Magenta tongue	
		Wabo	Ihu	Wabo	Ihu	Wabo	Ihu
		% (No.)	% (No.)	% (No.)	% (No.)	% (No.)	% (No.)
0 – 4 years	M	– (0)	5% (3)	– (0)	2% (1)	– (0)	– (0)
	F	– (0)	2% (0)	– (0)	2 (1)	– (0)	– (0)
5 – 9 years	M	17% (4)	19% (2)	9% (2)	14% (20)	13% (3)	5% (7)
	F	– (0)	16% (21)	7% (1)	9% (12)	7% (1)	5% (7)
10 – 14 years	M	40% (4)	23% (33)	– (0)	11% (15)	– (0)	4% (6)
	F	– (0)	19% (20)	10% (1)	12% (13)	– (0)	6% (6)
25 – 19 years	M	29% (2)	11% (1)	– (0)	– (0)	14% (1)	33% (3)
	F	– (0)	9% (3)	– (0)	– (0)	50% (0)	– (0)
Adults	M	20% (11)	9% (10)	22% (12)	15% (16)	18% (10)	6% (6)
	F	16% (11)	11% (20)	15% (10)	12% (21)	10% (7)	3% (5)
Total	M	19% (21)	19% (21)	16% (74)	11% (52)	13% (14)	4% (22)
	F	10% (11)	13% (65)	11% (12)	9% (47)	9% (9)	4% (18)

number of children who had died. Over 50% of the married women interviewed had lost one or more children aged less than 5 years, indicating high infant and toddler mortality rates. The mean number of living children per family was 1.87 for the Wabo population (this included a high proportion of childless couples) and 2.84 at Ihu. The Department of Health estimated that the total child mortality rate (0 to 4.99 years) was 202 per 1000 in 1973 (Papua New Guinea. Dept. of Health 1974).

Breast feeding appeared to be the norm for all babies, for all the children below 2 years of age in the Wabo sample were still being breast fed, as were 71% of the Ihu children below 2 years. There was a lack of adequate sanitary facilities in both areas with 41% of the Wabo families having access to a latrine, and only 19% of the Ihu families with this facility.

4. Discussion

At the time of the survey very little was known about the nutritional status of the people in the area. Korte (1975), in a national review of data from maternal and child health clinics, found that 45% of children attending these clinics were malnourished, the seventh highest provincial figure in the country. Due to the low population density, and problems with transport, 24% of the population were found to live more than two hours travelling time from the nearest aid-post. Health expenditure per capita in 1974 was K4.35, compared with a national average of K5.82 (Papua New Guinea. Dept. of Health 1974).

The results of the anthropometric data reveal a high incidence of subclinical malnutrition or undernutrition among both the Wabo and Ihu populations. Among pre-school children approximately 35% of the children in both areas were under 80% of the standard weight for age. Those children aged between 12 and 23 months were the worst nourished with at least 50% of the children below the 80% line. Above this age there appeared to be some improvement in nutritional status, a commonly observed trend, which may be due to the fact that older children are able to gather foods from the bush such as wild berries, nuts, fruits, and various insects like caterpillars and locusts. Given the very high child mortality rate, it is likely that many of the more vulnerable children do not survive beyond the first year or two of life. These results may be compared with the findings of a survey done at maternal and child health clinics in the province in 1975 (Papua New Guinea. Dept. of Health. Nutrition Section, undated), which revealed that in the Ihu area 51% of 1 to 5-year olds (69 out of 136) were under 80% weight for age. No figures were available for the Wabo area.

There was a further increase in subclinical malnutrition among school age children, rising to a maximum among the 10 to 14-year olds, probably as a result of the cumulative effects of chronic undernutrition. The differences between the two populations do not appear to be significant, although in general the children from Ihu appear to be less well nourished than those from the Wabo area, a somewhat surprising finding in view of better services and access to the sea at Ihu. The lower incidence of undernutrition among the children under 5 months indicates that, as with other Papua New Guinean populations, the potential for adequate growth is there, but that this is seldom realised.

This high incidence of subclinical malnutrition, compares unfavourably with some of the highland provinces. Several surveys in these provinces, including one by the author in 1975 (Lambert 1975), have shown much lower rates of subclinical malnutrition. This undernutrition, if severe, frequently leads to decreased mental performance, apathy and indifference, reduced work output and productivity, and increased susceptibility to infections with correspondingly high morbidity and mortality rates.

Research by Smith (unpublished manuscript) carried out in the Southern Highlands Province amongst infants attending maternal and child health clinics, has revealed the dire consequences of malnutrition in infancy. For those children aged 0 to 4.99 years who were clinically malnourished, 50% died before the next clinic was held. Among those children aged less than 30 months, mortality rates were three times higher for the subclinically malnourished (between 60 and 80% normal weight for age), than for those of adequate nutritional status.

Mid upper arm circumference and triceps skinfold measurements revealed a similar pattern with a high incidence of undernutrition. Generally the Ihu population appeared to have similar arm circumference but greater skinfold thickness than the Wabo population, indicating the superior musculature of the latter population, typical of highlanders.

Mean haemoglobin levels revealed very low blood haemoglobins for the Wabo children below the age of 10. Practically all the children in this age group had haemoglobin values below 12 g per 100 ml, the level established by the World Health Organization (1972), below which anaemia can be said to exist. This anaemia is probably a result of a high incidence of malaria and hookworm rather than primarily a dietary deficiency, however, the poor diet observed undoubtedly exacerbated the situation. Among older children and adults in Wabo the situation was improved, and in fact mean haemoglobin levels were higher than those of the Ihu population.

The dietary information reveals a very high dependence on sago with its high starch, low protein content, particularly in the Wabo area. Low intakes of protein-rich foods, fruit and vegetables were also observed. This added to the fact that less than 50% of the population consumed 3 or more meals a day, helps to explain the high incidence of malnutrition. Feeding experiments carried out by Binns (1976) in the Enga Province revealed that children aged four to six years, fed ad libitum on a balanced traditional diet, could not meet their energy needs unless some energy-rich food such as fat or oil was added to the diet. The existing diet is most probably inadequate in all essential nutrients, particularly energy and protein. Sago is a high bulk, low protein food that must be supplemented with high energy, high protein foods such as peanuts or fish. In addition, locally available fruits and vegetables such as pawpaw, mangoes, and all types of dark green leaves should be encouraged and consumed in view of the observed vitamin A and riboflavin deficiencies. There is a growing volume of evidence that the variety of the diet has declined over the last twenty years, as noted by Jackson (1977) in Western Province and Lambert (1975) in the Simbu Province.

The low population density and the undoubted fertility of the soil, given suitable cultivation techniques, together with the free availability of fish in the sea and all rivers, gives the area the potential to support the existing population at an acceptable level of nutrition with little difficulty. However, a number of basic health problems will have to be tackled at the same time, in particular malaria and other blood parasites and intestinal parasites. Clean water supplies and adequate waste disposal are also essential. What is needed is a co-ordinated education programme using all extension agents to inform the population of the nature and causes of malnutrition, backed up by a distribution scheme for seeds and planting materials.

5. Conclusions

The results of this survey indicate that both clinical and subclinical protein energy malnutrition exists in the Wabo and Ihu areas; in addition, there was evidence that over 10% of the population may have been suffering from deficiencies of vitamin A and riboflavin. Questioning of mothers revealed high infant and toddler mortality rates, and anaemia was found to be very widespread.

While it is accepted that the food recall method of assessing food intake is somewhat crude, all available indicators suggested that food intakes were, by any standards, inadequate. The great majority of the survey population appeared to consume a diet consisting almost entirely of sago. Very little fish was consumed, despite its easy availability, and intakes of fresh fruit and vegetables were very low. No legumes were reported as being consumed by any of the 1175 people covered in the survey, and none were seen growing.

The existing nutritional status of the population therefore leaves much to be desired and can be compared unfavourably with the inhabitants of other parts of the country. Should the Purari Hydroelectric Scheme go ahead there is no guarantee that nutritional status will improve. Experience gained from the current coffee flush in the highlands shows that economic advancement can lead to increased consumption of non-food items and foods of low nutritional value, social problems, and the neglect of food gardens and consequent food shortages. Experience from resettlement schemes in West New Britain and the East Sepik does not give reason for any optimism (Lambert 1979; Cox 1979).

The inundation of large areas of land above the dam site and the concentration of a large population around the proposed industrial and port complex will make the traditional sago harvesting very difficult. Consequently the population will be forced to change their principle staple food, unless supplies can be obtained from further afield.

The extent and duration of breast feeding can be expected to decline, as a result of increasing sophistication of the population. Unless this trend is matched by an improvement in the quality and quantity of supplementary feeding, dire nutritional consequences can be expected. Should some mothers cease breast feeding their infants altogether and feed their infants with a bottle, greatly increased rates of malnutrition and infantile diarrhoea can be expected (Lambert and Basford 1977; WHO 1981).

While it is accepted that there had been great improvements in health, malnutrition was still a major problem in the Purari at the time of the survey. A child mortality rate in excess of 200 per thousand per year provides further support for this. In order to reduce infant and toddler mortality rates, it is essential that the quality of the diet be improved, utilising local resources wherever possible.

References

Bailey, K.V., 1963. Dental development in New Guinea infants. J. Paediatrics 64: 97 – 100.

Binns, C., 1976. Does food volume limit the dietary intake of highland children. Nut. and Development 2: 4 – 10.

Cox, E., 1979. Gavien and Bagi: rubber/profit versus people/community. East Sepik Province. In: C.A. Valentine and B.L. Valentine (eds.), Going Through Changes: Villagers, Settlers and Development in Papua New Guinea, pp. 15 – 31. Institute of Papua New Guinea Studies, Port Moresby.

Jackson, R., 1977. Kiunga development study. Mimeographed report. University of Papua New Guinea, Geography Department, Port Moresby.

Jelliffe, D.B., 1966. The assessment of the nutritional status of the community. World Health Organization. Monograph Series, No. 53. Geneva.

Korte, R., 1975. Food and nutrition in Papua New Guinea. In: W. Wilson and R.M. Bourke (eds.), Proceedings of the 1st Food Crops Conference, pp. 15 – 18. Dept. of Primary Industry, Konedobu, Papua New Guinea.

Lambert, J.N., 1975. A study of nutritional status and economic development in the Chimbu. M.Sc. Thesis, University of London.

Lambert, J.N., 1979. 1978 national nutrition survey. Mimeographed report. National Planning Office, Waigani, Papua New Guinea.

Lambert, J.N., 1979. The relationship between cash crop production and nutritional status in Papua New Guinea. History of Agriculture Working Paper, No. 33. University of Papua New Guinea.

Lambert, J.N. and J. Basford, 1977. Port Moresby infant feeding survey. P.N.G. Med. J. 20: 175 – 179.

Papua New Guinea. Dept. of Health, 1974. National Health Plan 1974 to 1978. Dept. of Health, Konedobu, Papua New Guinea.

Papua New Guinea. Dept. of Health. Nutrition Section, undated. A survey of maternal and child health services in Papua New Guinea. Unpublished data.

Warrillow, C., 1983. A short history of the upper Purari and the Pawaia people. This volume, Chapter III, 1.

World Health Organization, 1972. The health aspects of food and nutrition. WHO Regional Offices, Manila and Bangkok.

World Health Organization, 1981. Impact of sales control on feeding bottles. Weekly Epidem. Rec., No. 16: 126.

10. Subsistence patterns and sago cultivation in the Purari River

S.J. Ulijaszek and S.P. Poraituk

1. Introduction

In 1924, F.E. Williams said about land ownership in the Purari Delta: 'Village land belongs normally to individuals. There are as a rule no boundaries except such as are formed by natural features, small creeks especially, or where in recent years fences have been made. In old established villages there is no uncertainty in the matter: a man builds on his father's land which has been in the family beyond living memory. Even the smaller mud creeks are under individual ownership, though if useful as waterways they may be used by any and everyone. The larger streams are regarded as common property. Bush land is acquired by inheritance, for the majority of land within reasonable distance has long since been taken up.'

This still held true in 1980 when the land use in the Purari Delta, with special reference to sago cultivation and management, was investigated in Koravake village in September. Clan territories are clearly defined, sago places, gardens and cultivated trees being owned by the males of individual families. The produce of these are shared with married sons, or joint ownership is held with male siblings. Inheritance of land is patrilineal. The subsistence economy is based on sago cultivation and fishing, root and tree crop culture being a subsidiary activity.

2. The people of the Purari Delta

Although sharing a common language, the people of the Purari Delta are made up of five clans (Fig. 1). The total population of the delta declined in the early 1950s, to reach presently somewhat lower numbers than those for the early 1920s (Table 1).

Petr, T. (ed.) The Purari – Tropical environment of a high rainfall river basin
© *1983, Dr W. Junk Publishers, The Hague / Boston / Lancaster*
ISBN-13: 978-94-009-7265-0

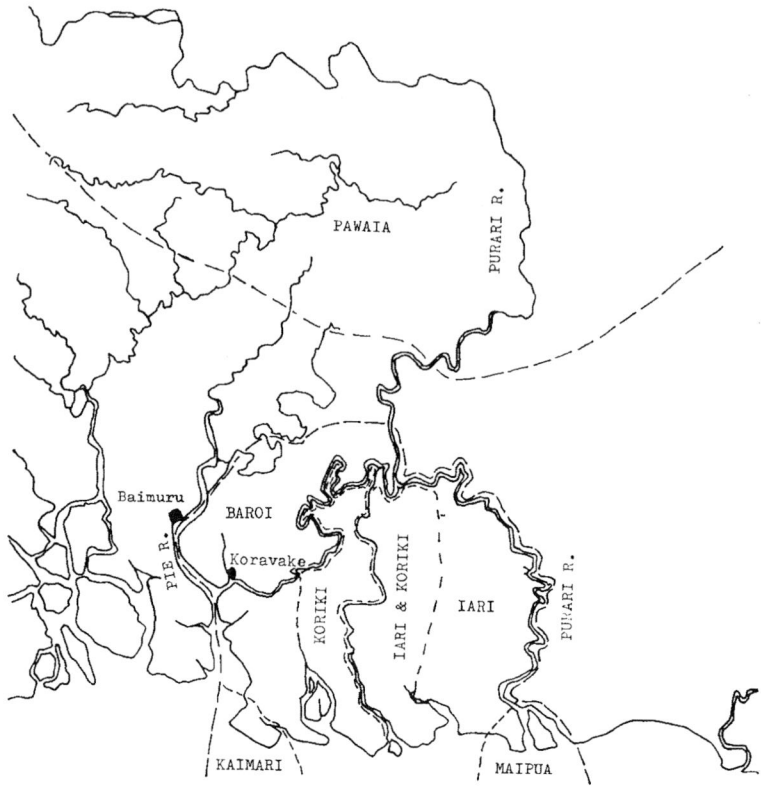

Fig. 1. Clan boundaries within the Purari Delta.

Amongst reasons for the reversal of the depopulation process can be listed improved health facilities and living standard.

The people are organised in large settled villages residing on sites determined by the immediate availability of fishing, crab collecting and gardening grounds. Sago may be cultivated anywhere in the swamps as long as the soil salinity is not too high (Petr and Lucero 1979), and is therefore not an impor-

Table 1. Total population of the Purari Delta.

	Population	Source
Early 1920s	8558	Maher (1961)
Early 1950s	5433	Maher (1961)
1980	7218	National Census Office

tant criterion for selection of village sites. The degree of sedentary life exhibited is made possible by the system of tidal waterways which allow easy access to all sites of food production. The contrasting life-style of the Pawaia (see Warrillow; Toft, this volume) who live above the delta, may be due in large part to the lack of such a transport system, the site of population units being dependent on the food availability in a particular area at a given time.

Interclan warfare was waged prior to administration, presumably over the availability of good gardening land, there being an abundance of fishing, crab collecting and potential sago grounds. This is reflected in the relative dearth of land resources available to the weaker clans, the Kaimari and Maipua. Significantly, the strongest and traditionally the most warlike clan, the Koriki, have the widest range of resources from which to shape their environment. There have been no disputes over land since the 1950s, all clans having settled into an equilibrium with each other and with their available resources.

In terms of work time allocated to different food-producing activities, sago making and fishing take equal priority, with hunting and the harvesting of coconuts and garden foods being of lesser importance.

Production of sago starch, crab and shellfish collection and fishing are mainly female activities, whilst harvesting tree crops and gardening are mainly male activities. However, there is considerable overlap of male and female roles where gardening is concerned. Hunting is totally a male activity producing infrequent kills, but important dietary contributions. The woman's work is most productive in bringing food to the pot. Observation in Koravake indicates that time is not a limiting factor in food production.

3. Time – space utilisation of the environment for subsistence

Three kinds of garden were distinguished by the authors:

Sago places in which the sago palm is cultivated, usually on tidal river banks. Normally growing on the perimeter of the sago garden are coconuts, banana, taro, sugarcane and betel nut.

Bush gardens which are areas cleared of nipa, burnt over and planted with taro, sugarcane and banana. These undergo an occasional tidal inundation, and there is no evidence to indicate that these plants thrive or produce crops of any substantial yield.

Village gardens which are made on drier ground within walking distance of the village.

At Koravake, each clan clears a communal area at the beginning of the dry season, planting cassava, bananas, sugarcane, pumpkin, taro, corn and

Diagramme of time – space utilisation of the Koravake (Baroi clan) environment.

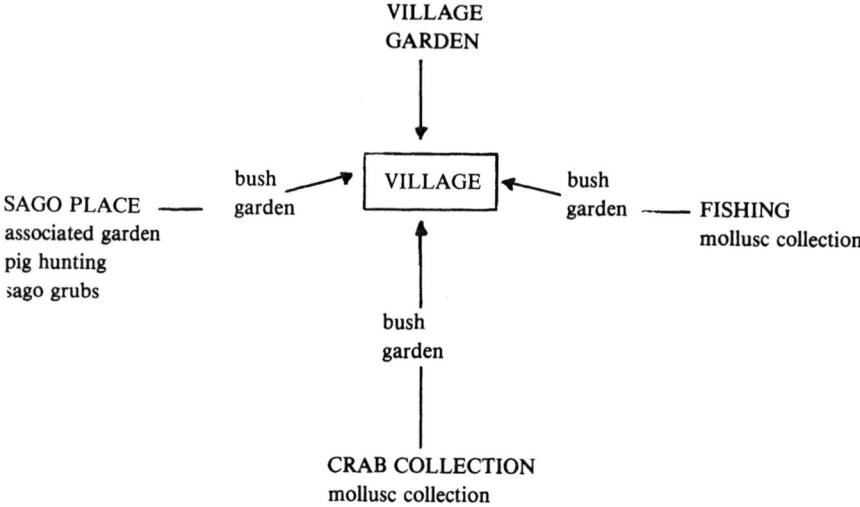

Primary subsistence activities in terms of dietary contribution and time spent are given in capital letters; secondary subsistence activities are given in small letters.

watermelon. These are left untended and are successively harvested as each crop comes to fruition. Initial crops of cassava may take three to four months to reach harvesting stage; the final crop of cassava may take up to three years to reach maturity, no replanting taking place. New gardens are made every year, produce being available from up to three gardens at any given time. With the exception of some areas of high fertility, gardens are abandoned after a single cropping and allowed to lie fallow, often long enough for tall secondary forest to develop.

The diagramme illustrates time – space utilisation of the environment for subsistence. All the gardens are clared during the dry season. For an individual family the sago places, crab collecting places and fishing grounds may be several, as are the bush gardens. Bush gardens are made at strategic places on the river banks en route to the sites of the primary subsistence activities where sago, fishing or crab and mollusc collection takes place. This ensures that small amounts of garden food can be harvested on the journey back to the village from a major food-getting expedition. Hunting of wild pig and gathering of sago grubs are subsidiary activities to crab collection and fishing.

Although the relative importance of the various subsistence activities cannot be precisely defined, there is a well organised system of land use which ensures readily available food supplies all the year round.

The list of vegetable foods cultivated and collected by the Baroi has not grown since 1947 (Ulijaszek and Poraituk 1981; Hipsley and Clements 1950). Following the 1947 New Guinea Nutrition Survey Expedition, the Department of Agriculture, Stock and Fisheries placed special emphasis on the improvement of native food supply. However, this took the form of up-grading of pig and poultry stock with the aim of making more protein available; this was clearly not a priority in light of the abundant waters of the delta. The promotion of new food crops by the Department of Primary Industry has been sporadic and fraught with difficulties. Perhaps the only foods to gain widespread popularity are corn, cucumber and watermelon, the latter two being of minimal nutritional value. In recent years high protein vegetables such as beans and peanuts have been vigourously promoted but these have yet to gain widespread acceptance; they remain very much the province of community schools' nutrition programmes, an important part of the food crop extension programme. Of greater importance is the establishment of some food cropping within the village (for easy access of supplementary foods during inclement weather) and the cultivation of green leafy vegetables, a fact appreciated by the Calverts in their rather more informal nutrition programme at Kapuna Christian Mission.

4. Sago cultivation

Of the eight species of the palm genus *Metroxylon* described in Oceania (Barrau 1958; Barrau 1959), only the *M. sagu* yields starch in any quantity. Some 150 000 ha of sago stands may be available in the Gulf Province (estimate based on Holmes and Newcombe 1980). This indicates great potential for the development of this resource. Estimates of productivity per hectare vary from 7 to 330 mature trunks per year (Morris 1953; Cavanaugh 1955), the lower range being under subsistence management by sago-collecting peoples relying totally on uncultivated stands, and the upper range being under well-organised commercial plantation management using highest yielding and fastest maturing varieties of the species, or cultivars. Flach (1977) suggests that a sustained yield of 135 mature trunks per hectare per year could be obtained under village management.

Although large stands of uncultivated sago palm in the Purari Delta are owned, these are a considerable distance from existing villages and are very rarely used. There is a history of trade in sago and considerable knowledge of sago cultivation. In addition to communal sago grounds at the sites of the previous clan villages, men possess several sago grounds at varying distances

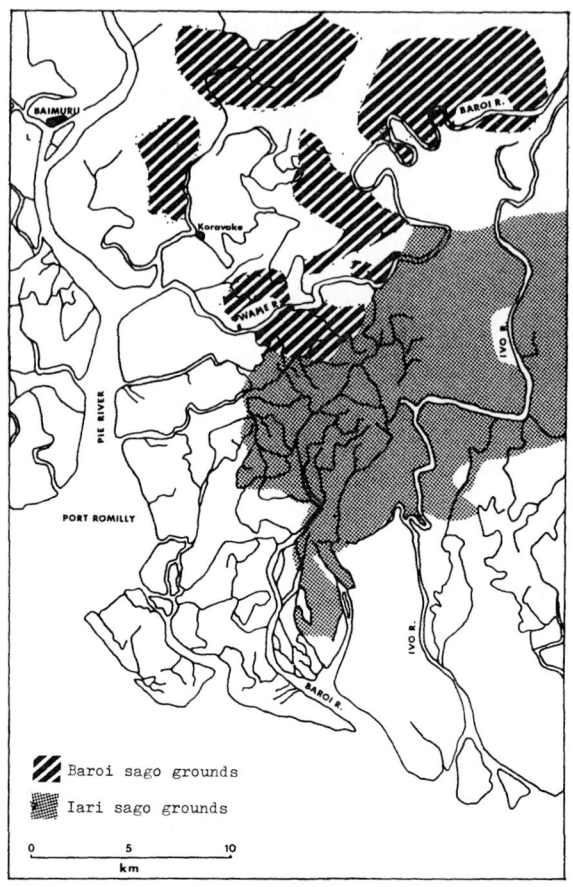

Fig. 2. The major cultivated grounds of the Baroi and Iari clans of the Purari Delta.

from the present village. These are river margin clearings often only 20 m² in area and having three or four sago palms at various stages of development. Progressive clearing is practiced, adjacent areas being cleared in subsequent years and more palms planted. Thus each family has adequate sago palms planted in various places to provide a constant supply of the main staple. Some individuals have created sago blocks similar to small plantations, anticipating exhaustion of their sago palms at the old clan village site. The major cultivated sago grounds of the Baroi are given in Fig. 2.

After planting, the Baroi clan practices very little palm management. When a palm is being harvested, dead leaves may be cleared from the surrounding palms. However, suckers are not removed from the growing palms, nor

582

is there thinning of clumps, accessory suckers dying off because of dominance of the main stand. When a palm is harvested, all suckers except the dominant one are removed; this is allowed to propagate and replace the harvested palm. Depending on a man's forecasted needs for sago and his cultivar preference, other well-formed suckers may be replanted in another sago place. Personal preference is an important factor in determining which cultivars are maintained.

In all, the Baroi have twelve names for the starch-yielding sago palm, distinguished by characteristics of spines, leaves, colour and height. A list of starch-yielding palms known to the Baroi is given in Table 2. Of the cultivars seen, only two were smooth petioled. Thorny cultivars may be preferred because the thorns are a deterrent to wild pigs who would otherwise eat the growing shoots.

Of the most common varieties, the most used in order of importance are *havea*, *pakeava*, *kauapa*, *opai* and *bauma*. The reasons given for the utilisation of these five are ease of sago making (low fibre to starch ratio), high yields, palatability, storage and colour of starch. It appears that in recent generations there has been increased propagation of high yielding, easily

Table 2. Cultivars of starch-yielding sago palms distinguished by the Baroi.

Local name	Smooth/thorny	Characteristics
Propagated		
*havea**	Thorny	Grows tall if cultivated on well-drained ground; slender
*pakeava**	Thorny	Grows tall and fat if cultivated on well-drained ground; short and fat on poorly drained ground
*opai**	Thorny	Grows tall with medium girth
*kauapa**	Thorny	Grows tall with medium girth
kairi-kairi	Thorny	Grows tall with medium girth
*aiameri**	Thorny	Grows short and fat; hard fibres make this a difficult tree to process
*kaivei**	Thorny	Medium height; slender
avei	Thorny	Mature palm not seen
*pirika**	Thorny	Mature palm not seen
calai	Thorny	Mature palm not seen
Propagated and wild		
*bauma**	Smooth	Grows tall and with medium girth on well drained ground; short and fat on poorly drained ground
ikipa	Smooth	Short with medium girth

* Asterisk signifies the most commonly used cultivars.

worked and more palatable cultivars to the neglect of poorer ones. This may be related to more recent clan migrations and the post-war attempt at business enterprise largely based on sago production (Maher 1961).

Unlike most subsistence activities, sago production per unit of labour is easier to measure than productivity per unit of land. Within the Sepik, yields vary from 2.2 to 3.7 kg of wet starch per hour of labour (Ruddle et al. 1978; Dornstreich 1973; Schindlbeck 1980). A more general measure of productivity is the number of man-hours needed to produce one million kilocalories of food energy (Carneiro 1957). Studies conducted by Edwards (1961), Lea (1964) and Townsend (1974) give values of 80, 154, and 157 man hours per million kilocalories respectively.

During the starch yield study, the number of hours worked and the starch produced by seven sago makers was measured. To obtain reasonably accurate estimates of starch energy obtained by each party, 100 gm samples of crude starch were taken and frozen for subsequent analysis of moisture content. An energy value of 400 kilocalories (1674 J) per 100 gm dry sago was used in calculating the number of hours of labour required to produce one million kilocalories sago starch energy. This value is a mean of the two energy analyses available (Peters 1957; FAO/USDHEW 1972).

The mean quantity of wet starch produced per hour's work is 3.5 kg, with a range of 2.0 to 8.6 kg per hour. This is within the range cited in the literature.

Using the figures for starch yields, hours of labour, water content and energy value, the number of hours of labour required to produce one million kilocalories was calculated at 133. This is equivalent to the mean of the means obtained by Edwards (1961), Lea (1964) and Townsend (1974).

As a measure of the efficiency of the subsistence economy, human output in sago-producing cultures of Papua New Guinea is similar to that of shifting cultivation, and higher than sedentary agriculture or hunting and gathering (Townsend 1974).

5. Yields of sago starch

The mean trunk length worked by the Baroi is 8.4 meters, which is greater than reported by Townsend (1974) and Flach (1981), who give values of 6.6 and 7.8 meters respectively. Mean outside diameter, however, is less than among the Sanio-Hiowe of the Sepik: 0.42 meters as opposed to 0.46 meters. Starch yields (wet weight) per trunk range from 38.4 to 358.6 kg with a mean of 133.9 kg. These yields can be compared with data obtained in other parts of Papua New Guinea and Irian Jaya presented in Table 3.

Table 3. Yield of sago starch (wet weight) from the Purari as compared with other areas.

Place	Size of sample	Mean yield (kg) (range in brackets)	Wild/cultivated	Source
Baroi (Purari Delta)	18	133.9 (38 – 359)	Cultivated	Present study
Abelam (East Sepik)	2	219	Cultivated	Lea 1964
Sanio-Hiowe (Upper Sepik)	5	86 (28 – 205)	Traditionally cultivated	Townsend 1974
Oriomo (Western Province)	8	66 (29 – 104)	Wild and cultivated	Ohtsuka 1975
Fly River	n.g.	(79 – 159)	n.g.	Riley 1925
Irian Jaya	n.g.	(110 – 160)	Cultivated	Barrau 1959

n.g. = not given.

Table 4. Yields of sago starch (dry weight) per cubic meter of pith.

Place	Yield/m^3 pith		Source
	Wet weight (kg) (range in brackets)	Dry weight (kg) (range in brackets)	
Baroi (Purari Delta)	106 (32.5 – 260.6)	57.0 (26.4 – 128.8)	Present study
Abelam (East Sepik)	230 (235.2 – 249.6)	n.d.	Lea 1964
Sanio-Hiowe (Upper Sepik)	101.4 (56.2 – 147.7)	n.d.	Townsend 1974
Gaikorobi (Middle Sepik)	212 (33.5 – 323.4)	n.d.	Schindlbeck 1980
Imbando, Kundima, Mamber (Sepik flood basin)	n.d.	158.9 (40.9 – 286.7)	Flach 1981
Aibom (Middle Sepik)	230	n.d.	Schuster 1965

n.d. = not data.

Mean yield (dry weight) per trunk was 74.3 kg, considerably less than the yields obtained in the Sepik, as estimated by Cavanaugh (1955), Flach (1981) and Toyo Menka Kaisha (1972). On a wet weight basis, starch yields in the Baroi exhibit the widest range of values of all similar studies conducted in Papua New Guinea.

Yield per cubic meter of pith is compared with other studies in Table 4. On a wet weight basis, yields per m^3 of pith are comparable to that of Sanio-Hiowe who use traditionally cultivated sago, and low compared with Gaikorobi, Abelam and Aibom. On a dry weight basis, yields are 35% of those estimated by Flach in the Sepik flood basin.

Although yields are of primary consideration in any commercial venture involving sago, this is clearly not so for the Baroi. The cultivar *pakeava* was found to give yields in the highest ranges reported for Papua New Guinea (Ulijaszek and Poraituk 1981), but other varieties are maintained for their other properties and cultural significance, respectively. Yields could be improved by proper management, particularly the pruning of suckers from the growing palm and the regular clearing away of dead leaves from the trunk.

6. Discussion and conclusions

The Purari Delta is one of the areas in Papua New Guinea where the subsistence staple, and the culture associated with it is based upon sago. The cultivars used are normally grown in specific sago areas and not collected from the wild. For these cultivars much consideration is given to selection for ease of processing, palatability and cooking properties.

Neither time nor land are limiting factors in food production, and there is little pressure upon the people to increase it, as there is no malnutrition sufficient for them to be aware of the need to do so. This would account for the rather mixed progress experienced by the Department of Primary Industry in food crop extension work.

The possibilities of using sago starch as an energy source through fermentation (as has been proposed for the Sepik) could change this situation. Yields would be increased through the organisation of sago areas, and traditional opposition to the mechanisation of sago production could weaken in light of its new-found commercial value.

The change in orientation would not necessarily impose marked differences upon the present pattern of land use. So one may suspect that excessive disturbance to custom may not ensue. However, the introduction of such commercial enterprise into the existing stable relationships between the clans might well be disturbing in other social ways, in that they might encourage traditional clan rivalries over available resources. Developers should be aware of this potential issue.

Commercial operations for growing sago as a food could well have to consider the additional factor of palatability. Apart from using a suitable cultivar, this much depends upon the technique of extraction and skill of the extractor. Traditional practices should be inspected closely in this regard.

The method of extracting sago from the trunk by the women appears to be fairly efficient in energetic terms compared to hunting – gathering or sedentary farming. Thus as a means of obtaining calories it is not to be decried.

Physical modifications in the upper part of the catchment, such as the building of dams like the one proposed for Wabo, are unlikely to affect the sago economy of the delta since the plant's distribution, cultivation, or method of use would not be directly influenced by changes in water regime or sediment transport. However, wage earning opportunities out of the delta area for the men could have some sociological repercussions. As men are responsible for clearing of garden lands and active sago cultivation, this could result in short-falls of total food supply and a more limited range of available foods some years into the future. It is questionable whether the income generated by this work would be used for the purchase of nutritious trade store foods. Such an emigration would put an extra burden on child-rearing women, resulting in increasing under-nutrition in pregnant and lactating women and their offspring.

References

Barrau, J., 1958. Subsistence agriculture in Melanesia. Bernice P. Bishop Museum Bull., Vol. 219. Honolulu.

Barrau, J., 1959. The sago palms and other food plants of marsh dwellers in the South Pacific islands. Economic Botany 13: 151 – 163.

Carneiro, R.L., 1957. Subsistence and social structure. Ph.D. Thesis. University of Michigan, Ann Arbor.

Cavanaugh, L.G., 1955. Sago flour production, Sepik River. Mimeo. Forest Products Research Centre, Boroko, Papua New Guinea.

Dornstreich, M.D., 1973. An ecological study of Gadio Enga (New Guinea) subsistence. Ph.D. Thesis. University of Michigan, Ann Arbor.

Edwards, E.T., 1961. The natural stands of sago palms, *Metroxylon* spp., in the Sepik River area of New Guinea and their possible use as a source of commercial starch. Mimeo. Report submitted to the Board of Directors of Geo. Fielder and Co., Ltd., Sydney.

FAO/USDHEW, 1972. Food composition table for use in East Asia. Dept. of Health, Education and Welfare. Publication No. (National Institute of Health) 75 – 465. Washington, D.C.

Flach, M., 1977. Yield potential of the sago palm and its realisation. In: K. Tan (ed.), Sago '76: Papers of the First International Sago Symposium, pp. 157 – 177. Kuala Lumpur.

Flach, M., 1981. Sago palm resources in the northeastern part the Sepik River flood basin. P.N.G. Dept. of Minerals and Energy. Energy Planning Unit. Report, No. 3/81. Konedobu, Papua New Guinea.

Hipsley, E.H. and F.W. Clements, 1950. Report of the New Guinea Nutrition Survey Expedition 1947. Government Printer, Sydney.

Holmes, E.B. and K. Newcombe, 1980. Potential and proposed development of sago (*Metroxylon* spp.) as a source of power alcohol in Papua New Guinea. In: W.R. Stanton and M. Flach (eds.), Sago. The Equatorial Swamp as a Natural Resource, pp. 164 – 174. Proc. 2nd Intern. Sago Symp., Kuala Lumpur, Sept. 15 – 17, 1979. Martinus Nijhoff, The Hague/Boston/London.

Lea, D.A.M., 1964. Abelam land and sustenance. Ph.D. Thesis. Australian National University, Canberra.

Maher, R.F., 1961. New men of Papua. University of Wisconsin Press, Madison.

Morris, H.S., 1953. Report on a Melanau sago-producing community in Sarawak. Colonial Research Studies, No. 9. Colonial Office, London.

Ohtsuka, R., 1975. The sago eaters: an ecological discussion with special reference to the Oriomo Plateau. Paper read at Sunda and Sahul Symposium, Pacific Science Congress, Vancouver, 1975.

Peters, F.E., 1957. Chemical composition of South Pacific foods. South Pacific Commission. Technical Paper, No. 115. Noumea, New Caledonia.

Petr, T. and J. Lucero, 1979. Sago palm (*Metroxylon sagu*) salinity tolerance in the Purari Delta. In: T. Petr (ed.), Purari River (Wabo) Hydroelectric Scheme Environmental Studies, Vol. 10: Ecology of the Purari River catchment, pp. 101 – 106. Office of Environment and Conservation, Waigani, Papua New Guinea.

Riley, E.B., 1925. Sago-making on the Fly River. Man 23: 145 – 146.

Ruddle, K., D. Johnson, P.K. Townsend and J.D. Rees, 1978. Palm sago, a tropical starch from marginal lands. The University Press of Hawaii, Honolulu.

Schindlbeck, M., 1980. Sago bei den Sawos (Mittelsepik, Papua New Guinea). Basler Beiträge zur Ethnologie 19: 87 – 92.

Schuster, M., 1965. Mythen aus dem Sepik-Gebiet. Basler Beiträge zur Ethnologie 2: 369 – 384.

Toft, S., 1983. The Pawaia of the Purari River: social aspects. This volume, Chapter III, 2.

Townsend, P.K., 1974. Sago production in a New Guinea economy. Human Ecology 2: 217 – 236.

Toyo Menka Kaisha Ltd., 1972. Feasibility study for the establishment of sago flour plant in the Territory of Papua New Guinea. Forest Products Research Centre. Report. Boroko, Papua New Guinea.

Ulijaszek, S.J. and S.P. Poraituk, 1981. The sago subsistence of the people of the Purari Delta. In: A.B. Viner (ed.), Purari River (Wabo) Hydroelectric Scheme Environmental Studies, Vol. 19: pp. 1 – 38. Office of Environment and Conservation, Waigani, Papua New Guinea.

Williams, F.E., 1924. The natives of the Purari Delta. Anthropological Report, No. 5. Government Printer, Port Moresby, Papua New Guinea.

Warrillow, C., 1978. The Pawaia of the upper Purari, Gulf Province, Papua New Guinea. In: T. Petr (ed.), Purari River (Wabo) Hydroelectric Scheme Environmental Studies, Vol. 4, pp. 1 – 88. Office of Environment and Conservation, Waigani, Papua New Guinea.

Warrillow, C., 1983. A short history of the upper Purari and the Pawaia people. This volume, Chapter III, 1.

11. Purari hydroelectric potential and possibilities for industrial development

K.W. Dyer

Papua New Guinea, because of its high rainfall and suitable topography, has great potential for hydroelectric development – probably in excess of 20 000 MW. Domestic and industrial demand within the country is small and limited. It increased from 75 MW to 210 MW when Bougainville Copper mining commenced in 1972, but since then has only increased marginally to about 250 MW. Industrial developments within the country would not require massive amounts of power and if concentrates were refined locally this would not require more than 15 MW.

Hydroelectric power, unlike oil, coal, mineral concentrates, timber and agricultural produce cannot be loaded into ships and transported to where there is a demand for it. Thus, if hydroelectric power is going to be developed on a very large or significant scale, then commercial users must be located in P.N.G. These industries would require imports of raw materials to be processed with P.N.G. power, the finished product being marketed overseas. The development would be massive and it is this scale of industry that is to be considered in this chapter.

The Purari River basin occupies a central position in Papua New Guinea. It drains the greater part of the 5 highland provinces before flowing through the Gulf Province to the Gulf of Papua. The highland catchment area is generally very rugged and mostly lies above 1250 metres with large areas above 2500 metres. The whole basin, including highlands and coastal plains, has an area of 33 670 sq. kilometres. The total length of the main stem of the Purari River is 630 km. A generally reliable but diversified rainfall pattern together with the topography indicates this river has total potential of about 10 000 MW with a very great number and range of potential hydroelectric power sites from the minihydro possibilities to those capable of substantial power output.

The collection of basic data on such a river system takes time, is expensive, and cannot be undertaken everywhere at the one time. Even pre-feasibility

Petr, T. (ed.) The Purari – Tropical environment of a high rainfall river basin
© *1983, Dr W. Junk Publishers, The Hague / Boston / Lancaster*
ISBN-13: 978-94-009-7265-0

assessments are expensive and a feasibility study much more so. For this reason potential sites are not closely examined until there is a special interest to do so.

Interest in the Wabo site on the Purari River came from the Japanese firm Nippon Koei Co. Ltd. who, in 1971 and 1972 with the P.N.G. Government Executive Council approval, undertook some pre-feasibility investigations. Very strict conditions were laid down by the government for this study. It was to be entirely at Nippon Koei's expense; the investigation permit was to be held by the Papua New Guinea Electricity Commission; and an engineer from that organisation was appointed to work with Nippon Koei, not only to ensure that conditions for the study were met but in particular to ensure that all information obtained was available to the Papua New Guinea government.

Before the report on these investigations was published in 1973 it became obvious that a hydroelectric project of very considerable magnitude was possible – also that P.N.G. had no resources to evaluate the feasibility and worth of such a vast project – nor the financial and other resources to implement it. Two matters of concern were already of great importance to the government. The first was the impact of such a scheme on the nation. Indications were that such a huge development would be several times the size of the then national budget. The government itself might well be overshadowed, its own priorities distorted, and its ability to control the nation threatened. Secondly it was concerned at possible environmental consequences. Whilst the government supported a serious study of possible development that may bring benefits to the nation, it was concerned to ensure its own interests were not swamped by foreign dominance. It had limited financial resources for any major study. It was already a large recipient of aid funds from the Australian government to which it might need to turn for more assistance if Purari potential was to be investigated.

The 1973 report by Nippon Koei sparked considerable interest with much activity and lobbying from consultants. A number of very large organisations offered their expertise to P.N.G. At least one very large company indicated it could not only do a feasibility study, environment study, regional study, build the power and industry complex but also arrange the industry and finance and deal with other governments. Furthermore, it was most pressing in its invitation for ministers and government officials to inspect its work overseas. These approaches increased the apprehension already felt that the country could lose control of its own destiny.

The government therefore went ahead cautiously, jealously guarding its right to make its own decisions and avoiding commitments that might do otherwise. Concerning the Wabo power site:

(i)   the assistance of the Australian government was sought;

(ii)   the Australian government agreed as a first step to review the work done by Nippon Koei and commissioned the Snowy Mountains Engineering Corporation (SMEC) to do this task;

(iii)   the P.N.G. government appointed its own special consultant, Mr T.A. Lang, then president of the American consultancy firm of Leeds, Hill and Jewitt, Inc. This was considered essential as the interests of Australia and Japan might not coincide with P.N.G. interests;

(iv)   all consultants agreed that a feasibility study should be undertaken and that Wabo was the appropriate dam site for investigation;

(v)   this led to a feasibility study to be conducted jointly by the Australian, Japanese and Papua New Guinea governments. An agreement, restricted to a technical study of the Wabo site only, was finalised late 1974 and a 2-year study was commenced January 1975;

(vi)   through the United Nations Development Program (UNDP) the P.N.G. government sought special advice about other required studies and in particular:
   − a regional study to pursue the possibilities of regional development over an area rather than just hydropower and industrial development (Barr 1974)
   − an environmental study assessment intended to advise what environmental aspects needed study, their relative importance and cost and how the work might be undertaken and funded (Goldman et al. 1975)
   − an industrial study assessment again primarily to advise the government what needed to be done (Spooner et al. 1974);

(vii)   through the Papua New Guinea Electricity Commission (ELCOM) a private consultant was commissioned to evaluate the potential of other major hydropower sites on the Purari River. It was foreseen that development of the Wabo site might affect the potential of other sites and not be in the best interests of the future development of the overall resource;

(viii) the government had special concern about the sketchy port proposals on the Papuan south coast so it commissioned expert international advice on the general practicability of a port in this region.

Thus during 1975 a feasibility study of the Wabo site was under way. By agreement between the parties this was restricted to a technical feasibility study primarily to determine physical feasibility of a dam at the Wabo site, an estimate of costs and an indication of the cost of power which was essential for any realistic discussion of its industrial use. This study progressed very well. The final 8-volume report (SMEC-NK 1977) was published in December 1977 and distributed early 1978. The indications were that it would be feasible to construct a dam at Wabo and that the power would be relatively cheap. It was evident however that the consultants had two basic problems in trying to arrive at power costs. Costs would depend to some extent on the source and cost of finance which could not be known; and overall costs would depend on port and industry locations and the number and nature of the power-consuming industries, matters on which the consultants received no guidance from the government.

The several reports from the UNDP consultants, whilst containing much valuable information, were not as helpful to the government as expected. Firstly the Barr report really proposed nothing new and had relied heavily on advice from government departments on development prospects for the Gulf Province which were extremely limited; and Barr had nothing novel except a further series of very elaborate and expensive studies which the government could not afford either in terms of money or human scientific resources. Even if the further studies were undertaken by scientific specialists from international aid sources, these would need to be serviced by local technical expertise which was very limited. If these limited resources were required to work solely on Purari, all other work in P.N.G. would cease or be deferred and national priorities distorted. Furthermore, a large influx of international experts might also provide much conflict with P.N.G. authorities and the further studies envisaged by Barr could not be justified at such an early and uncertain stage of the project. It was reaffirmed that regional considerations were extremely important but, without a major project, not justified.

The environmental report by E.R.A. Associates (Goldman et al. 1975) was vague in determining priorities, although it contained a comprehensive list of many possible environmental consequences. The list was useful but did not assist the government in determining priorities within ranges of financial constraints or how the studies and assistance with them might be organised. Various cost estimates of the proposals indicated a multimillion kina study which, once under way, would become difficult to control and limit financially. It was also evident that, even if funded and organised by international aid agencies, there would be a great strain on local technical expertise required to support Purari research to the detriment of important work going on elsewhere in Papua New Guinea.

As a practical approach, working on the consultant's list of tasks to be done, a government committee of mainly technical expertise worked through the E.R.A. report to determine matters of priority importance and timing. It also sought the appointment of a full-time environmental study manager.

When this manager was ultimately appointed, with the assistance, guidance and support of the Physical Environment Committee, the most important and essential studies to the limit of finance available were actively pursued. This monograph, together with reports already published underline the success of this programme which has used predominantly locally available expertise, was financially inexpensive and quite adequate to the present status of the Wabo project. More may need to be done, but the focus for this will only become clear when a more definite project is in view, industries determined, and the areas for port and industrial development defined.

Most studies have gone as far as is deemed prudent until a customer for the potential power is in view and the nature and location of potential industry is known. In summary: (i) the feasibility study of the Wabo site has been completed. Any further evaluations depend on finding a consumer; (ii) environmental studies should allow some overall appreciation of the impact of a dam on the area and provide a firm basis for further studies that might be required if and when a project is being considered; (iii) regional interests must be considered if a large industrial project came into the area; (iv) the government itself has decided there will be no further expenditure on the Wabo scheme until there is serious interest by a potential consumer. Such interest would need to be backed by a willingness to undertake a feasibility study and enter into discussions with the government.

There is no doubt that the Purari River basin has a huge potential for hydroelectric power development. It is also clear that large-scale power development will never proceed without a customer for the power. The remaining part of this chapter will therefore examine the possibilities for industrial development, which is not just a matter between industry and government but is also influenced by people and their life-styles.

When the preliminary studies were being undertaken by Nippon Koei in 1971/72, it was thought that an aluminium smelting industry might become the core user of the power. Projections of aluminium demand and supply at that time showed the need for an aluminium smelting capacity of an additional million tons per annum during the 1980s even at what was then considered a modest 10% growth rate for the industry. Aluminium smelting is a large consumer of power, and critical shortages of energy resulting from the world oil crisis of the early 1970s indicated an opportunity for aluminium production overseas.

It was evident that Wabo, the site being examined, could not be economically built by stages, and that it would produce considerably more power than would be required for what would be the world's largest smelter − up to 800 000 tonnes p.a. capacity. During preliminary investigations there was concern about how the power could be used. Many smaller industries would benefit if power on a large scale was available, but at least one large core user was essential to get any project off the ground.

During 1974 there was an appraisal by international consultants of P.N.G. energy resources in relation to uranium enrichment. This was part of a survey being carried out at that time throughout the world. Some effort was made by the P.N.G. government to obtain basic information about the uranium enrichment industry. This was part of a program to obtain information about as many power intensive industries as possible. The uranium enrichment industry probably could have paid higher prices for power, thereby making it a substantial income earner for the government. It was however never seriously considered as a potential consumer of Wabo power on obvious technical and political grounds. Although Wabo itself might not have had sufficient power for a uranium enrichment industry, there were additional adjoining sites which, developed with Wabo, would have provided ample power.

In August 1974 a large group of leading Japanese industrialists headed by the International Bank of Japan (referred to as the IBJ Mission) visited P.N.G. Prior to leaving Japan the P.N.G. government was asked what information the Mission could provide that would assist P.N.G. The opportunity was taken to ask for ideas on specific power-intensive industries that might be interested in using Wabo power, and information about such industries including basic requirements for land, power, water, labour, raw materials and other significant industry parameters or special requirements. In response to this request an excellent summary of information was provided listing 8 industries which together, at certain stated levels of production, could use about 1800 MW of power which was the expected size of Wabo. This list was incorporated in the IBJ Mission report presented early 1975. As the P.N.G. government at the conclusion of the Wabo feasibility study in 1977 had no industrial consumer in view, that hypothetical package of 8 industries was used as the basis of some cost assessments relating to that study.

During the visit of the IBJ Mission and in its report it was apparent that Japanese interests required not only cheap and stable power but other incentives such as a taxation holiday, a free-trade zone and other concessions. Cheap power was not enough. Paradoxically, whilst these other incentive requirements appear to have confirmed to Wabo critics that a definite package of industries was being considered, it showed quite clearly to government officials that the sale of Wabo power might not be able to provide any significant benefit to Papua New Guinea. Nevertheless the belief was still strongly held that the Wabo and Purari resource ought to be of value in an energy-starved world, and the government clearly indicated it wished to continue seeking a way to develop the resource for the benefit of the nation. The government was encouraged in the belief that the Wabo resource could be developed by continuing visits and approaches from governments and overseas commercial interests, but this interest always fell short of any practical proposal.

Other factors which dampened official optimism in P.N.G. were the known Japanese interests in hydropower resources of other countries, a changed outlook for aluminium growth prospects coupled with a world economic recession, and the very long lead time from any decision to go-ahead with a project until power was actually available – over 10 years in fact. These factors will now be examined more closely.

Whilst the Japanese continued to express interest in Wabo, it was no secret that they were also interested in similar resources in Indonesia and Brazil. The Indonesian government signed a contract with the Japanese for the Asahan hydroelectric power project mid 1975. Even though all funds were provided from Japanese sources it appeared that the Indonesian government had very little, if any, control over what was to become virtually a foreign enclave, and it was difficult to see what significant overall benefit would accrue to Indonesia. The type of development, though along lines sought by the Japanese, was not likely to be attractive to Papua New Guinea.

At about this time a reassessment of world aluminium demand and supply showed a slump in the industry and that predicted shortages were no longer evident. In fact demand and supply were now expected, even under the most favourable conditions to the industry, to be more or less in equilibrium for the next 15 years – or the foreseeable future. It was also known that many existing smelters were operating at considerably less than their potential output, and this situation in Japan appears to be continuing with the industry remaining in serious difficulties.

Indications from the Wabo feasibility study (SMEC-NK 1977) concerning the timing of any future project were that if and when a decision to go ahead was made, it would take an estimated 10½ years before the first two generators were installed and power was available. This represented 2½ years for detailed investigation, design and mobilisation, and 8 years construction. To arrive at a decision point would require several years for further investigations, discussion and agreement by industry and government. There would need to be a long-term agreement on the use and pricing of power. There is a vast array of problems to be overcome in agreeing to a whole range of issues with such a long lead time to the availability of power – especially in a world situation which is increasingly subject to rapid change.

There were also internal domestic factors that tended to dampen enthusiasm. The large-scale type of development envisaged, the massive industrial estate, was quite contrary to most of the government development strategy and its 8-point plan. It was becoming more apparent and perhaps better understood within P.N.G. that the industrial complex to use the power within P.N.G. would be enormous and concentrate large numbers of people into a large urban complex with all its related social and welfare problems, and a way of life or life-style not only alien but also not desired by most people. Opposition to the scheme was becoming more vocal, organised and highly emotional. The main opposition was channelled through the 'Purari Action Group' and the book 'Overpowering Papua New Guinea' (Pardy et al. 1978) published on their behalf. Whilst this publication contains much quoted factual information, the interpretation is distorted and misleading. Critics of Wabo have viewed the project as if a deal for development had already tacitly been agreed. When the Wabo feasibility studies commenced, many people believed work on the dam had started, and no amount of publicity and argument could convince them otherwise. Government would agree that more studies will need to be done before any project could go ahead. It has always asserted that from the first investigations what is being done is merely a study and no project will proceed unless it is in the overall interests of the nation and all environmental aspects are thoroughly examined.

Many people however remained unconvinced and referred to the reputedly high cost of the study, arguing that neither the Japanese nor Australian governments would provide such generous amounts of finance merely for a study. This underlines their misunderstanding of the size of the project, which even educated and sophisticated people find difficult to grasp. Eight million dollars is a considerable amount to establish technical feasibility and an order of costs, especially when there is no consumer in sight. However this is a relatively trivial amount compared to the $120 million that would be required for further investigations prior to design, and a construction project which, at 1976 prices, would cost $1500 million and probably three times that amount if a package of industries and their needs are included.

The crux of the government problem was that on the one hand it wished to develop a resource in the interests of the nation; it supported the feasibility study but had virtually no knowledge or expertise to guide it on a vast range of related matters. On the other hand, when it sought advice on regional, environmental and industrial study needs, it received grandiose proposals with open ended requirements for vast amounts of money. It had not initiated the interest in Wabo which may or may not prove important. There were many other areas of national importance and concern, and the government did not have unlimited funds or resources to pour into investigations that might prove wasteful.

How deeply should the government become involved in the pursuit of industrial knowledge, understanding and promotion of the Wabo power resource? As with the environment investigations, the government opted for a cautious approach to test what appeared to be the main possibilities. An official was appointed to collect and collate information and to institute certain minimum studies. He was supported by a committee of officials from departments with interests in trade, commerce and industry. In this way much useful information was gathered. Preliminary studies were undertaken into the aluminium smelting industry and the large-scale production of steel using electricity. The outcome was not promising as neither industry seemed likely to be able to afford power prices as would be necessary to bring benefits to P.N.G.

It is not surprising therefore that further studies were abandoned at least until some initiative and specific interest came from industry. The government, however, still made efforts through diplomatic and trade channels, and overseas visiting missions, to promote interest in the Wabo resource.

Despite the lack of specific proposals there was still continuing interest in the power resource by visiting consultants, financiers and trade missions from other countries. During one such visit from a Norwegian group it was evident that transmission of power by underwater cable was technically possible over relatively long distances and that power loss during transmission might not be a significant deterrent. The power would be transmitted as DC and accordingly high costs would be incurred with major transformers at each end of the transmission cable.

Fears about the adverse effects of massive industrial development necessary to consume Wabo power might be lessened if a significant amount of the power could be sold outside the country; if, for example, 800 MW could be sold overseas leaving 1000 MW for aluminium smelting and other industries within P.N.G., the project may be more manageable and perhaps more acceptable. It seemed to be a possible option worth at least some consideration.

Australia is the only country to which it might be possible to transmit power in this way. When these thoughts were germinating in 1976, oil prices were continuing to rise at an alarming rate. In the medium to long term it seemed possible that it might be in Australia's economic interest to convert coal to oil rather than use it for large-scale power generation if power was available from P.N.G. Other possible advantages foreseen were the possible lessening of dependence by P.N.G. on Australian financial aid, and the goodwill that could flow from a joint enterprise between neighbouring nations − provided both could establish that there would be mutual benefits.

The acceptance of such an idea in Australia was not likely to be easy, as the premiers of most Australian states were interested in developing their own power resources and providing cheap power to encourage industry. It would be difficult for Australia to appraise the worth of such an idea in a truly national and overall economic context. Nevertheless, the P.N.G. Cabinet decided the option worth pursuing and that an approach be made to the Australian authorities. This approach does not appear to have been handled well. It was widely interpreted as a specific approach to sell power, whereas it should have been an approach initially to test interest and to see if there was any acceptance that there may be mutual benefits which could lead to more formal discussions between the two governments. Hopefully these discussions could lead to a feasibility study of the submarine cable concept which would be essential before there would be any intergovernment agreement on power sharing.

Whilst the final outcome of this approach is not known, it seems that there is no interest or enthusiasm by Australian authorities at this stage. This does not necessarily mean that at some time in the future this situation could not change. The fact remains that there is a very considerable resource of both water and hydroelectric power right on Australia's doorstep which continues to remain undeveloped.

The prospects for industrial development and for Wabo at the present time and for the foreseeable future are not good if P.N.G. is to receive significant benefits. Enclave development on the pattern of Asahan in Indonesia could be of interest to Japanese industry and government, but this type of development is not likely to be acceptable in P.N.G. But the resource and potential remains. Circumstances can and will change. With existing economic parameters and technical knowledge, the changes necessary for renewed interest in Wabo are more likely to be a combination of the following: (i) a surge in demand and much higher prices for metals relative to raw material and other costs such that industry could pay much more for power; (ii) another serious oil shortage or crisis − or escalation of oil prices. This sort of event could influence energy policies in countries such as Japan and the U.S.A., with the phasing out of energy-intensive industries. This might lead to renewed interest in P.N.G. on more favourable terms; (iii) changes in attitude towards development by government and the people.

In the simplest terms development of Wabo will depend on how urgently P.N.G. may wish to develop the resource and how urgently industry needs it and is willing to pay for it. The basic studies are done so a renewal of interest could be assessed at any time − even perhaps 50 or more years hence. But more is involved than merely agreement between the government and industry. There needs to be acceptance − maybe even a demand − by the people.

Whether the nation is likely to receive an economic or commercial benefit

598

from any future proposal should be capable of precise analysis and assessment. The effects of industrial development on the people and their reaction are much more difficult to establish. Provincial and national interests do not necessarily coincide – nor do the interests of people between the various provinces. Furthermore, within the same province, what is seen by some as a gain and desirable is not acceptable to others. Thus a young family man may prefer urban living with its health and education services and the other pleasures that give rise to social problems, whilst his father or grandfather would be horrified at any intrusion into his age-old peaceful subsistence existence. With people, over time, attitudes can change.

For the past few years the government of P.N.G. has tried to pursue a policy of more even development throughout the nation and in particular to encourage people to remain in rural areas rather than migrate to the cities. This has much to commend it, but in practice will have serious limitations. There are many areas within the country for which nothing can be done simply because of the physical nature of the area and difficulties of access. This is not the exception but would apply to most areas, excluding the well-populated and accessible river valleys and coastal areas served by road and ship. A great number of people live in inaccessible areas. It is likely that an increasing number of them would welcome the opportunity of employment and the benefit of social services for themselves and families in a large urban environment. Such a migration from other provinces would not be acceptable to people now living on the Papuan coast. But their attitude may change. At present they fail to realise that such a migration will be essential for any significant development, and some catalyst for development is what the Gulf Province sorely needs.

The Gulf people say they do not want a major project such as Wabo – they would prefer a number of smaller projects. Small dams with the cultivation of rice was one example. It is difficult to imagine any people within P.N.G. adapting to the cultivation of paddy rice. There are not sufficient people in the Gulf Province to make a serious start. There is no bar to smaller projects going ahead under the existing provincial government system, without Wabo, provided suitable resources are made available and there is sufficient manpower to undertake the work. But suitable resources are hard to find and having them made available for development probably more difficult. Manpower would be inadequate. The burden of subsistence living and village responsibilities are such as to have already driven large numbers of able-bodied workers to urban areas where prospects are better. This in turn makes life more difficult for those remaining at home. They would be hard pressed to do more than meet existing responsibilities.

The agricultural resources of the Gulf Province are not attractive and with one or two possible exceptions are not favourable for development prospects. Thanks to the high and regular rainfall that falls on large areas of the highlands and the geographic accident that allows the surface water to flow out to the Gulf of Papua, the Gulf Province does have very substantial resources of hydroelectric power. It might well be that the development of this hydroelectric power along the lines possible with the Wabo project is the only chance for significant development in this deprived province. Whatever the development potential may be, nothing significant will be achieved until there is an acceptance of migrants from other areas. This would be vigorously opposed at the present time.

In conclusion it is worth noting that the Asahan scheme now coming to fruition in Indonesia was first surveyed in 1908 and has been the subject of numerous further surveys since then. Wabo was first investigated about 1956, which is comparatively recent. Whatever happens with Wabo, the Papua New Guinea government has in its possession valuable studies, not only of the technical feasibility but also environmental information as a starting point for assessment of development plans that might be presented should world industry and economic parameters change and the people and government of the country foresee substantial overall benefits to the nation from implementation of this project and be willing to implement it.

Acknowledgements

I wish to thank Professor Robert L. Weigel for his contributions, and his deep interest in the preparatory work for the Purari Hydroelectric Scheme.

References

Barr, R.L., 1974. United Nations Development Programme, Papua New Guinea. Comprehensive River Basin Development (Purari). Report to UNDP by Tippetts-Abbott-McCarthy-Stratton, Engineers and Architects. New York.

Goldman, R.C., R.W. Hoffman and A. Allison, 1975. Environmental studies design, Purari River development, Papua New Guinea. Ecological Research Associates, Davis, California.

Pardy, R., M. Parsons, D. Siemon and Ann Wigglesworth, 1978. Purari overpowering Papua New Guinea? International Development Action for Purari Action Group, Globe Press, Fitzroy, Victoria, Australia.

Snowy Mountains Engineering Corporation – Nippon Koei (SMEC-NK), 1977. Purari River Wabo Power Project feasibility report. 8 volumes.

Spooner, E.C.R., I.W. Snedden and A.H. Rintoul, 1974. Report in industrial study requirements for the Purari Development Committee. Office of Environment and Conservation, Waigani, Papua New Guinea.

Appendix A. Basic data on the Wabo hydroelectric scheme

Purari River catchment at Wabo	26 300 km²
Total Purari River catchment	33 670 km²
Average flow rate at Wabo	2360 m³/sec
Maximum recorded flow at Wabo	10 450 m³/sec
Minimum recorded flow at Wabo	487 m³/sec
Annual discharge at Wabo	74,4 km³
Sediment load at Wabo	40 × 10⁶ m³
Denundation rate of the catchment	1.4 mm per annum
Annual rainfall at the dam	8500 mm per annum

Wabo reservoir

Maximum depth (approximate)	90 m
Surface area	260 to 290 km²
Storage at full supply level	16 km³
Draw-down (maximum calculated)	20 m
Annual outflow to volume ratio	5:1
Calculated loss in storage capacity in 50 years	15%
100 years	27%
Installed capacity	6 × 360 MW

Petr, T. (ed.) The Purari – Tropical environment of a high rainfall river basin
© *1983, Dr W. Junk Publishers, The Hague / Boston / Lancaster*
ISBN-13: 978-94-009-7265-0

Appendix B. Reports in the series Purari River (Wabo) Hydroelectric Scheme Environmental Studies, published by the Office of Environment and Conservation and the Department of Minerals and Energy, Papua New Guinea.

Vol. I. Workshop 6 May 1977, edited by T. Petr, 1977.

Engineering feasibility − and what comes after it, by K. Dyer, pp. 4−6.

Environmental studies − general considerations, by L. Hill, pp. 7−12.

The effect of Wabo dam on sedimentation processes in the Purari River and sediment delivery to its delta, by G. Pickup, pp. 13−15.

Water quality − nutrient cycling, by T. Petr, pp. 16−20.

Notes on aquatic and semi-aquatic flora, by B. Conn, pp. 21−27.

Mangroves of the Gulf of Papua, by K.J. White, pp. 28−29.

Wood-boring molluscs and crustaceans of mangrove stands, by S. Rayner, pp. 30−31.

Fish and fisheries of the Purari River and delta, by A.K. Haines, pp. 32−36.

Prawn research in the Gulf of Papua, by C. McPadden, pp. 37−40.

Preservation of fish by salting and smoking in the delta and processing of sago, by R. Wanstall, pp. 41−42.

Wildlife and wildlife habitat in the area to be affected by the Purari scheme, by D.S. Liem pp. 43−45.

Purari scheme and provincial planning, by J. Morolla, pp. 46−47.

Social environment impact study, by C.S. Mero, pp. 48−51.

The malaria situation in the Gulf Province, by N.W. Tavil, pp. 52−55.

Purari nutrition survey, by J. Lambert, pp. 56−62.

Health aspects of the Purari scheme with particular reference to arthropod-borne diseases, by M. Alpers, pp. 63−64.

Ancestral and prehistoric sites at the Wabo dam site and the impoundment area, by B. Egloff, pp. 65−66.

Vol. 2. Computer simulation of the impact of the Wabo hydroelectric scheme on the sediment balance of the lower Purari, by G. Pickup, 1977.

Vol. 3. The ecological significance and economic importance of the mangrove and estuarine communities of the Gulf Province, Papua New Guinea, by D.S. Liem and A.K. Haines, 1977.

Vol. 4. The Pawaia of the upper Purari, Gulf Province, Papua New Guinea, C. Warillow, 1978.

Vol. 5. An archaeological and ethnographic survey of the Purari River (Wabo) dam site and reservoir, by B.J. Egloff and R. Kaiku, 1978.

Vol. 6. An ecological survey of fish of the lower Purari River system, Papua New Guinea, by A.K. Haines, 1979.

Vol. 7. Health and disease patterns of the Purari people, by A. Hall, 1979.

Vol. 8. Viral and parasitic infections of the people of the Purari River system and mosquito vectors in the area, edited by M. Alpers, 1980.
Mosquitoes of the Purari River area, by E.N. Marks, pp. 3 – 27.
Vector mosquitoes and human arbovirus infections of the Purari River drainage, by T.H. Work and M. Jozan, pp. 29 – 45.
Parasites of Purari people, by R.W. Ashford and D. Babona, pp. 47 – 53.
Red cell enzymes, serum proteins and viral antibodies among the Pawaia, by D.T.G. Hazlett, N.M. Blake and G.T. Nurse, pp. 55 – 65.
The Pawaia as agents and patients, by G.T. Nurse, pp. 67 – 84.

Vol. 9. The status and transport of nutrients through the Purari River, by A.B. Viner, 1979.

Vol. 10. Ecology of the Purari River catchment, edited by T. Petr, 1979.
A preliminary list of birds for the Purari River between Wabo and Baimuru (Gulf Province, Papua New Guinea), by R.W. Ashford, pp. 3 – 8.
Bird migration across Torres Strait with relevance to arbovirus dissemination, by R.W. Ashford, pp. 9 – 30.
Phragmites karka and the floating islands at Lake Tebera (Gulf Province), with notes on the distribution of *Salvinia molesta* and *Eichhornia crassipes* in Papua New Guinea, by B.J. Conn, pp. 31 – 36.
The vegetation of the lakes of Mt. Giluwe area (Southern Highlands Province), Papua New Guinea, by B.J. Conn, pp. 37 – 62.
Notes on the aquatic and semi-aquatic flora of the Lake Kutubu (Southern Highlands Province), Papua New Guinea, by B.J. Conn, pp. 63 – 90.
Marine wood borers in the Purari Delta and some adjacent areas, by S. Cragg, pp. 91 – 99.
Sago palm (*Metroxylon sagu*) salinity tolerance in the Purari River delta, by T. Petr and J. Lucero, pp. 101 – 106.

Vol. 11. Aquatic ecology of the Purari River catchment, edited by T. Petr, 1980.

A note on zooplankton from four Papua New Guinea lakes (Altitudinal range 538 – 3630 m), by I.A.E. Bayly and D.W. Morton, pp. 3 – 5.

A preliminary report on the zooplankton of the Purari estuary, by I.A.E. Bayly, pp. 7 – 11.

Organochlorine residues in the Purari River Delta, Gulf Province, Papua New Guinea, by R.W. Olafson and D. Mowbray, pp. 13 – 23.

Some aspects of the microbial ecology of the Purari River, Papua New Guinea, by H.W. Paerl and P.E. Kellar, pp. 25 – 39.

Bacterial densities and faecal pollution in the Purari River catchment, Papua New Guinea, by T. Petr, pp. 41 – 58.

A preliminary note on baseline levels of nine metals, including mercury, in freshly deposited sediments of the lower Purari River in Papua New Guinea, by T. Petr, pp. 59 – 68.

Vol. 12. Social study of the Pawaia, Papua New Guinea, by S. Toft, 1980.

Vol. 13. The agricultural and fishing development in the Purari Delta in 1978 – 1979, by R.N. Stevens, 1980.

Vol. 14. Census of crocodile population and their utilisation in the Purari area, by J. Pernetta and S. Burgin, 1980.

Vol. 15. Possible effects of the Purari Hydroelectric Scheme on subsistence and commercial crustacean fisheries in the Gulf of Papua, Workshop 12 December 1979, edited by D. Gwyther, 1980.

Some preliminary observations on the occurrence of juvenile Penaeidae in the Gulf of Papua, by C. McPadden, pp. 3 – 10.

The inshore prawn resource and its relation to the Purari Delta region, by S.D. Frusher, pp. 11 – 27.

The Gulf of Papua offshore prawn fishery in relation to the Wabo hydro-electric scheme, by D. Gwyther, pp. 29 – 52.

Distribution of penaeid prawn species within the trawling grounds of the Gulf of Papua prawn fishery, by C.D. Tenakanai, pp. 53 – 65.

Surface and bottom drift in Kerema, Orokolo and Deception Bays, by J.W. MacFarlane, pp. 67 – 81.

The mangrove crab *Scylla serrata*, by L.J. Opnai, pp. 83 – 91.

Vol. 16. The climate of the Purari River catchment above Wabo, by D.T. Evesson, pp. 1 – 15; Discharge data for the Purari River and some of its tributaries, by J.H. Carter, pp. 17 – 23, 1980

Vol. 17. Geomorphology of the Purari Delta, by B.G. Thom, and L.D. Wright, 1982.

Vol. 18. Geochemistry of the Purari catchment with special reference to clay mineralogy, by G. Irion and T. Petr, 1980.

Vol. 19. The sago subsistence of the people of the Purari River Delta, by S. Ulijaszek and S. Poraituk, 1981.

Vol. 20. Molluscs in the subsistence diet of some Purari Delta people, by S. Poraituk and S. Ulijaszek, 1981.

Vol.21. Anthropometry and nutritional status of some people of Purari Delta, by S. Ulijaszek and S. Poraituk, 1981.

Enquiries about the availability of these reports should be addressed to:
Office of Environment and Conservation,
Central Government Offices,
Waigani,
Papua New Guinea.

General index

609

611

616

Taxonomic index

619